T0076084

METHODS IN MOLECULAR BIOLOGY™

For further volumes:
http://www.springer.com/series/7651

Nanotoxicity

Methods and Protocols

Edited by

Joshua Reineke

Department of Pharmaceutical Sciences, Eugene Applebaum College of Pharmacy and Health Sciences, Wayne State University, Detroit, MI, USA

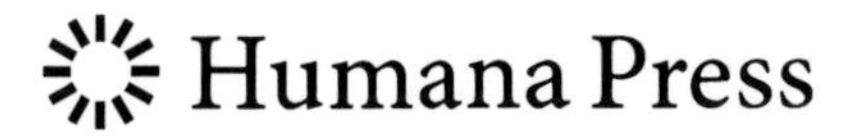

Editor
Joshua Reineke
Department of Pharmaceutical Sciences
Eugene Applebaum College of Pharmacy
and Health Sciences
Wayne State University
Detroit, MI, USA

ISSN 1064-3745 ISSN 1940-6029 (electronic)
ISBN 978-1-62703-001-4 ISBN 978-1-62703-002-1 (eBook)
DOI 10.1007/978-1-62703-002-1
Springer New York Heidelberg Dordrecht London

Library of Congress Control Number: 2012944369

Printed on acid-free paper

Humana Press is a brand of Springer
Springer is part of Springer Science+Business Media (www.springer.com)

Preface

I have the pleasure of introducing this edition of *Nanotoxicity: Methods and Protocols* in the successful *Methods in Molecular Biology* series. I must express my humble gratitude for the work done by the many contributors of this book and for the patience and assistance of the series editor, John Walker. Their many efforts have enabled the fruition of this project.

The field of nanotechnology has developed very rapidly over the past decade lending great promise to medical applications in drug delivery, therapeutics, and biological imaging. Additionally, broad arrays of consumer products have utilized nanomaterials including cosmetics, food products, textiles, and agriculture. Due to the great promise, rapid development, and broad application of nanomaterials, it is imperative that researchers from development through application seek an understanding of nanotoxicity. Many existing toxicology techniques have been applied to nanomaterials, and many newly developed methods to address the unique considerations of nanomaterials are continually emerging. The methods, protocols, and perspectives highlighted in *Nanotoxicity: Methods and Protocols* address the special considerations when applying toxicity studies to nanomaterials and detail newly developed methods for the study of nanotoxicity. These methods span in vitro cell culture, model tissues, in situ exposure, in vivo models, analysis in plants, and mathematical modeling. The diverse protocols covered are relevant to pharmaceutical scientists, material scientists, bioengineers, toxicologists, environmentalists, immunologists, and cellular and molecular biologists to name a few. This timely edition aims to diversify the capabilities of current researchers involved in nanotoxicology and to enable researchers in related fields to expand the knowledge of how nanomaterials interface with the biological environment. Expansion in the field of nanotoxicology will enable the progression of nanotechnology to its full potential.

Detroit, MI, USA *Joshua Reineke*

Contents

Contributors

SHOELEH ASSEMI • *Department of Metallurgical Engineering, University of Utah, Salt Lake City, UT, USA*
RICHARD D. BEGER • *Division of Systems Biology, National Center for Toxicological Research, US Food and Drug Administration, Jefferson, AR, USA*
PRIYANKA BHATTACHARYA • *Department of Physics and Astronomy, Clemson University, Clemson, SC, USA*
SHUPING BI • *School of Chemistry and Chemical Engineering, State Key Laboratory of Coordination Chemistry of China and MOE Key Laboratory for Life Science, Nanjing University, Nanjing, China*
JOHN P. BOHNSACK • *Department of Pharmaceutics and Pharmaceutical Chemistry, University of Utah, Salt Lake City, UT, USA*
PAUL L. CARPINONE • *University of Florida, Gainesville, FL, USA*
VINCENT CASTRANOVA • *Pathology and Physiology Research Branch, National Institute for Occupational Safety and Health, Morgantown, WV, USA*
NATARAJAN CHANDRASEKARAN • *Centre for Nanobiotechnology, VIT University, Vellore, India*
FANG CHANG • *School of Chemistry and Chemical Engineering, State Key Laboratory of Coordination, Chemistry of China and MOE Key Laboratory for Life Science, Nanjing University, Nanjing, China*
LOUIS W. CHANG • *Division of Environmental Health and Occupational Medicine, National Health Research Institutes, Zhunan, Taiwan*
YUNG CHANG • *Center for Infectious Disease and Vaccinology, The Biodesign Institute, Arizona State University, Tempe, AZ, USA*
RAN CHEN • *Department of Physics and Astronomy, Clemson University, Clemson, SC, USA*
YONGSHENG CHEN • *The School of Sustainable Engineering and the Built Environment, Arizona State University, Tempe, AZ, USAY. Chen*
School of Civil and Environmental Engineering, Georgia Institute of Technology, Atlanta, GA, USA
IN-HONG CHOI • *Department of Microbiology, College of Medicine, Yonsei University, Seoul, South Korea*
MARTIN J.D. CLIFT • *BioNanomaterials, Adolphe Merkle Institute, University of Fribourg, Fribourg, Switzerland*
TATIANA DA ROS • *Dipartimento Scienze Chimiche e Farmaceutiche, Piazzale EuropaTrieste, Italy*
PAOLO ELVATI • *Department of Mechanical Engineering, University of Michigan, Ann Arbor, MI, USA*
VINITA ERNEST • *Centre for Nanobiotechnology, VIT University, Vellore, India*
TRISHA EUSTAQUIO • *Weldon School of Biomedical Engineering, Purdue University, West Lafayette, IN, USA*
RUBYE H. FARAHI • *Oak Ridge National Laboratory, Oak Ridge, TN, USA; Department of Physics, University of Tennessee, Knoxville, TN, USA*

XAVIER FONT • *Department of Chemical Engineering, Escola d'Enginyeria, Universitat Autònoma de Barcelona, Bellaterra, Spain*
DARIN Y. FURGESON • *Department of Pharmaceutics and Pharmaceutical Chemistry, Department of Bioengineering, Department of Pediatrics, University of Utah, Salt Lake City, UT, USA*
ANA GARCIA • *Department of Chemical Engineering, Escola d'Enginyeria, Universitat Autònoma de Barcelona, Bellaterra, Spain*
HAMIDREZA GHANDEHARI • *Department of Pharmaceutics and Pharmaceutical Chemistry, Department of Bioengineering, Utah Center for Nanomedicine, Nano Institute of Utah, University of Utah, Salt Lake City, UT, USA*
KHALED GREISH • *Department of Pharmacology & Toxicology, Otago School of Medical Sciences, University of Otago, Dunedin, New Zealand*
SEISHIRO HIRANO • *Environmental Nanotoxicology Section, RCER, National Institute for Environmental Studies, Ibaraki, Japan*
EVANGELIA HONDROULIS • *Nanobioengineering/Bioelectronics Laboratory, Department of Biomedical Engineering, Florida International University, Miami, FL, USA*
YUKO IBUKI • *Institute for Environmental Sciences, University of Shizuoka, Shizuoka-shi, Japan*
JIYOUNG JANG • *Department of Microbiology, College of Medicine, Yonsei University, Seoul, South Korea*
PU CHUN KE • *Department of Physics and Astronomy, Clemson University, Clemson, SC, USA*
A. SAIRAM KISHORE • *International Institute for Biotechnology and Toxicology (IIBAT), Kancheepuram, Tamil Nadu, India*
MAMTA KUMARI • *Centre for Nanobiotechnology, VIT University Vellore, India*
JAMES F. LEARY • *School of Veterinary Medicine, Department of Basic Medical Sciences, Weldon School of Biomedical Engineering, Bindley Bioscience and Birck Nanotechnology Centers, Purdue University, West Lafayette, IN, USA*
CHEN-ZHONG LI • *Nanobioengineering/Bioelectronics Laboratory, Department of Biomedical Engineering, Florida International University, Miami, FL, USA*
MINGGUANG LI • *Optimum Therapeutics, LLC, San Diego, CA, USA*
DAE-HYOUN LIM • *Department of Microbiology, College of Medicine, Yonsei University, Seoul, South Korea*
CHIA-HUA LIN • *Division of Environmental Health and Occupational Medicine, National Health Research Institutes, Zhunan, Taiwan*
PINPIN LIN • *Division of Environmental Health and Occupational Medicine, National Health Research Institutes, Zhunan, Taiwan*
JAN D. MILLER • *Department of Metallurgical Engineering, University of Utah, Salt Lake City, UT, USA*
YIQUN MO • *Department of Environmental and Occupational Health Sciences, School of Public Health and Information Sciences, University of Louisville, Louisville, KY, USA*
ABDUL KHADER MOHAMMAD • *Department of Pharmaceutical Sciences, Eugene Applebaum College of Pharmacy and Health Sciences, Wayne State University, Detroit, MI, USA*
AMITAVA MUKHERJEE • *Centre for Nanobiotechnology, VIT University, Vellore, India*
P. BALAKRISHNA MURTHY • *International Institute for Biotechnology and Toxicology (IIBAT), Kancheepuram, Tamil Nadu, India*

JENNIFER F. NYLAND • *Department of Pathology, Microbiology, and Immunology, School of Medicine, University of South Carolina, Columbia, SC, USA*
MARICICA PACURARI • *Mary Babb Cancer Center, School of Medicine, West Virginia University, Morgantown, WV, USA*
WOLFGANG J. PARAK • *Fachbereich Physik and Wissenschaftliches Zentrum für Materialwissenschaften, Philipps Universität Marburg, Marburg, Germany*
SABINA PASSAMONTI • *Dipartimento Scienze della Vita, Universitá di Trieste, Trieste, Italy*
ALI PASSIAN • *Oak Ridge National Laboratory, Oak Ridge, TN, USA; Department of Physics, University of Tennessee, Knoxville, TN, USA*
KEVIN W. POWERS • *University of Florida, Gainesville, FL, USA*
TATSIANA A. RATNIKOVA • *Department of Physics and Astronomy, Clemson University, Clemson, SC, USA*
SONIA RECILLAS • *Department of Chemical Engineering, Escola d'Enginyeria, Universitat Autònoma de Barcelona, Bellaterra, Spain*
JOSHUA REINEKE • *Department of Pharmaceutical Sciences, Eugene Applebaum College of Pharmacy and Health Sciences, Wayne State University, Detroit, MI, USA*
PILAR RIVERA_GIL • *Fachbereich Physik and Wissenschaftliches Zentrum für Materialwissenschaften, Philipps Universität Marburg, Marburg, Germany*
BARBARA ROTHEN RUTISHAUSER • *BioNanomaterials, Adolphe Merkle Institute, University Of Fribourg, Fribourg, Switzerland*
ANTONI SÁNCHEZ • *Department of Chemical Engineering, Escola d'Enginyeria, Universitat Autònoma de Barcelona, Bellaterra, Spain*
ANNETTE SANTAMARIA • *Center for Toxicology and Mechanistic Biology, Exponent®, Inc, Houston, TX, USA*
LAURA K. SCHNACKENBERG • *Division of Systems Biology, National Center for Toxicological Research, US Food and Drug Administration, Jefferson, AR, USA*
KERRY N. SIEBEIN • *University of Florida, Gainesville, FL, USA*
JINCHUN SUN • *Division of Systems Biology, National Center for Toxicological Research, US Food and Drug Administration, Jefferson, AR, USA*
P. SUREKHA • *International Institute for Biotechnology and Toxicology (IIBAT), Kancheepuram, Tamil Nadu, India*
LAURENE TETARD • *Oak Ridge National Laboratory, Oak Ridge, TN, USA; Department of Physics, University of Tennessee, Knoxville, TN, USA*
GIRIDHAR THIAGARAJAN • *Department of Bioengineering, Utah Center for Nanomedicine, Nano Institute of Utah, University of Utah, Salt Lake City, UT, USA*
THOMAS THUNDAT • *Oak Ridge National Laboratory, Oak Ridge, TN, USA; Department of Physics, University of Tennessee, Knoxville, TN, USA*
TATSUSHI TOYOOKA • *Institute for Environmental Sciences, University of Shizuoka, Shizuoka-shi, Japan*
FEDERICA TRAMER • *Dipartimento Scienze della Vita, Universitá di Trieste, Trieste, Italy*
ANGELA VIOLI • *Department of Mechanical Engineering, University of Michigan, Ann Arbor, MI, USA*
BRYNN H. VOY • *Oak Ridge National Laboratory, Oak Ridge, TN, USA; Department of Animal Science, University of Tennessee, Knoxville, TN, USA*
RONG WAN • *Department of Environmental and Occupational Health Sciences, School of Public Health and Information Sciences, University of Louisville, Louisville, KY, USA*

JIANGXIN WANG • *Center for Infectious Disease and Vaccinology, The Biodesign Institute, Arizona State University, Tempe, AZ, USA*
NA WANG • *School of Chemistry and Chemical Engineering, State Key Laboratory of Coordination, Chemistry of China and MOE Key Laboratory for Life Science, Nanjing University, Nanjing, China*
EUN-JEONG YANG • *Department of Microbiology, College of Medicine, Yonsei University, Seoul, South Korea*
MO-HSIUNG YANG • *Department of Biomedical Engineering and Environmental Sciences, National Tsing Hua University, Hsinchu, Taiwan; Center for Nanomedicine Research, National Health Research Institutes, Zhunan, Taiwan*
FUPING ZHANG • *School of Chemistry and Chemical Engineering, State Key Laboratory of Coordination, Chemistry of China and MOE Key Laboratory for Life Science, Nanjing University, Nanjing, China*
QUNWEI ZHANG • *Department of Environmental and Occupational Health Sciences, School of Public Health and Information Sciences, University of Louisville, Louisville, KY, USA*
XIAOSHAN ZHU • *Graduate School at Shenzhen, Tsinghua University, Shenzhen, China*

Chapter 1

Historical Overview of Nanotechnology and Nanotoxicology

Annette Santamaria

Abstract

Although scientists have been studying nanoscience phenomena for many decades, technological developments in the second half of the twentieth century provided valuable tools that permitted researchers to study and develop materials in the nanoscale size range and helped formalize nanotechnology as a scientific field. This chapter provides a brief history of the field of nanotechnology, with an emphasis on the development of nanotoxicology as a scientific field. A brief overview of the worldwide regulatory activities for nanomaterials is also presented. The future development and safe use of nanomaterials in a diverse range of consumer products will be interesting, intellectually challenging, exciting, and hopefully very beneficial for the society.

Key words: Nanotechnology, Nanotoxicology, History, Nanomaterials, Consumer products

Nanotechnology is a means to develop and use materials, structures, devices, and systems that have novel properties and functions due to their small size. Nanotechnology is not a new field and scientists have been studying nanoscience phenomena for many decades in several fields of science and medicine. For example, Watson and Crick's discovery of the structure of DNA in 1953 can be considered nanoscience, just as much as the identification and research on carbon fullerenes in the 1980s. Nanoparticles having at least one dimension on the nanoscale (1–100 nm) are not new, as they have been around in the atmosphere since the dawn of the universe. The first observations and size measurements of nanoparticles was made during the first decade of the twentieth century in 1914 by Richard Adolf Zsigmondy, who was the first to use the term nanometer for characterizing a particle size of 1/1,000,000 of a millimeter, and he developed the first system classification based on particle size in the nanometer range (1). However, interest in nanoparticles, in particular "engineered nanoparticles," has increased significantly over the past several decades. Engineered nanomaterials are intentionally developed materials that have at least one dimension

Joshua Reineke (ed.), *Nanotoxicity: Methods and Protocols*, Methods in Molecular Biology, vol. 926,
DOI 10.1007/978-1-62703-002-1_1,

of 1–100 nm and exhibit novel properties compared to the non-nanoscale form of a material of the same composition.

The first time the idea of nanotechnology was introduced was in 1959, when Richard Feynman, a physicist at Caltech, gave a talk entitled, "There's Plenty of Room at the Bottom" (2). Dr. Feynman's talk has been viewed as the first academic talk that dealt with a main tenet of nanotechnology, the direct manipulation of individual atoms or molecular manufacturing. Although he did not use the term "nanotechnology," he suggested that it will eventually be possible to manipulate atoms and molecules to create "nanoscale" machines, through a cascade of billions of factories (3). Feynman proposed that these tiny "machine shops" would then eventually be able to create billions of tinier factories, and that as the scale got smaller and smaller, gravity would become more negligible, while both Van Der Waals attraction and surface tension would become more significant. The term "nanotechnology" was first used in a publication in 1974 by a student from the Tokyo Science University named Norio Taniguchi where he stated, "Nano-technology mainly consists of the processing of, separation, consolidation, and deformation of materials by one atom or one molecule" (4). However, nanotechnology did not develop into a field until the 1980s. In 1979, a scientist at the Massachusetts Institute of Technology named Eric Drexler expanded upon Feynman's vision of molecular manufacturing with contemporary developments in understanding protein function. Many believe that this is when the field of nanotechnology was created. In 1981, Drexler published his first article on the subject entitled "Molecular engineering: An approach to the development of general capabilities for molecular manipulation" in the journal "Proceedings of the National Academy of Sciences" (5). In this article, Drexler discussed the possibility of molecular manufacturing as a process of fabricating objects with specific atomic specifications, using designed protein molecules. Drexler took these concepts and expanded their potential in a book entitled, "Engines of Creation: The Coming Era of Nanotechnology" (6). Directly after the publication of this book, Drexler founded the Foresight Institute, whose stated goal was to "ensure the beneficial implementation of nanotechnology." As a result, due to the publicity generated by both Drexler's work and the Forsight Institute, scientists from all over the world began to have a vested interest in the field of nanotechnology. The idea that one could actually in some sense "touch" atoms and molecules came about in the 1980s, when scientists attempted to further study Drexler's proposed theory regarding the ability to manipulate atoms and molecules.

Just as Watson and Crick's discovery of the structure of DNA led to a biotechnology revolution, nanotechnology has been driven by the development of instrumentation and the availability of tools that allow scientists to see things that they were not able to see in the past. The development of sophisticated microscopes helped with the development of nanotechnology as a field. For example, the scanning tunneling microscope (STM) was invented by Binnig and Rohrer in 1981 and with this technology, individual atoms could be clearly identified for the first time, a breakthrough essential for the development of the field of nanotechnology because what had previously been concepts could be viewed and tested (3). Some of the limitations in STM were eliminated through the development of the Atomic Force Microscope (AFM) in 1986, a microscope that could image nonconducting materials such as organic molecules and could be used to study biological macromolecules and living organisms. AFM was integral for the study of carbon fullerenes (also known as buckyballs), which fall within the angstrom (Å) range (10 Å = 1 nm) and were first identified and produced in Richard Smalley's laboratory at Rice University in Texas in 1985 (7). Research with fullerenes marked the beginning of the current era of nanoscale science and technology and its unprecedented impacts across broad sectors of society. The buckyball, which is also known as a fullerene, was named after Richard Buckminster Fuller because of its resemblance to the late architect's geodesic domes. The structure is approximately 1 nm in diameter and consists of a 60 carbon atom cage that forms the shape of a soccer ball. There are also fullerenes that have been designed with up to 100 carbon atoms. This unique form of carbon was noted for its strength but did not find use in many applications until it led the way to the development of carbon nanotubes in the 1990s (8). However, the discovery of buckyballs was significant because it led to a host of nanotechnological developments and the incorporation of nanomaterials into products across many industries, particularly the chemical and semiconductor sectors.

While nanotechnology came into existence through Feynman's and then Drexler's vision of molecular manufacturing, the field has evolved to include research in several fields, including chemistry, materials science, medicine, toxicology, ecotoxicology, and industrial hygiene. With the development of microscopic technologies such as STM and AFM, nanotechnology could develop through the scientific method rather than through the conceptual visions of scientists such as Feynman and Drexler, and nanotechnology as a scientific field was established in a way that diverged from Drexler's original vision of molecular manufacturing. Whereas Drexler created interest in the field but also outlined a nanotech revolution, researchers around the world have brought nanotechnology to a more realistic and attainable level. The goal for nanotechnology research is not to create billions of assemblers that will revolutionize our world, but rather to explore the manufacturing and nonmanufacturing aspects of nanotechnology through a combination of chemistry, materials science, molecular biology, and molecular engineering (3).

There was a significant increase in the popularity of nanotechnology and the development of diverse nanomaterials in the 1990s and the 2000s, with the incorporation of nanomaterials into a wide variety of consumer products. Attempts to coordinate US federal work on nanotechnology began in November 1996 when staff members from several agencies met regularly to discuss nanotechnology plans and programs. In September 1998, the group was designated as the Interagency Working Group on Nanotechnology (IWGN) under the National Science and Technology Council. In August 1999, IWGN completed its first draft of a plan for an initiative in nanoscale science and technology, which went through an approval process involving the President's Council of Advisors on Science and Technology and the Office of Science and Technology Policy. In 2001, US federal funding for nanotechnology began under President Clinton following the approval of the federal initiative called the National Nanotechnology Initiative (NNI) with the allocation of half a billion dollars to several governmental agencies. Instead of focusing on molecular manufacturing, the NNI chose to fund nanoscale technology, which it defined as anything with a size between 1 and 100 nm with novel properties. This broad definition encompassed cutting-edge semiconductor research, several developing areas of chemistry, nanomedicine, and advances in nanomaterials.

Beginning in the early 2000s, concerns about the potential human and environmental health effects of nanomaterials were being expressed by many scientists, regulators, and nongovernmental agencies because particles and materials in the nanosize range may pose toxicological hazards due to their enhanced reactivity (e.g., chemical, electrical, magnetic) and potential for systemic availability and environmental occurrence (9–12). Because of the physicochemical properties of nanomaterials, they can modify cellular uptake, protein binding, translocation from portal of entry to the target site, and may have the potential for causing tissue injury (13). During this time, multidisciplinary research programs were initiated by the National Center for Environmental Research of the US Environmental Protection Agency, National Toxicology Program, National Institute of Environmental Health, and National Institutes of Health to address the impact of nanoparticles on human health and the environment (10).

There have been several toxicological and epidemiological studies conducted since the 1970s to evaluate the respiratory toxicity and pulmonary effects of "ultrafine" (<100 nm) particles versus larger particles (14–18). Much of this research was conducted to evaluate the behavior and potential health effects of ambient ultrafine particles that occur in the atmosphere as a result of natural and anthropogenic activities. The research on ultrafine particles provided a solid foundation regarding the understanding of how particles of different sizes are deposited in the respiratory tract that is useful for evaluating the deposition and behavior of engineered nanomaterials in the body. Beginning in the 1970s, research conducted to develop and evaluate the safety of nanoparticles for use in medical applications and in toxicological studies that addressed health and safety issues associated with the development of engineered nanomaterials were being conducted and published in the scientific literature (19–22). In 1990, two studies reported nanoscale-sized particles of titanium dioxide (TiO_2) and aluminum oxide (Al_2O_3) elicited a greater inflammatory response in the lungs of rats compared to the same mass of larger particles with the same chemical composition (15, 16). The theoretical ability for engineered nanomaterials to interact with biological systems in adverse ways created intense interest in the toxicology and ecotoxicology community that greatly accelerated in the 1990s. Besides the ongoing research in the pharmaceutical industry on the use of nanomaterials for drug delivery systems, there was a lot of research focusing on the evaluation of the potential for dermal penetration of nanosized titanium dioxide and zinc oxide because they were being used in dermally applied products such as sunscreen (23–25). With respect to the inhalation risks of engineered nanomaterials, three nanotoxicological studies that evaluated the effects on carbon nanotubes on the lungs in rodents created a lot of interest in the toxicology community in 2004 (26–28). Results from these three studies using intratracheal dosing of single or multiwalled carbon nanotubes indicated significant acute inflammatory pulmonary effects that either subsided in rats (26) or were more persistent in mice (27, 28). Around the same time, a study that evaluated the effects of carbon fullerenes on largemouth bass and reported lipid peroxidation in the brain and gills increased concern about nanomaterials in the field of ecotoxicology (29). The term "nanotoxicology" was first used in 2004 in an editorial by Donaldson et al. (30). A few members of the Society of Toxicology formed a Specialty Section entitled "Nanotoxicology" in 2007 and the journal called "Nanotoxicology" was launched in the same year. Since this time, there has been an explosion of in vitro and in vivo nanotoxicological research (e.g., (31–43)). There has also been an increasing amount of research to evaluate the toxicokinetics and toxicodynamics of nanomaterials (44, 45). The inhalation and dermal routes of exposure have been the primary focus for health effects research of nanomaterials; however, research on the ingestion of nanomaterials from food (46), the use of nanomaterials in medical devices (47, 48), diagnostics (49), and therapeutics (50–52) also increased significantly during the 2000s. The International Council on Nanotechnology (ICON) keeps a database of publications related to nanomaterial environmental health and safety that is available on their website (53). The number of technical research publications on nanomaterials identified with the terms "engineered" and "mammalian" has increased from approximately 15 publications from 1960 to 1980, to 89 from 1981 to 1990, to 510 from 1991 to 2000, and to more than 2,500 from 2001 to 2010.

During the early 2000s, several groups recommended frameworks and/or screening strategies for developing nanomaterials that may be used safely, including the International Life Sciences Institute (ILSI), the European Centre for Ecotoxicology and Toxicology of Chemicals (ECETOC), and a collaborative partnership between DuPont Corporation and Environmental Defense (13, 31, 54). The intent of these frameworks or screening strategies was to define a systematic process for identifying, managing, and reducing potential environmental, health, and safety risks of engineered nanomaterials from production through manufacture, use, disposal, and ultimate fate. The screening strategies are targeted toward companies and public and private research institutions that are actively working with nanomaterials and developing associated products and applications.

Companies around the world are currently harnessing the properties of nanomaterials for use in products across a number of sectors and are expected to continue to find new uses for these materials (55). Most nanoparticles that are currently in use today have been made from transition metals, silicon, carbon (single-walled carbon nanotubes; fullerenes), and metal oxides (zinc dioxide and titanium dioxide). The General Accounting Office (GAO) identified a variety of products that currently incorporate nanomaterials already available in commerce across the following eight sectors: automotive; defense and aerospace; electronics and computers; energy and environment; food and agriculture; housing and construction; medical and pharmaceutical; and personal care, cosmetics, and other consumer products (55). In addition, GAO predicted that the world market for nanotechnology-related products is growing and is expected to total between $1 trillion and $2.6 trillion by 2015.

The Woodrow Wilson Project on Emerging Nanotechnologies (PEN) developed an inventory of consumer products that are reported to contain nanomaterials and may be found on their website at http://www.nanotechproject.org/inventories/consumer/. The inventory grew from 212 products in 2006 to 1,015 products in 2009, a growth of almost 380%. The major product categories in the inventory include the following: health and fitness, home and garden, electronics and computers, food and beverage, automotive, and appliances. The largest number of nanotechnology products fall into the health and fitness sector, which includes cosmetics and personal care products. It is possible that at least some of the products claiming to have been derived from nanotechnology may in fact not really be, while additional products may contain nanomaterials even though it is not obvious from the label or marketing material that they contain such ingredients. Nanomaterials such as nano-titanium dioxide, nano-zinc oxide, and nanosilver are used in personal care products such as sunscreen, toothpaste, creams, hair products, and makeup. Carbon nanotubes are being used to reinforce a variety of sporting goods, such as bicycle frames, tennis rackets, baseball bats, and hockey sticks, because they offer greater strength and reduced weight, while increasing stiffness. Carbon nanotubes are also being used in a variety of other products, including automobile and airplane parts, in several biomedical applications, diagnostics, drug delivery systems, and in electrical and computer components. The continued miniaturization of electrical components using nanotechnology has facilitated enhancements to a number of consumer electronics including cell phones, personal computers, and MP3 players (56). Silver nanoparticles are being used in washing machines, clothing, personal care products, and food contact products (e.g., utensils, cutting boards) as antimicrobial agents. Many of the world's largest food and food packaging companies are reported to be actively exploring the potential of nanotechnology for use in food, dietary supplements, or food packaging (57). There are four major types of applications of nanotechnology in the food industry: (1) agriculture, (2) food processing, (3) food packaging, and (4) dietary supplements. Applications include the development of improved tastes, color, flavor, texture, and consistency of foodstuffs; increased absorption and bioavailability of nutrients and health supplements; new food packaging materials with improved mechanical, barrier, and antimicrobial properties; and nano-sensors for traceability and monitoring the condition of food during transport and storage (57). Nanomaterials are being developed to more efficiently and safely administer pesticides, herbicides, and fertilizers by controlling more precisely when and where they are released (55). Extensive research efforts are also underway to produce nanoscale drug delivery systems, diagnostics, and therapeutics (58). There are many benefits to formulating a pharmaceutical into a nanoparticulate including: increased bioavailability, faster onset of action, dose uniformity, reduction in fasted and fed variability, decreased toxicity, smaller dosage form and stable dosage forms of a drug that could not previously be formulated conventionally. Nanomaterials may be incorporated into a variety of building materials to increase energy efficiency, reduce aging due to sunlight, make steel stronger, make concrete more durable and more easily placed, and for antimicrobial purposes (55, 59). Functional textiles are being manufactured with nanomaterials that react to light to create power-generating clothing and nanosilver is being used in clothing and in antimicrobial wound dressings. Nanomaterials are also being used in coatings to make fabric and clothing stain and water-resistant and to provide nano-enabled surfaces that can remove scratches, stains, and scuff marks. Nanomaterials can be embedded on the surface of fabric fibers, creating a cushion of air around them and the fabric allows sweat to pass out, while also causing surface water to bead up and roll off.

In 2004, Lux Research predicted a global market share for nanotechnology products of 4% of general manufactured products in 2014, with nanotech in 100% of personal computers, in 85% of the consumer electronics, in 23% of pharmaceuticals, and in 21% of automobiles (60). In the same Lux report, the estimated global sales of products incorporating emerging nanotechnology were broken down by region of origin, and the largest regions of origin for the sale of nanotechnology products were Asia and the Pacific regions, followed by the USA and Europe on similar levels. Lux Research estimated the market for pure nanomaterials (carbon nanotubes, nanoparticles, quantum dots, dendrimers, etc.) to grow to approximately $3.6 billion by 2010 (from $413 million in 2005) and they forecasted the entire "nanotechnology impact" by 2010 to be approximately $1,500 billion (61).

As the fields of nanotechnology and nanotoxicology have been developing, federal, state, and global regulatory agencies have been struggling with how to classify and regulate nanomaterials. Several countries such as the United States, Australia, Canada, Japan, United Kingdom, and the European Union (EU) have begun to collect scientific data to understand the potential risks associated with nanomaterials and are reviewing their legislative authorities to determine the need to modify existing regulations. These countries have initiated the development of guidelines and regulations pertaining to nanomaterials, primarily in requesting the voluntary submission of scientific information about the nanomaterial. The US Environmental Protection Agency (EPA) developed the Nanoscale Materials Stewardship Program (NMSP) to "help provide a firmer scientific foundation for regulatory decisions by encouraging submission and development of information for nanoscale materials" (62). United Kingdom launched a voluntary reporting scheme for nanomaterials in 2008 that targeted manufacturers, importers, and users (63). This effort focused on free nanomaterials that are not enclosed in other materials because they were identified as having greater potential for environmental exposure. Information requested by the UK included chemical identity, dimensions and shape, size range, predictions of surface area, uses, available toxicological data, and certain physical and chemical characteristics, such as water solubility, stability, and flammability. The EU passed its chemical legislation in 2007, known as Regulation, Evaluation and Authorization of Chemicals (REACH), under which the EU requests information on all chemicals from companies. However because REACH requirements apply to chemicals with a production volume of greater than 1 metric ton per year, there has been concern that the provisions of REACH will not identify the risks of most nanomaterials if companies do not produce these materials at this level or volume. The EU is reviewing whether the provisions of REACH need to be modified to take into consideration the unique properties of nanomaterials by, for example, adjusting the volume-based requirement. In addition to efforts under REACH, the EU has developed regulations to require labeling on certain types of products containing nanomaterials. For example, an EU Cosmetics Regulation will require cosmetic products that contain nanoscale ingredients to be labeled as such (64). The regulation requires all nanomaterial ingredients be clearly indicated in the list of ingredients and the names of such ingredients to be followed by the word "nano" in brackets. The regulation would also require the manufacturers of new cosmetic products containing nanomaterials to notify regulators and provide them with safety information. Manufacturers of products containing nanoscale ingredients already being sold in the EU also would have to notify regulators and submit safety information. The regulation also calls for the EU to compile a publicly available catalogue of all nanomaterials used in cosmetic products placed on the market, including those used as colorants, UV filters, and preservatives, which will go into effect in July 2013. The EU has also begun to regulate nanomaterials in food and the regulations state that when there is a change in the particle size of a previously approved food additive, a new approval is required before the additive goes to market (65). In 2009, Australia's National Industrial Chemicals Notification and Assessment Scheme issued requests for companies to voluntarily provide information on nanomaterials (NICNAS 2009). In 2009, Health Canada proposed a requirement for companies to provide information on nanomaterials produced in or imported into Canada in excess of 1 kg. Health Canada stated that information required would include chemical and trade name, molecular formula, and any available information on the shape, size range, structure, physical and chemical properties such as solubility in water, toxicological data, quantity imported or manufactured, and known or predicted uses. Some states in the United States have also begun to address the potential risks from nanomaterials by collecting information from manufacturers on a limited number of nanomaterials in use in those states. Specifically, in January 2009, California required companies that manufacture or import carbon nanotubes into the state to submit certain readily available data on these materials to the California Department of Toxic Substances Control. Some municipalities have considered collecting information on nanomaterials such as the City of Berkeley, California, which issued a hazardous materials ordinance that requires companies to report the manufacture or use of nanomaterials. Berkeley's ordinance requires that facilities that manufacture or use nanoparticles submit a separate written disclosure of the material's known toxicology and how the facility will safely handle, monitor, contain, dispose, track, and mitigate the risks of such materials. Other states that have requested that companies submit scientific information on nanomaterials include Massachusetts, Maine, Washington, Pennsylvania, and South Carolina (55).

At this point, the research, development, and production of nanomaterials are greatly outpacing the speed by which toxicological information is being acquired on engineered nanomaterials. An understanding of the mammalian and ecotoxicological profiles of nanomaterials will be necessary to ensure that nanomaterials are safe for use and to establish appropriate safety procedures for handling those nanomaterials that may pose potential health hazards if there is sufficient exposure in the workplace, environment, or to consumers. For newly developed nanomaterials, it may be necessary to conduct a broad range of in vitro and in vivo studies to evaluate potential toxicological effects following oral, dermal, or inhalation exposure. The lessons we have learned from the chemical and biotechnology industries pave the way for the successful and safe incorporation of materials developed from technologies such as nanotechnology into products that can improve the quality of life. It is important that nanomaterials are developed responsibly, with optimization of benefits and minimization of risks, with international cooperation to identify and resolve gaps in knowledge. In the coming years, there is much research that needs to be completed to gather sufficient information for developing toxicological profiles for most nanomaterials that are currently being developed, manufactured, and incorporated into products. This is an exciting and interesting time for the fields of nanotechnology and nanotoxicology, and the safe development of nanomaterials for use in a broad range of consumer products will require the coordination of the efforts of a diverse group of engineers and scientists from academia, industry, and governmental agencies.

References

1. Zsigmondy R (1914) Colloids and the ultramicroscope. Wiley, New York
2. Feynman RP (1960) There's plenty of room at the bottom. Eng Sci 23(5):22–36
3. Fanfair D, Desai S, Kelty C (2007) The early history of nanotechnology. Produced by the Connexions Project, Module m14504, Rice University. http://cnx.org/content/m14504/latest/. Accessed 16 July, 2012
4. Taniguchi N (1974) On the basic concept of nano-technology. In: Proceedings of the international conference of production engineering. Japan Society of Precision Engineering, Tokyo
5. Drexler KE (1981) Molecular engineering: an approach to the development of general capabilities for molecular manipulation. Proc Natl Acad Sci USA 78(9):5275–5278
6. Drexler KE (1986) Engines of creation: the coming era of nanotechnology. Random House, New York
7. Kroto H et al (1985) C60: buckminsterfullerene. Nature 318:162–163
8. Iijima S (1991) Helical microtubules of graphitic carbon. Nature 354:56–58
9. Colvin VL (2003) The potential environmental impact of engineered nanomaterials. Nat Biotechnol 21(10):1166–1170
10. National Center for Environmental Research (NCER), US Environmental Protection Agency (2003) Impacts of manufactured nanomaterials on human health and the environment. http://epa.gov/ncer/rfa/current/2003_nano.html. Accessed 16 July, 2012
11. UK Royal Society and Royal Academy of Engineering (2004) Nanoscience and nanotechnologies: opportunities and uncertainties. http://www.nanotec.org.uk/finalReport.htm. Accessed 16 July, 2012
12. Hett A (2004) Nanotechnology. Small matter, many unknowns. SwissRe, Zurich
13. Oberdorster G et al (2005) Principles for characterizing the potential human health effects from exposure to nanomaterials: elements of a screening strategy. Part Fibre Toxicol 2:8

14. Wilson FJ Jr et al (1985) Quantitative deposition of ultrafine stable particles in the human respiratory tract. J Appl Physiol 58(1): 223–229
15. Ferin J et al (1990) Increased pulmonary toxicity of ultrafine particles. 1. Particle clearance, translocation, morphology. J Aerosol Sci 21: 381–384
16. Oberdorster G et al (1990) Increased pulmonary toxicity of ultrafine particles. 2. Lung lavage. J Aerosol Sci 21:384–387
17. Oberdorster G et al (1992) Role of the alveolar macrophage in lung injury: studies with ultrafine particles. Environ Health Perspect 97: 193–199
18. Churg A, Brauer M (2000) Ambient atmospheric particles in the airways of human lungs. Ultrastruct Pathol 24(6):353–361
19. Marty JJ, Oppenheim RC, Speiser P (1978) Nanoparticles – a new colloidal drug delivery system. Pharm Acta Helv 53(1):17–23
20. Couvreur P et al (1980) Tissue distribution of antitumor drugs associated with polyalkylcyanoacrylate nanoparticles. J Pharm Sci 69(2): 199–202
21. Kante B et al (1982) Toxicity of polyalkylcyanoacrylate nanoparticles I: free nanoparticles. J Pharm Sci 71(7):786–790
22. Kreuter J et al (1984) Toxicity and association of polycyanoacrylate nanoparticles with hepatocytes. J Microencapsul 1(3):253–257
23. Dussert AS, Gooris E, Hemmerle J (1997) Characterization of the mineral content of a physical sunscreen emulsion and its distribution onto human stratum corneum. Int J Cosmet Sci 19(3):119–129
24. Pflucker F et al (2001) The human stratum corneum layer: an effective barrier against dermal uptake of different forms of topically applied micronised titanium dioxide. Skin Pharmacol Appl Skin Physiol 14(Suppl 1): 92–97
25. Bennat C, Muller-Goymann CC (2000) Skin penetration and stabilization of formulations containing microfine titanium dioxide as physical UV filter. Int J Cosmet Sci 22(4): 271–283
26. Warheit DB et al (2004) Comparative pulmonary toxicity assessment of single-wall carbon nanotubes in rats. Toxicol Sci 77(1):117–125
27. Lam CW et al (2004) Pulmonary toxicity of single-wall carbon nanotubes in mice 7 and 90 days after intratracheal instillation. Toxicol Sci 77(1):126–134
28. Shvedova AA et al (2005) Unusual inflammatory and fibrogenic pulmonary responses to single-walled carbon nanotubes in mice. Am J Physiol Lung Cell Mol Physiol 289(5):L698–L708
29. Oberdorster E (2004) Manufactured nanomaterials (fullerenes, C60) induce oxidative stress in the brain of juvenile largemouth bass. Environ Health Perspect 112(10):1058–1062
30. Donaldson K et al (2004) Nanotoxicology. Occup Environ Med 61(9):727–728
31. Warheit DB et al (2007) Pulmonary toxicity study in rats with three forms of ultrafine-TiO2 particles: differential responses related to surface properties. Toxicology 230(1):90–104
32. Shvedova AA et al (2008) Inhalation vs. aspiration of single-walled carbon nanotubes in C57BL/6 mice: inflammation, fibrosis, oxidative stress, and mutagenesis. Am J Physiol Lung Cell Mol Physiol 295(4):L552–L565
33. Sayes CM, Reed KL, Warheit DB (2007) Assessing toxicity of fine and nanoparticles: comparing in vitro measurements to in vivo pulmonary toxicity profiles. Toxicol Sci 97(1): 163–180
34. Gopee NV et al (2007) Migration of intradermally injected quantum dots to sentinel organs in mice. Toxicol Sci 98(1):249–257
35. Park S et al (2007) Cellular toxicity of various inhalable metal nanoparticles on human alveolar epithelial cells. Inhal Toxicol 19(Suppl 1): 59–65
36. Baker GL et al (2008) Inhalation toxicity and lung toxicokinetics of C60 fullerene nanoparticles and microparticles. Toxicol Sci 101(1): 122–131
37. Ji JH et al (2007) Twenty-eight-day inhalation toxicity study of silver nanoparticles in Sprague–Dawley rats. Inhal Toxicol 19(10):857–871
38. Ryman-Rasmussen JP, Riviere JE, Monteiro-Riviere NA (2007) Variables influencing interactions of untargeted quantum dot nanoparticles with skin cells and identification of biochemical modulators. Nano Lett 7(5):1344–1348
39. Poland CA et al (2008) Carbon nanotubes introduced into the abdominal cavity of mice show asbestos-like pathogenicity in a pilot study. Nat Nanotechnol 3(7):423–428
40. Zhang LW, Monteiro-Riviere NA (2008) Assessment of quantum dot penetration into intact, tape-stripped, abraded and flexed rat skin. Skin Pharmacol Physiol 21(3):166–180
41. Muller J et al (2009) Absence of carcinogenic response to multiwall carbon nanotubes in a 2-year bioassay in the peritoneal cavity of the rat. Toxicol Sci 110(2):442–448
42. Robbens J et al (2010) Eco-, geno- and human toxicology of bio-active nanoparticles for biomedical applications. Toxicology 269(2–3):170–181
43. Porter DW et al (2010) Mouse pulmonary dose- and time course-responses induced by exposure to multi-walled carbon nanotubes. Toxicology 269(2–3):136–147

44. Riviere JE (2009) Pharmacokinetics of nanomaterials: an overview of carbon nanotubes, fullerenes and quantum dots. Wiley Interdiscip Rev Nanomed Nanobiotechnol 1(1):26–34
45. Yang RS et al (2010) Pharmacokinetics and physiologically-based pharmacokinetic modeling of nanoparticles. J Nanosci Nanotechnol 10(12):8482–8490
46. Kuzma J,VerHage P (2006) Nanotechnology in agriculture and food production. The project on emerging nanotechnologies. http://www.nanotechproject.org/publications/archive/nanotechnology_in_agriculture_food. Accessed 16 July, 2012
47. Williams D (2007) Carbon nanotubes in medical technology. Med Device Technol 18(2):8, 10
48. Christenson EM et al (2007) Nanobiomaterial applications in orthopedics. J Orthop Res 25(1):11–22
49. Yang X (2007) Nano- and microparticle-based imaging of cardiovascular interventions: overview. Radiology 243(2):340–347
50. Wang J et al (2008) Potential neurological lesion after nasal instillation of TiO(2) nanoparticles in the anatase and rutile crystal phases. Toxicol Lett 183(1–3):72–80
51. Bai S et al (2006) Recent progress in dendrimer-based nanocarriers. Crit Rev Ther Drug Carrier Syst 23(6):437–495
52. Lockman PR et al (2004) Nanoparticle surface charges alter blood–brain barrier integrity and permeability. J Drug Target 12(9–10):635–641
53. InternationalCouncilonNanotechnology(2010) Environment, Health, and Safety Database. http://cohesion.rice.edu/centersandinst/icon/report.cfm. Accessed 16 July, 2012
54. Environmental Defense and DuPont (ED/DuPont) (2007) Nano Risk Framework. A partnership of Environmental Defense Fund and DuPont. http://www.anoriskframework.com/. Accessed 16 July, 2012
55. Office GA (2010) Nanomaterials are widely used in commerce, but EPA faces challenges in regulating risk. In: Report to the Chairman, Committee on Environment and Public Works, US Senate
56. Thomas T et al (2009) Moving toward exposure and risk evaluation of nanomaterials: challenges and future directions. Wiley Interdiscip Rev Nanomed Nanobiotechnol 1(4):426–433
57. Chaudhry Q et al (2008) Applications and implications of nanotechnologies for the food sector. Food Addit Contam 25:241–258
58. Sandhiya S, Dkhar SA, Surendiran A (2009) Emerging trends of nanomedicine – an overview. Fundam Clin Pharmacol 23(3):263–269
59. Lee J, Mahendra S, Alvarez PJ (2010) Nanomaterials in the construction industry: a review of their applications and environmental health and safety considerations. ACS Nano 4(7):3580–3590
60. Hullmann A (2006) Who is winning the global nanorace? Nat Nanotechnol 1:81–83
61. Berger M (2007) Debunking the trillion dollar nanotechnology market size hype. NanoWerk. http://www.nanowerk.com/spotlight/spotid=1792.php. Accessed 16 July, 2012
62. U.S. Environmental Protection Agency (2009) Nanoscale Materials Stewardship Program, Interim Report. http://www.epa.gov/opptintr/nano/nmsp-interim-report-final.pdf. Accessed 16 July, 2012
63. UK Voluntary Reporting Scheme for engineered nanoscale materials (2008) Department for Environment, Food and Rural Affairs, London
64. European Parliament and of the Council on cosmotic products (2009) In: EC Regulation No. 1223/2009, C.o.t.E. Union (Editor)
65. European Parliament and of the Council of Food Additives (2008) In: EC Regulation No. 1333/2008, C.o.t.E. Union (Editor)

Chapter 2

Characterization of Nanomaterials for Toxicological Studies

Kevin W. Powers, Paul L. Carpinone, and Kerry N. Siebein

Abstract

The scientific community, regulatory agencies, environmentalists, and most industry representatives all agree that more effort is required to ensure the responsible and safe development of new nanotechnologies. Characterizing nanomaterials is a key aspect in this effort. There is no universally agreed upon minimum set of characteristics although certain common properties are included in most recommendations. Therefore, characterization becomes more like a puzzle put together with various measurements rather than a single straightforward analytical measurement. In this chapter, we emphasize and illustrate the important elements of nanoparticle characterization with a systematic approach to physicochemical characterization. We start with an overview describing the properties that are most significant to toxicological testing along with suggested methods for characterizing an as-received nanomaterial and then specifically address the measurement of size, surface properties, and imaging.

Key words: Particle characterization, Size distribution, Surface charge, Surface chemistry, Microscopy

1. Introduction

The increased emphasis on nanotechnology and its potential widespread commercial applications over the last two decades has sparked great interest in the potential human health and environmental effects of nanomaterials. Most of this attention is directed toward particulate nanomaterials as these are the most likely to be spread through the environment as aerosols or water-borne suspensions. If uncontrolled and unregulated, there is some potential in the minds of many that such particles can pose a human health hazard and/or environmental damage. Although the history of nanotechnology may trace its roots back to 1959 with Richard Feynman's "There is room at the Bottom" essay and popularized by Alex Drexler's 1991 book *Unbounding the Future: the Nanotechnology Revolution*, the first serious attention to the potential toxic effects of nanotechnology came much later (1, 2).

Joshua Reineke (ed.), *Nanotoxicity: Methods and Protocols*, Methods in Molecular Biology, vol. 926,
DOI 10.1007/978-1-62703-002-1_2, © Springer Science+Business Media, LLC 2012

In the spring of 2003, several groups presented findings at the annual American Chemical Society meeting illustrating cases where the size of nanomaterials affected their distribution and toxic behavior in animal studies (3). This brought national and international attention to the issue. The next year, the Royal Society published a report, *Nanoscience and nanotechnologies: opportunities and uncertainties*, that focused serious attention on the health and safety implications of nanotechnologies (4). In the Fall of 2004, the University of Florida, in conjunction with the National Toxicology Program, held the first US National Symposium on experimental approaches available to evaluate the toxicity of nanomaterials (5). Since that time, numerous workshops, conferences, organizations, and regulatory guidelines have been initiated to address this pressing issue. The journal, *Nanotoxicology*, was created in 2007 and the Society of Toxicology now has a specialty section focusing on nanotoxicity. Because of the uncertainty in the health and safety effects of nanotechnology, funding for nanotoxicity research has gradually increased with a US National Nanotechnology Initiative FY2011 (all agency) budgetary request of $117 million devoted to EH&S research (out of the 1.76 billion total request). Virtually all researchers in the field acknowledge the importance—and the difficulty—of sound characterization of nanomaterials as a necessary element in assessing their toxicity and/or other biological activity (6–9).

Particle characterization is an essential aspect of any attempt to assess potential biological effects of nanoparticulate systems. This may seem obvious, but to those unaccustomed to analyzing or quantifying particulate systems, the thorough characterization of their nanomaterials is a daunting task, especially in the context of a complex biological environment. Most traditional particle size analyzers have been designed to measure particle properties under controlled conditions with limited confounding factors (10, 11). In mineral or chemical systems, one often has the luxury of adjusting the environment by dilution, changing the solvent system, adjusting the pH, or adding surfactants to promote dispersion and aid the analysis. In biological systems, the presence of multiple components, high ionic strength, a limited temperature range, and potential toxic effects of dispersion aids hamper the investigator in his ability to measure important particle attributes. Properties such as size distribution, state of aggregation, surface charge, surface chemistry, translocation, and interaction with biological components can be very difficult to quantify and interpret in these complex environments (12). In the end, characterization becomes more like a puzzle put together with various measurements rather than a single straightforward analytical measurement. In this chapter we emphasize and illustrate the important elements of nanoparticle characterization with a systematic approach to physicochemical characterization. Subheading 2 is an overview describing the properties that are most significant to toxicological testing along with suggested methods for characterizing an as-received nanomaterial. Subheadings 3–5 address the measurement of size, surface properties, and imaging, respectively.

2. Measurements and Methodologies

A number of individual researchers and several National and International organizations have pooled resources to try and define which properties of nanomaterials are needed to best evaluate or predict their toxicological behavior (6, 8, 13, 14). Most researchers agree on the basic characterization parameters. These usually include those shown in Table 1.

There are many other properties that have been suggested for the complete characterization of Nanomaterials. Perhaps the most

Table 1
Important properties for characterization of nanomaterials

As-received	In vitro[a]	In vivo
Particle size distribution	Stability	ADME
Particle shape	Surface chemistry	Translocation/distribution
Bulk composition	Zeta potential	Agglomeration
Purity	Cytotoxicity	Immune response
Solubility	Hemolytic properties	Inflammation
Surface chemistry[b]	Surface adsorbed species	Toxicity
Surface area	ROS generation	
Surface morphology (crystallinity, shape, surface roughness)	Sterility	
Hydrophobicity		
Zeta potential		
Stability/agglomeration behavior		

[a]See NIST-NCL protocols at http://ncl.cancer.gov/

[b]Surface chemistry is often rather broadly defined (if defined at all). Most often it is meant to describe the chemical species on the surface, their concentration, and their effect on the interaction with the surrounding solvent system, surfaces, and other suspended solids. See Subheading 2 for greater detail

complete list has been compiled by OECD's Working Group on Manufactured Nanoparticles (15). This list includes a number of additional properties such as dustiness, water–octanol partition coefficient, a TEM micrograph, and number of other surface properties. Ultimately, it is up to the individual researcher to decide which properties are pertinent to his or her experiment. For example, the OECD list includes "dustiness" which may be important in the setting of aerosol exposure limits but may be irrelevant to an in vitro or parenteral exposure.

Once the properties of interest are identified, there are two issues that must be addressed: measuring the properties and expressing the measurement in some standard fashion acceptable to other scientists. The "dustiness" property is a good example because it is somewhat obscure in its definition (16). Intuitively, we may all understand that "dustiness" relates to the propensity for a dry powder to become aerosolized. However, there are different methodologies for quantifying this property oriented toward specific industries, size ranges, and concentrations (e.g., ASTM Standard D 7486-08). A dustiness measure designed for the coal industry likely will not adequately capture this property for dry agglomerated nanomaterials. Such a measurement for dry nanomaterial applications has yet to be standardized.

Fortunately, if we go back to the list in Table 1, we will find that for most (but not all) of these measurements there are established analytical methods (17). Perhaps the most notable exception is the "agglomeration behavior" of particulate systems. The quantification of this important property is a major issue as particles are introduced into biological environments and will be treated in more detail in the next section.

The bulk chemical composition and purity of the material is usually assessed by standard analytical means. For metals and metal oxides, the elemental composition can be measured by inductively coupled plasma (ICP-AES or ICP-MS), atomic absorption, or other quantitative techniques. Impurities can also be assessed by these means. For polymers or other organic constituents standard analytical techniques can be applied, sometimes requiring dissolution in appropriate solvents (18). A variety of methods such as Gel permeation chromatography, FTIR, NMR, LCMS, and others can be used to determine the structure and average MW of polymers. Analytical techniques can be highly specialized and often require sophisticated instrumentation and special expertise. If the nanomaterial is procured commercially, one of the most obvious sources of this information is the manufacturer. Manufacturer's technical support personnel are usually willing to provide additional information that may not be on the specification sheet for their materials. It is always useful, however, to conduct one's own analysis if possible, as the manufacturer may not quantify the same information that is relevant for toxicological studies. For example, a few ppm of copper or heavy metals in silver nanoparticles may not materially affect the intended antimicrobial properties of the silver but, if concentrated near the surface, may have a dramatic effect on its toxicity toward higher organisms. We once found 0.50 ppm mercury in a 50 ppm colloidal silver preparation that was intended for human consumption! The method of manufacture, transport, dispersion, storage, or even atmospheric exposure can sufficiently alter the surface composition of some nanomaterials so as to impart toxicological properties significantly different from the bulk material. Hence purity should always be a concern.

Solubility is often overlooked as being relatively straightforward. When it comes to toxicity, it is not. Looking up the "solubility" in a CRC handbook and ascribing "soluble" or "not soluble" is often insufficient to characterize the solubility of nanomaterials. Almost all metals and metal oxides have some solubility in aqueous systems and even very slight solubilities can have significant metabolic/toxic effects. Amorphous silica has an equilibrium solubility of 90 ppm at pH 7 which increases dramatically as particle diameter decreases below 10 nm (19). The previous example of silver nanoparticles is another case in point. The equilibrium solubility of silver in deionized water is low enough (only a few ppm) that it would generally be described as insoluble, yet its solubility is high enough to exert a deadly effect on microbes. Solubility can change dramatically with particle size due to increased surface energy, and changes in pH can have very dramatic effects on the dissolution and solubility of particles. Many metals are multivalent and amphoteric, and the hydrolysis and speciation of metal cations can be an important parameter (e.g., Cr^{3+} vs Cr^{6+}). At the very least, the equilibrium concentration of soluble constituents should be established and the aqueous chemistry of the nanomaterial should be well understood when conducting toxicological tests (20).

3. Size, Size Distribution, and Agglomeration

The determination of particle size seems very intuitive and simple to the layman but is far more challenging and complex than is generally recognized. Both ISO and ASTM have committees and numerous working groups devoted exclusively to particle characterization and particle size measurement (ISO TC24, SC4 particle characterization, and ASTM E-29 on particle and spray characterization). The complexity in particle characterization lies in two basic attributes: particles are generally small and hence they are generally numerous. Representative sampling is a crucial (often overlooked) issue. A large number of particles must be sampled and measured to adequately represent the size distribution, especially for broad size distributions. The second attribute is that particles have a natural tendency to agglomerate in solution or "coagulate" from an aerosol dispersion (21). Unless stabilized by some form of repulsive surface forces (typically electrostatic or steric), van der Waals forces will cause particles to stick together as they move in Brownian motion and randomly come into contact with each other. Hence, one is often dealing with a dynamic system where the particle (or at least agglomerate) size changes as a function of time or with changes in the surrounding environment.

Most individual particle sizing techniques do not clearly differentiate between primary particles and particle agglomerates. In addition, most measurement techniques interpret or report particle size based on the assumption that the particles are spheres with homogeneous properties (e.g., uniform density, refractive index). When agglomeration causes particle shape or density to depart from these assumptions, the assessment of the particle size distribution becomes less reliable. Thus, multiple analysis techniques and imaging are normally combined to develop a more complete picture of size distribution and state of dispersion. An example of this is provided in Figs. 1, 2, and 3 for a nominally spherical 50 nm aluminum sample. In this example the particle size was determined by three common methods: electron microscopy, BET surface area, and laser diffraction. Primary particle size appears to be qualitatively distributed in the 20–100 nm range by TEM/SEM. The aluminum powder has a specific surface area of 35.1 m^2/g and a median diameter by laser diffraction as recorded below in Table 2.

Figure 1 shows the laser diffraction size measurement of the Al powder dispersed and sonicated in deionized water. Note how different the laser diffraction data looks when plotted as a number distribution versus the volume weighted distribution. The number distribution is calculated to be the relative frequency of particles of a given diameter while the volume distribution is weighted by the volume or "cube" of the particle diameter. A particle system with a broad size range or with a high degree of agglomeration will be evident by a greater disparity between the two distributions when

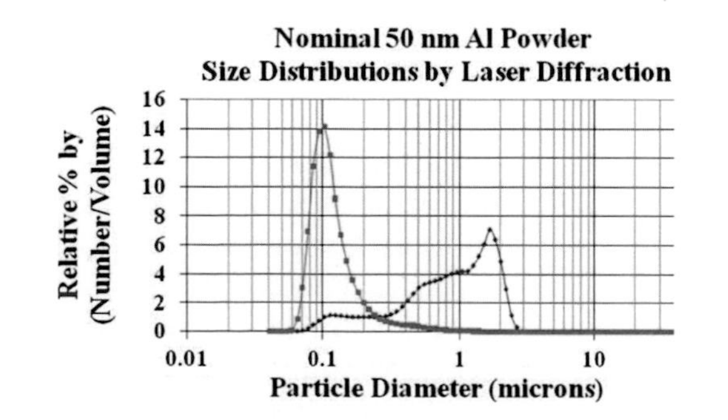

Fig. 1. Coulter LS13320 laser diffraction particle size for a nominal 50 nm aluminum sample (number and volume distributions).

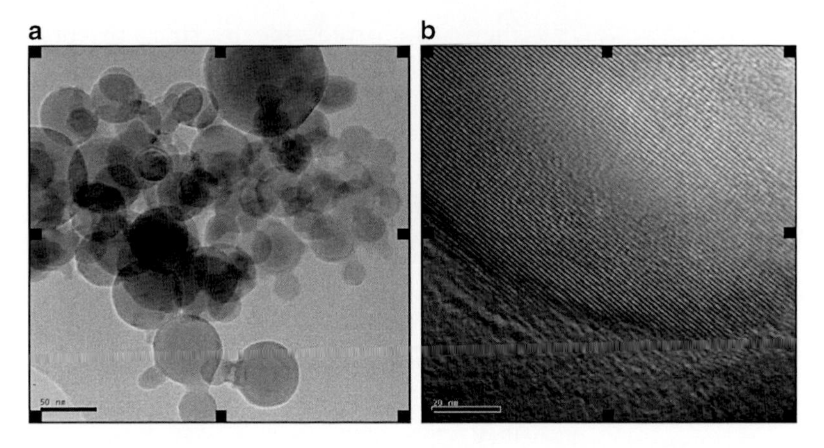

Fig. 2. TEM micrographs of aluminum 50 nm sample. (**a**) Aggregated particles, (**b**) high mag showing crystalline FCC lattice in particle interior and amorphous surface coating. Scale bars = 50 nm (**a**) and 20 nm (**b**).

Fig. 3. SEM micrographs of as-received aluminum 50 nm powder.

Table 2
Particle size data for a nominally 50 nm Al powder

50 nm Al powder	
Median diameter (number)	111 nm
Median diameter (volume)	1.02 μm
Specific surface area	35.1 m^2/g (~56.9 nm sphere)

they are plotted side by side. In this example, the additional information gained by BET surface area analysis and by SEM/TEM (Figs. 2 and 3) enables one to conclude that the primary particle size distribution lies predominantly between 20 and 100 nm, and that the powder is substantially agglomerated when dispersed in water.

In assessing particle size distribution for toxicity testing there are additional issues to be addressed. Since aggregate size is often dynamic, there is the question of when and under what conditions to make the measurements. Usually particles are characterized most thoroughly *as-received* from the manufacturer. Here the researcher has the maximum control over the environment and may freely manipulate the system to measure the desired properties. For example, the pH of the solution may be adjusted to promote dispersion, surfactants may be used, sonication or high shear can be applied to promote dispersion, and particles can be prepared for microscopy with fewer potential artifacts. Most measurement techniques are not designed for complex systems; therefore the best results are usually attained by keeping the particle-solvent system as simple as possible. Ensemble techniques usually measure properties averaged over a large number of particles in the measurement zone (e.g., dynamic light scattering (DLS), laser diffraction). Thus, assuming that the sampling is representative, the results are statistically robust. Once the properties of the material are thoroughly understood ex vivo, samples can be prepared appropriately for exposure with greater confidence and understanding.

For nanomaterials, the most common wet sizing techniques are DLS, laser diffraction, centrifugal sedimentation, acoustic techniques, Brownian motion analysis, electrozone sensing, and dark field, fluorescent, or confocal microscopy, albeit optical microscopy is limited to viewing relatively large nanoparticle agglomerates. For very small nanoparticles, large macromolecules, dendrimers and proteins, size exclusion chromatography (SEC), asymmetric field flow fractionation (AFFF), and ultracentrifugation are popular methods. For dry powders or aerosolized nanomaterials one can determine particle size from BET-specific surface area, dynamic mobility analysis (DMA), time of flight mass spectroscopy (TOFMS), light scattering, and electron microscopy. Atomic force microscopy (AFM) and cantilever resonance techniques (e.g., quartz balances, microfluidic cantilevers) can be conducted on either dry or wet samples.

All of these measurement techniques become more difficult to implement in complex heterogeneous systems such as cell culture media and biological fluids. The complexity increases in three ways. Additional components can influence the aggregation state of the particles (often by a change in pH, ionic strength, or surface adsorption). Secondly, they influence important physical properties of the surrounding medium such as viscosity, refractive index, or

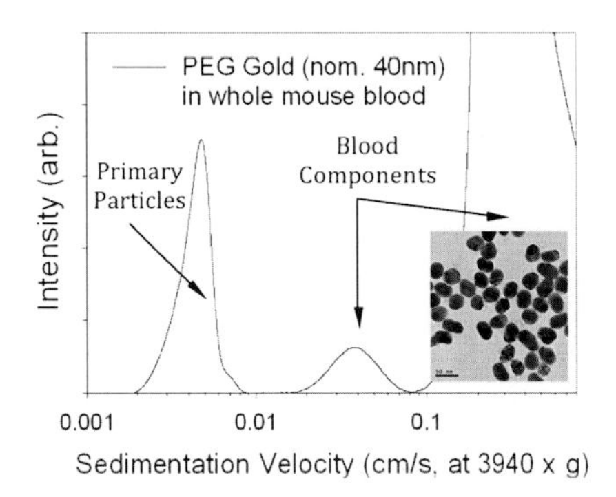

Fig. 4. Analytical centrifugation data for PEG-coated 40 nm gold in whole blood. *Inset*: TEM micrograph of primary particles (50 nm bar).

absorbance, and third, additional moieties in biological systems are often particles themselves (e.g., large proteins, biopolymers, cells, microparticles, etc.) that can confound or obscure the measurement. In Fig. 4, centrifugal analysis demonstrates these phenomena for a sample of PEG-coated 40 nm gold particles dispersed into whole blood. There appears a relatively distinct peak (at the equivalent of 40 nm) representing the primary gold particles with two peaks representing various cellular components of the lysed blood. The cellular components of the blood present a background which can easily obscure larger gold aggregates present in the sample. In this example, the analysis is further complicated by the fact that the various components have distinctly different densities.

Due to its broad popularity, DLS deserves a few comments of its own. DLS is based on the principle that particles diffuse under Brownian motion as a function of their hydrodynamic diameter (or size), fluid viscosity, and temperature. Smaller particles tend to diffuse more rapidly than larger particles. Mathematically, the diffusion of a particle in a fluid is described by the Stokes–Einstein equation. Again, in this basic equation the particle is assumed to be spherical and homogeneous.

$$d_\mathrm{h} = \frac{k_\mathrm{B} T}{3\pi\eta D_\mathrm{T}},$$

where d_h = hydrodynamic diameter, k_B = Boltzman constant, T = temperature (K), η = solvent viscosity, and D_T = translational diffusion coefficient. At a given temperature and viscosity, the hydrodynamic diameter can be calculated by experimentally determining the diffusion coefficient.

In DLS, the diffusion coefficient is usually derived through the use of photocorrelation or power spectrum analysis (Doppler shift). Without delving into the mathematical details, one can see that for uniform, well-dispersed, homogenous spheres or emulsions, there will be a single diffusion coefficient, and DLS is highly reliable and robust (22). Various advanced algorithms enable DLS to be used for polydisperse samples, but the accuracy suffers the further a sample deviates from the ideal. As with most other techniques, highly aggregated, contaminated, or heterogeneous samples can yield poor results if the researcher does not use appropriate care in making the measurement. Though not strictly defined as DLS, Brownian motion analysis can be applied to individual particles if their individual movement is captured by video with a dark field microscope and laser illumination. The motion of individual particles is captured and analyzed to determine their diffusion coefficient, and a statistical picture of the whole sample is developed by the automated analysis of many particles. This technique suffers from its own weaknesses including high sensitivity to sample concentration, clarity and depth of field, and a relatively smaller sample size. However, as a counting technique, it can handle multimodal or broad distributions.

The preceding discussion highlights the care that must be taken when conducting particle size analysis before, during, and after exposure in toxicological studies. Multiple techniques are almost always essential to develop a full understanding of primary particle size and state of agglomeration. The investigator should thoroughly understand the physical principles by which these measurements are made and give careful consideration to the composition of the surrounding fluid and how it affects these measurements. Collaboration with researchers trained in particle characterization is always a good idea.

4. Surface and Interfacial Properties

Particles are all about size, shape, and surface properties. The most obvious surface property for fine particles is the high surface area. For spheres, specific surface area increases inversely with particle diameter, thus for a given mass it increases dramatically as particle size approaches the nano range. The theoretical specific surface area is usually measured in meters squared per gram. For a spherical particle it can be calculated by the following formula where density (ρ) is described in g/cm^3 and diameter is given in microns.

$$\mathrm{SSA} = \frac{6}{\rho \times d_{\mathrm{microns}}}.$$

Thus, a 1 μm silica sphere (ρ=2.2 g/cm^3) has a SSA of 2.72 m^2/g, whereas a 10 nm silica particle has a 100-fold greater area of 272 m^2/g! A perfect sphere is the shape of minimum surface area, thus any shape deviation, surface roughness, or porosity serve to increase the specific surface area. This equation is most often used to calculate the spherical equivalent particle size from a BET surface area measurement (23). The BET measurement should be included in the particle characterization whenever there is sufficient dry powder available. For wet systems, or where there is insufficient material available to conduct BET analysis, surface area is most often approximated by measuring particle size and calculating the surface area using a spherical assumption (24). Occasionally, surface area is approximated in solution by dye adsorption or acid/base titration. For example, the Sears test is a standard acid base titration method used for determining surface area in the colloidal silica industry (25). There has been some success in measuring the in situ surface area of aerosol powders using the epiphaniometer and more recently by a diffusion charging (DC) technique (26, 27).

Surface area is important because it is the surface that interacts with the surroundings; however, it is not surface area alone that defines these interactions. Surface chemistry consists of the chemical structure of particle surfaces, development of surface charge, surface functional groups, surface active sites, and the propensity to adsorb moieties from the environment both pre- and postexposure. Surface adsorption and/or coating can completely change the nature of the surface. Surface chemical analysis is most easily conducted on as-received materials. Two principle measurements of surface chemistry are hydrophilicity and zeta potential. These will largely determine how the particles physically respond to dispersion in aqueous fluids. The two are related in that hydrophobic particles seldom develop a substantial surface charge and therefore may be difficult or impossible to disperse in solution. They may however have a greater affinity for lipids, membranes, or other lypophilic environments.

There is no easy way to quantify the contact angle of particles. Typically researchers will simulate the particle surface composition on a flat bulk surface and measure contact angle accepting the fact that this is an approximation of the highly curved nanoparticle surface. Another method is by powder capillary rise using the Laplace–Washburn equation; however, this technique suffers from insufficient precision in defining the effective interparticle pore radius (r_{eff}) for a nonspherical packed nanoparticle bed plus difficulty dealing with hysteresis effects (28).

$$\Delta P = \frac{2\gamma LV \cos\theta c}{\gamma_{\mathrm{eff}}}.$$

One will note that the OECD working group on manufactured nanomaterials suggests using the octanol–water partition coefficient to characterize hydrophilicity (15). This method is used routinely in the pharmaceutical industry and by environmental toxicologists to partition and identify hydrophobic species dissolved in aqueous solutions. Its usefulness for nanoparticles is still under debate. In its most simplified form, nanoparticles are suspended in water mixed with an organic solvent, and the concentration of nanoparticles that partition into each phase is quantified through a partition coefficient (log *P*). This parameter can be used in a comparative manner against standard materials to estimate the hydrophobicity of a powder (29).

$$\log P = \log\left(\frac{C_{\mathrm{o}}}{C_{\mathrm{W}}}\right).$$

There are several issues with this simplified technique. It fails to account for effects such as particle size, sedimentation, and surface tension effects at the water–octanol interface. Standardized methods for applying it to nanoparticles are still under consideration.

A more quantitative method of determining the hydrophilicity of particles is by AFM. In this method, particles are affixed to an AFM cantilever and force measurements are recorded as the AFM tip is lowered into a drop of water (30, 31). This is a rather specialized technique with its own set of issues such as how particles are affixed to the cantilever tip and their geometry as the tip approaches the water droplet. It has not yet been adapted as a standardized method but is very useful if the instrumentation and expertise are available.

Since almost all exposures in toxicology studies eventually involve aqueous biological fluids, it is important to determine the surface charge that develops on particles when introduced into aqueous solutions. Since the charge at the particle surface is difficult to measure directly, the zeta potential (net potential at the shear plane) and isoelectric point of the particles are typically used to quantify particle charge (32). A complete analysis should be made in both deionized water and in a buffered solution of similar ionic strength and pH relevant to the intended exposure route. Figure 5 shows a zeta potential titration of aluminum nanoparticles and demonstrates the sensitivity to the increase ionic strength of tap water and the adsorption of even small concentrations of polyvalent ions. When measuring (or attempting to measure) zeta potential, the dispersion stability of the particles in suspension often becomes quite evident and provides insight into the agglomeration behavior of the materials. For aqueous suspensions, zeta potential is generally positive below the isoelectric point (low pH) and negative above the IEP. One should be alert that as pH approaches very high or very low values solubility often increases dramatically.

Generally, zeta potentials of absolute value greater than about 30 mV are characteristic of well-stabilized colloids. As the pH approaches the isoelectric point, dispersions become less stable and more prone to agglomeration. Due to the high ionic strength and near neutral pH of many physiological fluids, it is common for nanoparticle systems to become totally unstable and agglomerate

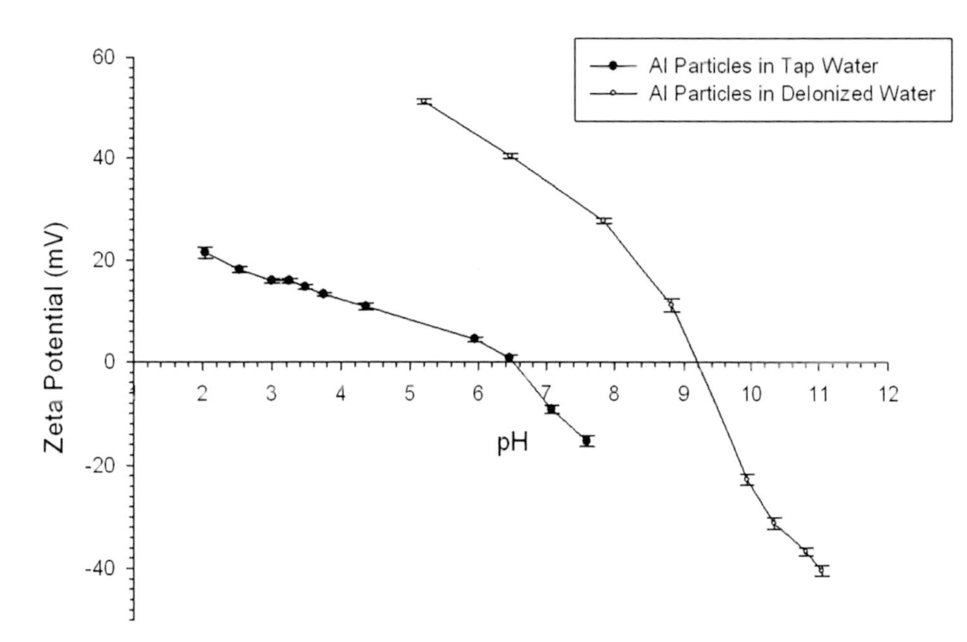

Fig. 5. Zeta potential titration of aluminum nanoparticles in both deionized and tap waters.

rapidly when introduced into the biological environment. The temptation at this point is to attempt to disperse the particles through the use of surfactants or other dispersion aids, high energy sonication, shear, coatings, other surface modifications, or by changing the solution conditions. The danger is that such treatments may promote unrealistic exposure protocols and begs the question, "Are the biological effects due to the particles or to the surface modifications used to disperse them?"

Zeta potential can be measured with several good techniques including electroacoustic, DLS, and electrophoretic methods (33–35).

The functional surface chemistry of nanomaterials consists of a broad range of interfacial properties that, like zeta potential, rely on the surroundings as well as the particle surface. Thus, surface composition and properties should be characterized as-received with the understanding that these properties may change once introduced into a biological setting. The elemental composition of the particle surface can be determined by surface sensitive techniques such as X-ray photoelectron spectroscopy (XPS), secondary ion mass spectroscopy (SIMS), energy dispersive spectroscopy (EDS), and auger electron spectroscopy (AES). These are all high vacuum techniques and vary in their capabilities and surface sensitivity. Surface chemistry may also change substantially when particles are handled, dried, fixed in a biological matrix, and placed under ultra-high vacuum for analysis. Molecular species can sometimes be deduced from these analyses (e.g., bond energies from XPS), but painting a complete picture of the surface atomic layers on particles is still an art. Standard spectroscopic techniques such as FTIR and Raman can be helpful in identifying surface groups, but suffer from issues such as sensitivity and (too high) depth of penetration. Fortunately, unless a specific surface chemistry is intentionally introduced, surface groups tend to be some combination of the bulk particle composition, oxides, hydroxyl groups, or contaminants introduced during synthesis or handling. Surface treatments, such as a peroxide baths (Fenton treatment) or plasma cleaning, can be used to clean surfaces of impurities with the understanding that one may be imparting an artificially "clean" surface to the particles prior to exposure (36).

The propensity for nanomaterials to catalyze or generate free radicals or reactive oxygen species is often of specific interest in assessing toxicity. Many semiconductor or transition metal oxides such as iron oxides, titanium dioxide, and cerium oxide can be catalytic toward free radical species under different conditions (37). Fullerenes, carbon nanotubes, graphene, along with various functional groups can also generate radicals (38, 39). While the mechanisms of ROS generation may or may not be a specific goal of the investigation, quantification of ROS generation is important in the toxicological evaluation of nanomaterials. Quantification of ROS in solution or in vitro is often accomplished using various colorimetric or fluorometric indicators such as dihydrorhodamine-123 (DHR) which is converted to the fluorescent form upon reaction with peroxide, singlet oxygen, and other reactive species (40, 41). Electron paramagnetic resonance (EPR) or spin resonance (ESR) is a specialized method that can be used to measure free radicals directly or by "spin trapping" radicals with a reporter molecule. It is a powerful in vitro technique and has seen increasing applications in in vivo detection and quantification of oxygen and ROS (42).

Clearly there are many other reactions that take place at the surface of particles in the biological environment. There may be specific uptake of proteins, specific cell membrane interactions, opsonization of particles, and numerous other interactions all of which might bear on the toxicity profile of these materials. Particles designed for therapeutic or imaging applications have specific surface properties engineered to chemically target them to the desired location or to control the lifetime in circulation. Consequently, the surface chemistry of nanoparticles can be as complex and as broad as the field of analytical chemistry itself.

5. Microscopy and Imaging

Microscopy is often considered the gold standard for measuring the properties of nanoparticles because it is visual, more intuitive, can be highly quantitative, and there is less ambiguity in the observed size, structure, and morphology (6). It is generally recognized as good practice to include micrographs at the appropriate magnification of any nanomaterial being characterized for toxicity testing. For nanomaterials, this normally will require scanning electron microscopy (SEM), transmission electron microscopy (TEM), or AFM. These techniques can contribute toward the measurement of the particle size, shape, morphology, particle size range, crystallinity, agglomeration state, purity, surface area, and surface morphology. The chemical composition of the nanoparticles and coatings can be determined using energy dispersive X-ray spectrometry (EDS) or electron energy loss spectroscopy (EELS). As with all particle characterization techniques, sampling and statistical reliability are two major issues when drawing conclusions from imaging alone. It is much more difficult to achieve and verify a representative sample for the very small samples required for microscopy, and there are many artifacts that are possible in the preparation and analysis of samples especially for in vacuo techniques such as SEM and TEM.

Optical techniques generally don't have the resolution to see individual nanoparticles but are useful to image the cells, tissue, and for pathology. Fluorescent probes can be embedded or surface attached to particles to track their behavior and location. Usually, TEM with the appropriate sample preparation and staining techniques is required to analyze the nanoparticles in the ultrastructure of cells and tissue. Morphological information is gained through comparison of images taken using multiple techniques over a wide range of magnifications (43, 44). For example, Fig. 6 shows a series of optical and TEM images at different magnifications that "home in" on the location, size, and aggregation state of gold nanoparticles deposited from the blood stream in the liver of a mouse. The EDS spectrum is necessary to positively identify the particles, which are often confused with staining artifacts or cellular structures.

A better understanding of the interactions between cells and nanoparticles is obtained using multiple microscopy techniques. Correlative microscopy includes any available microscopic techniques, including light, probe, laser, and electron microscope techniques. A few of the microscopy techniques that are typically used in biological sciences and the cellular uptake of nanoparticles are transmitted light, fluorescence microscopy, confocal microscopy, and TEM with elemental analysis such as EDS or EELS (44–47). The analysis can be in two or three dimensions. Three-dimensional analysis is more difficult to achieve, but can provide additional information on shape, particle distribution, and relationship to various cellular structures. Confocal imaging, stereo SEM, cryosectioning, AFM, and soft X-ray tomography are all techniques that provide 3D data. Typically, an optical technique is used to screen samples and to locate areas of interest. TEM is needed for ultrastructure information obtained through high resolution

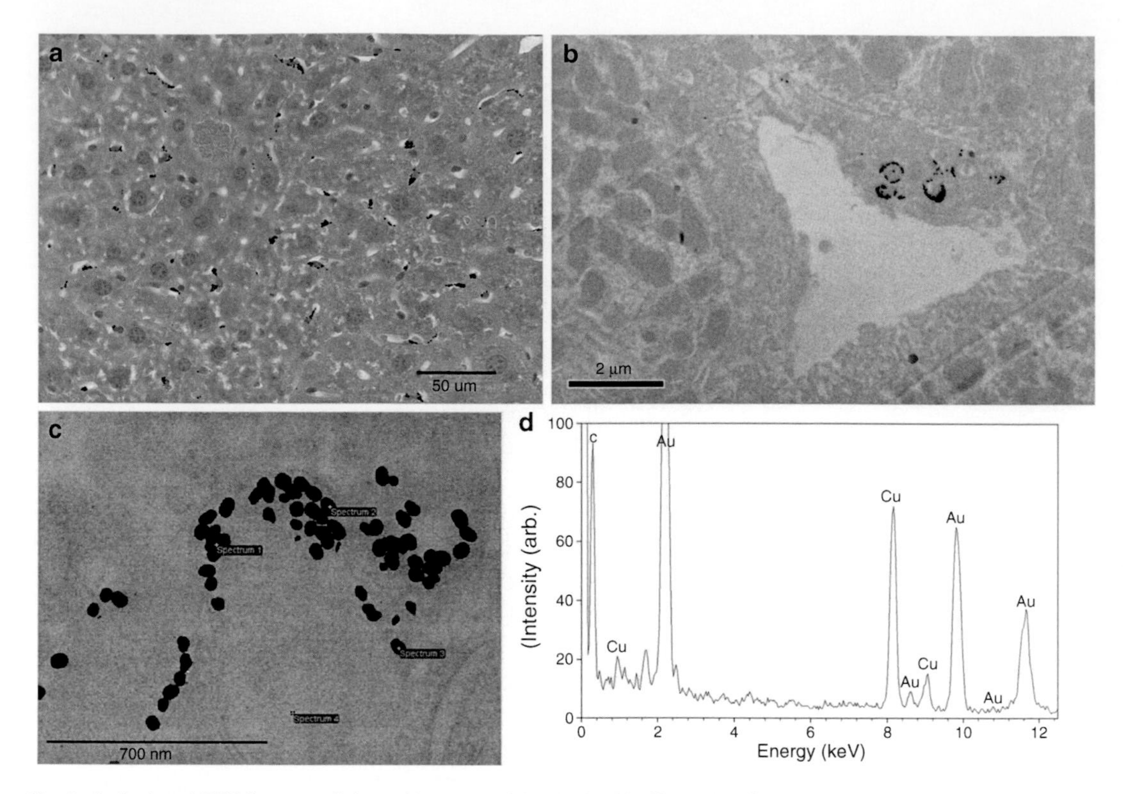

Fig. 6. Optical and TEM images of the gold nanoparticles in the Kupffer cells of a mouse liver with the EDS spectrum confirming elemental composition. (**a**) Optical micrograph of gold aggregates in a histological liver section. (**b**) TEM of particles in liver section at higher magnification. (**c**) TEM at still higher magnification showing primary gold particles. (**d**) EDS spectra taken by TEM confirming that the particles in the image are gold (copper peaks are from TEM grid). Scale bars = 50 μm (**a**), 2 μm (**b**), and 700 nm (**c**).

imaging and elemental analysis. The preparation of samples for TEM requires special expertise that is expensive and labor intensive. In many cases, the TEM results are necessary to aid in the interpretation of the optical results, after which the optical technique can be used as a screening process.

The latest advances in TEM analysis of biological materials are in cryo preparation, cryo TEM, and TEM tomography (48–50). Cryo preparation procedures, such as high pressure freezing, eliminate artifacts from the conventional preparation procedures of fixation, dehydration, and embedding. The possibility of washing away nanoparticles is also eliminated when high pressure freezing is used. The frozen samples can be imaged directly in the cryo TEM without staining. TEM tomography is based on a tilt series of two-dimensional projections of the object along different directions that is reconstructed into a three-dimensional projection of the original object.

Ion abrasion SEM (IA-SEM) or focused ion beam SEM (FIB-SEM) is being used to present three-dimensional views of cells and tissue (48–53). It is a relatively new technique to the world of biological sciences. It combines the removal of thin layers of samples by an ion beam and then imaging the newly exposed face of the sample. This slice and image technique combined with stereological image analysis software can be used to build the 3D image from SEM slices. The result is an image cube that shows the nanoparticle distribution in three dimensions.

Some of the more exciting recent advances in biological imaging include X-ray microscopy and soft X-ray tomography. The resolution of X-ray microscopy lies between that of the optical microscope and the electron microscope. X-ray microscopy using synchrotron soft X-ray sources such as the Advanced Light Source at Berkeley Labs has achieved resolutions as fine as 15 nm. The advantage of X-ray microscopy is that biological samples can be analyzed in their natural hydrated state, preserving the unaltered microstructure of the cell. Soft X-ray tomography provides three-dimensional imaging much like the medical computed tomography (CT) scan we are all familiar with, but at much higher resolutions. The biggest drawback to the widespread use of X-ray-based biological imagining has been the need for a synchrotron source of X-rays. However, commercial instruments with laser plasma X-ray sources are now being developed with submicron capabilities and the potential to do small animal imaging (54, 55).

Imaging is currently the only method available to assess the shape of nanoparticles. Although many nanoparticulate systems are nominally equiaxed or spherical, many of the most interesting structures are not. Carbon nanotubes (CNTs), nanowires, and nanoflakes might pose a significant risk if toxicity is linked directly to shape factors. High aspect fibers (e.g., asbestos) have been linked to mesothelioma at larger particle sizes and hence researchers and regulatory agencies are cautious. There have been a number of studies to try and determine if shape is a factor in the toxicity of CNTs and, although there are a number of studies showing pulmonary or cellular toxicity, it is still inconclusive (56, 57). Nonetheless, shape affects many other physical properties such as aerosolization, aggregation, and diffusion and should be quantified in any characterization protocol.

6. Conclusions

There has been a great deal of attention and research devoted to the issue of nanotoxicology over the last 10 years. The scientific community, regulatory agencies, environmentalists, and most industry representatives all agree that more effort is required to ensure the responsible and safe development of new nanotechnologies. Characterizing nanomaterials is a key aspect in this effort. There is no universally agreed upon minimum set of characteristics although certain common properties are included in most recommendations (see Table 1 and additional references). Reviewers for many professional journals have been sensitized to look for such information. Ultimately it is up to the researcher to decide the properties that are the most pertinent to the goals of the investigation and to characterize their nanomaterials accordingly. Research teams incorporating multidisciplinary personnel can be most helpful in this respect.

References

1. Feynman RP (1959) There's plenty of room at the bottom: an invitation to enter a new field of physics. American Physics Society, Caltech Engineering and Sciences
2. Drexler E, Peterson C, Pergamit G (1991) Unbounding the future: the nanotechnology revolution. William Morrow & Co., New York
3. Service RF (2003) Nanomaterials show signs of toxicity. Science 300:243
4. Society R (2004) Royal Society and Royal Academy of Engineering Report on nanoscience and nanotechnologies: opportunities and uncertainties (Society TR, ed), London
5. Bucher J, Masten S, Moudgil B, Powers K, Roberts S, Walker N (2004) Developing experimental approaches for the evaluation of toxicological interactions of nanoscale materials. Final Workshop Report, Gainesville
6. Oberdorster G, Maynard A, Donaldson K, Castranova V, Fitzpatrick J, Ausman K, Carter J, Karn B, Kreyling W, Lai D, Olin S, Monteiro-Riviere N, Warheit D, Yang H (2005) Principles for characterizing the potential human health effects from exposure to nanomaterials: elements of a screening strategy. Part Fibre Toxicol 2:8
7. Thomas K, Sayre P (2005) Research strategies for safety evaluation of nanomaterials, Part I: evaluating the human health implications of exposure to nanoscale materials. Toxicol Sci 87:316–321
8. Powers KW, Brown SC, Krishna VB, Wasdo SC, Moudgil BM, Roberts SM (2006) Research strategies for safety evaluation of nanomaterials. Part VI. Characterization of nanoscale particles for toxicological evaluation. Toxicol Sci 90:296–303
9. Borm PJA, Robbins D, Haubold S, Kuhlbusch T, Fissan H, Donaldson K, Schins R, Stone V, Kreyling W, Lademann J, Krutmann J, Warheit D, Oberdorster E (2006) Part Fibre Toxicol 3:11–46
10. Allen T (ed) (2004) Powder sampling and particle size measurement, vol I, 5th edn. Chapman & Hall, London
11. Allen T (ed) (2004) Surface area and pore size determination, vol 2. Chapman & Hall, London
12. Knapp J, Barber T, Lieberman A (1996) Liquid- and surface-borneparticle measurement handbook. Marcel Decker, New York
13. OECD (2008) OECD Environment, Health and Safety Publications Series on the safety of manufactured nanomaterials No 6, Document ENV/JM/MONO (development, O. f. e. c. a., Ed.), Paris
14. Scientific Committee on Emerging and Newly-Identified Health Risks (SCENIHR) (2007) Opinion on the appropriateness of the risk assessment methodology in accordance with the technical guidance documents for new and existing substances for assessing the risks of nanomaterials, European Commission
15. OECD (2010) Guidance manual for the testing of manufactured nanomaterials: OECD's sponsorship programme; first revision ENV/JM/MONO, (development, O. f. e. c. a., Ed.), Paris
16. Boundy M, Leith D, Polton T (2006) Method to evaluate the dustiness of pharmaceutical powders. Ann Occup Hyg 50:453–458
17. Nieman TA, Skoog DA, Holler FJ (2006) Principals of instrumental analysis. Brooks/Cole, Pacific Grove
18. Holler FJ, Skoog DA, West DM (1996) Fundamentals of analytical chemistry. Saunders College Publications, Philadelphia
19. Iler RK (1979) The chemistry of silica. Wiley, New York
20. Baes CEJ, Messmer RE (1976) The hydrolysis of cations. Krieger Publishing Co, Malabar
21. Hinds WC (1999) Aerosol technology: properties, behavior, and measurement of airborne particles. Wiley, New York
22. Berne B, Pecora R (1976) Dynamic light scattering. Wiley, New York
23. Brunauer S, Emmett PH, Teller E (1938) J Am Chem Soc 60:309–319
24. Maynard AD (2003) Estimating aerosol surface area from number and mass concentration measurements. Ann Occup Hyg 47:123–144

25. Sears GW (1956) Determination of specific surface area of colloidal silica by titration with sodium hydroxide. Anal Chem 28: 1981–1983
26. Baltensperger U, Gaggeler HW, Jost DT (1988) J Aerosol Sci 19:931–934
27. Jung HJ, Kittelson DB (2005) Characterization of aerosol surface instruments in transition regime. Aerosol Sci Technol 39:902–911
28. Forny L, Saleh K, Denoyel R, Pezron I (2010) Langmuir 26(4):2333–2338
29. Weisner MR, Bottero JY (eds) (2007) Environmental nanotechnology: applications and impacts of nanomaterials. McGraw-Hill, New York
30. Rabinovich YI, Yoon RH (1994) Use of atomic force micro-scope for the measurements of hydrophobic forces. Colloids Surf A Physicochem Eng Asp 93:263–273
31. Yakubov GE, Vinogradova OI, Butt HJ (2001) A study of the linear tension effect on the polystyrene microsphere wettability with water. Colloid J 63:518–525 (Translated from Kolloidnvi Zhurnal 63(4):567–575
32. Delgado AV, Gonzalez-Caballero F, Hunter RJ, Koopal LK, Lyklema J (2005) Measurement and interpretation of electrokinetic phenomena (IUPAC Technical Report). Pure Appl Chem 77:1753–1850
33. Booth F (1948) Theory of electrokinetic effects. Nature 161:83–86
34. Dukhin SS, Semenikhin NM (1970) Koll Zhur 32:366
35. O'Brien RW, Hunter RJ (1981) Can J Chem 59:1878
36. Martin R, Alvaro M, Herance JR, Garcia H (2010) Fenton-treated functionalized diamond nanoparticles as gene delivery system. ACS Nano 4:65–74
37. Nel A, Xia T, Madler L, Li N (2006) Toxic potential of materials at the nanolevel. Science 311:622–627
38. Donaldson K, Stone V, Tran CL, Kreyling W, Borm PJ (2004) Nanotoxicology. Occup Environ Med 61:727–728
39. Garza KM, Soto KF, Murr LE (2008) Cytotoxicity and reactive oxygen species generation from aggregated carbon and carbonaceous nanoparticulate materials. Int J Nanomed 3:83–94
40. Brehm M, Schiller E, Zeller WJ (1996) Quantification of reactive oxygen species generated by alveolar macrophages using lucigenin-enhanced chemiluminescence – methodical aspects. Toxicol Lett 87:131–138
41. Hanson KM, Clegg RM (2002) Observation and quantification of ultraviolet-induced reactive oxygen species in ex vivo human skin. Photochem Photobiol 76:57–63
42. Holley AE, Cheeseman KH (1993) Measuring free radical reactions in vivo. Br Med Bull 49:494–505
43. Jahn K, Barton D, Braet F (2007) Correlative fluorescence and Scanning transmission electron microscopy for biological investigation. Mod Res Educ Top Microsc 1:203–211
44. Porter AE, Muller K, Skepper J, Midgley P, Welland M (2006) Uptake of C60 by human monocyte macrophages, its localization and implications for toxicity: studied by high resolution electron microscopy and electron tomography. Acta Biomater 2:409–419
45. Kapp N, Studer D, Gehr P, Geiser M (2007) Electron energy-loss spectroscopy as a tool for elemental analysis in biological specimens. Methods Mol Biol 369:431–447
46. Stearns RC, Paulauskis JD, Godleski JJ (2001) Endocytosis of ultrafine particles by A549 cells. Am J Respir Cell Mol Biol 24:108–115
47. Yen HJ, Hsu SH, Tsai CL (2009) Cytotoxicity and immunological response of gold and silver nanoparticles of different sizes. Small 5: 1553–1561
48. Robinson JM, Toshihiro T, Pombo A, Cook PR (2001) Correlative fluorescence and electron microscopy on ultrathin cryosections: bridging the resolution gap. J Histochem Cytochem 49:803–808
49. Sartori A, Gatz R, Beck F, Rigort A, Baumeister W, Plitzko JM (2007) Correlative microscopy: bridging the gap between fluorescence light microscopy and cryo-electron tomography. J Struct Biol 160:135–145
50. van der Wel NN, Fluitsma DM, Dascher CC, Brenner MB, Peters PJ (2005) Subcellular localization of mycobacteria in tissues and detection of lipid antigens in organelles using cryo-techniques for light and electron microscopy. Curr Opin Microbiol 8:323–330
51. Heymann JA, Hayles M, Gestmann I, Giannuzzi LA, Lich B, Subramaniam S (2006) Site-specific 3D imaging of cells and tissues with a dual beam microscope. J Struct Biol 155:63–73
52. Heymann JA, Shi D, Kim S, Bliss D, Milne JL, Subramaniam S (2009) 3D imaging of mammalian cells with ion-abrasion scanning electron microscopy. J Struct Biol 166:1–7
53. Matthijs de Winter DA, Schneijdenberg CTWM, Lebbind MN, Lich B, Verkleij AJ, Drury MR, Humbel BM (2009) Tomography of insulating biological and geological materials using focused ion beam (FIB) sectioning and low kV imaging. J Microsc 223: 372–383

54. Yamamoto Y, Shinohara K (2002) Application of X-ray microscopy in analysis of living hydrated cells. Anat Rec 269:217–223
55. Parkinson DY, McDermott G, Etkin LD, Le Gros MA, Larabell CA (2008) Quantitative 3-D imaging of eukaryotic cells using soft X-ray tomography. J Struct Biol 162:380–386
56. Lam CW, James JT, McCluskey R, Arepalli S, Hunter RL (2006) A review of carbon nanotube toxicity and assessment of potential occupational and environmental health risks. Crit Rev Toxicol 36:189–217
57. Poland CA, Duffin R, Kinloch I, Maynard A, Wallace WA, Seaton A, Stone V, Brown S, Macnee W, Donaldson K (2008) Carbon nanotubes introduced into the abdominal cavity of mice show asbestos-like pathogenicity in a pilot study. Nat Nanotechnol 3:423–428

Chapter 3

Methods for Understanding the Interaction Between Nanoparticles and Cells

Pilar Rivera_Gil, Martin J.D. Clift, Barbara Rothen Rutishauser, and Wolfgang J. Parak

Abstract

A critical view of the current toxicological methods used in nanotechnology and their related techniques. Hereby, toxicological effects derived from the intracellular accumulation and uptake will be examined. Then advantages/disadvantages of these methods will be discussed. Additional analytical techniques necessary to implement the results will be reviewed.

Key words: Nanotechnology, Nanoparticle–cell interactions, Methodologies in nanotoxicology, In vitro methods, In vivo methods, Microscopy, Spectroscopy, Biochemical tests, Molecular biology, Nanoparticle internalization

1. Introduction

Nanotechnology is a rapidly emerging field with the amount of money invested into its research and development, as well as its economic value increasing yearly (1), as shown by the level of nanotechnology-related products manufactured and distributed for use in a wide and diverse range of consumer, industrial, and technological applications (2). Due to this constant influx of nanotechnology-related products, it is essential that rigorous assessment of their risk to both human health and the environment is performed (3, 4).

In the past decade, progress in the field of nanotoxicology has gained increased intensity (2–5). Despite this, the field has not been able to maintain an even balance in regard to the known effects of nanoparticles (NPs) and the production of NP-based products and applications (4). The inability to provide clear and

Joshua Reineke (ed.), *Nanotoxicity: Methods and Protocols*, Methods in Molecular Biology, vol. 926,
DOI 10.1007/978-1-62703-002-1_3, © Springer Science+Business Media, LLC 2012

up-to-date knowledge on the effects of newly, engineered NPs has caused public perception of nanotechnology to form upon a lack of and incorrect information, an issue that could repeat the problems which inadvertently ended the genetically modified food industry (6). Clear examples of such issues include the recently published paper by the European Respiratory Journal, which claimed that chronic exposure to NPs not only induced altered cardiovascular and pulmonary function, but also caused human fatalities (7). In this study, Song et al. suggested that seven females who worked within a poorly ventilated print plant were exposed to NPs (polyacrylate ester) for up to 13 months. A range of different biological and immunological tests were performed on the female workers from which it was noted that five of the workers showed signs of increased pleural effusion, pulmonary fibrosis, and signs of granuloma formation in the lung. It was also suggested that two of the female employees died as a result of the NP exposure, although further assessment is necessary to confirm the subjects' cause of death. Additionally, the study assessed the effects of one male employee who was exposed to the polyacrylate ester from the paint in the workplace for a total of 3 months. It was reported that he only suffered from asthmatic-like symptoms. There was no association of this health state with NP exposure, however, as well as if he may or may not already have been asthmatic prior to working in the print plant. The response to this publication has brought increased skepticism and caution from world-leading figures in the field of nanotoxicology, particularly in relation to the lack of confirmation that the employees were exposed to actual NPs (by definition), as well as a clear association of the detrimental health effects observed following exposure to this NP type. It is of the upmost importance therefore that it is deduced whether or not the NPs that the workers were exposed to in the occupational setting drove the adverse health effects observed.

Despite such instances, research into the effects of engineered NPs has continued to increase over the past 10 years in order to obtain information that will enable the proposed advantages of nanotechnology to be realized (2, 4, 8). In light of this however, the proposed basis that the testing strategies used to determine the effects of ambient air NPs could be used to determine the potential toxicity of new, engineered NPs is not sufficient (9). Concurrent with the ever-increasing number of studies published in regard to testing the toxicity of NPs, heightened attention has focused upon the novel characteristics of each different type of engineered NP and how these characteristics might affect their interactions with biological systems (2–4, 8–10).

The aim of this chapter is to therefore highlight the major aspects of research in nanotoxicology, discussing the many issues and problems of the field, in order to provide a foundation for high impact, good quality scientific research in, and understanding of nanoscience (11).

2. Engineered vs. Accidentally Produced NPs

NPs are comprised from a diverse number of materials such as carbon, titanium, gold, polystyrene, cadmium, iron and zinc, and can either be accidentally produced (also referred to as naturally occurring NPs), or specifically engineered (8).

Naturally occurring NPs have been experienced by humans throughout all evolutionary stages. Examples of naturally occurring NPs include particles derived from volcanoes and fires (8, 10). Over the past two centuries, the increase in industry, the invention of the car engine as well as other combustion processes have resulted in a rise in the level of air pollution, as well as the amount of unintentional or accidental NPs released into the atmosphere (8). Within the workplace, NPs have also been found to be present in the emissions from welding fumes (12) and natural gas-powered equipment (13), whereas in the environment, power plant emissions (independent of being powered by coal, oil or natural gas) have been observed to produce large amounts of NPs into the atmosphere. The major source of accidentally produced NPs in the environment, however, is from traffic exhaust emissions (8). Also known as combustion-derived NPs, these emissions can increase the number of particles in the air by up to ten million particles per cm^3 (14). The effects of exposure to increased levels of particulate present within environmental air pollution have been well documented over the past 20 years. Such research has demonstrated accidental NPs to be more potent in causing toxic and adverse health effects, such as cardiovascular disease and chronic obstructive pulmonary disease (COPD) (15), than their larger particle counterparts at the same mass dose (8). It is the findings from studies (investigating the toxicity of NPs present within environmental air pollution) which have prompted concerns as to the possible hazardous health effects associated with exposure to engineered NPs on human health and the environment (2, 16).

Engineered NPs are defined as particles engineered or manufactured on the nanoscale with specific physicochemical compositions and structures to exploit properties and functions associated with their dimensions (8, 10). Furthermore, engineered NPs include particles that have either a homogeneous or heterogeneous chemical composition or structure, or are multifunctional and can be different shapes, for example spheres, fibers, tubes, rings, and planes (8). Examples of these include: (1) carbon nanotubes (CNTs)—graphite-like carbon tubes with capped ends; can be either a single graphite-like carbon tube, known as single-walled CNTs (SWCNTs), or a series of SWCNTs, which can be either (1) stacked inside one another, or (2) a single CNT rolled inside of itself, known as multiwalled CNTs (MWCNTs); (5), (2) C_{60} fullerenes—complex, semi-conductor, spherical 3D carbon structures

(17), (3) superparamagnetic iron oxide NPs (SPIONs)—spherical NPs consisting of an iron (Fe) core, coated with a shell material (usually polysaccharides or polymers (18) but it can also be gold (19)), (4) other inorganic NPs made out of an inorganic core (like metallic, semi-conductors elements surrounded by a polymer shell) (20), and (5) dendrimers—a symmetrical arrangement of macromolecule branches arising from a multifunctional core (21). In addition, the specific physicochemical properties of engineered NPs can be manipulated to create different surface coatings/charges surface chemistries and surfactants, such as organic (neutral charge), carboxylated (COOH) (negative charge), and amino (NH_2) or amine (positive charge), as well as the ability to attach polymer groups and biologically based structures (22–25).

Despite the vast array of manufactured NPs already in use throughout a wide range of applications (4), newly engineered NPs are constantly being created to incorporate the increased demand for smaller, lighter, faster, more efficient, and durable products (4). The potential toxic and adverse health effects of these "new" NPs, however, are not fully understood (3, 4, 26). It is vital that efficient, model ways of analyzing the diverse physiochemical components of manufactured NPs is determined to comprehend both the potential detrimental and advantageous effects of engineered NPs (2).

3. Human Exposure to NPs and Their Subsequent Cellular Interaction

Due to the many different forms of NPs being produced in a wide variety of consumer, industrial, and technological applications, the exposure routes for which humans will inevitably be exposed to NPs are numerous. The specific routes by which NPs may enter the human body, and potentially elicit adverse effects, are understood to include inhalation, injection, ingestion, and permeation through (diseased) skin (Fig. 1) (10, 27). Therefore, primary exposure to these organs (lung/blood/gut/skin) must be extensively investigated, although it is also essential that the concerns regarding NP translocation and secondary toxicity are also studied.

Currently, research into the effects of nanomaterial exposure via ingestion is negligible. The understanding of the effects of NPs following injection is also limited. Analysis of the effects of nanomaterials on the skin, however, although also being in its infancy, has widely suggested that the optimal opportunity for uptake by the epidermis is when the outer skin cells are broken, such as when the skin is (sun)burned or diseased (28). Tinkle et al. (29) hypothesized that when broken skin is flexed it is more sensitive to penetration by NPs, such as titanium dioxide (TiO_2), as it forms a more

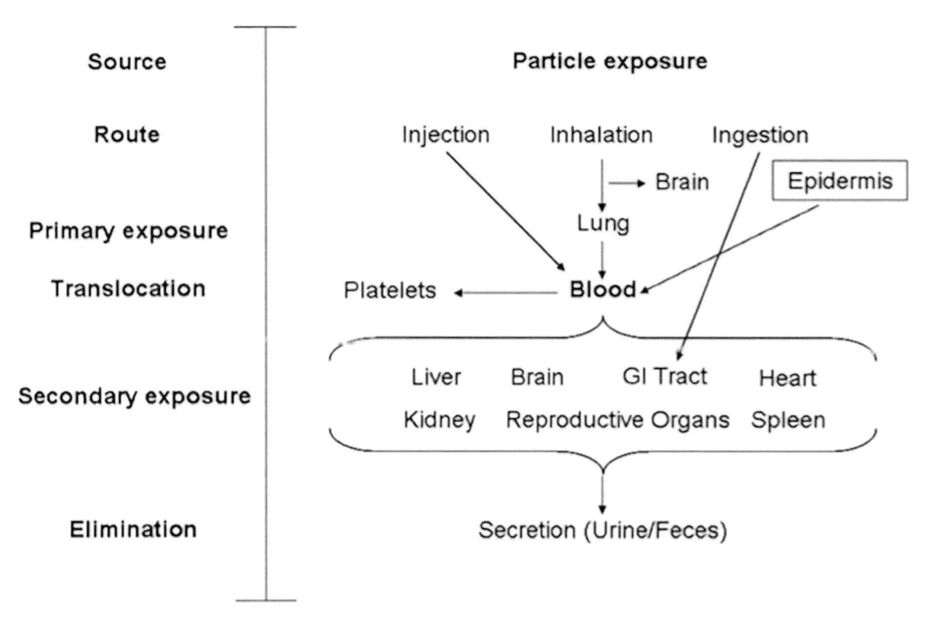

Fig. 1. Overview of the exposure routes of NPs into the human body and the subsequent fate (adapted from Oberdorster et al. (10)). In this figure, the epidermis is highlighted to denote that only the diseased skin is at risk to NPs, and not healthy skin.

permeable environment, allowing translocation of the nanomaterials to the lymph nodes and then subsequently into the blood circulation. In addition there is ongoing discussion if certain NPs can penetrate the epidermis through hair follicles and/or sweat glands (28). Further investigation is necessary in order to support this hypothesis, as well as to understand the interaction of NPs with both healthy and diseased/injured skin.

Although information relevant to the interaction of NPs with the gut, blood, and skin are relatively unknown, the interaction of NPs with the lung have been well documented over the past 20 years (8, 30, 31). Indeed, exposure to most forms of NPs will initially be via inhalation, especially when considering occupational exposure, and thus will affect the respiratory system (8). The exposure of NPs to biological systems of the lung, specifically cell (mono)cultures, has been a basis for increased discussion recently (32, 33). In most nanotoxicology studies, particles are applied in suspension (e.g. suspended in cell culture medium). If NPs are to translocate from the lung to secondary target organs (such as the liver or brain), then it is possible that the interaction between NPs and cells could be mimicked via NP suspension exposure to cell cultures (due to the NPs being translocated within blood). This aspect, however, as previously highlighted, requires further, in-depth analysis to (1) determine if and how NPs are translocated from the lung and (2) at what dose the NPs are translocated in and thus interact with the secondary organs. In addition, it is paramount that the specific characteristics of the translocated NPs are determined, as it is also possible that they are no longer in the same format as the NPs that entered the lung. Therefore, in order to achieve a more advantageous correlation between the in vitro and the in vivo situation, regarding specific NP-lung interactions, cells should be cultivated at an air–liquid interface and the particles have to be applied from the air. In order to study particle–cell interactions in cell culture systems representing the airway epithelial barrier, it is important to mimic the in vivo interactions of particles with cells as closely as possible. The particles have to maintain their shape as well as their chemical and physical properties when they come into contact with the cell surface to realistically clarify which particle properties contribute to toxic effects. As highlighted previously, the majority of existing cell culture models used to assess lung cell–particle interactions use cells immersed in medium, with particles added to the cell cultures as a suspension in liquid (30, 34–37). This does not reflect the physiological condition of lung epithelial cells, which are exposed to air, and separated from it by only a thin liquid lining layer with a surfactant film at the air–liquid interface (36, 38, 39). In order to rectify this issue, a number of exposure systems have been constructed to enable the assessment of various different NP types on lung cell cultures at the air–liquid interface (32, 40, 41).

The availability and toxicity of any substance to a biological organism are determined by both the concentration/dose that the organism is exposed to and the "toxicokinetics" of the substance. These include the uptake, transport, metabolism, and sequestration to different compartments by the organism, as well as the elimination of the substance from the biological organism. These parameters are essential, since the potential toxicity of substances is dependent upon the specific organs or cell types exposed, which form the substance is in (e.g., bound to serum protein, aggregated, dissolved, oxidized), as well as the period of time the substance interacts/remains at the site of primary and secondary exposure.

These factors are important to NP toxicology since the realization that NP localization and fate are not only restricted to their portal of entry (Fig. 1). Recently, it has been reported that NPs can be distributed to organs distal to their site of exposure, so that potential NP toxicity can occur in any number of secondary targets (i.e., liver/kidney/brain) (9). Research into the possible secondary toxicity of NPs (i.e., in distal organs from their original exposure route) is extremely limited and is very much in its infancy. Despite this, the limited number of studies investigating the effects of NP translocation to secondary organs have shown that NPs can elicit negative effects to the liver, brain, GI tract (following inhalation), spleen, reproductive systems, and the placenta (42–45).

As mentioned before, exposure of these particles is often unavoidable. Directly (e.g., medicines, pollution) or indirectly (e.g., electronics) they become in contact with the human body. Depending on the way of exposure, NPs can distribute systemically and can translocate to different organs/tissues. Herein NPs are internalized or enter through different uptake mechanisms into the different cell types. Due to particle internalization, cells or tissues in contact with NPs will be exposed to the particles for extended periods of time. As a result of their intracellular accumulation, NPs may cause cytotoxicity by interfering with the regular cell cycle (46). Furthermore, upon uptake most NPs agglomerate intracellularly due to a loss of stability because of the acidic environment (47), which can cause cellular damage (48), but might not necessarily induce cell death (49). These effects can be accentuated by the fact that most NPs are not biodegradable, and the lack of metabolization/excretion leads to toxic bioaccumulation. Concerning cytotoxic effects one has to distinguish between effects related to the nature of the material and effects common to NPs of even inert materials. Also, for inert particles such as gold, inflammatory effects in tissues caused by particles have been demonstrated. However, in cell culture experiments Au NPs are regarded as biocompatible, and acute cytotoxicity has not been observed so far. In particular, no release of toxic ions as in the case of cadmium-based NPs has been reported. On the other hand, there are few examples of toxic effects related to the nature of Au, which might depend on the cell line, on surface chemistry, and on the NP size. Actin fibers inside the cell, for example, can be affected by the presence of NPs, and very small Au-clusters have been demonstrated to fit into the grooves of DNA molecules and thus cause cytotoxic effects (50). The assessment of nanotoxicity is intrinsically linked to the characterization of NPs uptake because uptake provides strong evidences about NP–cell interactions. For example, the whole NP uptake process as observed by detection of changes in the zeta potential values after cell incubation with NPs for different time lengths affects the surface charge of cells over time (51).

NP–cell interactions depend on the surface aspects of materials, which may be described according to their chemistry or lyotropic characteristics but also depend on the physics behind, which may be described thermodynamically according to their colloidal state. Both parameters determine how the particles will adsorb to the cell surface and specially determine the cell behavior on contact and thus cytotoxicity. NPs can strongly interact with cell membranes. In case of charged NPs, NP–cell interactions tend to be caused by van der Waals and other electrostatic interactions, which can generate stresses that, in turn, result in deformations of the contacting membrane. Due to their hydrophobic tendencies, uncharged or hydrophobically modified NPs of small sizes (~1–10 nm comparable to the cell membrane bilayer) can incorporate into cell membranes resulting in membrane swelling (52). However, strongly charged or more hydrophilic NPs have attractive interaction with the zwitterionic headgroups and thus can pull the phospholipids away from the membrane, forming a hybrid micelle (association colloids made from a NP core and phospholipid shell). These processes can lead to the rupture of the original membrane and the formation of holes (52). The mechanism of the changes in the membrane structure as a result of the interaction with NPs can be explained thermodynamically. Association colloids are made of amphiphilic molecules (like in the lipid bilayer) which strongly reduce the interfacial or the system free energy so that the formation of these hybrids micelles is energetically favorable (53). In case of larger particles, the activation barriers for these processes increase proportionally to the particle surface area. For large neutral particles, these barriers may become difficult to overcome and the membrane conformation might remain intact. For large, strongly charged particles, and in case the adhesion strength to the membrane is above a threshold value, the cellular membrane first wraps around the particle and then the particle internalized in a vesicle can detach form the original membrane. In this way, charged particles could overcome the activation barriers and pull the phospholipids away from the membrane in the form of a vesicle (endocytosis) without significantly disrupting the integrity of the cellular membrane (54, 55). A thermodynamic model based on the "wettability criterion" (56) suggested that in case of a particle radius less than the contact radius, the entire particle can become engulfed by the membrane (57). Even if the particle radius equals the contact radius, in case NP–cell membrane interactions engulfment is also energetically favorable since the interfacial energy is sufficiently low due to the amphiphilic nature of the cell membrane. In this sense, the initial NP–cell interactions are of passive nature although further cellular uptake still relies on endocytic processes. Beside the passive uptake there is also active internalization processes, which are mediated by the functional groups attached to the surface of NPs and/or by the protein corona that may be formed upon interaction of the NPs with biological fluids. Active uptake is triggered mainly by ligand–receptor interactions that lead to specific interactions with the cell membrane. Signal-mediated transport via pores occurs when the size of the complex is less than 0.1 μm (58). Otherwise there is a vesicle-mediated transport through the cytoskeleton, i.e., macropinocytosis, clathrin-mediated endocytosis, caveolin-mediated endo- and transcytosis, and clathrin- and caveolin-independent endocytosis. Remarkably, the uptake mechanism of the ligand can be perturbed upon conjugation to the NP. In this sense, the normal route of internalization and thus the following activation of signaling cascades may be disrupted (59, 60). Intracellular accumulation of a ligand-NP bioconjugate can induce alterations in the normal intracellular transport of unconjugated ligands mostly due to the increase size of the complex (61). The nature of these ligands is predominantly protein. Upon adsorption to the NPs, these proteins might change their conformation and consequently their function. As a result of conformational changes, the immunological system may then not recognize these proteins as native but rather as foreign objects and may try to eliminate them, inducing autoimmunity. Additionally, adsorbed molecules that are normally excluded from cells could enter the cells with the help of the NPs.

In order to correlate any toxic reaction with a NP type, it is indispensable to investigate if the particles are adsorbed to the cell surface or if they enter cells. If NPs are found inside the cells, their localization in different compartments such as endosomes, lysosomes, mitochondria, the nucleus, or the cytosol may also provide some answers regarding their potential toxicity. In order to elucidate how to maximally accumulate NPs in cells, without killing or altering cellular function is indispensable to quantitatively and qualitatively characterize uptake and distribution.

3.1. Optical Microscopy

Laser scanning microscopy (LSM), also described as confocal LSM, is a valuable tool for obtaining high spatial resolution of specific fluorescently labeled structures at the light microscopic level. The LSM was developed to overcome some of the limitations of conventional (i.e., widefield) microscopy. In the latest, the whole sample is illuminated from the light source and the fluorescence is collected from the whole specimen. A laser scanning microscope collects information from a thin, focal plane and ignores out of focus information (62). The information obtained by the scanning process is free of out of focus blur resulting in a greater depth of field. Nowadays, LSM is routinely performed by using multilabeled specimens, allowing a comparison of the spatial arrangement of various proteins or cell structures. The LSM provides advantages over other microscope methods, as the specimen preparation is fast and nondestructive. However, the resolution of optical microscopy is limited by light and therefore is unable to resolve single NPs.

Video-enhanced differential interference contrast (VEDIC) techniques are able to visualize NPs as small as 10 nm, although the image resolution remains at the limits of optical microscopy. This hybrid technique combines the high resolution of color video-enhanced microscopy and the polarized light conditions of DIC. VEDIC enables to track the trajectories of NPs inside cells and easily distinguish NPs from organelles and cellular granules (63).

3.2. Electron Microscopy

Transmission electron microscopy (TEM) offers adequate resolution to visualize NPs at a single particle level as well as the ability to determine their localization in different cellular compartments. However, only few particle types, such as gold NPs, show unique characteristics like particle shape and electron density that can be easily recognized within cellular compartments. For NPs which are not that electron dense or are the same size as organelles (i.e., ribosomes), or can be mistaken with high-density cellular granules, or are made up of numerous different metals—such as quantum dots—additional elemental analysis of the NP compositions could help to confirm uptake and their intracellular localization. This can be performed at the TEM level by energy filtered TEM, since each chemical element shows a characteristic electron energy loss spectrum (64). Electron energy loss analysis in combination with TEM employed in the spectrum mode (electron energy loss spectroscopy, EELS) and/or in the image mode (electron spectroscopic imaging, ESI) is an alternative approach that can offer advantages in terms of spatial resolution and NP identification (65, 66). With this technique, however, the need to use very thin specimens has represented a severe limitation.

The question for electron microscopy-derived techniques arises from the fact that if during exposure to aqueous fixatives and dehydration solutions, loss and/or redistribution of the NPs from the original location take place.

3.3. Spectroscopy

The analytical technique conductively coupled plasma atomic (or optical) emission spectroscopy (ICP-A(O)ES), can add important quantitative information to optical and electron imaging methods such as LSM, VEDIC, and TEM regarding the fate of NPs in cells. It is a type of emission spectroscopy that uses inductively coupled plasma as a radiation source to dissociate molecules. The ionized atoms emit electromagnetic radiation at wavelengths characteristic of a particular element. The intensity of this emission is indicative of the concentration of the element within the sample. The most crucial limitation is the possible interferences between some elements and the different matrices. ICP-AES have been widely used to study the uptake of different NPs quantitatively (67, 68). However, one has to keep in mind that ICP-AES does not discriminate the size and shape within a heterogeneous sample (69).

4. Methods to Assess Nanotoxicity

4.1. Cytotoxicity and Cell Death

When studying the potential toxicity of NPs, researchers have predominantly used an assessment of the permeability or integrity of the cell membrane in order to assess the cytotoxicity of NPs. This has predominantly been achieved by measuring the amount of lactate dehydrogenase (LDH) present within the cell supernatants following NP exposure. LDH is an enzyme present within the cytoplasm of cells and can be subsequently released following a loss in cell membrane integrity (70). LDH release from cells has been shown to be a good indicator of cytotoxicity (71), as well as associated with necrotic cell death (70). It must be stressed, however, that assessment of LDH release from cells does not indicate necrosis, nor any form of cell death. Quantification of the level of LDH enzyme release from particulate-treated cells into the external cell culture medium (cell supernatant) (this process is also known as membrane leakage) can be achieved by assessing the catalytic interconversion of pyruvate to lactate, concomitantly with an investigation of the interconversion of nicotinamide adenine nucleotide (NADH) to its oxidized form nicotinamide adenine dinucleotide (NAD^+). Predominantly LDH is assessed via spectrophotometry and absorbance.

The measurement of the level of LDH release to denote the cytotoxicity of a NP is not sufficient. Numerous colorimetric tests are also available to assess the potential cytotoxicity of NPs. These include the MTT, WST-1, and MTS assays. These assays are based upon the ability of living, metabolically active cells to cleave a tetrazolium salt. These assays enable the determination of cell survival, proliferation, and activation within a wide variety of cells, including macrophage-like tumor and lymphoma cell lines, as well as stimulated myeloma cells. Similar to the LDH assay, this is achieved via quantification by spectrophotometry of the formazan generation by active mitochondria, indicative of the energy metabolism of the mitochondria (levels of mitochondrial succinate dehydrogenase) within living cells (72). Generally, it is accepted that an observed increase in absorbance is indicative of cell proliferation, whilst a decrease in absorbance suggests either a reduction in metabolic activity and/or a decrease in viable cell number.

Although the above-mentioned assays take into consideration the potential cytotoxicity of NPs on the mitochondria and the cell membrane, in order to determine the absolute toxicity, otherwise known as the lethal dose (LD_{50}) of a substance, an assessment of the substance's ability to cause cell death is necessary. Cell death can be attributed to numerous cell cascades, resulting in different cell death processes. There are, however, two specific forms of cell death, which have received increased attention in relation to NP exposure to cells. The first of these processes is apoptosis, also known as controlled cell death. The second, necrosis, is a pathological process, as it has been shown to occur in response to externally induced toxicity, including inflammation. Cell death is commonly assessed by the use of specific fluorescent staining solutions/antibodies for both apoptosis and necrosis. Although there are many different kits and tests available, assessment of NP-induced cell death is generally achieved by performing the Annexin V assay. When cells undergo apoptosis, one part of this process includes the transport of inner cell membrane lipids (such as phosphatidylserine) to the outer side, serving as a marker for macrophages to eliminate these cells. In the Annexin-V assay, the fluorescent dye Annexin-V-Fluorescein binds to phosphatidylserine, thus marking the apoptopic cells. The necrotic cells are also distinguished via the use of a fluorescent dye, propidium iodide. Staining with the fluorescent dye propidium iodide identifies the DNA of cells. This method can be achieved, similar to the mitochondrial membrane potential assay, both qualitatively and quantitatively, thus the result can be observed via both CLSM and FACS™ analysis.

4.2. Reactive Oxygen/ Nitrogen Species and Oxidative Stress

If the NPs exposed to cells do not cause an LD_{50}, then there are a variety of possible adverse, sub-lethal effects that they (the NPs) can elicit upon the cell. It can be argued that the oxidative stress paradigm (15) is the foundation for nanotoxicology research and for studying such potential sub-lethal effects. There are numerous assays to assess such endpoints contained within this paradigm. The most common are ones associated with the assessment of the formation of reactive oxygen species (ROS) and oxidative stress.

In regard to the production of ROS, the most common parameter used is the fluorescent dye 2′,7′ dichlorodihydrofluorescein-diacetate (DCFH-DA), a reliable fluorogenic marker for ROS in living cells. This method assesses the production of ROS via the increase in fluorescence. As upon the production of ROS, the DCFH-DA molecule is cleaved, resulting in a DCFH molecule which is highly fluorescent. The fluorescence signals can be both imaged via CLSM and quantified using FACS™ analysis and fluorescent spectophotometry. A similar form of testing procedure is also based upon the fluorescent dye dihydroethidium (DHE). Similar to the DCFH-DA reaction, in which ROS production is determined by the level of fluorescence, DHE detects the extent of oxidative stress and ROS production in cells, by oxidizing to ethidium, which is highly fluorescent. These methods are extremely advantageous, as not only can they be performed via multiple approaches, but they can also be used in a cell-free environment. This is important, as it provides important information pertaining to the NPs alone, and thus their ability to cause ROS independent of a cellular environment.

The production of ROS is concomitant with a reduction in antioxidants, thus causing an oxidant imbalance within the cell and therefore oxidative stress (73). It is therefore pertinent that an assessment of the antioxidant capacity is also performed in order to obtain a thorough and clear understanding of the oxidant-related effects of NPs. This form of analysis has commonly been performed in regard to an assessment of intracellular glutathione levels. Glutathione is an intracellular nonprotein, thiol readily available within a wide range of living cells, and is a key factor in a number of biological functions, including oxidative stress (74). Usually present in cells as GSH in an oxidative environment, it exists as oxidized glutathione (GSSG), although is rapidly converted to GSH via an enzymatic reaction using glutathione reductase. In the method originally published by Hissin and Hilf (74), the levels of both GSH and GSSG were determined via their pH-sensitive reactions with the fluorescent reagent ophthaldehyde. There are of course many alternative ways to determine glutathione levels within cells, including diagnostic kits (via either absorbance or fluorescence) and microscopy related assays. In addition, there are numerous ways to assess the antioxidant capacity of cells, such as the Trolox Equivalent Antioxidant Capacity (TEAC) method; however, this is extremely laborious and difficult to optimize.

In addition to determining the antioxidant capacity, it is possible to determine the level of oxidants present within the biological sample. This is notoriously difficult and time consuming, but provides extremely valid and optimal results in regard to the oxidative potential of NPs. Such testing strategies include assessment of the hydroxyl radical (•OH), superdismutase 2 (SOD2), and heme oxygenase-1 (HO-1). Due to the intrinsic nature of these radicals within a cellular environment, the most effective method to determine their presence is polymerase chain reaction (PCR), despite the laborious and expensive nature of this technique.

In addition to the measurement of ROS and oxidative stress, it is prudent to determine the ability for nitric oxide (NO) to be produced. Often, however, NO is not assessed in relation to the radical formation potential of NPs. Increased research should be directed toward this aspect though, due to the fundamental input of NO to cellular function. A radical gas, NO is derived from the guanidino nitrogen of the amino acid l-arginine and molecular oxygen in a reaction catalyzed by the enzyme nitric oxide synthase (NOS). There are at least two types of NOS. The calcium-dependent form is present constitutively in a variety of tissues and produces the physiological concentration of NO needed for maintenance of blood pressure. The other, calcium-independent form is inducible NOS (iNOS), present in a number of different cell types including macrophages, hepatocytes, neutrophils, and endothelial cells, by a variety of immunological stimuli such as interferon-γ (IFN-γ) and bacterial lipopolysaccharide (LPS). iNOS is a high output enzyme and nitrite concentrations are easily detected. It is possible to measure NO (or iNOS) via spectrophotometry using a colorimetric test known as the Greiss reaction, for which numerous diagnostic kits are available. It is important to note, however, that problems have been reported when assessing NO in primary cells, as well as some cell lines (e.g., A549 epithelial cells). This test is, however, advantageous in assessing the ability for NPs only to produce NO (i.e., in a cell-free environment). In order to assess NO in cells therefore, it is also possible to use PCR analysis.

4.3. Inflammation

Assessment of inflammation in vitro is usually achieved via the use of enzyme-linked immunosorbant assay (ELISA) diagnostic kits. Although expensive, these diagnostic kits provide valuable and reproducible testing strategies to measure the plethora of different pro-inflammatory cytokines and chemokines that have been associated with NP exposure to culture cells. In addition to ELISA kits, it is possible to measure cytokine/chemokine production via either PCR analysis or FACS™ analysis; however, these methods are again laborious as well as expensive, and in comparison to the ELISA method provide no real additional advantages.

4.4. Cell Signaling and Gene Expression

The influence of NPs on cell signaling, specifically upon Ca^{2+} signaling in cells, is a significant factor relating to the oxidative stress paradigm (15). In relation to this, Ca^{2+} signaling is not as well researched as in comparison to the ability for NPs to stimulate ROS and inflammatory cytokines/chemokines. Although possible to measure intracellular calcium levels via a number of means, again including diagnostic kits, the most commonly used indicator of changes in intracellular calcium is the fluorescent marker FURA 2-AM (75). The function of this dye is similar to DCFH-DA, as during the incubation period with the cells (and xenobiotic), if there is a change in Ca^{2+}, intracellular esters are able to remove the acetoxymethyl (AM) group from the Fura 2-AM molecule, revealing the membrane impermeable and Ca^{2+} sensitive fluorescent dye, Fura 2. It is then possible to determine the level of Ca^{2+} present in samples via spectrophotometry, as well as by CLSM.

In addition to this, PCR is a very useful tool in assessing the effects of cell signaling, as well as gene expression. As mentioned, however, although this method can identify specific proteins and genes (ribose nucleic acid (RNA) transcription/isolation), it is laborious and requires increased optimization time in order to obtain valid and reliable results.

4.5. Genotoxicity

As with cytotoxicity/cell death, oxidative stress and inflammation, when investigating the genotoxicity of NPs there are numerous different protocols that can be performed. Recent research, however, has focused upon only a few specific tests which measure the ability for NPs to inflict DNA damage, mutagenicity, and cell proliferation.

DNA damage is commonly investigated via the comet assay. In addition to measuring DNA damage, this assay assesses other DNA alterations in cells (76). The extent of DNA damage is revealed via electrophoresis of the agarose embedded single cell samples, showing both DNA fragments or damaged DNA to migrate away from the nucleus (forming a comet shape). After staining with a DNA-specific fluorescent dye, such as propidium iodide, the fluorescent signal of the gel is then measured in the head and tail and length of tail. The extent of DNA liberated from the head (to the tail) of the comet is directly proportional to the amount of DNA damage. Quantification of the liberated DNA is generally achieved via the use of specific computer software, but can also be determined by eye by giving comets a series of numbers relating to the severity of DNA damage. This latter method, although original, is flawed, specifically due to the impact of human error, and so is not advisable despite the expense of the computer software. In addition to the original comet assay, alternative testing protocols have been developed in which oxidative DNA damage can be assessed simultaneously. For example, by inclusion of the formamido pyrimidine glycosylase (FPG) enzyme, the assay also allows for specific detection of oxidative DNA lesions (77). Although this form of the comet assay requires double the samples, it is extremely advantageous to obtain an understanding of the oxidative/genotoxic potential of NPs.

Assessment of a NP's ability to cause mutagenicity is also a prominent testing strategy when determining the genotoxic potential of NPs. A well-defined testing protocol for mutagenicity is the micronucleus assay. This test has been proven to detect both clastogenic and aneugenic events, as well as to be valid for genotoxicity testing of micron- and nano-sized particles (78, 79). In addition to the micronucleus test, a recent testing protocol for NP genotoxicity is the Ames test. This test is based on strains of bacteria that have point mutations on the histidine (His) operon. Due to the position of the mutation on the HIS-gene, the bacteria are incapable of producing histidine and therefore unable to proliferate unless histidine is present (80). If chemicals are mutagenic, then additional mutations occur on the His-gene, known as base substitutions or frameshift mutations, which enable the bacteria to produce histidine and subsequently proliferate. The mutagenicity of the chemical is then determined via the quantification of the ability of the histidine-free bacteria's ability to replicate. Previously, this test has been used within industry to screen chemicals; however, recently it has all been successfully used to study the mutagenicity of NPs, specifically DEPs (81). There is debate, however, to the specificity of this mutagenicity test, due to the interaction of NPs with bacteria and not cells. The understanding of the interaction between NPs and bacteria is extremely limited, if not unknown, and therefore it is not understood whether or not the NPs will interact with the bacteria in a similar manner as to a cell, and thus whether or not use of this protocol will provide specific, realistic mutagenic effects of NPs in relation to human health.

Cell proliferation is also a key factor in the genotoxic potential of NPs (82). As previously highlighted, this can be achieved via a number of different assays, such as the MTT, WST-1, and MTS assays. Due to the reported concerns and difficulties of using these assays, an alternative test for determining the ability of cells to proliferate is the BrdU Assay. The BrdU assay is an immunochemistry technique which assess newly synthesized (or partially denatured double stranded) DNA strands in cells that are undergoing active proliferation via fluorescent microscopy. This technique enables

Assay type	Category	Cellular property/process probed	Assays
Viability			
	Proliferation	Metabolic activity	MTT, XTT, WST-1, Alamar Blue
		DNA synthesis	$[^3H]$Thymidine incorporation
		Colony formation	Cologenic
	Necrosis	Membrane integrity	LDH, Trypan Blue, Neutral Red, propidium iodide
	Apoptosis	Membrane structure	Annexin-V
	LIVE-DEAD	Esterase activity/membrane integrity	Calcein acetoxymethyl/ethidium homodimer
Mechanistic			
	DNA damage	DNA fragmentation	Comet, CSE
		DNA double-strand breakage	TUNEL
	Oxidative stress	Presence of reactive oxygen species (ROS)	DCFDA, Rhodamine123
		Lipid peroxidation	C11-BODIPY
		Lipid peroxidation	TBA assay for malondialdehyde
		Presence of lipid hydroperoxides	Amplex Red
		Antioxidant depletion	DTNB
		Superoxide dismutase (SOD) activity	Nitro blue tetrazolium
		SOD expression	Immunoblotting

Fig. 2. Summary of common in vitro toxicological techniques used in the assessment of nanotoxicity. Reprinted with permission from (83). Copyright 2009, RSC Publishing.

the assessment of cells that have been treated with NPs to actively synthesize and proliferate. Similar to the MTT, WST-1, and MTS assays, this test can be performed via purchasing a diagnostic kit and using spectrophotometry, as well as CLSM, therefore, providing a valid and reliable testing strategy for cell proliferation in relation to NP genotoxicity.

4.6. Other Analytical Techniques

In general in vitro toxicological techniques used to assess nanotoxicity can be divided according to the assay type into mechanistic and viability (Fig. 2). The first referred to cellular processes related to DNA damage and oxidative stress. The latter rather involves cellular processes like proliferation (e.g., XTT, colony formation, Alamar blue), necrosis/apoptosis (e.g., LDH, Neutral red, Annexin-V, etc.), or sterase activity/membrane integrity assays (83).

Alternatively, there are novel techniques that can implement the characterization of nanotoxicity. Gene expression analysis has been used to compare at the level of RNA NP-exposed cells to control cells (84, 85). High-content screening assays have also been recently applied to nanomaterials to detect toxicity of different parameters simultaneously (85, 86). In addition, a computer-based optical counting method was developed (87) and used to sensibly and quantitatively detect the toxic effects of even very low NP concentrations based on the adhesion behavior of living cells (88).

5. Use of Appropriate Controls When Assessing the Toxicity of NPs

In addition to using the correct (and most realistic) concentration of NPs (89), the correct form of exposure system as well as fully characterizing the NPs in their buffer, water, and specific cell culture media, (90) it is also vital that additional and appropriate controls are performed.

When choosing a toxicology testing strategy, it is important that a relevant positive control is also chosen. Although negative controls are easily identified (usually cell culture medium/buffer only), determination of the correct positive control must be given detailed thought. Also, the ability for proteins to adsorb to the surface of NPs is realistic and can significantly affect the ability for NPs to interact with cellular systems. It is therefore also essential that the ability for proteins to adsorb to the surface of particles during toxicological tests is assessed (such as when using an ELISA). This is also true of enzymes, such as LDH. It is known that this enzyme can also adsorb to the surface of NPs, masking their toxicity and thus providing a false-negative toxic result. As with protein adsorption, enzyme adsorption must also be investigated to correlate valid and representative toxicity data. Additionally, it is necessary to determine the ability for the NPs used to interact with the assays in regard to the fluorescent dyes or formazans that are used. An example of this was reported by Worle-Knirsch et al. (91), where CNTs were found to interact with the MTT formazan (tetrazolium salt) used and subsequently provided a false-negative toxicity. In addition to this, it is pertinent that an assessment of the toxicity of the suspension media/buffer is performed. Increasingly, NPs are suspended in buffers (or surfactants for optimal dispersion) such as Pluronic F127 in order to obtain a well-dispersed and characteristic NP suspension (92). If, however, a toxic response is observed following cellular exposure with NPs suspended in such buffers, it is essential that the toxicity of these buffers is known, in order to assess the specific effects of the NPs only (92).

6. Alternative Test Systems for Nanotoxicology

In vivo (specifically mice and rat models, however, also considering research on hamster, guinea pig, and monkey models) research is known to enable scientific research to observe almost "first-hand" the effects of a substance as they would occur in a *Homo sapiens*, in vitro models provide the possibility to investigate toxic effects on human cells extensively, which cannot be conducted in vivo (93). Cultured human and animal cells can be better controlled and

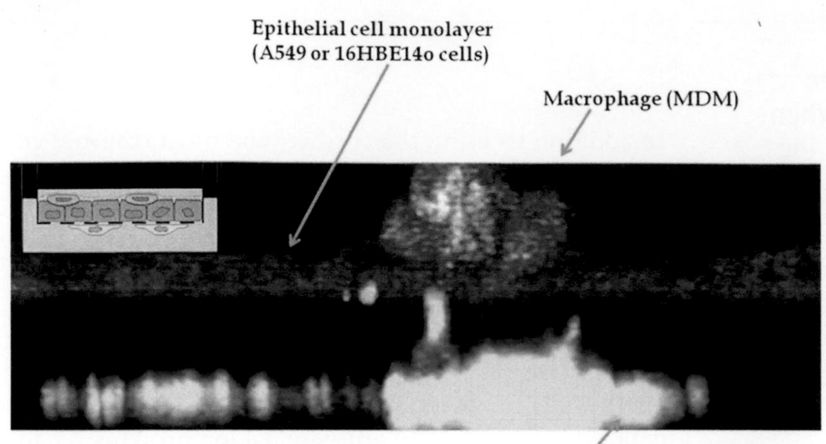

Fig. 3. A three-dimensional (3D) cellular model of the human respiratory tract to study the interaction with particles. The 3D triple cell coculture model system is composed of epithelial cells (A549 or 16HBE14o), human monocyte-derived macrophages (MDM), and dendritic cells (MDDC). Each cell type has been stained for specific markers (CD14 for the MDM and CD86 for the MDDC) as well as for F-actin at the side of the epithelial monolayer (where the MDM are placed) to allow visualization under a confocal laser scanning microscope. A schematic representation of the 3D coculture is presented on the *upper left corner*. The figure has been adapted from (93, 95) (Color figure online).

therefore yield more reproducible data than in vivo systems; however, they require a high standardization to maximize reproducibility.

Although monoculture systems provide the basis for high-throughput analysis for nanotoxicology (94), they do not represent a realistic model of how NPs will interact with the specific organ of the body in vitro. Over the past 5 years, there have been increased efforts to establish more realistic models to study the toxic potential of NPs. An example of such an effort is a 3D triple cell coculture model system composed of epithelial cells (Fig. 3), human monocyte-derived macrophages, and dendritic cells which has been established, simulating the most important barrier functions of the epithelial airway (95). This model provides a clear basis for investigating the interaction of NPs with the lung (93, 95) as well as at the air–liquid interface (96). One aspect emphasized by the model of Rothen-Rutishauser et al. (95) is that the architecture is specific to that as it is in the human lung (i.e., macrophages on the luminal side, a layer of epithelial cells and dendritic cells on the basal side). This type of detail is essential, as it provides a clear sign as to the interaction of NPs at the epithelial airway wall.

Studies using coculture cell systems have reported different effects from the combined cell culture system, compared to monoculture analysis (97, 98). Such reactions observed, however, from a culture containing two, three, or four different types of cells merely cultured in the same dish, although providing data showing that NPs interact with different cells in an opposite manner to each cell monoculture, do not specifically highlight the effects as they would occur in the human body. Thus, architecture of such coculture models is essential. To implement correct architecture in in vitro coculture models, a number of issues need to be taken into consideration. These include: (1) understanding how the organ works, (2) performing series of baseline tests, (3) understanding the manner in which the cells interact with one another, (4) knowing and applying the correct ratio of different cells to each other cell type in the coculture, and finally (5) making sure that the coculture system can remain viable. These are all essential points that must be thoroughly investigated prior to the setup or use of any coculture system for in vitro (nanotoxicology) testing strategies.

Another example of the efforts employed to establish in vitro models to study the toxic potential of NPs comes with the hand of the microfabrication and microfluidics. These techniques are based on a dynamic exposure that mimics the physiological conditions to assess toxicity. By using an integrated multicompartmented microfluidic device in order to develop a flow exposure condition, an in vitro cell array system can be established (99). The different channels in these platforms are individually addressable even in both directions, which demonstrates the abilities of these platforms as a multiplexed live cell array for cell cytotoxicity screening (100).

7. Final Remarks

In conclusion, the field of nanotoxicology is complex and requires increased research in order to obtain essential and adequate amounts of information to enable the advantages of the nanotechnology industry to be realized.

Whereas first investigations concerning the toxicity of NPs were based on in vivo experiments (i.e., inhalation studies, etc.), the ability to design a large variety of different NPs led to a huge body of work based on in vitro studies. Due to their presumed lower complexity, in vitro studies with well-defined model NPs enable the identification of conceptual models for interactions of NPs with cells. In vitro high-dose toxicology and mechanistic studies should be viewed as proof-of-principle studies, though, that ultimately require validation in vivo. A major issue that needs to be carefully considered is the relevancy of the doses applied in vitro for predicting in vivo outcomes. With the necessity of in vivo validation kept in mind, several properties of NPs have been demonstrated to change the in vitro (and partly also in vivo) toxicity of NPs as compared to the bulk state. First, in comparison with bulk materials, NPs possess a higher surface-to-volume ratio and thus an enhanced contact area with their surroundings than do bulk materials at the same mass. Second, due to their small size, NPs are retained in many cells and organs to a larger extent than are larger particles. Third, in comparison with unstructured bulk material, the shape of NPs can play a crucial role in a determining response.

It is important though that by performing toxicity tests, increased controls and supporting tests must be performed concomitantly, as well as numerous tests (at least three) performed in regard to the same endpoint (such as cytotoxicity) in order to obtain a clear, definitive, and realistic understanding of the risk that NPs might pose.

References

1. Service RF (2004) Nanotoxicology: nanotechnology grows up. Science 304:1732–1734
2. Maynard AD, Aitken RJ, Butz T, Colvin V, Donaldson K, Oberdorster G, Philbert MA, Ryan J, Seaton A, Stone V, Tinkle SS, Tran L, Walker NJ, Warheit DB (2006) Safe handling of nanotechnology. Nature 444:267–269
3. Hoet PH, Bruske-Hohlfeld I, Salata OV (2004) Nanoparticles – known and unknown health risks. J Nanobiotechnol 2:12
4. Maynard AD (2007) Nanotechnology: the next big thing, or much ado about nothing? Ann Occup Hyg 51:1–12
5. Donaldson K, Aitken R, Tran L, Stone V, Duffin R, Forrest G, Alexander A (2006) Carbon nanotubes: a review of their properties in relation to pulmonary toxicology and workplace safety. Toxicol Sci 92:5–22
6. Gaskell G, Allum N, Bauer M, Durant J, Allansdottir A, Bonfadelli H, Boy D, de Cheveigne S, Fjaestad B, Gutteling JM, Hampel J, Jelsoe E, Jesuino JC, Kohring M, Kronberger N, Midden C, Nielsen TH, Przestalski A, Rusanen T, Sakellaris G, Torgersen H, Twardowski T, Wagner W (2000) Biotechnology and the European public. Nat Biotechnol 18:935–938
7. Song Y, Li X, Du X (2009) Exposure to nanoparticles is related to pleural effusion, pulmonary fibrosis and granuloma. Eur Respir J 34:559–567
8. Oberdoerster G, Oberdoerster E, Oberdoerster J (2005) Nanotoxicology: an emerging discipline evolving from studies of ultrafine particles. Environ Health Perspect 113: 823–839
9. Oberdorster G, Maynard A, Donaldson K, Castranova V, Fitzpatrick J, Ausman K, Carter J, Karn B, Kreyling W, Lai D, Olin S, Monteiro-Riviere N, Warheit D, Yang H (2005) Principles for characterizing the potential human health effects from exposure to nanomaterials: elements of a screening strategy. Part Fibre Toxicol 2:8
10. Oberdörster G, Stone V, Donaldson K (2007) Toxicology of nanoparticles: a historical perspective. Nanotoxicology 1:2–25
11. The Royal Society and The Royal Academy of Engineers Report (2004) Nanoscience and Nanotechnologies: Opportunites and Uncertainties
12. Zimmer AT, Baron PA, Biswas P (2002) The influence of operating parameters on number-weighted aerosol size distribution generated from a gas metal arc welding process. J Aerosol Sci 33:519–531
13. Rundell KW (2003) High levels of airborne ultrafine and fine particulate matter in indoor ice arenas. Inhal Toxicol 15:237–250
14. Kittelson DB, Watts WF, Johnson JP (2004) Nanoparticle emissions on Minnesota highways. Atmos Environ 38:9–19
15. Donaldson K, Stone V, Borm PJA, Jimenez LA, Gilmour PS, Schins RPF, Knaapen AM, Rahman I, Faux SP, Brown DM, MacNee W (2003) Oxidative stress and calcium signaling in the adverse effects of environmental particles (PM10). Free Radical Biol Med 34:1369–1382
16. Colvin VI (2003) The potential environmental impact of engineered nanomaterials. Nat Biotechnol 21:1166–1170
17. Jensen AW, Wilson SR, Schuster DI (1996) Biological applications of fullerenes. Bioorg Med Chem 4:767–779
18. Pellegrino T, Manna L, Kudera S, Liedl T, Koktysh D, Rogach AL, Keller S, Rädler J, Natile G, Parak WJ (2004) Hydrophobic nanocrystals coated with an amphiphilic polymer shell: a general route to water soluble nanocrystals. Nano Lett 4:703–707
19. Gupta AK, Gupta M (2005) Cytotoxicity suppression and cellular uptake enhancement of surface modified magnetic nanoparticles. Biomaterials 26:1565–1573

20. Rivera Gil P, Hühn D, del Mercato LL, Sasse D, Parak WJ (2010) Nanopharmacy: inorganic nanoscale devices as vectors and active compounds. Pharmacol Res 62:115–125
21. Duncan R, Izzo L (2005) Dendrimer biocompatibility and toxicity. Adv Drug Deliv Rev 57:2215–2237
22. Sperling RA, Pellegrino T, Li JK, Chang WH, Parak WJ (2006) Electrophoretic separation of nanoparticles with a discrete number of functional groups. Adv Funct Mater 16: 943–948
23. Lin C-AJ, Sperling RA, Li JK, Yang T-Y, Li P-Y, Zanella M, Chang WH, Parak WJ (2008) Design of an amphiphilic polymer for nanoparticle coating and functionalization. Small 4:334–341
24. Fernández-Argüelles MT, Yakovlev A, Sperling RA, Luccardini C, Gaillard S, Medel AS, Mallet J-M, Brochon J-C, Feltz A, Oheim M, Parak WJ (2007) Synthesis and characterization of polymer-coated quantum dots with integrated acceptor dyes as FRET-based nanoprobes. Nano Lett 7:2613–2617
25. Stehr J, Hrelescu C, Sperling RA, Raschke G, Wunderlich M, Nichtl A, Heindl D, Kürzinger K, Parak WJ, Klar TA, Feldmann J (2008) Gold nano-stoves for microsecond DNA melting analysis. Nano Lett 8:619–623
26. Nel A, Xia T, Madler L, Li N (2006) Toxic potential of materials at the nanolevel. Science 311:622–627
27. Knol AB, de Hartog JJ, Boogaard H, Slottje P, van der Sluijs JP, Lebret E, Cassee FR, Wardekker A, Ayres JG, Borm PJ, Brunekreef B, Donaldson K, Forastiere F, Holgate ST, Kreyling WG, Nemery B, Pekkanen J, Stone V, Wichmann HE, Hoek G (2009) Expert elicitation on ultrafine particles: likelihood of health effects and causal pathways. Part Fibre Toxicol 6:19
28. Baroli B (2010) Penetration of nanoplarticles and nanomaterials in the skin: fiction or reality? J Pharm Sci 99:21–50
29. Tinkle SS, Antonini JM, Rich BA, Roberts JR, Salmen R, DePree K, Adkins EJ (2003) Skin as a route of exposure and sensitization in chronic beryllium disease. Environ Health Perspect 111:1202–1208
30. Rothen-Rutishauser B, Mühlfeld C, Blank F, Musso C, Gehr P (2007) Translocation of particles and inflammatory responses after exposure to fine particles and nanoparticles in an epithelial airway model. Part Fibre Toxicol 4:9
31. Muhlfeld C, Rothen-Rutishauser B, Blank F, Vanhecke D, Ochs M, Gehr P (2008) Interactions of nanoparticles with pulmonary structures and cellular responses. Am J Physiol Lung Cell Mol Physiol 294:L817–L829
32. Rothen-Rutishauser B, Grass RN, Blank F, Limbach LK, Muehlfeld C, Brandenberger C, Raemy DO, Gehr P, Stark WJ (2009) Direct combination of nanoparticle fabrication and exposure to lung cell cultures in a closed setup as a method to simulate accidental nanoparticle exposure of humans. Environ Sci Technol 43:2634–2640
33. Stone V, Johnston H, Schins RPF (2009) Development of in vitro systems for nanotoxicology: methodological considerations. Crit Rev Toxicol 39:613–626
34. Duffin R, Tran L, Brown D, Stone V, Donaldson K (2007) Proinflammogenic effects of low-toxicity and metal nanoparticles in vivo and in vitro: highlighting the role of particle surface area and surface reactivity. Inhal Toxicol 19:849–856
35. Gurr JR, Wang ASS, Chen CH, Jan KY (2005) Ultrafine titanium dioxide particles in the absence of photoactivation can induce oxidative damage to human bronchial epithelial cells. Toxicology 213:66–73
36. Limbach LK, Li YC, Grass RN, Brunner TJ, Hintermann MA, Muller M, Gunther D, Stark WJ (2005) Oxide nanoparticle uptake in human lung fibroblasts: effects of particle size, agglomeration, and diffusion at low concentrations. Environ Sci Technol 39:9370–9376
37. Stearns RC, Paulauskis JD, Godleski JJ (2001) Endocytosis of ultrafine particles by A549 cells. Am J Respir Cell Mol Biol 24:108–115
38. Gil J, Weibel ER (1971) Extracellular lining of bronchioles after perfusion-fixation of rat lungs for electron microscopy. Anat Rec 169:185–199
39. Schürch S, Gehr P, Hof VI, Geiser M, Green F (1990) Surfactant displaces particles toward the epithelium in airways and alveoli. Respir Physiol 80:17–32
40. Lenz AG, Karg E, Lentner B, Dittrich V, Brandenberger C, Rothen-Rutishauser B, Schulz H, Ferron GA, Schmid O (2009) A dose-controlled system for air-liquid interface cell exposure and application to zinc oxide nanoparticles. Part Fibre Toxicol 6:32
41. Muller L, Comte P, Czerwinski J, Kasper M, Mayer ACR, Gehr P, Burtscher H, Morin JP, Konstandopoulos A, Rothen-Rutishauser B (2010) New exposure system to evaluate the toxicity of (scooter) exhaust emissions in lung cells in vitro. Environ Sci Technol 44: 2632–2638
42. Kreyling WG, Semmler M, Erbe F, Mayer P, Takenaka S, Schulz H, Oberdorster G, Ziesenis A (2002) Translocation of ultrafine

insoluble iridium particles from lung epithelium to extrapulmonary organs is size dependent but very low. J Toxicol Environ Health A 65:1513–1530

43. Kreyling WG, Semmler-Behnke M, Möller W (2006) Ultrafine particle–lung interactions: does size matter? J Aerosol Med 19:74–83
44. Wick P, Malek A, Manser P, Meili D, Maeder-Althaus X, Diener L, Diener PA, Zisch A, Krug HF, von Mandach U (2010) Barrier capacity of human placenta for nanosized materials. Environ Health Perspect 118: 432–436
45. Perry MJ, McAuliffe ME (2007) Are nanoparticles potential male reproductive toxicants? A literature review. Nanotoxicology 1:204–210
46. Xia T, Kovochich M, Liong M, Zink JI, Nel AE (2008) Cationic polystyrene nanosphere toxicity depends on cell-specific endocytic and mitochondrial injury pathways. ACS Nano 2:85–96
47. Nativo P, Prior IA, Brust M (2008) Uptake and intracellular fate of surface-modified gold nanoparticles. ACS Nano 2:1639–1644
48. Murphy CJ, Gole AM, Stone JW, Sisco PN, Alkilany AM, Goldsmith EC, Baxter SC (2008) Gold nanoparticles in biology: beyond toxicity to cellular imaging. Acc Chem Res 41:1721–1730
49. Cheng JX, Huff TB, Hansen MN, Zhao Y, Wei A (2007) Controlling the cellular uptake of gold nanorods. Langmuir 23:1596–1599
50. Sperling RA, Rivera Gil P, Zhang F, Zanella M, Parak WJ (2008) Biological applications of gold nanoparticles. Chem Soc Rev 37:1896–1908
51. Ozkan CS, Zhang Y, Yang M, Portney NG, Cui DX, Budak G, Ozbay E, Ozkan M (2008) Zeta potential: a surface electrical characteristic to probe the interaction of nanoparticles with normal and cancer human breast epithelial cells. Biomed Microdev 10:321–328
52. Stellacci F, Verma A (2010) Effect of surface properties on nanoparticle-cell interactions. Small 6:12–21
53. Ginzburg VV, Balijepalli S (2007) Modeling the thermodynamics of the interaction of nanoparticles with cell membranes. Nano Lett 7:3716–3722
54. Deserno M, Gelbart WM (2002) Adhesion and wrapping in colloid-vesicle complexes. J Phys Chem B 106:5543–5552
55. Smith KA, Jasnow D, Balazs AC (2007) Designing synthetic vesicles that engulf nanoscopic particles. J Chem Phys 127:084703
56. Chen HM, Langer R, Edwards DA (1997) A film tension theory of phagocytosis. J Colloid Interface Sci 190:118–133
57. Rimai DS, Quesnel DJ, Busnaina AA (2000) The adhesion of dry particles in the nanometer to micrometer-size range. Colloids Surf A Physicochem Eng Asp 165:3–10
58. Geiser M, Rothen-Rutishauser B, Kapp N, Schurch S, Kreyling W, Schulz H, Semmler M, Hof VI, Heyder J, Gehr P (2005) Ultrafine particles cross cellular membranes by nonphagocytic mechanisms in lungs and in cultured cells. Environ Health Perspect 113:1555–1560
59. Diagaradjane P, Orenstein-Cardona JM, Colon-Casasnovas NE, Deorukhkar A, Shentu S, Kuno N, Schwartz DL, Gelovani JG, Krishnan S (2008) Imaging epidermal growth factor receptor expression in vivo: pharmacokinetic and biodistribution characterization of a bioconjugated quantum dot nanoprobe. Clin Cancer Res 14:731–741
60. Zhang H, Zeng X, Li Q, Gaillard-Kelly M, Wagner CR, Yee D (2009) Fluorescent tumour imaging of type I IGF receptor in vivo: comparison of antibody-conjugated quantum dots and small-molecule fluorophore. Br J Cancer 101:71–79
61. Tekle C, van Deurs B, Sandvig K, Iversen TG (2008) Cellular trafficking of quantum dot-ligand bioconjugates and their induction of changes in normal routing of unconjugated ligands. Nano Lett 8:1858–1865
62. Pawley JB (2006) Handbook of biological confocal microscopy
63. Tkachenko AG, Xie H, Liu Y, Coleman D, Ryan J, Glomm WR, Shipton MK, Franzen S, Feldheim DL (2004) Cellular trajectories of peptide-modified gold particle complexes: comparison of nuclear localization signals and peptide transduction domains. Bioconjugate Chem 15:482–490
64. Brandenberger C, Clift MJD, Vanhecke D, Muhlfeld C, Stone V, Gehr P, Rothen-Rutishauser B (2010) Intracellular imaging of nanoparticles: is it an elemental mistake to believe what you see? Part Fibre Toxicol 7:15
65. Egerton RF (2009) Electron energy-loss spectroscopy in the TEM. Rep Prog Phys 72:016502
66. Ottensmeyer FP, Andrew JW (1980) High-resolution microanalysis of biological specimens by electron-energy loss spectroscopy and by electron spectroscopic imaging. J Ultrastruct Res 72:336–348
67. Derfus AM, Chan WCW, Bhatia SN (2004) Probing the cytotoxicity of semiconductor quantum dots. Nano Lett 4:11–18

68. Feldheim DL, Ryan JA, Overton KW, Speight ME, Oldenburg CM, Loo L, Robarge W, Franzen S (2007) Cellular uptake of gold nanoparticles passivated with BSA-SV40 large T antigen conjugates. Anal Chem 79:9150–9159
69. Chithrani BD, Ghazan AA, Chan CW (2006) Determining the size and the shape dependence of gold nanoparticle uptake into mammalian cells. Nano Lett 6:662–668
70. Tarloff JB, Kendig DM (2007) Inactivation of lactate dehydrogenase by several chemicals: implications for in vitro toxicology studies. Toxicol In Vitro 21:125–132
71. Henderson RF, Benson JM, Hahn FF, Hobbs CH, Jones RK, Mauderly JL, Mcclellan RO, Pickrell JA (1985) New approaches for the evaluation of pulmonary toxicity - bronchoalveolar lavage fluid analysis. Fundam Appl Toxicol 5:451–458
72. Mosmann T (1983) Rapid colorimetric assay for cellular growth and survival: application to proliferation and cytotoxicity assays. J Immunol Methods 65:55–63
73. MacNee W (2001) Oxidative stress and lung inflammation in airways disease. Eur J Pharmacol 429:195–207
74. Hissin PJ, Hilf R (1976) A fluorometric method for determination of oxidized and reduced glutathione in tissues. Anal Biochem 74:214–226
75. Grynkiewicz G, Poenie M, Tsien RY (1985) A new generation of Ca-2+ indicators with greatly improved fluorescence properties. J Biol Chem 260:3440–3450
76. Wallin H, Jacobsen NR, Moller P, Cohn CA, Loft S, Vogel U (2008) Diesel exhaust particles are mutagenic in FE1-Muta (TM) Mouse lung epithelial cells. Mutat Res 641:54–57
77. Wallin H, Jacobsen NR, Pojana G, White P, Moller P, Cohn CA, Korsholm KS, Vogel U, Marcomini A, Loft S (2008) Genotoxicity, cytotoxicity, and reactive oxygen species induced by single-walled carbon nanotubes and C-60 fullerenes in the FE1-Muta (TM) mouse lung epithelial cells. Environ Mol Mutagen 49:476–487
78. Singh NP, Mccoy MT, Tice RR, Schneider EL (1988) A simple technique for quantitation of low-levels of DNA damage in individual cells. Exp Cell Res 175:184–191
79. Schins RPF, Li H, Van Berlo D, Shi T, Spelt G, Knaapen AM, Borm PJA, Albrecht C (2008) Curcumin protects against cytotoxic and inflammatory effects of quartz particles but causes oxidative DNA damage in a rat lung epithelial cell line. Toxicol Appl Pharmacol 227:115–124
80. Ames BN, Lee FD, Durston WE (1973) An improved bacterial test system for the detection and classification of mutagens and carcinogens. Proc Natl Acad Sci USA 70:782–786
81. Zhao HW, Barger MW, Ma JKH, Castranova V, Ma JYC (2004) Effects of exposure to diesel exhaust particles (DEP) on pulmonary metabolic activation of mutagenic agents. Mutat Res 564:103–113
82. Schins RPF, Knaapen AM (2007) Genotoxicity of poorly soluble particles. Inhal Toxicol 19:189–198
83. Marquis BJ, Love SA, Braun KL, Haynes CL (2009) Analytical methods to assess nanoparticle toxicity. Analyst 134:425–439
84. Hauck TS, Ghazani AA, Chan WCW (2008) Assessing the effect of surface chemistry on gold nanorod uptake, toxicity, and gene expression in mammalian cells. Small 4:153–159
85. Zhang TT, Stilwell JL, Gerion D, Ding LH, Elboudwarej O, Cooke PA, Gray JW, Alivisatos AP, Chen FF (2006) Cellular effect of high doses of silica-coated quantum dot profiled with high throughput gene expression analysis and high content cellomics measurements. Nano Lett 6:800–808
86. Jan E, Byrne SJ, Cuddihy M, Davies AM, Volkov Y, Gun'ko YK, Kotov NA (2008) High-content screening as a universal tool for fingerprinting of cytotoxicity of nanoparticles. ACS Nano 2:928–938
87. Rivera Gil P, Yang F, Thomas H, Li L, Terfort A, Parak WJ (2011) Development of an assay based on cell counting with quantum dot labels for comparing cell adhesion within cocultures. Nano Today 6:20–27
88. Kirchner C, Liedl T, Kudera S, Pellegrino T, Javier AM, Gaub HE, Stölzle S, Fertig N, Parak WJ (2005) Cytotoxicity of colloidal CdSe and CdSe/ZnS nanoparticles. Nano Lett 5:331–338
89. Oberdorster G (2010) Safety assessment for nanotechnology and nanomedicine: concepts of nanotoxicology. J Intern Med 267:89–105
90. Bouwmeester H, Lynch I, Marvin HJ, Dawson KA, Berges M, Braguer D, Byrne HJ, Casey A, Chambers G, Clift MJ, Elia G, Fernandes TF, Fjellsbo LB, Hatto P, Juillerat L, Klein C, Kreyling WG, Nickel C, Riediker M, Stone V (2011) Minimal analytical characterization of engineered nanomaterials needed for hazard assessment in biological matrices. Nanotoxicology 5:1–11
91. Worle-Knirsch JM, Pulskamp K, Krug HF (2006) Oops they did it again! Carbon nanotubes hoax scientists in viability assays. Nano Lett 6:1261–1268

92. Wick P, Manser P, Limbach LK, Dettlaff-Weglikowska U, Krumeich F, Roth S, Stark WJ, Bruinink A (2007) The degree and kind of agglomeration affect carbon nanotube cytotoxicity. Toxicol Lett 168:121–131
93. Rothen-Rutishauser B, Blank F, Muhlfeld C, Gehr P (2008) In vitro models of the human epithelial airway barrier to study the toxic potential of particulate matter. Expert Opin Drug Metab Toxicol 4:1075–1089
94. Sauer UG (2009) Animal and non-animal experiments in nanotechnology – the results of a critical literature survey. Altex-Alternativen Zu Tierexperimenten 26:109–134
95. Rothen-Rutishauser BM, Kiama SG, Gehr P (2005) A three-dimensional cellular model of the human respiratory tract to study the interaction with particles. Am J Respir Cell Mol Biol 32:281–289
96. Blank F, Rothen-Rutishauser B, Gehr P (2007) Dendritic cells and macrophages form a transepithelial network against foreign particulate antigens. Am J Respir Cell Mol Biol 36:669–677
97. Müller L, Riediker M, Wick P, Mohr M, Gehr P, Rothen-Rutishauser B (2010) Oxidative stress and inflammation response after nanoparticle exposure: differences between human lung cell monocultures and an advanced three-dimensional model of the human epithelial airways. Interface Focus 7:27–40
98. Lehmann AD, Blank F, Baum O, Gehr P, Rothen-Rutishauser BM (2009) Diesel exhaust particles modulate the tight junction protein occludin in lung cells in vitro. Part Fibre Toxicol 6:26
99. Rhee SW, Mahto SK, Yoon TH (2010) A new perspective on in vitro assessment method for evaluating quantum dot toxicity by using microfluidics technology. Biomicrofluidics 4
100. Thorsen T, Wang ZH, Kim MC, Marquez M (2007) High-density microfluidic arrays for cell cytotoxicity analysis. Lab Chip 7:740–745

Chapter 4

Single-Cell Gel Electrophoresis (Comet) Assay in Nano-genotoxicology

Maricica Pacurari and Vincent Castranova

Abstract

The Comet assay, or single-cell gel electrophoresis assay, is an easy and simple yet reliable method to evaluate DNA damage in cells. Under the alkaline conditions of this method, DNA strand breaks and alkaline-labile sites are detected. Here we describe the alkaline version of the Comet assay with applications in testing nanoparticles.

Key words: Comet assay, DNA damage, Nanoparticles

1. Introduction

With the advent of nanotechnology, nanoparticles can be synthesized with unique physicochemical properties such as diameter, length, weight, and surface area, which present great prospects for numerous applications. However, these very same characteristics could have detrimental effects on human health. The genotoxic effect of many materials can be identified by a number of techniques that detect DNA damage. However, those techniques sometimes are difficult. The Comet assay is a method of choice for the analysis of DNA damage in toxicology testing (1). The Comet assay or single-cell gel (SCG) electrophoresis detects DNA damage at the level of one single cell. The term "Comet" describes the pattern of migration of DNA from a single cell if there is DNA damage (1, 2). The method was first developed by Östling and Johanson in 1984 (3) as a microgel

Joshua Reineke (ed.), *Nanotoxicity: Methods and Protocols*, Methods in Molecular Biology, vol. 926,
DOI 10.1007/978-1-62703-002-1_4, © Springer Science+Business Media, LLC 2012

electrophoresis technique for detecting DNA damage of a single cell under neutral conditions. Under this method, the cells are embedded in agarose, deposited on microscope slides, and exposed to lysing solutions. Then the slides are electrophoresed under neutral conditions. Cells with DNA double strand breaks displayed increased DNA migration toward the anode. The migrating DNA was detected by staining with ethidium bromide, and quantitative analysis was performed by measuring the intensity of fluorescence of the migrated DNA using a microscope photometer. Several years later in 1988, Singh et al. (4) introduced alkaline conditions (pH > 13) to the microgel electrophoresis. Under these conditions, increased DNA migration is associated with DNA damage at several levels of the DNA strands, such as single strand breaks (SSB) associated with incomplete excision repair, and alkaline-labile sites including apurinic and apyrimidinic sites due to loss of sugar group from the DNA strand. This method offered increased sensitivity. The term of Comet assay was coined by Olive et al. in 1992 (2) when this group introduced another version of the alkaline method under which electrophoresis is performed at a pH of ~12.3. Since then, the alkaline version of the Comet assay has been used by numerous investigators, and with the advent of technology, automated image analysis has been commercially available making the use of Comet assay simple and effective (1). Due to its ease of use and reliability, the Comet assay in testing genotoxic effects of metal- and carbon-based nanoparticles has been extensively used by numerous investigators including our own laboratory (5–7). In this chapter, we describe the Comet assay alkaline version to detect DNA damage in cells as a result of genotoxic effects of nanoparticles (Fig. 1). Under this method, the cells are embedded with low-melting point agarose, placed on specially formulated microscope slides, and electrophoresed under alkaline conditions followed by DNA staining, visualization, and quantitative image analysis.

The principle of the Comet assay is based on the ability of DNA, either denatured or fragmented, to migrate out of the cell upon cell lysis under the influence of an electric field. The extent of DNA migration depends on the degree of DNA damage present in the cells. Undamaged DNA migrates slowly and usually remains within nucleoid, whereas damaged DNA migrates out of the cell in the direction of anode, thus appearing like a "Comet" when visualized (Fig. 2). The size and shape of the Comet and the distribution of DNA within the Comet correlate with the extent of DNA damage (8).

In the past few years, the Comet assay described in this chapter has been extensively used to study genotoxic effects of nanoparticles (6, 9). The method is sensitive, fast, and easy to use.

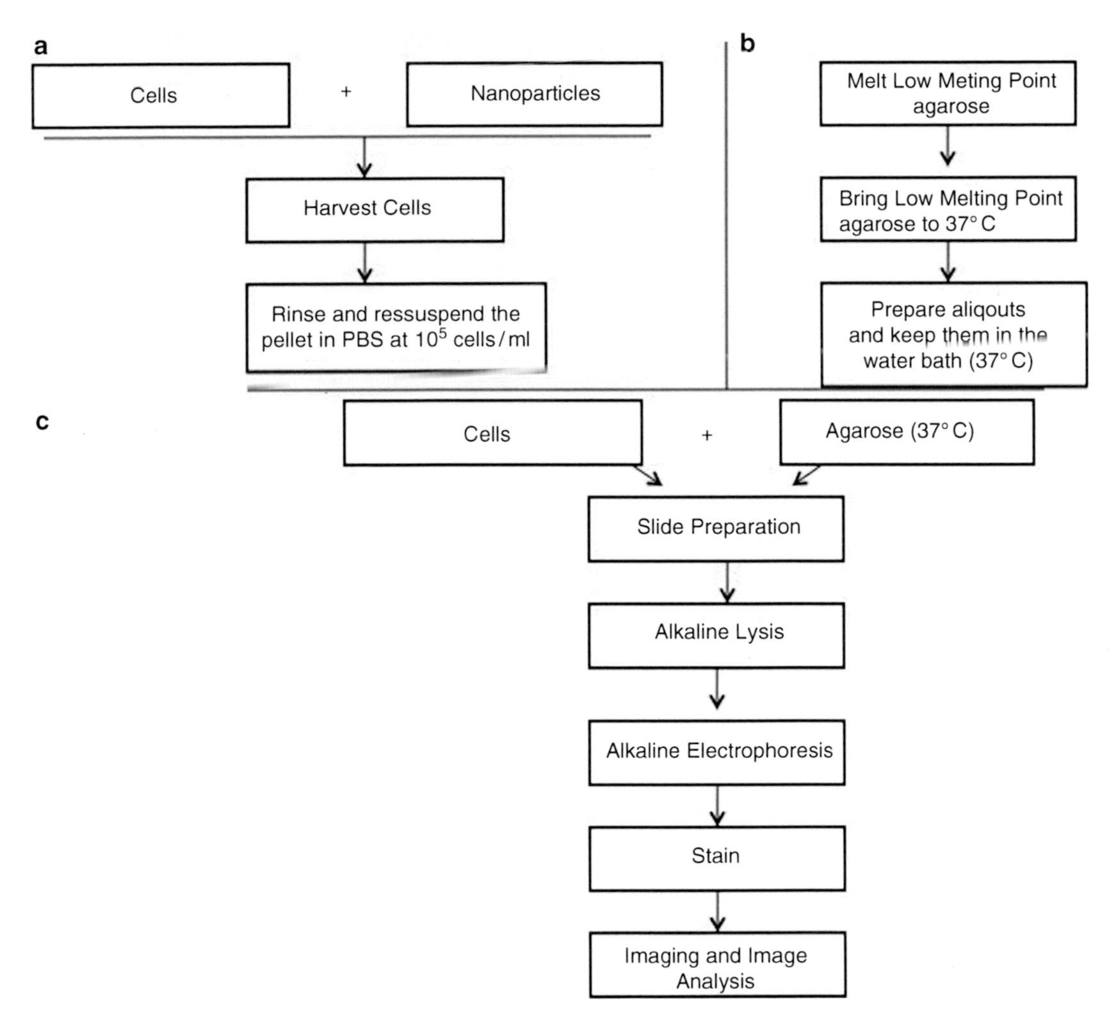

Fig. 1. Schematic representation of the Comet assay alkaline version. The scheme is separated in three parts: part (**a**) shows preparation of cells for the Comet assay; part (**b**) shows the preparation of low-melting point agarose; and part (**c**) shows the steps for the running of the Comet assay.

2. Materials

Prepare all solutions using ultrapure water (i.e., deionized water filtered through a MiliQ column at 18 MΩ cm at 25°C). Several materials required for this method are commercially available.

2.1. Equipment

1. Microscope slides especially formulated (Trevigen, Gaithersburg, MD).
2. Pipette tips (1–20, 20–200, 200–1,000 μl).
3. Serological pipettor and pipettes.

Fig. 2. DNA Comets from alkaline version of the Comet assay. Particle-treated rat cells analyzed for DNA damage according to the Comet assay protocol described above. This representative micrograph indicates three levels of DNA damage: total DNA damage indicated by the largest Comet (*solid arrow*); intermediate DNA damage indicated by a smaller Comet (*broken arrow*); and minimal DNA damage indicated by a minor Comet (*dotted arrow*) (Color figure online).

4. Boiling water bath and 37°C water bath.
5. Horizontal electrophoresis apparatus.
6. Epifluorescence microscope equipped with appropriate filter.
7. If using silver stain, need a light transmission microscope.
8. Graduated cylinder.
9. Optional: peristaltic pump for recirculation of buffer.

2.2. Reagents

Several reagents must be prepared fresh immediately before each use. Follow good laboratory safety rules by wearing gloves, lab coat, eye protection, using a chemical hood, and properly disposing waste.

1. 1× PBS, Ca^{2+}/Mg^{2+} free: Dilute 10× PBS with deionized water.
2. Optional: dimethylsulfoxide.
3. 10× TBE buffer (if using neutral electrophoresis).
4. 0.5 M EDTA pH 8.0.
5. Ethanol.
6. TE buffer: 10 mM Tris-HCl pH 7.5, 1 mM EDTA.
7. Lysis solution for Comet assay (Trevigen), must be stored at 4°C. Chill lysis solution at 4°C or on ice before use for at least 20 min.
8. Low-melting point agarose (Trevigen).

9. Alkaline solution pH > 13: for 100 ml total volume, 1.2 g NaOH pellets, 500 μl 200 mM EDTA (pH 8.0), 99.5 ml ddiH_2O (see Note 1).
10. Electrophoresis solution pH > 13: for 1.0 l total volume, 12.0 g (300 mM) NaOH pellets, 2.0 ml (1 mM) 500 mM EDTA pH 8.0, bring up to 1.0 l with ddiH_2O (see Note 2).
11. SYBR® Green I staining solution: 1 μl SYBR® Green I concentrate (10,000× in DMSO) (Trevigen), 10 ml TE buffer pH 7.5. Mix well and store at 4°C in the dark up to 3–4 weeks (*see* Note 3).
12. Antifade solution (see Note 4): For 50 ml total volume, 500 mg *p*-phenylenediamine dihydrochloride, 4.5 ml 1× PBS, *adjust* pH with 10 N NaOH (~400 μl) until the solution is at pH 7.5–8.0. Bring the volume to 5 ml with 1× PBS. Add 45 ml glycerol and vortex the solution until thoroughly mixed. Store at −20°C up to a month. Use 10 μl of the antifade solution per slide.

3. Methods

3.1. Sample Preparation

Sample preparation as well as preparation of solutions must be a coordinated process in order to obtain reliable results and to maintain the normal flow of steps of the assay.

1. All the steps of the Comet assay that involve cells must be carried out under dimmed light or yellow light in order to protect cells from ultraviolet light damage.
2. Prepare the low-melting point agarose. Melt the low-melting point agarose by placing it in a small beaker of boiling water for 5 min with the container loosely capped (see Note 5). Transfer the heated agarose to a 37°C water bath, and let it stand in the water bath for 20 min to cool down to 37°C. Aliquot molten agarose into tubes and keep them in the water bath until combining with the cells (see Note 6).
3. Prepare cell samples just before the start of the assay.
4. All buffers, except where specified, should be chilled to 4°C.

3.2. Cell Sample Preparation

After performing cell treatments:

1. Collect cells by scraping the cell layer with a rubber policeman and transfer the cell suspension to a tube. Do a cell count. Also, perform a cell viability assay (see Note 7).

2. Collect cells by centrifugation at 700 rpm for 5 min. Rinse the pellet with cold 1× PBS Ca^{2+}/Mg^{2+} free.
3. Remove the supernatant and resuspend cell pellet at a density of 1×10^5 cells/ml in ice-cold 1× PBS Ca^{2+}/Mg^{2+} free.
4. Keep the cells on ice until ready to perform the assay.

3.3. Slides Preparation

1. Determine the number of slides needed. Usually one sample per slide (see Note 8).
2. Label the slides with a pencil.
3. Place the slides in a tray that will hold a certain amount of solution in such a way as the slides will be completely submerged into solution.
4. The slides should be at room temperature before use.

3.4. Comet Assay

For good results, the cell samples should be prepared the same day as the Comet assay is run, as cell preservation may affect the results.

1. Mix gently the cells (1×10^5 cell/ml) prepared as specified in Subheading 3.2 with aliquots of warm agarose at a ratio of 1:10 (see Note 9).

Low-melting agarose (molten at 37°C from step 3.1.2)	500 µl
Cells at 1×10^5/ml 1× PBS (Ca^{2+}/Mg^{2+} free)	50 µl

2. Immediately transfer 75 µl of the mixture onto the slide. If the aliquot is not spreading, gently spread it with pipette tip.
3. Place the slides flat at 4°C in the dark for 10 min. Move the slides gently (see Note 10).
4. Place the slides in prechilled lysis solution and keep the slides on ice, or at 4°C for 30–60 min.
5. Gently remove the slides from the lysis solution and tap off the excess solution, and immerse the slides into freshly prepared alkaline solution $pH > 13$.
6. Leave the slides in the alkaline solution for 20–60 min at room temperature in the dark.
7. Remove the slides from the alkaline solution and place them onto the gel tray of a horizontal electrophoresis apparatus. The slides should be aligned equidistant from the electrodes and from each other.
8. Gently pour the alkaline solution on both sides of the electrophoresis apparatus until the level of the solution just covers the samples. Make sure that the top of each slide is immersed into solution.

9. Set the voltage to about 1 V/cm. The current should be approximately 300 mA and perform electrophoresis for 20–40 min (see Note 11).
10. At the end of electrophoresis, wear gloves and gently take out the slides from the electrophoresis solution, tap off the excess of electrophoresis solution, rinse the slides several times in ddiH_2O, and then immerse the slides in 70% ethanol for 5 min.
11. Take out the slides from the ethanol solution and air dry them for 15–30 min. The slides can be stored at room temperature with desiccant prior to scoring.
12. If silver staining is used, note that silver will permanently stain the samples providing the advantage for samples archiving and long storage.
13. Any dye that will stain DNA can be used, such as ethidium bromide, propidium iodine, or SYBR® Green I. If SYBR® Green I is used, then place 50 μl of SYBR® Green working solution (prepared as described in Subheading 2.2, step 11) on each blob of dried agarose. Let it stand for 5–10 min and then visualize the slides.
14. Visualize the slides using an epifluorescence microscopy. The microscope should be equipped with appropriate filters depending on the fluorochrome used. If using SYBR Green I, set excitation/emission at 494/521 nm (see Notes 12 and 13).
15. Capture the images using automated image analysis systems (microscope, camera, and software) (see Notes 14 and 15).
16. Analyze captured images using commercially available software for the Comet assay (see Notes 16 and 17).
17. Comet analysis can be scored qualitatively and quantitatively (see Note 18).

3.5. Qualitative Analysis

1. Score the intensity of DNA in the tail. Scoring is determined according to nominal, medium, or high intensity of DNA content in the Comet tail. A minimum of 75 cells should be scored per each sample (see Note 19).

3.6. Quantitative Analysis

1. Captured Comet assay images are used for analysis of DNA damage using Comet parameters according to software packages. There are different commercially available software packages; however, the principle of software packages for image analysis is the same (see Note 19). Many parameters can be determined from the Comet assay. A typical Comet assay micrograph is shown in Fig. 3, and it shows the delineation of the Comet parts: the head and the tail.
2. Primary Comet assay measurements include: (a) tail length, (b) tail DNA, and (c) DNA distribution profile in the tail.

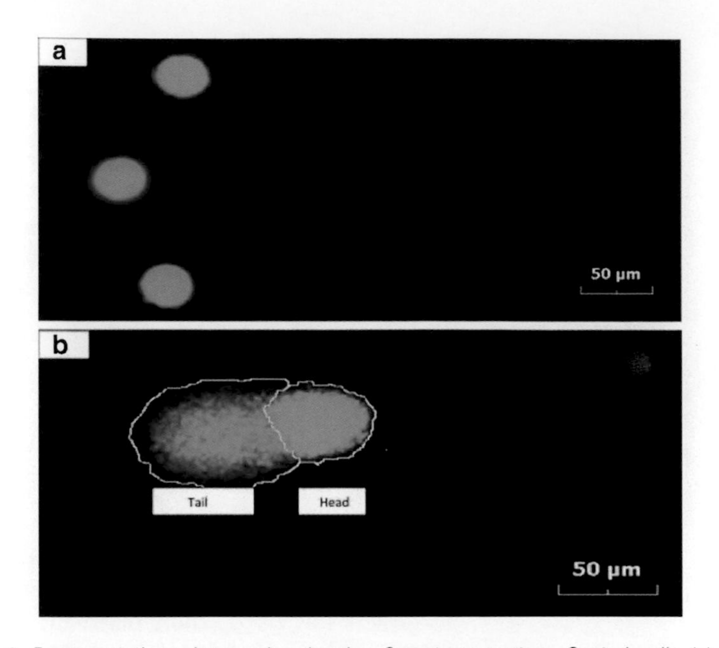

Fig. 3. Representative micrographs showing Comet parameters. Control cells (**a**) and treated cells (**b**) were prepared and analyzed according to the Comet assay protocol described above. In the control cells, there is no DNA Comet; whereas in the treated cells, there is a large DNA Comet. DNA Comet components, the head and the tail, are delineated in (**b**) (Color figure online).

3. The most accepted Comet measurements include: (a) tail moment (TM), and (b) percentage of DNA in the tail (% Tail DNA) (10). However, as to what Comet parameters are to be determined, it is at laboratory's choice.
4. TM is calculated as a product of % Tail DNA and the DNA tail length along the *x*-axis of the Comet (see Note 20).
5. % Tail DNA is given by the ratio of total intensity of the tail and the total intensity of the Comet (head and tail) (see Note 21).
6. Tail length might be a good parameter, but only for low levels of DNA damage.
7. % tail DNA is the most common measurement as it accounts for a wider range of damage and is related to the numbers of DNA breaks in a cell.
8. Analyze the captured images without bias. Thus, blind label the slides, and by using systematic random sampling steps in *x* and *y* coordinates, or any other different method of sampling, score all the cells from the sampling area. Score a high number of cells per slide (11). If possible, score all the cells from a slide to reduce data variability.

4. Notes

1. This solution must be prepared fresh immediately before each use. Must wear gloves when preparing the solution. Stir until the pellets are totally dissolved. The solution must be at room temperature before use.
2. The volume of water must not exceed the specified volume. Therefore, the pellets must be dissolved and then the final volume must be adjusted accordingly.
3. Prepare the solution in a brown bottle.
4. This solution is usually not necessary if the slides are analyzed immediately.
5. The agarose container should not be submerged into the boiling water, but it should be placed in the boiling water in such a way as not to get water into the agarose container.
6. This step is helpful when dealing with a large number of samples, and also it shortens the time for the assay.
7. Before performing the Comet assay, determined cell viability as Comet assay should be performed with at least 75% viable cells (10).
8. Comet assay requires the use of slides that are formulated specifically for this assay (Trevigen).
9. Mix the cells by pipetting up and down 1–2 times or by gently swirling. This step has to be gentle but fast as agarose will solidify fast.
10. Immerse the slides in a prechilled lysis solution and keep the slides on ice, or at 4°C for 30 min up to 60 min. The slides should be uniformly submerged into the lysis solution.
11. Alkaline electrophoresis solution is a non-buffered system. Therefore, temperature control is highly recommended. Using a larger electrophoresis apparatus along with recirculation of the electrophoresis solution will greatly improve the quality of electrophoresis. Alternatively, if the use of a larger electrophoresis apparatus is not possible, performing electrophoresis at cooler temperatures (e.g., 16 or 4°C) provides several advantages such as diminished background, increased sample adherence to the slide, and reproducible results. However, choose the method that is the most convenient in your laboratory. For reproducibility always use the same conditions, apparatus, and power supplies.
12. Antifade solution can be added to slides, since it reduces the rate of signal quenching.

13. There are image analysis systems (microscope, camera, and computer analysis software). These image analysis systems can be set up to measure Comet characteristics based on fluorescence intensity profiles of the Comet (12).
14. For good reproducible results from one experiment to another, keep the same camera settings for image acquisition and always use same camera.
15. Capture images of high quality, as good Comet images should show many details such as good light intensity in the tail and in the head of a Comet.
16. Avoid capturing saturated images, capture only images that show variation of light intensity in the Comet and in the background. A good image should have a good contrast but not saturated.
17. For image analysis, keep the same image analysis settings for each experiment. Before image analysis, spatial calibration must be performed and each image must have a scale bar.
18. Always run a positive control along with your experiment.
19. Select random areas for scoring, and score a higher number of cells to reduce variability of analysis (11, 13).
20. For good results of image analysis from experiment to experiment, follow the same protocol for image analysis (14).
21. For this measurement, the absolute measurements units are in μm (10, 12).
22. % Tail DNA is a good Comet assay parameter (11).

References

1. Tice RR, Agurell E, Anderson D, Burlinson B, Hartmann A, Kobayashi H, Miyamae Y, Rojas E, Ryu JC, Sasaki YF (2000) Single cell gel/comet assay: guidelines for in vitro and in vivo genetic toxicology testing. Environ Mol Mutagen 35:206–221
2. Olive PL, Banath JP, Durand RE (1990) Heterogeneity in radiation-induced DNA damage and repair in tumor and normal cells measured using the "comet" assay. Radiat Res 122:86–94
3. Ostling O, Johanson KJ (1984) Microelectrophoretic study of radiation-induced DNA damages in individual mammalian cells. Biochem Biophys Res Commun 123:291–298
4. Singh NP, McCoy MT, Tice RR, Schneider EL (1988) A simple technique for quantitation of low levels of DNA damage in individual cells. Exp Cell Res 175:184–191
5. Pacurari M, Yin XJ, Zhao JS, Ding M, Leonard SS, Schwegier-Berry D, Ducatman BS, Sbarra D, Hoover MD, Castranova V, Vallyathan V (2008) Raw single-wall carbon nanotubes induce oxidative stress and activate MAPKs, AP-1, NF-kappa B, and Akt in normal and malignant human mesothelial cells. Environ Health Perspect 116:1211–1217
6. Karlsson HL (2010) The comet assay in nanotoxicology research. Anal Bioanal Chem 398:651–666
7. Petersen EJ, Nelson BC (2010) Mechanisms and measurements of nanomaterial-induced oxidative damage to DNA. Anal Bioanal Chem 398:613–650
8. Fairbairn DW, Olive PL, O'Neill KL (1995) The comet assay: a comprehensive review. Mutat Res 339:37–59
9. Collins AR, Oscoz AA, Brunborg G, Gaivao I, Giovannelli L, Kruszewski M, Smith CC,

Stetina R (2008) The comet assay: topical issues. Mutagenesis 23:143–151

10. Kumaravel TS, Jha AN (2006) Reliable Comet assay measurements for detecting DNA damage induced by ionising radiation and chemicals. Mutat Res 605:7–16
11. McArt DG, Wasson GR, McKerr G, Saetzler K, Reed M, Howard CV (2009) Systematic random sampling of the comet assay. Mutagenesis 24:373–378
12. Kumaravel TS, Vilhar B, Faux SP, Jha AN (2009) Comet assay measurements: a perspective. Cell Biol Toxicol 25:53–64
13. McArt DG, McKerr G, Howard CV, Saetzler K, Wasson GR (2009) Modelling the comet assay. Biochem Soc Trans 37:914–917
14. Vilhar B (2004) Help! There is a comet in my computer! http://botanika.biologija.org/exp/comet/comet_guide01.pdf

Chapter 5

Single-Cell Nanotoxicity Assays of Superparamagnetic Iron Oxide Nanoparticles

Trisha Eustaquio and James F. Leary

Abstract

Properly evaluating the nanotoxicity of nanoparticles involves much more than bulk-cell assays of cell death by necrosis. Cells exposed to nanoparticles may undergo repairable oxidative stress and DNA damage or be induced into apoptosis. Exposure to nanoparticles may cause the cells to alter their proliferation or differentiation or their cell–cell signaling with neighboring cells in a tissue. Nanoparticles are usually more toxic to some cell subpopulations than others, and toxicity often varies with cell cycle. All of these facts dictate that any nanotoxicity assay must be at the single-cell level and must try whenever feasible and reasonable to include many of these other factors.

Focusing on one type of quantitative measure of nanotoxicity, we describe flow and scanning image cytometry approaches to measuring nanotoxicity at the single-cell level by using a commonly used assay for distinguishing between necrotic and apoptotic causes of cell death by one type of nanoparticle. Flow cytometry is fast and quantitative, provided that the cells can be prepared into a single-cell suspension for analysis. But when cells cannot be put into suspension without altering nanotoxicity results, or if morphology, attachment, and stain location are important, a scanning image cytometry approach must be used. Both methods are described with application to a particular type of nanoparticle, a superparamagnetic iron oxide nanoparticle (SPION), as an example of how these assays may be applied to the more general problem of determining the effects of nanomaterial exposure to living cells.

Key words: Nanotoxicity, Single cell, Superparamagnetic iron oxide nanoparticles, SPIONs, Cytotoxicity, Apoptosis, Necrosis, Flow cytometry, Scanning image cytometry, LEAP

1. Introduction

1.1. Superparamagnetic Iron Oxide Nanoparticles for Nanomedicine

Nanomedicine is a field of nanotechnology that involves the engineering of nanoparticles and other nanostructures for dual diagnostic and therapeutic purposes ("theragnostics" or "theranostics"). These nanomedical systems are developed and produced at the nanometer level (<100 nm) and can be programmed to perform

Joshua Reineke (ed.), *Nanotoxicity: Methods and Protocols*, Methods in Molecular Biology, vol. 926,
DOI 10.1007/978-1-62703-002-1_5,

a precise order of molecular functions. For example, nanoparticles can be programmed to detect diseased cells or tissues and deliver targeted therapy, where the therapeutic release can also be designed to be responsive to the extent of the diseased state (1). Superparamagnetic iron oxide nanoparticles (SPIONs), the focus of this chapter, are widely used for enhanced magnetic resonance imaging (MRI) contrast due to their higher relaxation rates (2). The formulation of the magnetic core material is usually iron-based, including compounds such as Fe_2O_4 (3–7), Fe_3O_4 (8, 9), Fe with Fe_2O_3/Fe_3O_4 shell (10), $FeCo/Fe_2O_3$ (11), and FeNi (12). Such iron-based cores are commonly coated with polymers, such as dextran, to improve their physical stability in solution (13–15). In addition to the polymer coating, SPIONs can be functionalized with biomolecules for active targeting and/or therapeutic purposes. Thus, an engineered SPION can be highly unique based on its core material, size, geometry, surface chemistry, and biomolecular coatings. Given the extensive number of combinations possible, it is obvious that certain SPION formulations possess distinct physicochemical properties and ensuing differences in cell–nanoparticle interactions. Moreover, the small size of SPIONs can induce toxic effects that are not observed in their bulk material states due to the increase in surface area per unit mass (16). This means that SPIONs may be much more active due to their capability of potentially acting as catalytic agents for other reactions on these surfaces. For these reasons, there is an urgent need to establish in vitro nanotoxicity screening protocols to ensure the safety of SPION formulations for clinical use and to assess their effects on the environment. It is also important that we have quantitative measures of this nanotoxicity, so public policy decisions can be made on the basis of scientific data rather than fear and ignorance.

1.2. Nanotoxicity via Necrosis and Apoptosis

One of the initial steps in an in vitro nanotoxicity study is to evaluate the cytotoxic potential of the SPION formulation in question. A common method is the use of fluorescent dyes to measure cell death via necrosis and apoptosis (17). Necrosis is cell death due to injury and is characterized by rupture of the cell membrane and spilling of cellular contents that may pose as a risk factor to surrounding cells, which can bring on an immune response that may create additional complications. Apoptosis, the safer and more environmentally friendly way for a cell to die, is "programmed cell death," whereby the cell maintains its membrane integrity until late-stage apoptosis and methodically breaks down its constituents into precursor amino acids and nucleotides that can be reused by neighboring cells. SPIONs can cause a cell to trigger its own apoptosis, and this action will not be measured by traditional necrosis assays until very late in the process when cell membrane integrity is compromised. For this reason, it is not sufficient to measure only necrosis but also apoptosis.

The advantage of fluorescent labeling of nonviable cells is that fluorescent dyes specific for either necrosis or apoptosis can be used in combination to allow for simultaneous detection of both modes of cell death within the same cell population. For example, a marker for early-stage apoptosis is the translocation of phosphatidylserine (PS) from the inner to the outer leaflet of the plasma membrane (18). The exposure of PS to the extracellular environment allows identification of early apoptotic cells via annexin V, a 35–36 kDa protein that has a strong affinity for PS. Here, annexin V is conjugated to a fluorophore or biotin to facilitate detection with fluorescence-based methods. On the other hand, necrotic and late-stage apoptotic cells can be easily detected based on the decreased integrity of their plasma membrane. The "leakiness" of the plasma membrane renders it permeable to small-molecule fluorophores, such as propidium iodide (PI). PI is a red fluorescent molecule (668–669 Da) that labels DNA by intercalation (19). The choice of fluorescent probes for necrosis/apoptosis assays is not limited to fluorescent annexin V and PI. Kits for labeling necrotic and apoptotic cells may include fluorescent probes of distinct emission and excitation spectra and are commercially available from several vendors.

1.3. Caveats of Bulk-Cell Analyses of Nanotoxicity

Certainly, there are many other types of cytotoxicity assays available to study nanotoxicity. A popular assay to measure cell viability is the MTT (3-(4,5-dimethylthiazol-2-yl)-2,5-diphenyltetrazolium bromide) assay or variations of this assay [MTS (3-(4,5-dimethylthiazol-2-yl)-5-(3-carboxymethoxyphenyl)-2-(4-sulfophenyl)-2H-tetrazolium), XTT (2,3-bis-(2-methoxy-4-nitro-5-sulfophenyl)-2H-tetrazolium-5-carboxanilide), WSTs (water-soluble tetrazolium salts), etc.]. The MTT assay measures the cellular level of mitochondrial dehydrogenase activity, where the formation of a colored product (e.g., a purple formazan) is proportional to the number of viable cells. Cells are lysed with a detergent [e.g., radioimmunoprecipitation assay (RIPA) buffer] before measuring the absorbance via spectrophotometry. The MTT assay and other similar spectrophotometric assays are classified as bulk-cell assays due to the measurement of the response from a large cell population to a stimulus. Since the overall absorbance is measured from a cell lysate, the well-known heterogeneous behavior of cells in a population is not taken into account. A common caveat for bulk-cell approaches is that it is difficult to discern whether the observed effect results from a small, homogeneous response from all cells or a large response from a small subpopulation of cells. This is especially apparent in the MTT assay since the cell lysate is comprised of cells that exhibit differing levels of metabolic activity. Thus, measurements of cell viability are averaged for the entire population and may be misleading.

1.4. Advantages of Single-Cell Analysis of Nanotoxicity

A better alternative to bulk-cell assays is single-cell analysis. The major advantage to single-cell assays is the ability to acquire dynamic information from individual cells, such as differences in gene and protein expression, proliferation, cell cycle, and drug response. Moreover, single-cell approaches can acquire and correlate data from multiple parameters (e.g., protein expression v. drug response) via the use of appropriate fluorescent probes. In the scope of this chapter, data about necrosis (e.g., PI) and apoptosis (e.g., Alexa Fluor® 488 annexin V) in a SPION-treated population can be collected simultaneously, which is not possible using the MTT assay. The MTT assay and other bulk-cell assays do not provide information about the SPION-induced mode of cell death in individual cells. In addition, smaller numbers of cells can be used (even cells isolated from primary tissue sources), along with a reduced amount of reagents required. This is critical since SPIONs are often synthesized on a small scale, making it difficult to submit the SPIONs to an extensive battery of nanotoxicity tests. This chapter describes two common single-cell analysis techniques for measuring SPION-induced apoptosis and necrosis—flow cytometry and scanning image cytometry.

1.4.1. Flow Cytometry

Flow cytometry is a technology which takes cells in a single-cell suspension and hydrodynamically runs them past a light source (typically, but not always, a laser). While the cells are in the light source, they scatter light which gives some information about cell size and integrity as well as the excited fluorescence of any fluorescent probes that identify subcellular molecular constituents. Conventional flow cytometry is a "zeroth order" measurement, meaning it measures the fluorescence associated with a single cell but does not typically provide information about subcellular localization. The advantages of flow cytometry are speed (many thousands of cells per second can be measured) leading to large statistically significant samplings of the total cell population, the ability to determine the presence of cell subpopulations with thorough differences in scattering or fluorescence (or their combination), and the ability to look simultaneously at multiple fluorescence colors to provide highly sophisticated and correlated molecular measurements. While there are now some commercially available instruments that can provide real-time single-cell imaging, in general, cells must be sorted (usually by a process similar to inkjet printing) and then subsequently analyzed by some form of microscopy or image analysis.

1.4.2. High-Speed Scanning Image Cytometry

Alternatively, single cells can be analyzed by high-speed scanning image cytometry. Here, cells are seeded and assayed in microplates or on microscope slides, where nanotoxicity measurements from individual cells can be correlated to location (cellular and subcellular) and morphology. Scanning image cytometry is the preferred method for adherent cells that cannot easily be put into suspension without affecting nanotoxicity results and for limited samples isolated from primary cell sources (e.g., ex vivo tumor tissue). There are a wide variety of such instruments on the market. In this chapter, we will describe a particular image scanning cytometer available in our laboratory. However, most of the assays described with this instrument can be more generally performed by other such scanning instruments.

The scanning image cytometer used for high-speed single-cell nanotoxicity assays in our laboratory is called LEAP™ (Laser Enabled Analysis and Processing, Cyntellect, Inc., San Diego, CA). LEAP is a microplate-based cytometry platform that enables automated in situ imaging of entire wells of microplates (96-well and 384-well) and rapid quantification of individual cells in fluorescence-based assays. Descriptions of LEAP technology and operation can be found in preceding patents (20–22). Unlike flow cytometry, LEAP does not require additional cell manipulation (e.g., detachment from growth substrate) that may significantly alter cell function. Furthermore, analysis by LEAP is performed in a closed system, eliminating potential issues with biohazard risks and sterility. Another advantage of LEAP is that it permits the use of small cell samples (e.g., <10^4 cells), which reduces the amounts of reagents necessary. Finally, LEAP permits the monitoring of long-term nanotoxicity of the same samples since measurements are collected in situ, which is not feasible with flow cytometry. Besides whole-well imaging and automated fluorescence quantification, LEAP also has applications for laser-mediated cell transfection and in situ cell purification (23–26).

1.5. Experimental Design Considerations

1.5.1. Selection of Cell Types

The selection of cell types is critical to mimic the in vivo environments that the SPIONs may encounter. Both targeted and non-targeted cell types must be included to study differences, if any, in SPION–cell interactions. For example, a panel of human breast cancer and normal cell lines would be included if the SPIONs are designed to deliver targeted therapy to breast tumors. Normal cells lines may consist of fibroblast, endothelial, and macrophage cell types. The inclusion of normal cell types may give rise to information about SPION "mistargeting" events. Often, the targeting mechanism of the SPION formulation exploits the overexpression of a cell surface receptor, and it is important to include cell types that express different levels of the receptor. For example, EphA2 is an extracellular receptor tyrosine kinase that is highly expressed on several aggressive cancer types, and its activation induces endocytosis of the receptor (27). Previously, EphA2-targeted SPIONs were designed for preferential uptake in EphA2-expressing breast cancer cells (28). Two human breast adenocarcinoma lines were used for this study (MDA-MB-231 and MDA-MB-468). MDA-MB-231 highly expresses EphA2,

(~1×10^5 receptors per cell), while MDA-MB-468 express little to no EphA2 (<10^4 receptors per cell). Transmission electron microscopy (TEM) showed that EphA2-targeted SPIONs preferentially accumulated in MDA-MB-231 cells and had little to no interaction with MDA-MB-468 cells.

1.5.2. Applied SPION Concentration and Exposure Time

SPION concentration and exposure time are known to play a role in SPION-derived cytotoxicity. It is recommended that a wide range of SPION concentrations are tested with exposure times of at least 4 h. SPION concentrations and exposure times can be adjusted in subsequent experiments, depending on results of initial nanotoxicity studies. Short exposure times provide quick results of direct cytotoxicity, but long-term effects may remain unseen. One approach to long-term nanotoxicity studies is changing the growth medium postexposure and monitoring cytotoxicity for an extended period of time. This method would enable the study of secondary cytotoxic effects that may arise from products released from the SPIONs upon dissolution in the intracellular environment.

1.5.3. Nanotoxicity Assay Controls

SPIONs are often surface functionalized to increase their stability, confer targeting ability, and deliver a therapeutic agent. It follows that a multifunctional SPION system should be analyzed as individual parts and as a whole. To determine if the conjugation of targeting biomolecules lead to increased cell uptake, for example, SPIONs with and without the conjugated biomolecules should be tested side-by-side.

Positive controls for apoptosis and necrosis should also be included to calibrate settings for fluorescence detection. These are prepared by the addition of appropriate inducing agents to the cells. For example, camptothecin can be used to induce apoptosis and 70 % methanol can be used to induce necrosis. Negative untreated controls are also included to validate the viability of the initial cell source. Finally, vehicle controls are used to check the biocompatibility of the SPION suspension buffer. Here, SPION-free suspension buffer is added at the same volume as is applied to SPION-containing samples.

1.5.4. Confirmation of Cellular Uptake of SPIONs

In order to evaluate the potential cytotoxicity of internalized SPIONs, actual SPION uptake must first be established. Common methods of verifying the presence of SPIONs inside cells include TEM and Prussian blue staining. Due to the iron-based, electron-dense composition, TEM can be utilized to detect small numbers of SPIONs and to verify their intracellular location and degree of uptake. Prussian blue staining is a histological method, of limited sensitivity, that relies upon the reaction of potassium hexacyanoferrate trihydrate ($K_4Fe(CN)_6 \cdot H_2O$) with Fe(III) from SPIONs in an acidic environment. It is critical to confirm positive SPION uptake with the aim of associating intracellular SPION concentration to cytotoxicity effects. For example, the addition of a biomolecule coating may appear to increase SPION biocompatibility but may be merely due to poor internalization or cell interaction. Without confirmation of cellular uptake, nanotoxicity data may be misleading.

2. Materials

2.1. Fluorescent Labeling of Apoptotic and Necrotic Cells

1. Phosphate-buffered saline (PBS), pH 7.4 (~4°C) (see Note 1).
2. Deionized water.
3. Alexa Fluor® 488 annexin V (Invitrogen, Carlsbad, CA): solution in 25 mM HEPES, 140 mM NaCl, 1 mM EDTA, 0.1 % BSA, pH 7.4.
4. 5× annexin V binding buffer (Invitrogen, Carlsbad, CA): 50 mM HEPES, 700 mM NaCl, 12.5 mM $CaCl_2$, pH 7.4.
5. Propidium iodide (PI; Invitrogen, Carlsbad, CA): 1 mg/ml (1.5 mM) solution in water.
6. Hoechst 33342 (Invitrogen, Carlsbad, CA): 10 mg/ml (16.2 mM) solution in water (for scanning image cytometry only).

Fluorescent probes may be purchased separately, but Invitrogen also offers combinations of these dyes in kits such as the Alexa Fluor® 488 Annexin V/Dead Cell Apoptosis Kit, which contains PI as its dead cell stain. In addition, Invitrogen's Vybrant® Apoptosis Assay line of products offers a wide array of other fluorescent probes used to label apoptotic and necrotic cells. For example, fluorescein isothiocyanate (FITC)-annexin V can be used instead of Alexa Fluor® 488 annexin V since they have similar excitation and emission spectra. Hoechst 33342 is a blue-fluorescent, cell-permeable dye used for labeling nucleic acids. Hoechst 33342 is used to label all cells in the population, regardless of live, necrotic, or apoptotic states. It was the first ever fluorescent dye that could reversibly stain A–T base pairs of DNA in live and dead cells (29). It does require a near-UV excitation (around 365 nm) and emits with a broad emission spectrum with a peak around 450 nm and an emission spectrum tail extending out to more than 600 nm. The more recent dyes excitable in the visible part of the spectrum were invented partly to allow their excitation by less expensive non near-UV light sources. For LEAP, staining with Hoechst 33342 is necessary for the quantification of total cell number and automatic calculation of percentages of necrotic and apoptotic cells in the population. For flow cytometry, Hoechst 33342 is not necessary since cells can be easily detected by forward and/or side scatter.

2.2. Induction of Apoptosis and Necrosis for Positive Nanotoxicity Controls

1. Camptothecin, for induction of apoptosis.
2. 70 % methanol, for induction of necrosis.

Camptothecin-treated cells serve as a positive control for apoptosis, while cells treated with 70 % methanol serve as a positive control for necrosis. Concentrations and exposure times of inducing agents should be optimized per cell type. Other inducing agents may be utilized. For example, a number of anticancer therapeutics are designed to induce apoptosis, such as paclitaxel, and can be used in lieu of camptothecin. To induce necrosis, 0.1 % saponin or 0.1–0.5 % digitonin may also be applied.

2.3. Instrumentation

2.3.1. Flow Cytometers

The following represent fluorescence-based flow cytometers capable of performing, with appropriate adaption of instrument settings and configurations, the nanotoxicity assays described in this chapter. There has been a recent flurry of acquisitions by Beckman Coulter (acquired Cytomation), Becton Dickinson (BD, acquired Cytopeia), and Sony (acquired i-Cyt). Beckman Coulter (http://www.beckmancoulter.com/) produces a wide array of both research and clinical instruments ranging from blood analyzers to cell sorters. Becton Dickinson (http://www.bd.com/) produces a wide array of both research and clinical instruments ranging from blood analyzers to cell sorters. They also provide a wide array of requisite reagents. Sony recently acquired i-Cyt (http://www.i-cyt.com/), manufacturer of a multi-modular flow cytometer/cell sorter housed in a biosafety level 2 (BSL-2) hood for live cell sorting. Partec (http://www.partec.com) produces a variety of small and portable flow cytometer analyzers (some using arc lamps, others using small lasers) that are also suitable for microbiology applications. A relatively recent arrival is Accuri (http://www.accuricytometers.com/) who produces small portable flow cytometers.

2.3.2. Scanning Image Cytometers

While there are many image analyzers on the market, the following three manufacturers are oriented toward single-cell fluorescence measurements as described in this chapter. Amnis (http://www.amnis.com/) produces an imaging in flow cytometer that can give spatial location to cellular fluorescence. It is effectively a hybrid flow and image cytometer. CompuCyte (http://www.compucyte.com/) produces a scanning image cytometer that can produce flow cytometry-like data and perform data sorting with cellular images. Cyntellect (http://www.cyntellect.com/) produces interactive scanning image cytometers that not only provide high field of view for high-throughput scanning image cytometry but also provide for elimination of undesired cells by laser ablation sorting. The nanotoxicity assay protocol in Subheading 3.2 is written for Cyntellect's LEAP system, but scanning image cytometers from the aforementioned manufacturers may also work with appropriate modifications.

3. Methods

3.1. Detection of Apoptotic and Necrotic Cells by Flow Cytometry

Flow cytometry can rapidly enumerate the relative numbers of live, necrotic, and apoptotic cells. However, these measurements on individual cells are done only once and, unlike scanning image cytometry, cannot be repeated on the same cells to study the time-dependency of nanotoxicity. Instead, we can only measure sub-populations of cells from a total population sampled at different time points on the flow cytometer. The following is a protocol that has been adapted from Invitrogen (30):

1. Seed 2×10^5 to 1×10^6 cells/ml in an appropriate tissue culture vessel (see Subheading 1.5.1 and Note 2) in the appropriate growth medium. For adherent cells, incubate plate at 37°C/5% CO_2 overnight to allow cells to settle and attach. For primary cultures at a small starting cell number, cells may be allowed to grow to 70–80 % confluence over several days.
2. Prepare working concentrations of SPION suspensions in serum-free growth medium (see Subheading 1.5.2). Also, prepare working solutions of camptothecin (apoptosis), 70 % methanol (necrosis), and SPION suspension buffer (vehicle) for nanotoxicity controls (see Subheading 1.5.3). Add working solutions to each cell sample. For negative control samples, add the same volume of serum-free growth medium (see Subheading 1.5.3 and Note 3). Expose cells for ≥4 h (see Subheading 1.5.2 and Note 4).
3. After the exposure period, wash cells with cold PBS to remove SPIONs and harvest cells using the appropriate method. Centrifuge at $200 \times g$ for 5 min, discard the supernatant, and resuspend cells in cold PBS (see Note 1).
4. Prepare 1× annexin-binding buffer by diluting 5× annexin-binding buffer 1/5 with deionized water. Prepare enough 1× annexin-binding buffer for all samples, approximately 500 μl per sample.
5. Prepare a 100 μg/ml working solution of PI in 1× annexin-binding buffer (see Note 5).
6. Recentrifuge the washed cells at $200 \times g$ for 5 min, discard the supernatant, and resuspend the cells in 1× annexin-binding buffer to a final cell concentration of 1×10^6/ml with a minimum total volume of 100 μl.
7. Add 5–25 μl Alexa Fluor® 488 annexin V (see Note 6), and 1 μl PI working solution (1 μg/ml final concentration) per 100 μl of cell suspension in 1× annexin-binding buffer.
8. Incubate the cells at room temperature for 15 min in the dark to avoid photobleaching the fluorophores.

9. After incubation, add 400 μl 1× annexin-binding buffer (to a final volume of 500 μl of cell suspension), mix gently, and keep on ice at ~4 °C (see Note 1).
10. As soon as possible, analyze the cells by flow cytometry.
    (a) Illuminate with blue light (488-nm laser line or BG12 excitation filter).
    (b) Measure green fluorescence of Alexa Fluor® 488 annexin V at 530 ± 20 nm or 530/40 nm (sometimes referred to as FL1) (see Note 7).
    (c) Measure red fluorescence of PI at >600 nm (sometimes referred to as FL3) (see Note 7).
11. Flow cytometry results should be confirmed under a fluorescent microscope using the appropriate filters for Alexa Fluor® 488 annexin V and PI (see Note 8).

When cells are labeled with Alexa Fluor® 488 annexin V and PI and analyzed in a flow cytometer, the data will look something like the data shown in Fig. 1. Each point on the two-dimensional scattergram represents one or more cells with that (x, y) coordinate pair of values of annexin V (x-value) and PI (y-value). The data of Fig. 1 shows camptothecin-treated MCF-7 (breast cancer) cells as a positive control. Camptothecin induces apoptosis in most cell types. In the case of SPION-treated cells, the numbers of necrotic and apoptotic cells should be relatively low and perhaps not much above background levels unless there is additional nanotoxicity from biocompatible coatings and conjugated biomolecules on the SPIONs.

3.2. Detection of Apoptotic and Necrotic Cells by LEAP Scanning Image Cytometry

A similar staining protocol is used for scanning image cytometry by LEAP except that the vital DNA stain Hoechst 33342 serves as a "cell marker source" and is used to locate nuclei of cells in order to quantify the total number of cells present in a sample for automatic calculation of percentages of necrotic and apoptotic cells in the sample. Hoechst 33342 is not necessary for measurement via flow cytometry since cells can be detected easily by forward and/or side scatter.

1. Seed 300–500 cells/10 μl/well in a 384-well C-lect™ microplate or 3,000–5,000 cells/50 μl/well in a 96-well C-lect microplate (see Subheading 1.5.1 and Note 9) in appropriate growth medium. Incubate plate at 37°C/5% CO_2 overnight to allow cells to settle and attach. For 384-well C-lect plates, gently centrifuge plates to remove bubbles before overnight incubation (see Note 10). For primary cultures at a small starting cell number, cells may be allowed to grow to 70–80 % confluence over several days.
2. Prepare 2× working concentrations of SPION suspensions in serum-free growth medium (see Subheading 1.5.2 and Note 11).

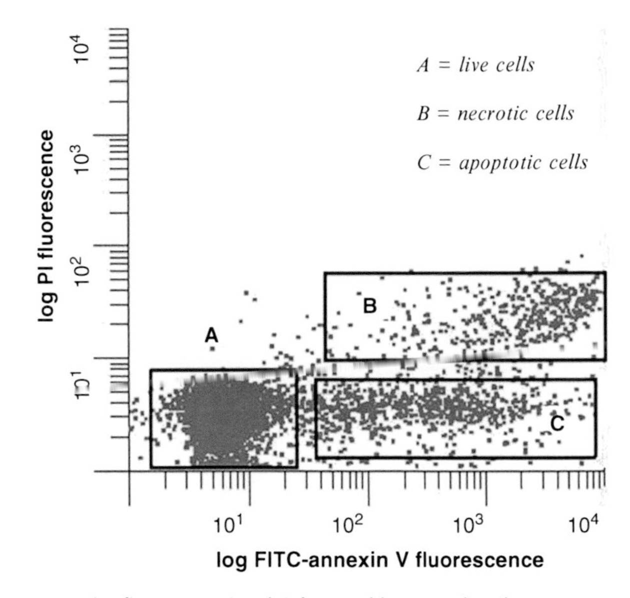

Fig. 1. Representative flow cytometry plot. As a positive control on the assay, camptothecin-treated human MCF-7 cells were stained with FITC-annexin V and propidium iodide (*PI*) and analyzed by flow cytometry. Three cell populations are evident: (*A*) live cells, (*B*) necrotic cells, and (*C*) apoptotic cells.

Also, prepare 2× working solutions of camptothecin (apoptosis), 70 % methanol (necrosis), and SPION suspension buffer (vehicle) for nanotoxicity controls (see Subheading 1.5.3). Pipet 10 μl (384-well) or 50 μl (96-well) of working solutions on top of cells. For negative controls, pipet 10 μl (384-well) or 50 μl (96-well) of serum-free growth medium on top of cells. Expose cells for ≥4 h (see Subheading 1.5.2 and Note 4).

3. After the exposure period, wash cells three times with cold PBS to remove SPIONs (see Notes 1 and 12).
4. Prepare 1× annexin-binding buffer by diluting 5× annexin-binding buffer 1/5 with deionized water. Prepare enough 1× annexin-binding buffer for all samples, including washes (10 μl per sample for 384-well and 50 μl per sample for 96-well).
5. Prepare 2× staining solution of Hoechst 33342, Alexa Fluor® 488 annexin V, and PI. Add 1 μl Hoechst 33342 (10 μg/ml 2× concentration), 10 μl Alexa Fluor® 488 annexin V, and 2 μl PI (2 μg/ml 2× concentration) per 1 ml of 1× annexin-binding buffer (see Notes 5 and 10).
6. Wash cells three times with 1× annexin-binding buffer. Add 10 μl (384-well) or 50 μl (96-well) of 2× staining solution to each well. Incubate the cells at room temperature for 15–30 min in the dark to avoid photobleaching the fluorophores. Keep microplate on ice until ready to load onto LEAP (see Note 1).

Table 1
Excitation, dichroic, and emission filter sets for Alexa Fluor® annexin V, PI, and Hoechst 33342

Fluorescent probe	Excitation wavelength (nm)	Dichroic wavelength (nm)	Emission wavelength (nm)
Alexa Fluor® 488 annexin V (apoptotic)	482/35	506LP	536/40
Propidium iodide (PI, necrotic)	562/40	593LP	624/40
Hoechst 33342 (total)	377/50	409LP	438/24

"___/___," mean wavelength/full width at half maximum (FWHM) transmission. *LP* long pass (see Note 6)

7. Turn on the LEAP instrument and its computer. Launch the LEAP application. Allow the camera, laser, and excitation lamp to warm up for 10 min.
8. Click on the Load Plate icon in the Protocol menu. Enter the Plate Label and select the type of plate (384-well or 96-well C-lect plate). Secure the plate onto LEAP stage with well A1 placed in the upper left-hand corner. Ensure the plate is seated flat. After loading, the plate is automatically aligned according to selected plate type.
9. Under the Machine Control tab, choose 3× magnification. At 3× magnification, an entire well in a 384-well C-lect plate is visible in one field of view (FOV). At 3× magnification, 16 images are acquired and stitched together to generate a whole-well image (see Note 13).
10. In the Protocol Options menu, select the Cell Viability protocol under Protocol Settings. For Viability Calculation, select Live/Total. In the LEAP Image Display window, Channel 1 is now labeled Live Cells and Channel 2 is now labeled Total Cells.
11. Under Auto Focus, check box to Enable Auto Focus to be performed on each well. For Focus Settings, select both Live Cell and Total Cell channels (see Note 14).

    The following steps are for acquiring data for necrotic cells. The same steps are repeated for acquiring data for apoptotic cells with some modifications (see step 17).
12. Under Optical Path in the LEAP Display window, select Fluorescence as the Illumination Source and select the appropriate Excitation, Dichroic, and Emission Wavelengths for the Live Cell and Total Cell channels (Table 1). To acquire necrosis data, Channel 1 (Live Cell) is set for viewing PI-positive cells. Channel 2 (Total Cell) is always used to identify Hoechst 33342-positive cells (see Note 15).

13. Navigate to a positive control well (treated with 70 % methanol) and select Autofocus to bring the cells into focus. In the Live Cell channel, calibrate imaging settings under the Image Preprocessing and Image Segmentation tabs. Ensure all PI-positive cells are efficiently identified and marked (see Note 16).
14. Navigate to a negative-control well (treated with serum-free growth medium only) and select Autofocus to bring the cells into focus. In the Total Cell channel, calibrate imaging settings under the Image Preprocessing and Image Segmentation tabs. Ensure all Hoechst 33342-positive cells are efficiently identified and marked (see Note 16).
15. After imaging settings have been set, select the wells to be processed in the Plate Map. Click Run Protocol to begin the data acquisition process.
16. After processing, data can be viewed in the Results Viewer. The percentages of necrotic cells are displayed on each well in the plate map. Data can be exported for further analysis.
17. Repeat steps 13–17 to acquire data for apoptotic cells with the following modifications. In step 13, set Channel 1 (Live Cells) for viewing Alexa Fluor® 488 annexin V-positive cells. In step 15, use the camptothecin-treated cells as the positive control for apoptosis (see Note 17).

4. Notes

1. Cold PBS (~4°C) is critical in order to hinder metabolic cellular processes and active transport across the cell membrane. It is important to NEVER freeze cells which will not only cause cell death but also denature proteins and antibodies. Maintaining a temperature of ~4°C is relatively easy to do with cell suspensions for flow cytometry by keeping the samples on ice in a common ice bucket. Cooling surfaces for scanning image cytometry (e.g., Peltier-effect thermoelectrical cooling) is very difficult and frequently impractical.
2. Tissue culture vessels in a variety of formats can be used, depending on the growth characteristics of the chosen cell type. For primary cell cultures, smaller formats may be used (e.g., 96-well microplates). In this case, several sample replicates may be prepared and then pooled in order to achieve the large numbers of cells necessary for measurement by flow cytometry.
3. The volume of working solutions can vary, depending on the chosen tissue culture format. Small volumes of working solutions

may be prepared, as long as the desired concentrations are achieved upon dilution in the cell samples. However, the added volumes should be uniform across all cell samples, including the negative controls.

4. Initial SPION exposure times may range from 4 to 24 h. Longer exposure times may be used if growth medium is changed over the course of the exposure period.
5. Extreme caution should be exercised when handling PI because it is a potent mutagen.
6. Invitrogen recommends 0.2–5 μg/ml final concentration of Hoechst 33342 for live mammalian cells. For annexin V, Invitrogen recommends 5–25 μl annexin V conjugate per 100 μl of cell suspension (1× concentration). Higher volumes of annexin V conjugate tend to generate better results. The staining concentrations should be optimized for each cell type.
7. Optical filter wavelengths are designated two ways. The first nomenclature is "mean wavelength/full width at half maximum (FWHM) transmission." The second nomenclature is "mean wavelength ± half width at half maximum (HWHM) transmission." For example, the designation 530 ± 20 nm is equivalent to 530/40 nm. Designations such as "FL1, FL2, FL3" are Becton Dickinson's (BD) instrument-specific nomenclature of photodetectors in their flow cytometers, corresponding approximately to green (530/30), orange (575/20), and red (630/30), but researchers really should not use this FL designation but rather use either the "mean wavelength ± HWHM" or "mean wavelength/FWHM" terminologies.
8. For quick confirmation of flow cytometry results via fluorescence microscopy, a few drops of cell suspension may be added to a glass slide with coverslip and viewed using an upright fluorescent microscope with the appropriate filters. To capture microscope images, however, cell suspensions should be transferred to a format with a set volume (e.g., 96-well or 384-well microplate) for more accurate qualitative representation of live, necrotic, and dead cell subpopulations. Since the samples lack DNA-specific (e.g., Hoechst 33342) fluorescent staining for live cells, brightfield images should also be taken to view these cells.
9. C-lect microplates are custom-made for LEAP. These plates exhibit good cell adhesion and growth, along with superior flatness and low autofluorescence for imaging. In addition, the software algorithms on LEAP are designed for these specific plates. The use of microplates of other types may not be optimal for LEAP.

10. Bubbles may occur when plating cells in 384-well C-lect plates. If bubbles are not removed from the bottom of the wells, the cells may not attach properly, resulting in cell death. Moreover, the occurrence of bubbles at the bottom of the wells may result in inefficient imaging of the cells by LEAP. Use a swinging plate holder for short, gentle centrifugation (100–200 × *g*).
11. For a 100 μg/ml final concentration, for example, prepare a 200 μg/ml working solution. The minimum volume needed per sample concentration is calculated from the plating volume and number of replicates. For example, three replicates in a 384-well C-lect plate require a minimum of 3 × 10 μl = 30 μl per sample. Prepare extra volume to account for pipetting error. Serum-free growth medium is used to provide cell nutrients without the possibility of SPIONs interacting with serum proteins (the so-called "protein corona") and aggregating.
12. To avoid the loss of cells, wash cells by aspirating approximately half the medium from each well and replacing with fresh PBS. Repeat this at least twice. If suspended cells are used, gently centrifuge plate at 100–200 × *g* (without brake) between each wash step.
13. Magnification at 3× is considered sufficient to quantify fluorescence-positive cells. Higher magnifications (5×, 10×, 20×) may be used, but this will increase the image acquisition and processing time.
14. Selecting both channels for focus ensures an accurate focus regardless of the level of viability. If the extremes of viability (i.e., 0 % or 100 %) are not present, one fluorescent channel can be selected for focus, allowing for a faster scan speed. Since all cells should stain positive for Hoechst 33342, it is recommended that the Total Cell Channel is selected for Auto Focus in this case.
15. The calculation for Cell Viability is a percentage of fluorescence-positive cells in the population (e.g., Live/Total × 100 %), regardless if the stain used is a live cell stain. In the case of viewing necrotic and apoptotic cells, Channel 1's designation of the "Live Cell" channel is a misnomer. Thus, the Live Cell channel could be designated for apoptotic (annexin V-positive) or necrotic (PI-positive) cells to compute % apoptotic and % necrotic, respectively. The Total Cell channel is designated for Hoechst 33342-positve cells for both cases.
16. Settings for Image Preprocessing and Image Segmentation are highly dependent on cell type and fluorophore. Channel settings may be saved for use in later experiments. Details about Image Preprocessing and Image Segmentation can be found in the latest LEAP user manual.

17. For future nanotoxicity studies, other fluorescent probes may be used in the same manner to acquire quantitative data about a different mode of cytotoxicity. For example, dihydroethidium (DHE) is a red fluorescence molecule that detects reactive oxygen species (ROS) (31). Cellular uptake of SPIONs may result in a transient generation of ROS which can impair normal cell functions and ultimately lead to cell death.

References

1. Haglund E, Seale-Goldsmith M-M, Leary JF (2009) Design of multifunctional nanomedical systems. Ann Biomed Eng 37:2048–2063
2. Corot C, Robert P, Idee J-M, Port M (2006) Recent advances in iron oxide nanocrystal technology for medical imaging. Adv Drug Deliv Rev 58:1471–1504
3. Ito A, Kuga Y, Honda H, Kikkawa H, Horiuchi A, Watanabe Y, Kobayashi T (2004) Magnetite nanoparticle-loaded anti-HER2 immunoliposomes for combination of antibody therapy with hyperthermia. Cancer Lett 212:167–175
4. Kim D, Lee S, Kim K, Kim K, Shim I, Lee M, Lee Y-K (2006) Surface-modified magnetite nanoparticles for hyperthermia: preparation, characterization, and cytotoxicity studies. Curr Appl Phys 6:e242–e246
5. Kumar CS, Leuschner C, Doomes EE, Henry L, Juban M, Hormes J (2004) Efficacy of lytic peptide-bound magnetite nanoparticles in destroying breast cancer cells. J Nanosci Nanotechnol 4:245–249
6. Zhang Y, Sun C, Kohler N, Zhang M (2004) Self-assembled coatings on individual monodisperse magnetite nanoparticles for efficient intracellular uptake. Biomed Microdev 6: 33–40
7. Zhang Y, Yang M, Portney NG, Cui G, Budak E, Ozbay M, Ozkan M, Ozkan CS (2008) Zeta potential: a surface electrical characteristic to probe the interaction of nanoparticles with normal and cancer human breast epithelial cells. Biomed Microdev 10:321–328
8. Berry CC, Wells S, Charles S, Aitchison G, Curtis AS (2004) Cell response to dextran-derivatised iron oxide nanoparticles post internalisation. Biomaterials 25:5405–5413
9. Mondalek FG, Zhang YY, Kropp B, Kopke RD, Ge X, Jackson RL, Dormer KJ (2006) The permability of SPION over an artificial three-layer membrane is enhanced by external magnetic field. J Nanobiotechnol 4:4
10. Bomati-Miguel O, Morales MP, Tartaj P, Ruiz-Cabello J, Bonville P, Santos M, Zhao X, Veintemillas-Verdaguer S (2005) Fe-based nanoparticulate metallic alloys as contrast agents for magnetic resonance imaging. Biomaterials 26:5695–5703
11. Moller W, Takenaka S, Buske N, Felten K, Heyder J (2005) Relaxation of ferromagnetic nanoparticles in macrophages: in vitro and in vivo studies. J Magn Magn Mater 293: 245–251
12. Yin H, Too HP, Chow GM (2005) The effects of particle size and surface coating on the cytotoxicity of nickel ferrite. Biomaterials 26:5818–5826
13. Boutry S, Brunin S, Mahieu I, Laurent S, VanderElst L, Muller RN (2008) Magnetic labeling of non-phagocytic adherent cells with iron oxide nanoparticles: a comprehensive study. Contrast Media Mol Imaging 3: 223–232
14. Geraldes CF, Laurent S (2009) Classification and basic properties of contrast agents for magnetic resonance imaging. Contrast Media Mol Imaging 4:1–23
15. Kustermann E, Himmelreich U, Kandal K, Geelen T, Ketkar A, Wiedermann D, Strecker C, Esser J, Arnhold S, Hoehn M (2008) Efficient stem cell labeling for MRI studies. Contrast Media Mol Imaging 3:27–37
16. Oberdorster G, Oberdorster E, Oberdorster J (2005) Nanotoxicology: an emerging discipline evolving from studies of ultrafine particles. Environ Health Perspect 113:823–839
17. Stone V, Johnston H, Schins RPF (2009) Development of in vitro systems for nanotoxicology: methodological considerations. Crit Rev Toxicol 39:613–626
18. van Engeland M, Nieland LJ, Ramaekers FC, Schutte B, Reutelingsperger CP (1998) Annexin V-affinity assay: a review on an apoptosis detection system based on phosphatidylserine exposure. Cytometry 31:1–9
19. Waring MJ (1965) Complex formation between ethidium bromide and nucleic acids. J Mol Biol 13:269–282

20. Koller MR, Palsson BO, Eisfeld TM (2003) Method for inducing a response in one or more targeted cells. US Patent 6,642,018
21. Palsson BO, Koller MR, Eisfeld TM (2003) Method and apparatus for selectively targeting specific cells within a cell population. US Patent 6,514,722
22. Palsson BO, Koller MR, Eisfeld TM (2003) Method and apparatus for selectively targeting specific cells within a mixed cell population. US Patent 6,534,308
23. Clark IB, Hanania EG, Stevens J, Gallina M, Fieck A, Brandes R, Palsson BO, Koller MR (2006) Optoinjection for efficient targeted delivery of a broad range of compounds and macromolecules into diverse cell types. J Biomed Opt 11:014034
24. Hanania EG, Fieck A, Stevens J, Bodzin LJ, Palsson BO, Koller MR (2005) Automated in situ measurement of cell-specific antibody secretion and laser-mediated purification for rapid cloning of highly-secreting producers. Biotechnol Bioeng 91:872–876
25. Koller MR, Hanania EG, Stevens J, Eisfeld TM, Sasaki GC, Fieck A, Palsson BO (2004) High-throughput laser-mediated in situ cell purification with high purity and yield. Cytometry A 61:153–161
26. Rhodes K, Clark I, Zatcoff M, Eustaquio T, Hoyte KL, Koller MR (2007) Cellular laserfection. Methods Cell Biol 82:309–333
27. Abraham SD, Knapp DW, Cheng L, Snyder PW, Mittal SK, Bangari DS, Kinch MS, Wu L, Dhariwal J, Mohammed SI (2006) Expression of EphA2 and ephrin A-1 in carcinoma of the urinary bladder. Clin Cancer Res 12:353–360
28. Seale M-M (2009) Design of targeted nanoparticles for multifunctional nanomedical systems. Ph.D. Thesis, Purdue University, Biomedical Engineering
29. Arndt-Jovin DJ, Jovin TM (1977) Analysis and sorting of living cells according to deoxyribonucleic acid content. J Histochem Cytochem 25:585–589
30. Invitrogen (2010, Sept) Alexa Fluor® 488 annexin V/Dead Cell Apoptosis Kit with Alexa® Fluor 488 annexin V and PI for Flow Cytometry. http://probes.invitrogen.com/media/pis/mp13241.pdf, [June 20, 2012]
31. Gallop PM, Paz MA, Henson E, Latt SA (1984) Dynamic approaches to the delivery of reporter reagents into living cells. Biotechniques 3:32–36

Chapter 6

Western Blot Analysis

Seishiro Hirano

Abstract

Electrophoresis and the following western blot analysis are indispensable to investigate biochemical changes in cells and tissues exposed to nanoparticles or nanomaterials. Proteins should be extracted from the cells and tissues using a proper method, especially when phosphorylated proteins are to be detected. It is important to select a good blocking agent and an appropriate pair of primary and peroxidase-tagged secondary antibodies to obtain good results in western blot analysis. One thing that may be specific to nanomaterials, and that you should keep in mind, is that some proteins may be adsorbed on the surface of particulate nanomaterials. In this chapter the whole process of western blot analysis, from sample preparation to quantitative measurement of target proteins, is described.

Key words: Proteins, Electrophoresis, SDS-PAGE, Nitrocellulose, PVDF, Antibody, Chemiluminescence, Densitometry

1. Introduction

Most nanomaterials are insoluble particulate substances. Exposure to nanomaterials either in vitro or in vivo leads to interaction of cell membranes with the surface of materials, which may be modified with surfactants in the alveolar lining fluid, serum, and other biomolecules. The particulate substances are taken up by phagocytosis or endocytosis into the cells and have chance to react with proteins in plasma membrane, cytosol, and other cellular compartments. Thus, it is important to select appropriate detergents or denaturing agents when proteins are extracted from the cells. For example, ionic detergents like sodium dodecylsulfate (SDS) should be used at a concentration of 0.1% to obtain the whole lysate proteins including nuclear proteins. On the contrary, noninonic detergents

Joshua Reineke (ed.), *Nanotoxicity: Methods and Protocols*, Methods in Molecular Biology, vol. 926,
DOI 10.1007/978-1-62703-002-1_6, © Springer Science+Business Media, LLC 2012

are to be used to obtain cytoplasmic proteins. Once you obtain the supernatant of the lysate which contains proteins (1–5 mg protein/mL), it is not difficult to quantitatively measure the amount of target proteins by western blot analysis, because a variety of agents and measurement kits are commercially available and, in general, good results will be obtained just by following the manufacture's instruction.

In nanotoxicology studies the "size effect" is usually discussed (1, 2). Generally smaller particles generated more intense oxidative stress than larger particles, probably because smaller particles have larger surface area than larger particles on a unit mass basis and the reaction of cells or biomolecules with particles occurs primarily on the surface of the particles. Thus, detection of stress proteins such as heme oxygenase-1 and heat shock proteins such as HSP70 and HSP27 as well as evaluation of apoptosis with western blot method is a useful approach to evaluate responses of the cells to manomaterials and nanoparticles. Also investigating changes in phosphorylation levels of MAP kinases such as SAPK (JNK) and p38 as an outcome of oxidative stress would be interesting.

2. Materials

1. Solubilization buffer: most frequently used one to obtain the whole lysate is radio immunoprecipitation assay (RIPA) buffer which contains 25 mM Tris–HCl or HEPES (pH 7.6), 150 mM NaCl, 1% Nonidet P-40 (see Note 1), 0.5–1% sodium deoxycholate, and 0.1% SDS. Omit 0.1% SDS to keep the nuclear membrane intact and to obtain cytosolic proteins.
2. Protease inhibitors: add the following to the lysate buffer; 1 mM phenylmethylsulfonyl fluoride (PMSF, prepare 100× solution in methanol and add to the solubilization buffer just before use), 10 μg/mL aprotinin, 5 mM benzamidine, 20 μM leupeptin, 10 μM pepstatin, and 5 mM EDTA. Protease inhibitor cocktails (usually provided as a 100× concentrate) are commercially available from Sigma, Thermo Fisher, and other chemical companies.
3. Phosphatase inhibitors: add the followings to the lysate buffer (final concentration); 1 mM *p*-nitrophenyl phosphate, 50 mM NaF, 1 mM orthovanadate (Na_3VO_4), and 20 nM calyculin A. Phosphatase inhibitor cocktails (usually provided as a 100× concentrate) are commercially available from Sigma, Thermo Fisher, and other chemical companies.
4. Polyacrylamide gel: when you prepare SDS-PAGE gels, acrylamide and bis-methyleneacrylamide solutions, 0.1 g/mL ammonium peroxydisulfate, and tetramethylethylenediamine

(TEMED) are required. Precast gels of various polyacrylamide concentrations and gradient gels (e.g., 4–12%) are commercially available (see Note 2).

5. Power supply: the power supply is necessary for both electrophoresis and western blotting. A special western blot apparatus (iBlot, Life technologies) can also be used for dry blotting.
6. Electrophoresis buffer: dissolve 6 g TRIZMA-Base, 22.8 g glycine, and 1 g SDS in 1 L of deionized water (pH 8.3) (see Note 3).
7. Western transfer buffer: dissolve 12.1 g TRIZMA-Base, 14.4 g glycine in 950 mL deionized water and then add 50 mL methanol.
8. Membrane/adsorbent paper filter: cut nitrocellulose or polyvinylidene difluoride (PVDF) membranes and adsorbent paper filters to the gel size.
9. Blocking solution: dissolve 5% skim milk or 2–3% bovine serum albumin (BSA) in PBS or 150 mM Tris–HCl (TBS) buffer (pH 7.2). Nonprotein blocking agents are commercially available.
10. Antibodies: dilute primary antibody to 1/200–1/1,000 and the peroxidase (POD)-tagged secondary antibody to 1/1,500–1/5,000 by blocking solution. Dilution solutions for the primary and secondary antibodies are commercially available.
11. Membrane wash solution: add Tween 20 (see Note 1) at a concentration of 0.05% to phosphate buffer or Tris–HCl buffer.
12. Substrate for antibody signal detection: enhanced chemiluminescence (ECL) solution is commonly used in western blot analysis (Fig. 1). Diaminobenzidine (DBA) can be used to stain the membrane directly.
13. Stripping buffer: the stripping buffer (62.5 mM Tris–HCl (pH 6.8) containing 2% SDS and 100 mM β-mercaptoethanol) is used to remove antibodies from the membrane after western analyses. The membrane can be reprobed with another set of antibodies.
14. Others: SDS sample buffer with loading dyes, molecular weight markers, western markers, X-ray films, shakers.

3. Methods

3.1. Exposure to Nanomaterials

It is important to disperse nanomaterials before an exposure experiment, because the level of agglomeration may dominate the effects of nanomaterials (3, 4). Hydrophilic particles can be dispersed easier and better than hydrophobic ones in the culture medium.

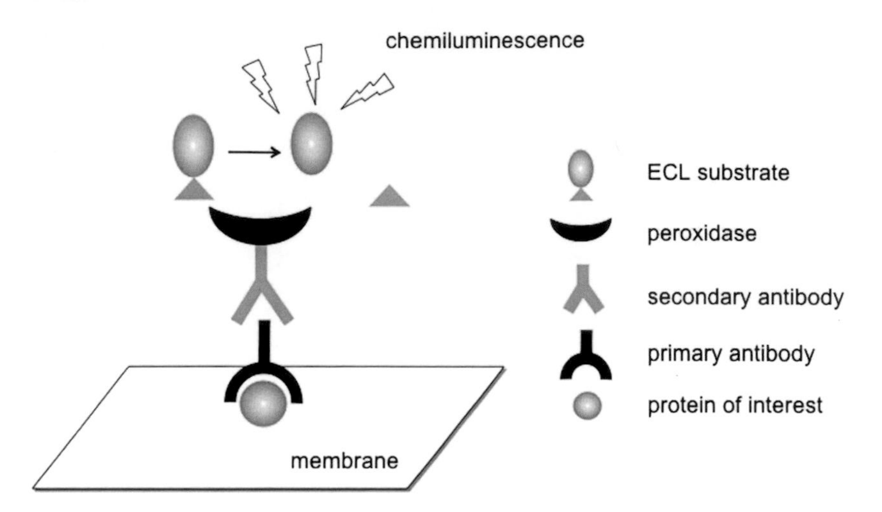

Fig. 1. A schematic presentation for the detection of proteins on the western blot membrane by ECL. ECL substrate is processed by the enzyme (peroxidase) tagged to the secondary antibody and generates chemiluminescence where the primary antibody reacted with the antigen on the membrane.

The following is a typical procedure to expose cells to hydrophobic nanomaterials such as carbon nanotubes.

1. Prepare a stock suspension of nanomaterials in nontoxic detergent vehicle such as Pluronic F68 of the cell culture grade.
2. Disperse the suspension by ultrasonication.
3. Dilute the sonicated nanomaterial suspension with culture medium. Pluronic F68 is not cytotoxic up to 1%.
4. Aspirate the conditioned medium and add the diluted nanomaterial suspension to the cells.
5. In parallel experiment measure the particle size distribution in the culture medium using dynamic light scattering (DLS) (see Note 4).
6. Remove the medium and wash the cells with HBSS or PBS for the following sample preparation.

3.2. Sample Preparation for Electrophoresis

1. First choose an appropriate method for extraction of proteins from cells or tissues according to your experiment purpose. Freezing and thawing several times, sonication using Bioruptor®, and direct lysis using solubilization buffer are commonly used to extract proteins from cells for electrophoresis–western blotting experiments.
2. Add protease inhibitors in protein extraction buffer to prevent proteolysis. If phosphorylated proteins are to be detected, add also phosphatase inhibitors.
3. Extract proteins from the samples. Keep the sample below 4°C during the extraction process.

4. Centrifuge the samples to remove cell debris and collect the supernatant (see Note 5).
5. Measure protein concentration of the supernatants (see Note 6) and adjust the protein concentration at 1–5 mg/mL.
6. It is important to add protease inhibitors and prepare the lysates on ice to prevent proteolysis.

3.3. Electrophoresis (SDS-PAGE) (5) (See Note 7)

1. Use an appropriate gel suitable for the detection of your target proteins. The concentration of acrylamide in commonly used gels varies 7–15%. If you need to analyze the whole proteins, try to use gradient gels (e.g., 4–12%).
2. Assemble the electrophoresis apparatus and set the gels.
3. Fill both inner and outer chamber of the electrophoresis assembly with the electrophoresis buffer and rinse each well of the gels by flushing it several times with the electrophoresis buffer.
4. Add SDS sample buffer to your samples and heat-treat your samples at 95°C for 3 min or at 75°C for 10 min to denature proteins.
5. Load the samples and markers to the wells of SDS-PAGE gel. You may want to load prestained molecular marker and western marker to empty wells. Prestained molecular markers electroblotted on the membrane are visible and you will easily trim the blotted area out of the membrane. The western markers are designed to react with the secondary antibody (see Note 8) and you will see the molecular marker on X-ray film or display of the gel image analyzer (vide infra).
6. Run the samples at a constant volt (e.g., 200 V) until bromophenol blue (BPB) usually added to the SDS sample buffer as a loading dye comes to the bottom of the gel.
7. Disassemble the electrophoresis apparatus and take out the gels.

3.4. Western Blotting

There are three types in western blotting: wet, semi-dry, and dry types. The dry-type western blotting can be done using a western blotting kit. For example, iBlot is commonly used for dry blotting worldwide. In the iBlot system no water is required because the necessary buffers are provided as packed gel plates. In the semi-dry and wet blotting you need to prepare western transfer buffer. The difference between semi-dry blotting and wet blotting is the amount of buffer to be used. Less amount of buffer and instead more number of absorbent filter papers are required in semi-dry blotting compared to wet blotting so that the membrane will be kept wet until the end of blotting. The following procedure is for wet-type western blotting and the typical wet blotting assembly is shown in Fig. 2.

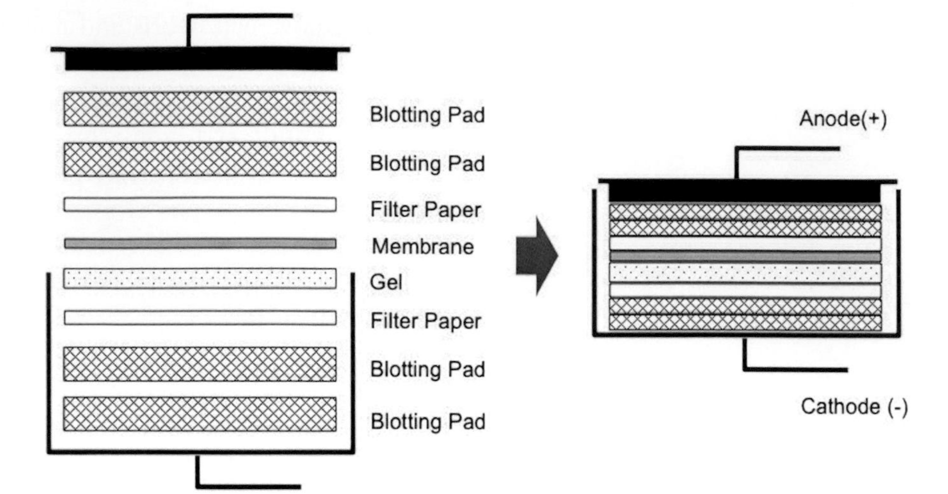

Fig. 2. Assembly of a wet-type western blotting. Proteins in the gel migrate onto the membrane where they are adsorbed and bound. The blotting pads are not used and instead 3–5 sheets of adsorbent paper are used on both sides of the membrane in semi-dry western blotting.

1. Choose either PVDF or nitrocellulose for blotting membrane on which proteins are to be adsorbed (see Note 9).
2. After SDS-PAGE electrophoresis trim the gel to fit the membrane. Handle the gel gently in the SDS running buffer, because the gels are fragile.
3. You need to cut the membrane and adsorbent filter papers (high absorbing capacity filter paper such as Whatman 3MM should be used) to the size of the gel.
4. Because PVDF membrane is water repellent, presoak the membrane shortly in 100% methanol before soaking in the buffer. This procedure is not necessary for nitrocellulose membrane.
5. Soak the membrane, filter papers, and blotting pads (in case of wet blotting) in western transfer buffer.
6. When assembling the blotting layers, avoid bringing in air bubbles in between the membrane and the gel. Keep the membrane and gels wet until the western blot analysis is completed.
7. Do not soak the gels in other buffers such as PBS, because proteins are negatively charged in the gels by dodecyl sulfate anion and it is important to keep the proteins negatively charged to be electroblotted on the membrane.
8. The current and time for western blotting are about 2 mA/ cm^2 gel and 60 min, respectively.
9. After electroblotting mark the membrane with a ball-point pen at one corner for identification of the membrane if you have two or more membranes.

3.5. Blocking

Several agents have been used for blocking the membranes. The blocking process is important to prevent nonspecific binding of antibodies, especially a POD-tagged secondary antibody, to the membrane and to reduce the background signal in the detection of proteins of interest.

1. Rinse the membrane in PBS or Tris–HCl buffer after electroblotting.
2. Add a blocking solution in a shallow container so that the membrane can be soaked. Either protein solutions (5% skim milk or 2–3% BSA) or nonprotein solutions can be used for blocking the membrane (see Note 10). In case that phosphorylated proteins are detected, try to use nonprotein blocking agents because some proteins in the blocking proteins may be phosphorylated.
3. Remove excess buffer on the membrane and place the membrane in the container.
4. Soak the membrane in the blocking solution with continuous gentle shaking or rocking at room temperature for 1 h.

3.6. Reaction with Antibodies

1. Prepare the primary antibody solution.
2. Usually the primary antibody is diluted to ×1/200 to ×1/1,000 with the blocking solution in the container (see Note 11).
3. Transfer the membrane to the primary antibody solution and shake or rock the container continuously at room temperature for 1 h or at 4°C overnight.
4. Wash the membrane three times with the membrane wash solution for 3–5 min.
5. Prepare the POD-tagged secondary antibody solution. Usually the secondary antibody is diluted to ×1/1,500 to ×1/5,000 with the blocking solution (see Note 11).
6. Transfer the membrane to the secondary antibody solution and shake or rock the container at room temperature for 1 h.
7. Wash the membrane four times with PBS-Tween20 for 3–5 min.

3.7. Detection/ Enhancement/ Quantitation

1. Rinse the membrane in PBS or Tris–HCl buffer and immerse the membrane in ECL solution.
2. Place the membrane on a thin flat plate and cover the membrane with transparent wrapping film (see Note 12).
3. Expose the X-ray film to the wrapped membrane in a X-ray film cassette. Handle X-ray films in a dark room, because they are sensitive to the light (see Note 13).
4. If a gel image analyzer (chemiluminescence detector) is available in the laboratory, place the wrapped membrane inside of

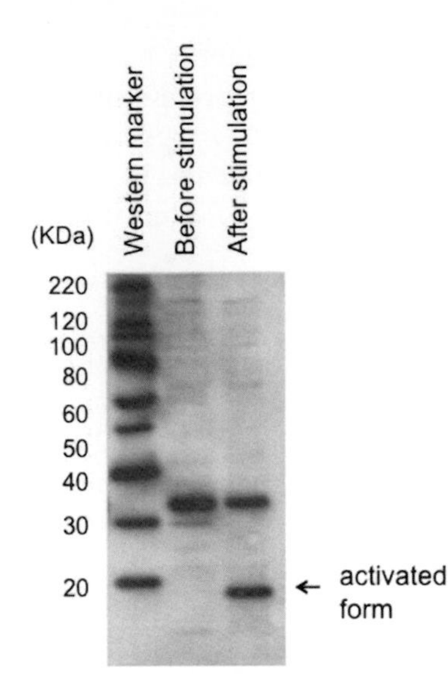

Fig. 3. A typical western blotting analysis. The cell lysate was obtained from control and MG132-treated macrophages in the presence of protease inhibitors. The proteins were resolved by SDS-PAGE (4–12%) and electroblotted onto a nitrocellulose membrane. The membrane was blocked with skim milk (BlockAce®). The membrane was probed with anti-Caspase 3 (×1/750) and then POD-tagged anti-mouse IgG (×1/1,500). The membrane was soaked in ECL solution and then wrapped. After exposure to the wrapped membrane in a film cassette, an X-ray film (Hyperfilm®, GE Healthcare) was developed in a dark room.

the instrument and take a picture using a CCD camera (Figs. 3 and 4).

5. It is also possible to visualize the proteins of interest using an appropriate substrate for peroxidase such as DAB instead of ECL. In this case the detection sensitivity can be enhanced by using an ABC (avidin–biotin complex) kit.
6. Evaluate the band using a densitometry software, if you need to do quantitative analysis.

3.8. Reprobing the Membrane

After completion of the first western blotting analysis, the membrane can be used for the detection of other proteins by stripping the antibodies from the membrane and reprobing the membrane using another set of antibodies. The reprobing of the membrane is necessary to normalize the target immunoprecipitated protein by IgG (antibody used in the immunoprecipitation). The reprobing is also necessary for the detection of a phosphorylated protein to normalize it by the total protein (non-phosphorylated and phosphorylated).

1. Remove the antibodies on the membrane by incubating it in the stripping buffer at 56°C for 15 min.
2. Wash the membrane three times with the membrane wash buffer for 2–3 min at room temperature.

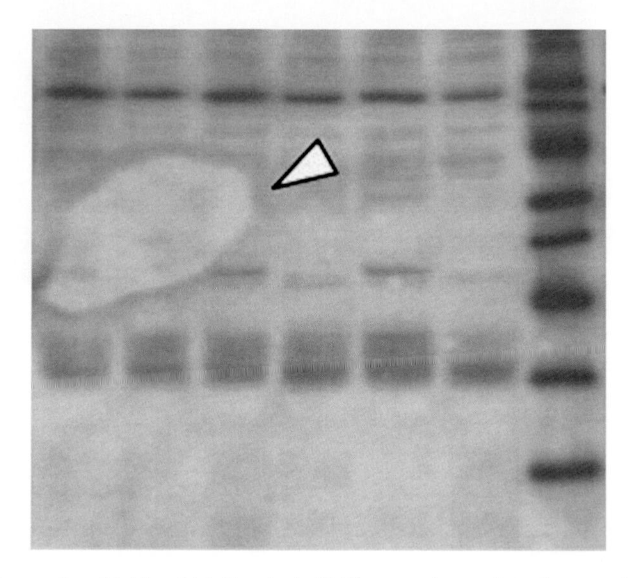

Fig. 4. The western blot in which the electroblotting was incomplete. An *arrowhead* indicates "non-blotted" area. The incomplete blotting was brought about probably because the membrane was not layered well on the gel and air bubbles were trapped between the membrane and the gel during the electroblotting.

3. The removal of antibodies can be confirmed by exposing an X-ray film to the washed membrane using ECL substrate. If no signal is detected, proceed to step 4. Otherwise repeat steps 1 and 2 again.
4. Block the membrane and follow the procedure in Subheading 3.5 using another set of antibodies and then follow the procedure described in Subheadings 3.6 and 3.7.

3.9. Antibody Arrays (Optional)

The antibody array is a reverse way for the detection of proteins of interest. Antibodies blotted in array on membranes are commercially available and you can semiquantitatively measure protein levels of your samples as shown in the following procedure.

1. Block membranes on which "capture antibodies" against a series of target proteins are immobilized.
2. React proteins in your sample with biotinylated detection antibodies.
3. Probe the membrane with the protein-biotinylated antibody complexes.
4. The target protein are finally detected by streptavidin-POD and a POD substrate like a sandwitch ELISA system.
5. Tens of proteins in the sample can be semiquantitatively measured in one experiment using the antibody array (Fig. 5).

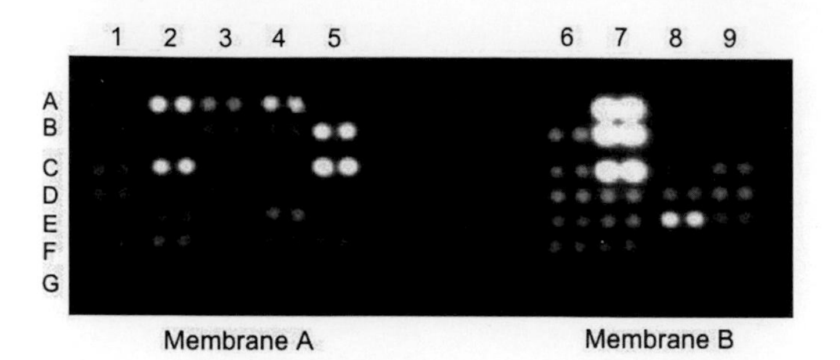

Fig. 5. Detection of phosphorylated proteins in BEAS-2B cells using an antibody array (human phospho-kinase array kit, R&D) (6). The antibodies are spotted on the membrane in duplicate. The phosphorylation levels of kinases were visualized by ECL and Lumino Imaging Analyzer (FAS-1100, TOYOBO).

4. Notes

1. Nonidet P-40 and Tween 20 are sticky liquids. Avoid using a micropipette to take an exact amount of those liquid detergents.
2. Some commercially available precast gels are stable for up to 1 year at 4°C.
3. Add SDS after filling the water to avoid bubble and foam generation.
4. DLS is used for particle size distribution as well as zeta-potential measurement. Both particle size distribution and zeta-potential are considered to be important determinants for toxicity of nanomaterials.
5. As most nanomaterials are insoluble particulate materials possessing a large specific surface area, some proteins are adsorbed on the surface of nanomaterials (7, 8). Boil the nanomaterial-containing pellets with SDS sample buffer after removal of the supernatant to extract the remaining proteins, when you need to investigate the adsorbed proteins.
6. BCA (bicinchoninic acid) assay method and Bradford method are commonly used for measurement of protein concentration. When detergents are present in the sample, avoid Bradford method.
7. Western blot analyses are also applicable for 2D gels. Once you complete the second dimension slab gel electrophoresis, the following western blotting process for the 2D gel is the same as that for the common SDS-PAGE.
8. The western makers are designed to react with secondary antibody of most animal species (human, mouse, rabbit, goat, rat,

chicken, etc.). However, the reactivity is different among the species.

9. The PVDF membrane can hold a larger amount of proteins than the nitrocellulose membrane. However, the background signal can be reduced when the nitrocellulose membrane is used. It is recommended to use nitrocellulose membranes first because reducing the background is more difficult than getting strong signals.
10. Tween 20 is usually added to nonprotein blocking solutions to the final concentration of 0.05%. However, Tween 20-containing nonprotein blocking solution is also commercially available (e.g., T20, Thermo Sci.).
11. Pre-made diluents for the primary and secondary antibodies are commercially available. The diluents are maximized to get clear signals with low background (e.g., Can Get Signal®, TOYOBO Co.). Omit the secondary antibody process if the primary antibody is already tagged with POD.
12. Saran wrap® easily wrinkles, which deteriorates the quality of picture. A little harder transparent plastic sheet is preferable.
13. X-ray films are less sensitive to red light. A pencil-type red light torch (GE Healthcare) may be useful to set an X-ray film in the dark room. When a tracker tape (GE Healthcare) is used, you can record "memo" directly on the X-ray film.

References

1. Fenoglio I, Croce A, Di Renzo F, Tiozzo R, Fubini B (2000) Pure-silica zeolites (Porosils) as model solids for the evaluation of the physicochemical features determining silica toxicity to macrophages. Chem Res Toxicol 13: 489–500
2. Oberdorster G, Oberdorster E, Oberdorster J (2005) Nanotoxicology: an emerging discipline evolving from studies of ultrafine particles. Environ Health Perspect 113:823–839
3. Tabet L, Bussy C, Amara N, Setyan A, Grodet A, Rossi MJ, Pairon JC, Boczkowski J, Lanone S (2009) Adverse effects of industrial multiwalled carbon nanotubes on human pulmonary cells. J Toxicol Environ Health A 72:60–73
4. Wick P, Manser P, Limbach LK, Dettlaff-Weglikowska U, Krumeich F, Roth S, Stark WJ, Bruinink A (2007) The degree and kind of agglomeration affect carbon nanotube cytotoxicity. Toxicol Lett 168:121–131
5. O'Farrell PH (1975) High resolution two-dimensional electrophoresis of proteins. J Biol Chem 250:4007–4021
6. Hirano S, Fujitani Y, Furuyama A, Kanno S (2010) Uptake and cytotoxic effects of multi-walled carbon nanotubes in human bronchial epithelial cells. Toxicol Appl Pharmacol 249:8–15
7. Hirano S, Kanno S, Furuyama A (2008) Multi-walled carbon nanotubes injure the plasma membrane of macrophages. Toxicol Appl Pharmacol 232:244–251
8. Jiang X, Weise S, Hafner M, Rocker C, Zhang F, Parak WJ, Nienhaus GU (2010) Quantitative analysis of the protein corona on FePt nanoparticles formed by transferrin binding. J R Soc Interf 7:S5–S13

Chapter 7

Application of Reverse Transcription-PCR and Real-Time PCR in Nanotoxicity Research

Yiqun Mo, Rong Wan, and Qunwei Zhang

Abstract

Reverse transcription-polymerase chain reaction (RT-PCR) is a relatively simple and inexpensive technique to determine the expression level of target genes and is widely used in biomedical science research including nanotoxicology studies for semiquantitative analysis. Real-time PCR allows for the detection of PCR amplification in the exponential growth phase of the reaction and is much more quantitative than traditional RT-PCR. Although a number of kits and reagents for RT-PCR and real-time PCR are commercially available, the basic principles are the same. Here, we describe the procedures for total RNA isolation by using TRI Reagent, for reverse transcription (RT) by M-MLV reverse transcriptase, and for PCR by GoTaq® DNA Polymerase. And real-time PCR will be performed on an iQ5 multicolor real-time PCR detection system by using iQ™ SYBR Green Supermix.

Key words: RNA isolation, Reverse transcription (RT), Polymerase chain reaction (PCR), Agarose gel electrophoresis, Real-time fluorescent quantitative PCR

1. Introduction

The study of gene expression in a cell or tissue at a particular moment gives an insight into the capacity of the cell for protein synthesis. Gene expression assays, for example, gene profiling, are an important tool and are widely used in nanotoxicity studies. There are several methods available to determine gene expression, such as northern blot analysis, ribonuclease protection assay (RPA), serial analysis of gene expression (SAGE), reverse transcription-polymerase chain reaction (RT-PCR), quantitative real-time polymerase chain reaction (qRT-PCR), PCR arrays, and microarrays. Among these techniques, Northern blot analysis remains a standard method for detection and quantitation of mRNA levels despite the advent of more robust techniques. Northern blotting involves the use of electrophoresis to separate RNA samples by size, then detect the

Joshua Reineke (ed.), *Nanotoxicity: Methods and Protocols*, Methods in Molecular Biology, vol. 926,
DOI 10.1007/978-1-62703-002-1_7, © Springer Science+Business Media, LLC 2012

mRNA with a hybridization probe complementary to part of the target sequence. RPA is an extremely sensitive technique for the quantitation of specific RNAs in solution. It can be performed on total cellular RNA or poly(A)-selected mRNA as a target. SAGE method, as well as PCR arrays and microarrays, is used to study partial or global gene expression in cells or tissues in various experimental conditions. In this chapter, we will describe the methods to determine gene expression by using RT-PCR and real-time PCR. RT-PCR as a relatively simple, inexpensive, extremely sensitive and specific tool to determine the expression level of target genes. Real-time PCR is a quantitative method for determining copy number of PCR templates, such as DNA or cDNA, and consists of two types: probe-based and intercalator-based. Probe-based real-time PCR, also known as TaqMan PCR, requires a pair of PCR primers and an additional fluorogenic oligonucleotide probe with both a reporter fluorescent dye and a quencher dye attached. The intercalator-based (SYBR Green) method requires a double-stranded DNA dye in the PCR which binds to newly synthesized double-stranded DNA and generates fluorescence. Both methods require a special thermocycler equipped with a sensitive camera that monitors the fluorescence in each sample at frequent intervals during the PCR. The principle techniques underlying both RT-PCR and real-time PCR are total RNA isolation, reverse transcription (RT), and PCR. Reverse transcription involves the synthesis of DNA from RNA by using an RNA-dependent DNA polymerase. PCR can amplify a single or a few copies of a piece of DNA across several orders of magnitude, generating thousands to millions of copies of a particular DNA sequence. Here we will introduce detailed procedures for RT-PCR and real-time PCR.

2. Materials

2.1. Total RNA Isolation

1. TRI Reagent. Store at 2–8°C.
2. Chloroform. Store at room temperature.
3. Isopropanol. Store at room temperature.
4. 75% Ethanol (40 ml): in a sterile and RNase-free 50 ml tube, add 10 ml of molecular grade and nuclease-free water, and 30 ml of 100% ethanol to make 40 ml of 75% ethanol. Store at −20°C.
5. Molecular grade and nuclease-free water. Store at 2–8°C.

2.2. Reverse Transcription

1. M-MLV reverse transcriptase with 5× M-MLV reaction buffer. Store at −20°C.
2. Recombinant RNasin® Ribonuclease Inhibitor. Store at −20°C.

3. dNTP mixture, 10 mM. Store at −20°C.
4. Oligo $(dT)_{18}$ primer: synthesized by SIGMA (The Woodlands, TX) and prepared at a concentration of 0.5 μg/μl with molecular grade and nuclease-free water. Aliquot and store at −20°C.
5. Molecular grade and nuclease-free water. Store at 2–8°C.
6. Thermal cycler (Mastercycler, Eppendorf, Westbury, NY).

2.3. Polymerase Chain Reaction

A basic PCR setup requires several components and reagents include the DNA template, two primers, Taq polymerase or another DNA polymerase, buffer solution, dNTPs, and divalent cations such as Mg^{2+}. The following reagents are used routinely for RT-PCR in our laboratory.

1. Template cDNA from the above RT. Store at −20°C.
2. Upstream and downstream primers: synthesized by SIGMA (The Woodlands, TX). Diluted to 5 μM with molecular grade and nuclease-free water. Store at −20°C.
3. GoTaq® DNA Polymerase with 5× reaction buffer. Store at −20°C.
4. dNTP mixture, 10 mM. Store at −20°C.
5. Molecular grade and nuclease-free water. Store at 2–8°C.
6. Agarose. Store at room temperature.
7. Ethidium bromide solution (10 mg/ml): dilute to 0.5 mg/ml with molecular grade and nuclease-free water. Store at room temperature.
8. 50× TAE: to make 1 l of 50× TAE buffer, need 242 g Tris base, 57.1 ml glacial acetic acid, and 100 ml of 0.5 M EDTA (pH 8.0). Add enough molecular grade H_2O to dissolve solids to a final volume of 1,000 ml. Store at room temperature.

2.4. Real-Time PCR

1. iQ™ SYBR Green Supermix. Store at −20°C.
2. 96-Well PCR plate. Store at room temperature.
3. Microseal® "B" Film. Store at room temperature.
4. Molecular grade and nuclease-free water. Store at 2–8°C.

3. Methods

3.1. Isolation of Total RNA

Obtaining high quality and intact RNA is the first and often the most critical step in performing RT-PCR and real-time PCR. RNA is easily degraded since RNase is very hard to inactivate. Several precautions need to be taken to prevent RNA from degradation. People should always wear a clean lab coat, disposable gloves, and change gloves frequently. The bench should be clean. Any aqueous solutions, tubes, and pipettes used for the procedure should be sterile and RNAse-free. To avoid contaminating your sample with RNases, do not talk while processing RNA extraction.

Currently, there are a number of RNA isolation kits commercially available. Although it is convenient, time-saving, and avoids contact with phenol/chloroform using commercially available kits, those kits using silica-membrane spin columns may not be ideal for studies of insoluble nanoparticle since the nanoparticles may clog the membrane pore of the spin column. Therefore, it is important to choose the right reagents or kits for total RNA isolation according to different experiments and specific characteristics of different nanoparticles. In our laboratory, we use TRI Reagent to isolate total RNA for nanoparticle studies. TRI Reagent is a mixture of guanidine thiocyanate and phenol in a monophase solution, which can effectively dissolve DNA, RNA, and protein after homogenization or lysis of tissue samples. It performs well with large or small amounts of tissue or cells. Here, we describe the procedures for isolating total RNA using TRI Reagent according to the manufacturer's instructions (1).

3.1.1. Sample Preparation

1. Lyse or homogenize the sample (see Notes 1–3).
   (a) Tissue (see Note 4): homogenize tissue samples in TRI Reagent (1 ml per 50–100 mg of tissue) in an appropriate homogenizer (see Notes 5 and 6). The volume of the tissue should not exceed 10% of the volume of the TRI Reagent.
   (b) Monolayer cells: lyse cells directly on the culture dish or plate (see Notes 7 and 8). Use 1 ml of the TRI Reagent per 10 cm^2 of glass culture plate surface area. After addition of the reagent, the cell lysate should be passed several times through a pipette to form a homogenous lysate (see Note 9).
   (c) Suspension cells: isolate cells by centrifugation at 1,000 rpm for 5 min and then lyse in TRI Reagent by repeated pipetting. One milliliter of the reagent is sufficient to lyse $5–10 \times 10^6$ animal, plant, or yeast cells, or 10^7 bacterial cells (see Notes 10 and 11).
2. In order to minimize the possibility of DNA contamination in the RNA extracted by TRI Reagent, after homogenization, centrifuge the homogenate at $12,000 \times g$ for 10 min at 2–8°C to remove the insoluble material (extracellular membranes, polysaccharides, and high molecular mass DNA). The supernatant contains RNA and protein. If the sample had a high fat content, there will be a layer of fatty material on the surface of the aqueous phase that should be removed. Transfer the clear supernatant to a fresh tube.

3. To ensure complete dissociation of nucleoprotein complexes, allow samples to stand for 5 min at room temperature.
4. Add 0.2 ml of chloroform (see Note 12) per ml of TRI Reagent used. Cover the sample tightly, shake vigorously for 15 s, and allow to stand for 2–15 min at room temperature.
5. Centrifuge the resulting mixture at 12,000 × *g* for 15 min at 2–8°C.

3.1.2. RNA Isolation

1. Transfer the colorless upper aqueous phase to a fresh tube and add 0.5 ml of isopropanol per ml of TRI Reagent used in Subheading 3.1.1, step 1 and mix. Allow the sample to stand for 5–10 min at room temperature (see Note 13).
2. Centrifuge at 12,000 × *g* for 10 min at 2–8°C. The RNA precipitate will form a pellet on the side and bottom of the tube.
3. Remove the supernatant and wash the RNA pellet by adding a minimum of 1 ml of 75% ethanol per 1 ml of TRI Reagent used in sample preparation (see step 1 of Subheading 3.1.1). Vortex the sample and then centrifuge at 7,500 × *g* for 5 min at 2–8°C (see Notes 14 and 15).
4. Briefly dry the RNA pellet for 5–10 min by air-drying or under a vacuum (see Note 16). Add an appropriate volume of molecular grade water to the RNA pellet. To facilitate dissolution, mix by repeated pipetting with a micropipette at 55–60°C for 10–15 min (see Note 17).
5. Measure the concentration of total RNA: in a sterile and RNase-free tube, add 48 μl of molecular grade and nuclease-free water and 2 μl of total RNA from step 3 to make a total volume of 50 μl. Mix well.
6. Measure the absorbance at 260 and 280 nm with a Spectrophotometer (DU 730 Spectrophotometer, Beckman Coulter, Fullerton, CA) (see Notes 18–21). 1 A_{260} unit/ml = 40 μg/ml.

3.2. Reverse Transcription

Reverse transcription involves the synthesis of DNA from RNA by using an RNA-dependent DNA polymerase, the *reverse* of normal transcription, which is from RNA to DNA. Although there are many kits commercially available for RT, the reverse transcriptase used in those kits usually is M-MLV reverse transcriptase from the Moloney murine leukemia virus or AMV reverse transcriptase from the avian myeloblastosis virus. M-MLV reverse transcriptase is the preferred reverse transcriptase in cDNA synthesis for long messenger RNA (mRNA) templates (>5 kb) because the RNase H activity of M-MLV reverse transcriptase is weaker than the commonly used AMV reverse transcriptase (2). Since M-MLV reverse transcriptase is less processive than AMV reverse transcriptase, therefore, more units of the M-MLV enzyme are required to generate the same amount of cDNA as in the AMV reaction (2). The following are the basic procedures for RT using M-MLV reverse transcriptase according to the manufacturer's instruction (2).

1. Preheat a water bath to 70°C.
2. In a sterile RNase-free 0.2 ml PCR tube, add 2 μl of Oligo $(dT)_{18}$ primer (0.5 μg/μl) and 2 μg of total RNA in a total volume of 15 μl with molecular grade and nuclease-free water (total volume is 17 μl).
3. Put the PCR tubes which contain the primer and total RNA in a plastic PCR rack (see Note 22), then put the rack in the 70°C water bath for 5 min to melt secondary structures within the template.
4. Cool the samples immediately by putting the tubes on ice to prevent secondary structures from reforming.
5. In a new sterile RNase-free 0.5 ml tube, mix the following reagents according to the number of samples. Each reaction should contain: 1 μl M-MLV reverse transcriptase, 1.25 μl of 10 mM dNTP, 0.75 μl Recombinant RNasin Ribonuclease Inhibitor, and 5 μl of 5× M-MLV reaction buffer (see Notes 23 and 24). Mix gently by flicking the tube, then spin briefly.
6. Spin the PCR tubes from step 4 briefly to collect the solution at the bottom of the tube and put the tubes back onto the PCR rack.
7. Add 8 μl of the above mixture (step 5) into each PCR tube, mix gently by flicking the tube, then spin briefly. The total volume in the PCR tube should now be 25 μl.
8. Put the PCR tubes on to a thermal cycler and incubate at 42°C for 60 min (see Note 25), at 94°C for 5 min, then keep at 4°C (see Note 26).
9. When finished, the samples can be stored at –20°C for later PCR experiments.

3.3. Polymerase Chain Reaction

Almost all PCR applications employ a heat-stable DNA polymerase, such as Taq polymerase. There are many DNA polymerases commercially available. Although their efficiency may be different, they are suitable for regular RT-PCR to determine the expression level of mRNA. Some experiments which will use PCR products for cloning purposes, especially those for cloning of promoter region with high G-C content, need to use high fidelity DNA polymerase. The PCR is commonly carried out in a reaction volume of 10–200 μl in small reaction tubes (0.2–0.5 ml volumes) in a thermal cycler. The following is an example of a PCR performed in our laboratory.

1. In a sterile nuclease-free microcentrifuge tube, mix the following reagents on ice. Each reaction contains: 1 μl of 5 μM each primer, 0.5 μl of 10 mM dNTP, 5 μl of 5× Green GoTaq Buffer, 0.25 μl GoTaq DNA polymerase (5 U/μl), and 16.25 μl

of nuclease-free water (total 24 μl) (see Notes 27–29). Mix gently by flicking the tube, then spin briefly.

2. In each PCR tube, add 24 μl of the above mixture to the bottom of the tube.
3. Add 1 μl of cDNA sample in each PCR tube. Mix gently by flicking the tube, then spin briefly.
4. Put the PCR tube onto a thermal cycler.
5. According to the primers and the length of PCR product, set up parameters for PCR (see Notes 30–34), then run.
6. Separate the PCR products by agarose gel electrophoresis and visualize with ethidium bromide (see Notes 35–37).
7. After separation on an agarose gel, PCR products are visualized by Gel Doc XR (Fig. 1) and analyzed by Quantity One.
8. To semi-quantify the expression level of mRNA, intensities of target gene products are normalized to that of housekeeping gene to obtain the relative densities.

3.4. Real-Time PCR

Traditional PCR uses agarose gel for detection of PCR amplification at the final phase or endpoint of the PCR (plateau). However, real-time PCR allows for the detection of PCR amplification in the exponential growth phase of the reaction. Theoretically, there is a quantitative relationship between the amount of starting target sample and the amount of PCR product at any given cycle number. Real-time PCR detects the accumulation of amplicon during the reaction. The data is then measured at the exponential phase of the

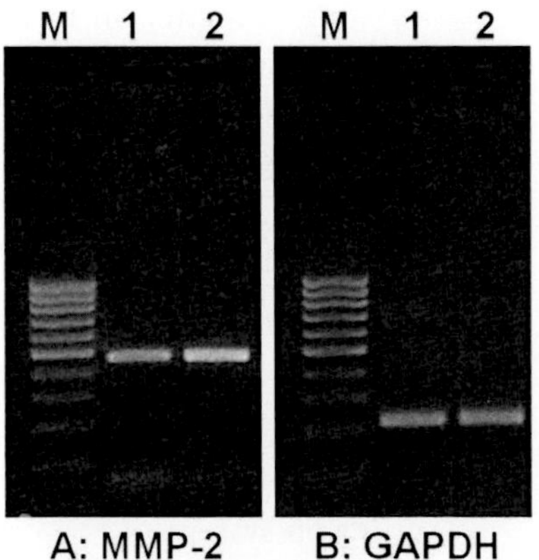

Fig. 1. mRNA expression level of matrix metalloproteinase-2 (MMP-2) by RT-PCR. Total RNA was extracted from human monocytes U937 by using TRI Reagent and reverse transcripted to cDNA by using M-MLV reverse transcriptase (Promega). PCR was performed on a thermal cycler (Mastercycler, Eppendorf) by using GoTaq DNA polymerase (Promega). The PCR products were separated on 1% agarose gel. Housekeeping gene GAPDH was used as internal control. M, 100 bp DNA Ladder (Fisher); (*1*) cells without any treatment were used as control; (*2*) cells were treated with 5.0 μg/ml of cobalt nanoparticles for 24 h.

PCR, which makes quantitation of DNA and RNA easier and more precise than traditional methods. There are two methods, which are often used in the laboratory. One is the 5′ nuclease assay in which an oligonucleotide called a TaqMan® Probe is added to the PCR reagent master mix. This probe is designed to anneal to a specific sequence of template between the forward and reverse primers and is also designed with a high-energy dye termed a Reporter at the 5′ end, and a low-energy molecule termed a Quencher at the 3′ end. When this probe is intact and excited by a light source, the Reporter dye's emission is suppressed by the Quencher dye as a result of the close proximity of the dyes. When the probe is cleaved by the 5′ nuclease activity of the enzyme, the distance between the Reporter and the Quencher increases causing the transfer of energy to stop. The fluorescent emissions of the reporter increase and the quencher decrease. An increase in Reporter fluorescent signal is directly proportional to the number of amplicons generated. Another real-time PCR method is by using SYBR Green Dye, which can bind the minor groove of any double-stranded DNA molecule. When SYBR Green dye binds to double-stranded DNA, the intensity of the fluorescent emissions increases. As more double-stranded amplicons are produced, SYBR Green dye signal will increase. The following is an example of real-time PCR by using iQ™ SYBR Green Supermix and performed on an iQ5 multicolor real-time PCR detection system.

1. Thaw all components used in step 2 at room temperature. Mix vigorously, and centrifuge to collect contents to the bottom of the tubes, then put the tubes on ice.
2. In a sterile nuclease-free microcentrifuge tube, mix the following reagents on ice. Each reaction contains: 1 μl of 5 μM of each primer, 10 μl of 2× iQ™ SYBR Green Supermix, and 7 μl of nuclease-free water (total 19 μl) (see Notes 38 and 39). Mix gently by flicking the tube, then spin briefly.
3. In each well of 96-well PCR plate, add 19 μl of the above mixture to the bottom of the tube.
4. Add 1 μl of cDNA sample in each well (see Note 40). Cover the plate with Microseal® "B" Film (see Note 41).
5. Mix gently by flicking the tube, then spin the plate briefly.
6. Put the PCR plate in an iQ5 multicolor real-time PCR detection system.
7. According to the primers and the length of PCR product, set up parameters for real-time PCR, then run (see Notes 42–44).
8. Analyze the data. If using SYBR Green, there should be only one peak in the melting curve (Fig. 2) (see Note 45).
9. In our laboratory, the relative expression level of each gene is calculated as fold dilution (see Note 46) by using a standard

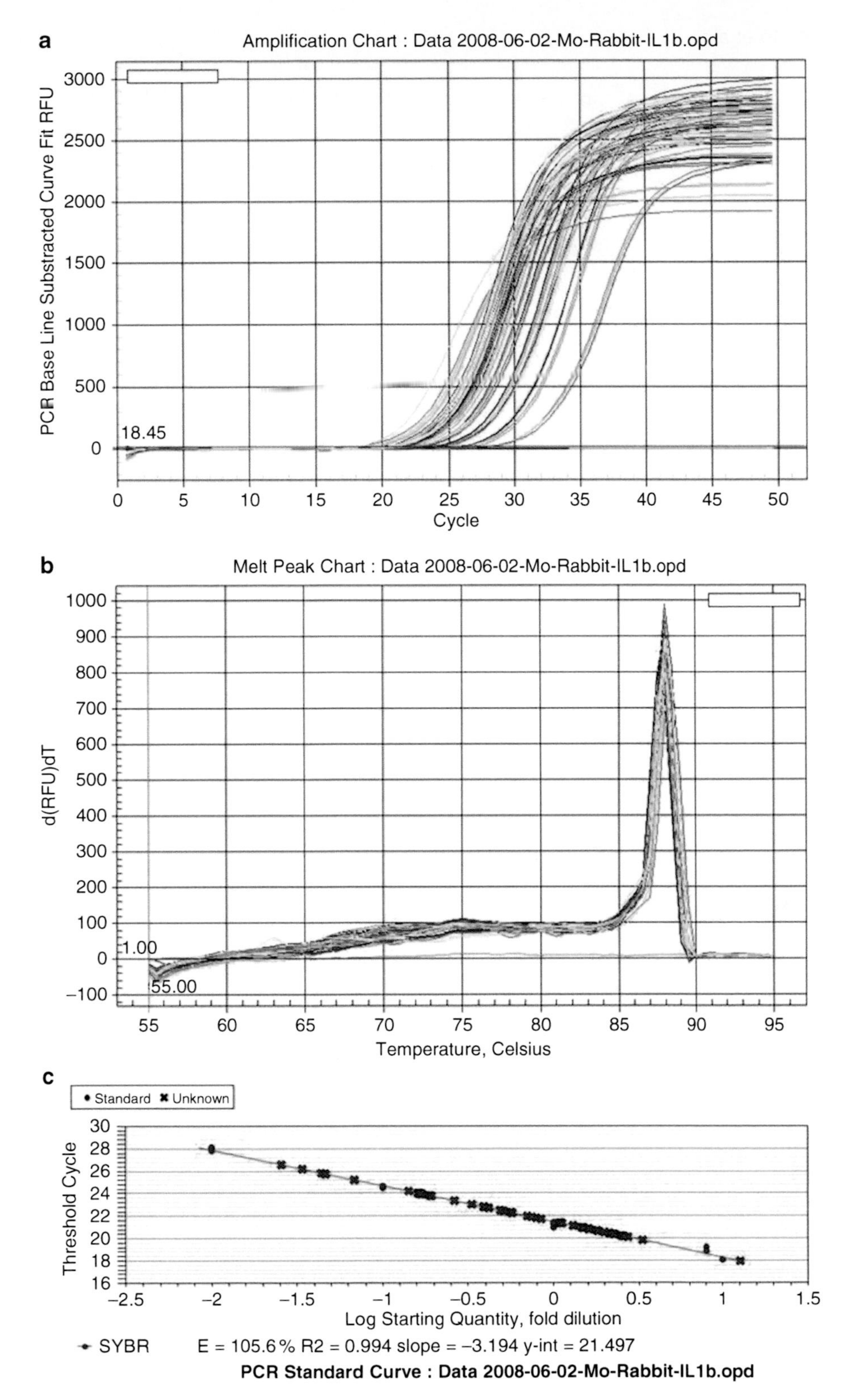

Fig. 2. Real-time PCR for rabbit IL-1β. Rabbit ear skin wound tissues were used for total RNA isolation by TRI Reagent (Sigma) and reverse transcripted to cDNA by using M-MLV reverse transcriptase (Promega). Real-time PCR was performed on an iQ5 multicolor real-time PCR detection system (Bio-Rad) by using 2× iQ™ SYBR Green Supermix (Bio-Rad). (**a**) The amplification curves; (**b**) the melting curves. Single peak in the melting curve represents that only the real target gene is amplified; (**c**) the PCR standard curve (Color figure online).

curve for each gene. Standard curves are obtained by real-time PCR using 3, 1, and 1 μl of 10-, 100-, and 1,000-fold dilution, respectively, of cDNA obtained from sample without any treatments (see Note 47). The expression level of each gene is then normalized to the relative expression level of housekeeping gene in the same sample.

4. Notes

1. Use enough TRI Reagent for the sample homogenization. Too small volume of TRI Reagent may result in DNA contamination.
2. If samples used for the isolation contain organic solvents (ethanol, DMSO), strong buffers, or alkaline solution, DNA contamination may occur.
3. Incomplete homogenization or lysis of samples may result in low yield of RNA.
4. The tissues need to be processed or frozen in liquid N_2 immediately after removing from the animal to prevent RNA from degradation. If not used immediately, the samples need to be stored at −70°C.
5. We usually put the tube which contains tissues and TRI Reagent on ice for 2–5 min after homogenizing for 10 s each time. Repeat homogenization for several times or until no tissues are visible.
6. The homogenizer we used is from Glas-Col, Terre Haute, IN.
7. TRI Reagent is *not* compatible with plastic culture plates.
8. Trypsin digestion of cells may result in RNA degradation.
9. After addition of the TRI Reagent, we usually put the dish or plate on a shaker for 5–10 min to let cells lyse totally.
10. Some yeast and bacterial cells may require a homogenizer.
11. After the cells have been homogenized or lysed in TRI Reagent, samples can be stored at −70°C for up to 1 month.
12. The chloroform used for phase separation should not contain isoamyl alcohol or other additives.
13. The mixture also can be put in −4 or −20°C for 1 h.
14. If the RNA pellets float, after vortexing, centrifuge at 12,000 × *g* for 5 min at 2–8°C.
15. Samples can be stored in ethanol at 2–8°C for at least 1 week and up to 1 year at −20°C.

16. A more complete evaporation of ethanol is required when RNA samples are to be used in RT-PCR. This is especially critical for small volume samples, which may contain a relatively high level of ethanol if not adequately dried. However, do not let the RNA pellet dry completely, as this will greatly decrease its solubility. Do not dry the RNA pellet by centrifugation under vacuum (Speed-Vac).
17. Incompletely dissolving the final RNA pellet may result in low RNA yield.
18. Final preparation of RNA is free of DNA and proteins. It should have a A260/A280 ratio of ≥1.7.
19. If the A260/A280 ratio is <1.65, it may be because: (a) the amount of sample used for homogenization may have been too small; (b) samples may not have been allowed to stand at room temperature for 5 min after homogenization; (c) there may have been contamination of the aqueous phase with the phenol phase; (d) the final RNA pellet may not have been completely dissolved; or (e) the water used for dilution of RNA may be acidic. Acidic pH can affect the A280 reading and lowers absorbance ratios. Try to dilute the RNA samples in TE buffer and measure again.
20. Typical yields from tissues (μg RNA/mg tissue): liver and spleen, 6–10 μg; kidney, 3–4 μg; skeletal muscle and brain, 1–1.5 μg; placenta, 1–4 μg.
21. Typical yields from cultured cells (μg RNA/10^6 cells): epithelial cells, 8–15 μg; fibroblasts, 5–7 μg.
22. The rack should be floated on the water surface. If a plastic PCR rack cannot be found, the plastic racks for 200 μl tips or other racks can be used.
23. Completely thaw and thoroughly vortex the buffer prior to use.
24. Spin the tubes containing reagents briefly before opening them.
25. The extension temperature may be optimized between 37 and 42°C.
26. Water bath can also be used in stand of a thermal cycler.
27. It is better to add water first and add DNA polymerase last. Put the DNA polymerase back in the −20°C freezer as soon as possible after using.
28. Completely thaw and thoroughly vortex the buffer prior to use.
29. Spin the tubes containing reagents briefly before opening them.
30. Initial denaturation of longer than 2 min at 95°C is unnecessary and may reduce the yield.
31. Annealing temperature should be optimized for each primer set based on the primer Tm.

32. The extension time should be at least 1 min/kb target length.
33. Housekeeping gene such as β-actin, GAPDH, and so on is also run to demonstrate equal loading.
34. An example profile of PCR parameters is given as follows.
    (a) Initial denaturation: at 95°C for 2 min for 1 cycle
    (b) Amplification (25–35 cycles)
        Denaturation: at 95°C for 0.5–1 min
        Annealing: at 42–65°C for 0.5–1 min
        Extension: at 72°C for 1 min/kb
    (c) Final extension: at 72°C for 5 min for 1 cycle
    (d) Soak at 4°C indefinite for 1 cycle
35. Make 1% (w/v) agarose gel: weigh 1 g of agarose and pour it into a beaker with 100 ml of 1× TAE or 1× TBE. Put the beaker into a microwave oven. Heat 30 s, take it out to mix, then put it back to the microwave oven. Repeat this step several times until all agarose is melted down. Let it stand at room temperature for several minutes to cool it down until your hand can hold the beaker. Then add ethidium bromide (EtBr) into the gel. *EtBr is a potent mutagen that must be handled carefully to avoid skin contact and contamination of the lab.* The amount of EtBr to add is as follows: of a 0.5 mg/ml stock solution, add 1/1,000 to your gel. For example, if we go back to our 100 ml gel, then you would add 100 μl of EtBr. Mix gently, then pour the gel onto the casting trays. Let the gel harden at room temperature before using.
36. The percentage of gel you run mainly depends on the size fragment of PCR product. See Table 1 for reference (3).
37. For reactions containing the 5X Green GoTaq® Reaction Buffer, load amplification reaction onto the gel directly after

Table 1
The optimal concentrations of agarose gel for dsDNA separation

% Agarose (w/v)	Range of resolution Linear dsDNA (kbp)
0.7	0.8–10
0.9	0.5–7
1.2	0.4–6
1.5	0.2–3
2.0	0.1–2

amplification. Do not need to add any more DNA loading buffer.

38. 2× iQ™ SYBR Green Supermix contains 2× reaction buffer with dNTPs, 50 U/ml iTaq DNA polymerase, 6 mM $MgCl_2$, SYBR Green I, 20 nM fluorescein, and stabilizers.
39. We use 20 μl as a total volume of one reaction to reduce the cost of experiments. It works well.
40. Directly add 1 μl of cDNA into the mixture in the wells and make sure no liquid clings to the tip when taking the tip out of the well.
41. Make sure that the edges of the plate are sealed securely. The film on the four corners is very easily detached.
42. The parameters for traditional PCR can also be used for real-time PCR.
43. For each gene to be determined, a test using a few samples is needed before performing many samples to make sure that the parameters are suitable and only one peak is observed in the melting curve.
44. Usually the experimental protocol consists of four programs:
    (a) Initial denaturation: at 95°C for 3 min for 1 cycle
    (b) Amplification (40 cycles)
        Denaturation: at 95°C for 10 s
        Annealing: at 42–65°C for 0.5–1 min
        Extension: at 72°C for 1 min/kb
    (c) Analysis of the melting curve to confirm the single product amplification during the PCR assay.
    (d) Cooling the rotor and thermal chamber at 25°C.
45. If there is more than one peak in the melting curve, you need to either optimize the PCR parameters or redesign the primers.
46. The expression level of each gene can also be calculated as copy number or others such as nanomoles, nanograms, and so on. However, a standard curve is necessary for either method.
47. It is better to have 4–5 points in a standard curve if possible.

Acknowledgments

This work was partly supported by American Lung Association (RG-872-N), American Heart Association (086576D), KSEF-1686-RED-11, Health Effects Institute (4751-RFA-05-2/06-12), CTSPGP 20018 from University of Louisville, T32-ES011564, and ES01443.

References

1. Product information, Technical Bulletin MB-205. TRI Reagent™ (Sigma, St. Louis, MO), Cat. #T9424
2. Usage information. M-MLV reverse transcriptase (Promega, Madison, WI), Cat. #M1701
3. Sambrook J, Russel DW (2001) Molecular cloning: a laboratory manual, 3rd edn. Cold Spring Harbor Laboratory Press, Cold Spring Harbor

Chapter 8

Deriving TC_{50} Values of Nanoparticles from Electrochemical Monitoring of Lactate Dehydrogenase Activity Indirectly

Fuping Zhang, Na Wang, Fang Chang, and Shuping Bi

Abstract

Nanotoxicity assessment methods for nanoparticles (NPs) such as carbon nanotubes (CNTs), nano-Al_2O_3, and tridecameric aluminum polycation or nanopolynuclear (nano-Al_{13}), particularly lactate dehydrogenase (LDH) assays are reviewed. Our researches on electrochemically monitoring the variations of LDH activity indirectly in the presence of multiwalled carbon nanotubes (MWCNTs), nano-Al_{13}, and nano-Al_2O_3 separately to derive toxic concentrations of NPs altering LDH activity by 50% (TC_{50}) values are discussed. TC_{50} values indicated that the toxicity order was Al (III) > MWCNTs > nano-Al_{13} > nano-Al_2O_3. Zeta potentials (ζ) data of these NPs in the literature proved that the surfaces of these NPs are charged negatively. Negatively charged surfaces might be a main cause in the reduction of LDH activity. Therefore, the classic LDH assays are doubtful to underestimate the nanotoxicities when they are applied to those NPs with negatively charged surfaces. These observations highlight and reconcile some contradictory results at present such as medium-dependent toxicity of NPs among the literature and develop novel analytical methods for evaluation of toxicities of NPs.

Key words: LDH assay, TC_{50}, NPs, MWCNTs, Nano-Al_2O_3, Nano-Al_{13}, Differential pulse voltammetry (DPV)

1. Introduction

Nanomaterials (NMs) and nanotechnologies (NTs) are widely employed in biomedical applications such as artificial organs, tissue materials, pharmacy delivery vehicles, therapeutic and diagnostic applications, optical sensors, chemical sensors, immunosensors, biosensors, imaging agents, catalysts, and water purification agents (1–20). With the developments of NMs and NTs, NMs are dispersed more and more widely in the environment. Consequently, the occupational and public incidental exposures to NMs increased steadily. NPs are capable of entering the human body through epidermical exposure, respiration, drinking polluted water, and eating

Joshua Reineke (ed.), *Nanotoxicity: Methods and Protocols*, Methods in Molecular Biology, vol. 926,
DOI 10.1007/978-1-62703-002-1_8, © Springer Science+Business Media, LLC 2012

contaminated food (21). NPs are observed to pierce into cells, transfer into and accumulate at both tissues and organs such as lung, liver, and kidney, ingress nervous system through the blood stream, disturb the immune system, and bind various biological components (18, 22–31). However, the knowledge about possible toxic effects of NPs on these systems and organs as well as their related components is still limited. Even more, some of the observations contradict each other (25, 32–36). Except for the complexities of the biosystems and the varieties of NPs themselves, these controversial observations are probably owing to the quite different evaluation methods for toxic effects and experiments designed for exploring toxic effects (35–42). Analytical developments and challenges of the nanotoxicological assessments for the characterization of NMs or NPs and their impacts on in vitro and in vivo functions were recently reviewed (23, 25, 30, 43–51). The contradictory data about the toxic effects of NPs highlighted the need for alternative ways to study their uptake by cells and effects on biological substances as well as plants and to develop new methods for full and appropriate evaluations of general and special impacts of NPs. In this chapter, we would like to focus on reviewing the recent studies on the nanotoxicity assessments, especially lactate dehydrogenase (LDH) assays (see Note 1), and our researches on electrochemically monitoring the variations of LDH activity indirectly in the presence of various NPs such as MWCNTs, nano-Al_{13}, and nano-Al_2O_3 to derive the important TC_{50} values (see Note 2).

2. Materials

All chemicals are of analytical reagent grade unless stated otherwise. The quartz double-distilled water was used for the preparation of all solutions.

2.1. Chemicals

1. MWCNTs: MWCNTs are ordered from Shenzhen Nanotech Port Co., Ltd, China with purity ≥ 99.5%, amorphous carbon < 3%, catalyst metals ≤ 0.02 wt%, special surface area of 40–300 m^2/g.
2. LDH: bovine heart LDH of concentration 10 mg/mL is obtained from Sigma Co. (St. Louis, MO) and stored in a refrigerator of 4°C (see Note 3).
3. β-NAD^+: β-NAD^+ of above 90% purity is purchased from Shanghai Bio Life Science & Technology Co., Ltd, China (see Note 3).
4. NADH: NADH of above 90% purity is purchased from Shanghai Bio-Life Science & Technology Co., Ltd, China (see Note 3).

5. P^-: P^- of 98.50% purity is purchased from Shanghai Chemical Reagent Factory.
6. Al^{3+}: preparation of Al^{3+} solution was referred to (52).
7. Nano-Al_2O_3: nano-Al_2O_3 with particle sizes of 50 and 1,000 nm is obtained from Sigma Co. (St. Louis, MO).
8. Nano-Al_{13}: preparation of nano-Al_{13} solution was referred to (52).
9. Tris–HCl buffer solution: values of pH were controlled by 0.1 M Tris–HCl buffer solution
10. Supporting electrolyte: 0.15 mol/L KCl was used as a supporting electrolyte.

2.2. Equipment

1. Three-electrode sample cell: a hanging mercury drop working electrode (HMDE, Jingsu Electroanalytical Instrument Factory, China), a platinum foil counter electrode, and a saturated calomel reference electrode.
2. CHI660B electrochemical system: CHI660B electrochemical system was bought from CH Instruments Inc., Shanghai, China (see Note 4).
3. PHS-2F digital ion meter: PHS-2F digital ion meter was bought from Shanghai Precision and Scientific Instrument Company, China. Values of pH are read on this meter.

3. Methods

3.1. LDH Assay

1. 25 mL of supporting electrolyte with an appropriate quantity of MWCNTs (see Note 5) is put into the three-electrode sample cell and sonicated afterward and degassed with nitrogen for 10 min.
2. Then, 30 mL of diluted LDH (see Note 6) is added into the cell along with stirring for $t_c = 5$ min (t_c, the precontact time between LDH and MWCNTs).
3. Subsequently, 100 μL of both 0.05 M NADH and 0.2 M P^- is separately spiked into the solution (see Note 7).
4. After the motionless of the solution (see Note 8), the DPV model (see Note 4) is applied according to a designed t_r (the reaction time of enzymatic reaction) and the signals are recorded.
5. The recording is terminated when the I_{pNAD^+} (cathodic peak current of NAD^+) gets flat or goes down continuously longer than 30 min (see Note 9).

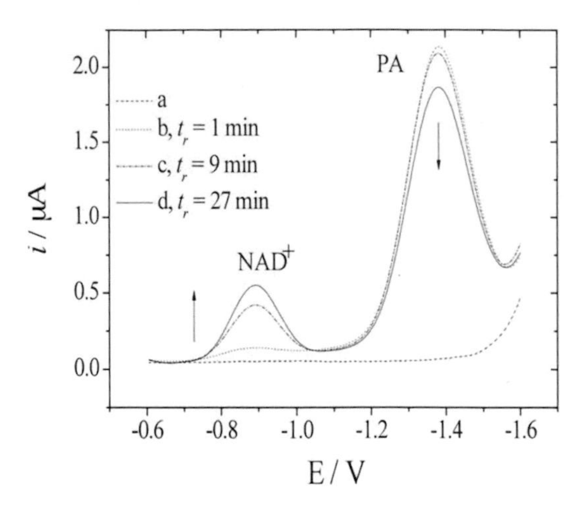

Fig. 1. DPV responses of NAD^+ with t_r in the absence and presence of MWCNTs (main diameters $d = 10–20$ nm, main lengths = 1–2 μm, purity ≥ 99.5%, amorphous carbon < 3%, catalyst metals ≤ 0.02 wt%, special surface area 40–300 m^2/g, Shenzhen Nanotech Port Co., Ltd). (**a**) Supporting electrolyte: 0.1 M pH 7.5 Tris–HCl buffer solution + 0.15 M KCl. (**b**), (**c**), (**d**) are the responses after the additions of 8.0×10^{-4} M PA, 2.0×10^{-4} M NADH, 1.2×10^{-4} mg/mL LDH to (**a**) and in the presence of 20 mg/L MWCNTs for $t_r = 1$, 9 and 27 min, respectively.

3.2. Deriving TC_{50} Values of NPs from Electrochemical Indirect Monitoring LDH Activity

Figure 1 shows that MWCNTs drastically inhibit LDH activity in the enzymatic reaction (1) (see Note 1). In the interest of deriving TC_{50} of NPs, the relative LDH activity of the samples in the absence of NPs is defined as:

$$\text{Relative LDH activity} = I_{\text{pNAD}^+(\text{mp})} / I_{\text{pNAD}^+(\text{ma})} \times 100\%$$

$I_{\text{pNAD}^+(\text{ma})}$ and $I_{\text{pNAD}^+(\text{mp})}$ are the maximal cathodic peak currents of NAD^+ of the samples in the absence and presence of NPs, respectively.

The relative LDH activity of the sample in the absence of NPs is 100%. Figure 2 reveals the variations of relative LDH activity with the quantity of MWCNTs (curve is simulated by Sigmoidal fitting). TC_{50} (toxic concentration of NPs to decrease LDH activity by 50%) of MWCNTs derived from Fig. 2 is about 40 mg/L. By the similar plots of our previous results, TC_{50} values of Al(III), nano-Al_{13}, Al_2O_3 (50 nm), and Al_2O_3 (1,000 nm) are achieved and listed in Table 1. TC_{50} values of nano-Al_2O_3 are in good agreement with those reported by S. Lanone and et al. (33). We conclude that the toxicity order is Al(III) > MWCNTs > nano-Al_{13} > nano-Al_2O_3.

3.3. Influence of MWCNTs Size on LDH activity

Figure 3 shows the variations of relative LDH activity to catalyze reaction (1) (see Note 1) with the sizes of MWCNTs. It indicates that the LDH activity decreases with both the increase of the main diameters and decrease of the main lengths. Namely, the shorter the length and the larger the diameters of MWCNTs are, the stronger the relative LDH activity is inhibited.

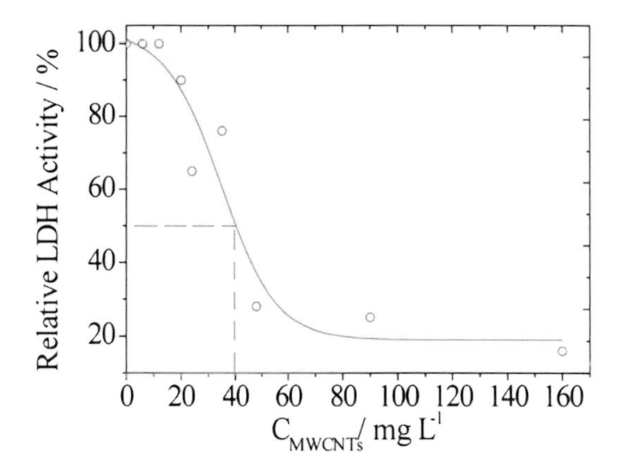

Fig. 2. The variations of relative LDH activity with the quantities of MWCNTs. Other conditions were the same as Fig. 1.

Table 1
TC_{50} values of MWCNTs, Al(III), nano-Al_{13}, and nano-Al_2O_3

Samples	TC_{50} (mg/L) 15°C	25°C	37°C	References
MWCNTs	–	40	–	This work
Al(III)	>5.5	5.5	1.2	(52)
Nano-Al_{13}	>120	50	>120	(52)
50 nm-Al_2O_3	–	>76.5	–	(53)
1000 nm-Al_2O_3	–	>76.5	–	(53)
Al_2O_3		82.2–866[a]		(37)
ZnO		1.7–4.1[a]		
TiO_2		369–845[a]		
CuO		3–31.1[a]		
SnO_2		3.4–174[a]		
Cu		1.7–6.6[a]		
Ag		19.3–1408[a]		

[a]mg/kg body wt

3.4. Influence of pH Values on LDH Activity

Figure 4 shows the variations of relative LDH activity with pH values with and without MWCNTs. Obviously, without MWCNTs, LDH maintains high activity between pH 7.0 and 9.6 and decreases sharply beyond this scope. However, in the presence of MWCNTs, pH range of high LDH activity shifts to lower pH value and LDH activity is inhibited by MWCNTs.

3.5. Discussion About the Profound Significance of Our Observation

The above TC_{50} data displays that Al(III), nano-Al_{13}, nano-Al_2O_3, and MWCNTs evidently reduce LDH activity to catalyze reaction (1) (see Note 1). With regard to diagnose whether a chemical component is a potential toxicant, the classic LDH assessment methods

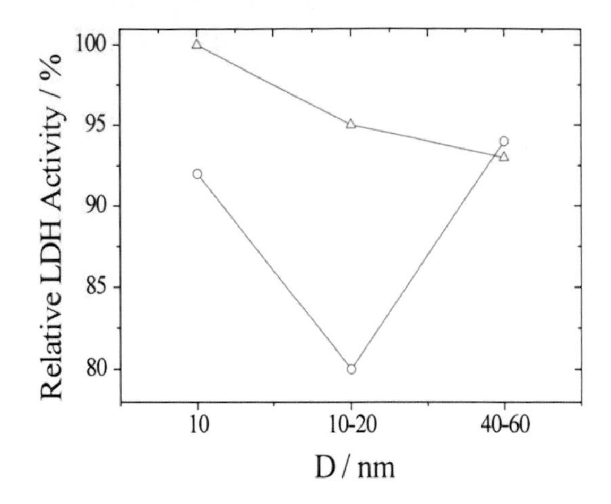

Fig. 3. The variations of relative LDH activity with the sizes of MWCNTs (35 mg/L; *D*, main diameters; *circle*, main lengths = 1–2 μm; *triangle*, main lengths = 5–15 μm). Other conditions were the same as Fig. 1.

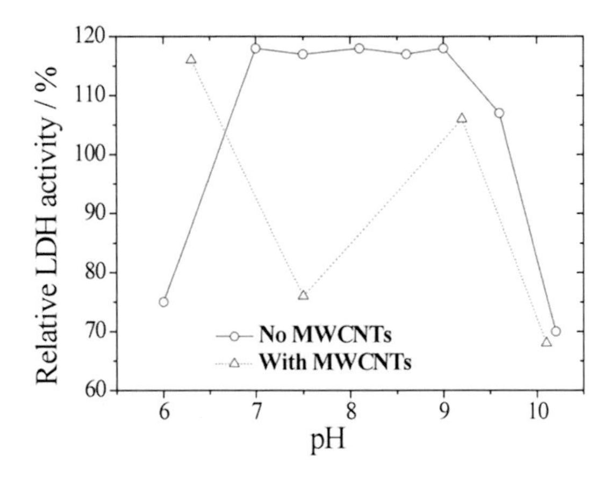

Fig. 4. Variations of relative LDH activity with pH values with (*triangle*) and without (*circle*) 35 mg/L of MWCNTs. Other conditions were the same as Fig. 1.

are applied to monitor the increase of LDH activity induced by the component. This means that the classic LDH assessment methods underestimate the nanotoxicities of these NPs. However, which kinds of NPs whose toxicities would also be underestimated by the classic LDH assays?

Let's check the detailed mechanisms of reaction (1) (see Note 1) firstly. The production of NAD^+ involved an intermediate step that NADH transferred its hydrogen to the negatively charged P^- which bound to the positively charged Arg-171, Arg-109, and His-195 residues of LDH (54, 55). Secondly, it is identified that the surfaces of graphite, C_{60}, SWCNTs, raw MWCNTs, MWCNT-OH, and MWCNT-COOH charged negatively at a wide value of pH, definite ionic strengths and after sonication (56–59). The oxidized and polished areas of MWCNTs also contain negatively charged carboxylates along the sidewalls of the defect sites (60). MWCNTs with negatively charged surface would electrostatically repel the P^- to close to the reactive sites of LDH. Thirdly, the surface of nano-Al_2O_3 at neutral pH is negatively charged too (61). Therefore, the negatively charged surfaces of NPs might be a main cause in the reduction of LDH activity. Consequently, the classic LDH assays underevaluate the nanotoxicities of those NPs with negatively charged surfaces. This observation also highlights some existing contradictory results such as medium-dependent toxicity of MWCNTs among the literatures (62) and develop novel analytical methods. Because ζ data symbolized the surface charge characteristics of NPs, Table 2 gives the summarization for ζ data of some NPs in different media. According to the above deduction, the toxicity of those NPs with negative ζ is expected to be underevaluated.

3.6. Conclusions

Although much progress has been achieved toward the development of classic spectral LDH assessment methods, there are still some intrinsic disadvantages such as insensitivity and nonspecificity when they are applied to evaluate the potential toxicities of NPs. Measurement of LDH activity indirectly by monitoring cathodic electrochemical signal of NAD^+ shows that NPs such as MWCNTs, nano-Al_{13}, and nano-Al_2O_3 decrease LDH activity. From the measured electrochemical signals, it is capable to derive important TC_{50} values of NPs rapidly and inexpensively. The TC_{50} values indicate that the toxicity ordered as: Al(III) > MWCNTs > nano-Al_{13} > nano-Al_2O_3. The negatively charged surfaces of NPs might be a main cause in the reduction of LDH activity, so that the classic LDH assays underestimate the nanotoxicities when they are applied to NPs with negatively charged surfaces.

4. Notes

1. Nanotoxicity is often evaluated by measurements of indexes such as half maximal effective concentration (EC_{50}), half lethal concentration (LC_{50}), TC_{50}, half minimal cell toxicity concentration (CC_{50}), half inhibitory concentration (IC_{50}), and half lethal dose (LD_{50}). Table 3 lists the definitions of these indexes and related assessment methods. It shows that LDH assays are the most important and populous assessment methods for evaluating the nanotoxicity. LDH is distributed widely in mammalian tissues and cytoplasm of all cells, as well as in yeasts and bacteria. The overall LDH enzyme activity is not markedly different between tissues.

   LDH is released from the cytosol upon cell lysis. Therefore, the assessment of the variation of LDH activity was capable to trace the alternations in cell integrities such as impermeability

Table 2
ζ data of some NPs in different media

$NPs_{(nm)}$	Media	pH	ζ (mV)	ZNSC[a] pH	Reference
$TiO_{2(15)}$	1 mM NaCl DI water[b]	<5.2 >7.8	>28 >−32	6.2	(63)
$Al_2O_{3(11)}$	ultra pure water	5.5	120	8.4	(64)
$TiO_{2(9)}$			−50		
$TiO_{2(12)}$			20	6.4	
$TiO_{2(17)}$			12	5.6	
$TiO_{2(21)}$			0		
$TiO_{2(25)}$			44	7	
$TiO_{2(142)}$			−33	5.2	
$TiO_{2(707)}$			1	5.4	
$TiO_{2(100)}$	DI water		−21		(65)
$TiO_{2(51)}$	DI water		−22		
$TiO_{2(50)}$	DI water		−14		
$TiO_{2(40)}$	DI water		−21		
$TiO_{2(39)}$	DI water		−20 to −23		
$TiO_{2(10)}$	DI water		−3		
$TiO_{2(6)}$	DI water		−29		
$TiO_{2(63)}$	1 mM NaCl DI water		6		(66)
$CuO_{(42)}$			31		
$ZnO_{(71)}$			27		
$Fe_2O_{3(29)}$			−17		
$Fe_3O_{4(20–30)}$			2		
$Graphite_{(5)}$	0.01 M KCl	2 9	40 −70	4.5–5.5	(57)
$C_{60(<0.22)}$	0.0001–0.1 M NaCl, KCl, Na_2SO_4	7	−50 to −20		(58)
	0 to 0.1 M NaCl	0–11	0 to −65	0	
$SWCNTs_{(118–326)}$	1.0 mg/L Tween 80 water		−5 to −10		(59)
$MWCNTs_{(209–223)}$	Water		−23 to −25		(56)
$MWCNTs\text{-}OH_{(181–187)}$			−23 to −24		
$MWCNTs\text{-}COOH_{(181–185)}$			−20 to −23		
$MWCNTs\text{-}O_{(187–196)}$			−22 to −25		
$MWCNTs_{(30–70)}$[c]			−46		(67)
MWCNTs-PEI[d]			35		
MWCNTs-PEI-Ac[e]			−1		
MWCNTs-PEI-SAH[f]			−21		

(continued)

Table 2 (continued)

$NPs_{(nm)}$	Media	pH	ζ (mV)	ZNSC[a] pH	Reference
Crystal $Si_{(452)}$	Water		-62		(68)
	PBS[g]		-16		
	Media		-1		
Amorphous $Si_{(35)}$	Water		-48		
	PBS		-27		
	Media		-5		
$ZnO_{(89)}$	Water		-55		
	PBS		-29		
	Media		-6		
$ZnO_{(111)}$	Water		-56		
	PBS		-15		
	Media		-9		

ZNSC[a], an isoelectric point pH with zero net surface charge; *DI water*[b], deionized water; *MWCNTs*[c], acid-treated *MWCNTs*; *MWCNTs-PEI*[d], polyethyleneimine (PEI) covalently bonded to acid-treated MWCNTs; *MWCNTs-PEI-Ac*[e] *and MWCNTs-PEI-SAH*[f], MWCNTs-PEI treated with acetic and succinic anhydrides respectively; *PBS*[g], phosphate-buffered saline

Table 3
Definitions of the nanotoxicity indexes and related assessment methods

Index	Definitions	Assays	References
EC_{50}	Median effect concentrations for immobilization or effect concentrations of NPs to reduce cell viability or to cause inhibition in untreated controls by 50%	LDH, MTT[a], neutral red, alamar blue	(33, 41, 42, 56, 69–71)
LC_{50}	Median lethal concentrations of NPs or concentrations of NPs causing 50% mortality compared to the control group	LDH, MTT, neutral red, XTT	(56, 69–75)
TC_{50}	Toxic concentrations of NPs inducing 50% cell mortality or altering LDH or MTT activity by 50%	LDH, MTT, neutral red	(37, 76)
CC_{50}	50% cytotoxic concentrations of NPs	LDH, MTT	(77, 78)
IC_{50}	Concentrations of NPs causing 50% inhibition of metabolic activity or cell reduction/death	LDH, neutral red, WST-1	(75, 79, 80)
LD_{50}	Median lethal doses of NPs		(38, 81)

[a]*MTT* 3-(4,5-dimethylthiazol-2-yl)-2,5-diphenyl tetrazolium bromide, *XTT* 2,3-bias (2-methoxyl-4-nitro-5-sulfophenyl)-2H-tetrazolium-5-carboxanilide, *WST-1* water-soluble tetrazolium salt 1 or 2-(4-iodophenyl)-3-(4-nitrophenyl)-5-(2,4-disulfophenyl)-2H-tetrazolium

and damages induced by potential toxicants. Accordingly, LDH assays are widely adopted to quantify the death of cells, the infiltration of cell membrane, and the cell proliferation induced by potential toxins such as NPs (36, 82–84). LDH assays detect the presence of LDH in both the intracellular and extracellular media indirectly by monitoring the variations of substrates in enzymatic reactions. Reaction (1) is a widely employed reversible enzymatic reaction catalyzed by LDH:

$$NADH + P^- + H^+ \overset{LDH}{\rightleftharpoons} NAD^+ + L^- \tag{1}$$

NADH, NAD^+, P^-, and L^- are abbreviations of 1,4-nicotinamide adenine dinucleotide, β-nicotinamide adenine dinucleotide, pyruvate or pyruvic acid, and lactate, respectively. Both the left and right reactions are adopted to estimate LDH activity indirectly (85–88). Conventionally, monitoring the variations of the UV absorbance of NADH at 340 nm ($\varepsilon = 6.22 \times 10^3\ M^{-1}cm^{-1}$) was frequently applied as a standard method for indirect evaluation of LDH activity varied by potential toxicants in clinic diagnoses (89–98). The concentration of LDH varied linearly with the decrease in the left reaction or increase in the right reaction of absorbance of NADH: LDH (U/L) = (ΔA_S (min^{-1}) − ΔA_C (min^{-1})) × 1000 $V_T/(6.3 \times V_S)$, ΔA_S and ΔA_C are the absorbance variation rates of NADH in sample and the control, respectively; V_T and V_S are the volumes of total media and sample, respectively (99). The percentage of LDH released from a treated sample can also be calculated from the absorbance of treated (A_T) and control (A_C) or untreated samples as $A_T/A_C \times 100\%$ (82, 100). Experimentally, when the left reaction was used for assessment, excessive P^- was added to inhibit the right reaction, otherwise excessive L^- was added. The disadvantages of this method are that it is time-consuming due to the measurement of a number of samples and expensive for the excessive depletion of NADH. A fluorescence assay that is based on the fluorescent responses of NADH by an excitation and an emission at 360 and 460 nm respectively required less quantity of NADH and run rapidly (101, 102).

During recent decades, many improvements were achieved for LDH assays. For examples, enzymatic reaction (1) coupled with the presence of diaphorase, NADH was oxidized to NAD^+ reversibly by tetrazolium salt, 2-(4-iodophenyl)-3-(4-nitrophenyl)-5-phenyl-2H-tetrazolium chloride (INT), following the reaction (2):

$$INT + NADH \overset{diaphorase}{\rightleftharpoons} INTH + NAD^+ \tag{2}$$

UV absorbencies of INTH (reduced INT) at 500 nm ($\varepsilon = 1.508 \times 10^4\ M^{-1}cm^{-1}$) and INT at about 490 nm are the alternative signals for monitoring LDH activity: LDH (U/L) = $0.3895\Delta A$(min^{-1}), ΔA is the absorbance variation rate of INTH or INT (103–105). This modified method was more sensitive and rapid and has been popularly applied for LDH assay to evaluate the potential toxicities of NPs (4, 33, 36, 84, 106–111). However, this assay is disturbed by phenol red indicator which is widely distributed in most tissue culture media and absorbed at 490 nm. A similar fluorescence method by employing rezasurin as an oxidative agent to convert NADH (generated in the left direction of reaction (1)) to NAD^+ in the presence of diaphorase like reaction (3) and detecting the fluorescent response of resorufin at 600 nm ($\varepsilon = 6 \times 10^4\ M^{-1}\ cm^{-1}$) was also established for monitoring LDH activity (112, 113).

$$\text{Rezasurin} + \text{NADH} \overset{\text{diaphorase}}{\rightleftharpoons} \text{NAD}^+ + \text{Resorufin} \quad (3)$$

Another improved LDH assay is the combined usage of MTT as an alternative indicator and measuring the reduction product of MTT, formazan (MTTF), at 570 nm along with a reference wavelength at 655 nm as shown in reactions (4) and (5) (MPMS, 1-methoxyphenazine methosulfate; RMPMS is the reduced MPMS) (114). These reactions are conveniently stopped by the addition of acid dimethylformamide/sodium dodecyl sulfate solution. This improved method is more economical than the standard one (114).

$$\text{NADH} + \text{MPMS} \rightleftharpoons \text{NAD}^+ + \text{RMPMS} \quad (4)$$

$$\text{RMPMS} + \text{MTT} \rightleftharpoons \text{MPMS} + \text{MTTF} \quad (5)$$

2. Due to the high sensitivity and selectivity, electrochemical measurements of the concentration variations of redox couples such as NAD^+/NADH and P^-/L^- catalyzed by LDH are capable of being applied as alternative methods for indirect determination of LDH activity. However, the usage of P^-/L^- redox couple for the evaluation of LDH activity has no practical meaning because of the requirement to expend excessive expensive NAD^+ or NADH as an inhibitor to restrain reverse reaction. Under an appropriate condition, the rate of the left enzymatic reaction (1) is proportional to the decrease of the electrochemical signal generated by NADH, whereas that of the right one was proportional to the increase of what generated by NADH. For example, an amperometric technique based on detecting a decrease in the anodic peak current that corresponded to the oxidation of NADH at 0.7 V vs SCE at a glassy carbon electrode (GCE) modified by adsorption of Mg^{2+} is used to measure LDH activity in serum, whole blood, lipemic sera, and homogenates of liver tissue without sacrificing sensitivity and accuracy compared with those of the conventional spectrophotometric methods (115). The oxidation of NADH at a carbon paste electrode occurs at 0.95 V vs Ag/AgCl, and related peak current is also observed to decrease linearly with the increase of the concentration of LDH (116).

   Owing to the direct oxidation of NADH encountering unfavorably high potential and fouling of the electrode surface, many works aimed to solve these problems. For instance, at an untreated GCE the oxidation peak of NADH is a broad irreversible uncatalyzed wave at 0.85 V vs Ag/AgCl, while after the electrode being activated at 1.2 V in 1 M NaOH, the related oxidation peak became much more defined and shifted down to 0.50 V (117). The oxidation of NADH at polyethyleneimine and adenosine diphosphate modified graphite electrodes at pH 9.0 appeared at 0.05 V vs Ag/AgCl and is applied in a flow injection-amperometric system to indirectly monitor LDH activity in human serum with a sampling rate of 40 h^{-1} and a response time of 12 s and a limit of detection (LD) of 8×10^{-9} M for NADH (the linear response between the enzymatic reaction rate and LDH activity is: v (nA/min) = 0.068(LDH) (U/L) + 0.04) (118). At a pretreated carbon fiber electrode, an electrocatalytic oxidative wave of NADH generated at 0.474 V vs Ag/AgCl and by monitoring this wave a LD of 10^{-6} M for NADH is achieved (119). However, this wave is produced only when NADH is introduced before ascorbate and the temperature must be maintained above 38°C (119). So far as we know, up to now the above-mentioned electrochemical methods are not accepted as routine techniques for the assessments of LDH activity variations induced by potential toxicants, especially by NPs.

In recent years, our research group has developed a series of novel electrochemical methods by immobilizing LDH in silica sol–gel film which was modified on gold electrode surface both without and with the addition of nano-TiO_2 (50 nm) and nano-Al_{13} for indirect monitoring L^- through the oxidation of NADH and for direct detection of resorcinol and *p*-xylene respectively (10, 17, 120). These studies indicate that LDH reacted with both nano-TiO_2 and nano-Al_{13}. Moreover, based on the tracing of the time-dependent increment of the DPV cathodic peak current signals of NAD^+ (I_{pNAD^+}) which is produced in the left direction of the enzymatic reaction (1) at −0.89 V vs SCE at a hanging mercury drop electrode (Fig. 1), Al(III), nano-Al_2O_3, and nano-Al_{13} are also observed to severely inhibit LDH activity (52, 53).

3. LDH, β-NAD^+, and NADH will lose bioactivity when they are stored at temperature higher than 4°C.

4. All electrochemical experiments are performed with this CHI660B electrochemical system. The differential pulse voltammetry (DPV) models are applied with the following parameters, scan rate 20 mV/s, pulse amplitude 50 mV, pulse width 50 ms.

5. Actually, MWCNTs did not dissolve in water solution. They often dispersed or spread highly uniformly in water to form a suspension after being sonicated and refused to deposit within 24 h due to their small sizes and surface charge.

6. The molecular weight of LDH is quite different from different kinds of sources. The merchant indicates the molecular weight of LDH is 1.5×10^5 g/mol. The calculated final concentrations of LDH is of 8.0×10^{-10} M.
7. The final concentrations of NADH and P^- are, 2.0×10^{-4} and 8.0×10^{-4} M, respectively.
8. If the solution is not motionless in the recording process, the obtained signals will be consisted of not only the responses of NAD^+, but also that of convective current.
9. When the enzymatic reaction reaches equilibrium the concentration of NAD^+ will reach a maximum amount. After this, due to the gradual oxidation of NADH by oxygen, the equilibrium will reverse, which will result in the slow decrease of the concentration of NAD^+.

Acknowledgments

This project is supported by the NSFC (20975049), State Key Laboratory of Electrochemistry of China in Changchun Applied Chemistry Institute (2008008) and Analytical Center of Nanjing University.

References

1. Bi SP, Zhang J, Cheng JJ (2009) Call from China for joint nanotech toxicity-testing effort. Nature 461:593
2. Huang XL, Teng X, Chen D, Tang FQ, He JQ (2010) The effect of the shape of mesoporous silica nanoparticles on cellular uptake and cell function. Biomaterials 31:438–448
3. Jonaitis TS, Card JW, Magnuson B (2010) Concerns regarding nano-sized titanium dioxide dermal penetration and toxicity study. Toxicol Lett 192:268–269
4. Bhardwaj V, Ankola DD, Gupta SC, Schneider M, Lehr CM, Kumar MNVR (2009) PLGA nanoparticles stabilized with cationic surfactant: safetystudies and application in oral delivery of paclitaxel to treat chemical-induced breast cancer in rat. Pharm Res 26:2495–2503
5. Pastorin G (2009) Crucial functionalizations of carbon nanotubes for improved drug delivery: a valuable option? Pharm Res 26:746–769
6. Azarmi S, Roa WH, Lobenberg R (2008) Targeted delivery of nanoparticles for the treatment of lung diseases. Adv Drug Deliv Rev 60:863–875
7. Kostarelos K, Bianco A, Prato M (2009) Promises, facts and challenges for carbon nanotubes in imaging and therapeutics. Nat Nanotechnol 4:627–633
8. Heller DA, Jin H, Martinez BM, Patel D, Miller BM, Yeung TK, Jena PV, Hoebartner C, Ha T, Silverman SK, Strano MS (2009) Multimodal optical sensing and analyte specificity using single-walled carbon nanotubes. Nat Nanotechnol 4:114–120
9. Krauss TD (2009) Nanotubes light up cells. Nat Nanotechnol 4:85–86
10. Cheng JJ, Huang DQ, Zhang J, Yang WJ, Wang N, Sun YB, Wang KY, Mo XY, Bi SP (2009) Electrochemical behavior of lactate dehydrogenase immobilized on "silica sol–gel/nanometre-sized tridecameric aluminium polycation" modified gold electrode and its application. Analyst 134:1392–1395
11. Plashnitsa VV, Elumalai P, Fujio Y, Miura N (2008) Sensing performances of zirconia-based NH3 sensor utilizing nano-Au sensing electrode. Chem Sensors 24:124–126
12. Liao YH, Yuan R, Chai YQ, Zhuo Y, Yang X (2010) Study on an amperometric immunosensor based on Nafion-cysteine composite membrane for detection of carcinoembryonic antigen. Anal Biochem 402:47–53

13. Su HL, Yuan R, Chai YQ, Zhuo Y, Hong CL, Liu ZY, Yang X (2009) Multilayer structured amperometric immunosensor built by self-assembly of a redox multi-wall carbon nanotube composite. Electrochim Acta 54:4149–4154
14. Sharma MK, Rao VK, Agarwal GS, Rai GP, Gopalan N, Prakash S, Sharma SK, Vijayaraghavan R (2008) Highly sensitive amperometric immunosensor for detection of Plasmodium falciparum histidine-rich protein in serum of humans with malaria: comparison with a commercial kit. J Clin Microbiol 46: 3759–3765
15. Lin JH, Zhang LJ, Zhang SS (2007) Amperometric biosensor based on coentrapment of enzyme and mediator by gold nanoparticles on indium-tin oxide electrode. Anal Biochem 370:180–185
16. Lin ZY, Huang LZ, Liu Y, Lin JM, Chi YW, Chen GN (2008) Electrochemiluminescent biosensor based on multi-wall carbon nanotube/nano-Au modified electrode. Electrochem Commun 10:1708–1711
17. Cheng JJ, Di JW, Hong JH, Yao KA, Sun YB, Zhuang JY, Xu Q, Zheng H, Bi SP (2008) The promotion effect of titania nanoparticles on the direct electrochemistry of lactate dehydrogenase sol–gel modified gold electrode. Talanta 76:1065–1069
18. Erdely A, Hulderman T, Salmen R, Liston A, Zeidler-Erdely PC, Schwegler-Berry D, Castranova V, Koyama S, Kim YA, Endo M, Simeonova PP (2009) Cross-talk between lung and systemic circulation during carbon nanotube respiratory exposure. potential biomarkers. Nano Lett 9:36–43
19. Wang XG, Kawanami H, Dapurkar SE, Venkataramanan NS, Chatterjee M, Yokoyama T, Ikushima Y (2008) Selective oxidation of alcohols to aldehydes and ketones over TiO_2-supported gold nanoparticles in supercritical carbon dioxide with molecular oxygen. Appl Catal A 349:86–90
20. Bekyarova E, Ni Y, Malarkey EB, Montana V, McWilliams JL, Haddon RC, Parpura V (2005) Applications of carbon nanotubes in biotechnology and biomedicine. J Biomed Nanotechnol 1:3–17
21. Singh S, Nalwa HS (2007) Nanotechnology and health safety – toxicity and risk assessments of nanostructured materials on human health. J Nanosci Nanotechnol 7:3048–3070
22. Duan YM, Liu J, Ma LL, Li N, Liu HT, Wang J, Zheng L, Liu C, Wang XF, Zhao XY, Yan JY, Wang SS, Wang H, Zhang XG, Hong FS (2010) Toxicological characteristics of nanoparticulate anatase titanium dioxide in mice. Biomaterials 31:894–899
23. Shvedova AA, Kagan VE, Fadeel B (2010) Close encounters of the small kind: adverse effects of man-made materials interfacing with the nano-cosmos of biological systems. Annu Rev Pharmacol Toxicol 50:63–88
24. Kim KT, Klaine SJ, Lin S, Ke PC, Kim SD (2010) Acute toxicity of a mixture of copper and single-walled carbon nanotubes to Daphnia magna. Environ Toxicol Chem 29: 122–126
25. Aillon KL, Xie Y, El-Gendy N, Berkland CJ, Forrest ML (2009) Effects of nanomaterial physicochemical properties on in vivo toxicity. Adv Drug Deliv Rev 61:457–466
26. Belyanskaya L, Weigel S, Hirsch C, Tobler U, Krug HF, Wick P (2009) Effects of carbon nanotubes on primary neurons and glial cells. Neurotoxicology 30:702–711
27. Deng XY, Wu F, Liu Z, Luo M, Li L, Ni QS, Wu MH, Liu YF (2009) The splenic toxicity of water soluble multi-walled carbon nanotubes in mice. Carbon 47:1421–1428
28. Elder A (2009) How do nanotubes suppress T cells? Nat Nanotechnol 4:409–410
29. Walker VG, Li Z, Hulderman T, Schwegler-Berry D, Kashon ML, Simeonova PP (2009) Potential in vitro effects of carbon nanotubes on human aortic endothelial cells. Toxicol Appl Pharmacol 236:319–328
30. Yu YM, Zhang Q, Mu QX, Zhang B, Yan B (2008) Exploring the immunotoxicity of carbon nanotubes. Nanoscale Res Lett 3:271–277
31. Elgrabli D, Floriani M, Abella-Gallart S, Meunier L, Gamez C, Delalain P, Rogerieux F, Boczkowski J, Lacroix G (2008) Biodistribution and clearance of instilled carbon nanotubes in rat lung. Part Fibre Toxicol 5:1–13
32. Grainger DW (2009) Nanotoxicity assessment: all small talk? Adv Drug Deliv Rev 61: 419–421
33. Jos A, Pichardo S, Puerto M, Sanchez E, Grilo A, Camean AM (2009) Cytotoxicity of carboxylic acid functionalized single wall carbon nanotubes on the human intestinal cell line Caco-2. Toxicol In Vitro 23:1491–1496
34. Elgrabli D, Abella-Gallart S, Robidel F, Rogerieux F, Boczkowski J, Lacroix G (2008) Induction of apoptosis and absence of inflammation in rat lung after intratracheal instillation of multiwalled carbon nanotubes. Toxicology 253:131–136
35. Monteiro-Riviere NA, Inman AO (2006) Challenges for assessing carbon nanomaterial toxicity to the skin. Carbon 44:1070–1078
36. Woerle-Knirsch JM, Pulskamp K, Krug HF (2006) Oops they did It again! carbon

nanotubes hoax scientists in viability assays. Nano Lett 6:1261–1268

37. Lanone S, Rogerieux F, Geys J, Dupont A, Maillot-Marechal E, Boczkowski J, Boczkowski J, Lacroix G, Hoet P (2009) Comparative toxicity of 24 manufactured nanoparticles in human alveolar epithelial and macrophage cell lines. Part Fibre Toxicol 6: 1–12
38. Monteiro-Riviere NA, Inman AO, Zhang LW (2009) Limitations and relative utility of screening assays to assess engineered nano particle toxicity in a human cell line. Toxicol Appl Pharmacol 234:222–235
39. Pulskamp K, Diabate S, Krug HF (2007) Carbon nanotubes show no sign of acute toxicity but induce intracellular reactive oxygen species in dependence on contaminants. Toxicol Lett 168:58–74
40. Casey A, Herzog E, Davoren M, Lyng FM, Byrne HJ, Chambers G (2007) Spectroscopic analysis confirms the interactions between single-walled carbon nanotubes and various dyes commonly used to assess cytotoxicity. Carbon 45:1425–1432
41. Davoren M, Herzog E, Casey A, Cottineau B, Chambers G, Byrne HJ, Lyng FM (2007) In vitro toxicity evaluation of single walled carbon nanotubes on human A549 lung cells. Toxicol In Vitro 21:438–448
42. Weyermann J, Lochmann D, Zimmer A (2005) A practical note on the use of cytotoxicity assays. J Intern Med 288:369–376
43. Marquis BJ, Love SA, Braun KL, Haynes CL (2009) Analytical methods to assess nanoparticle toxicity. Analyst 134:425–439
44. Shvedova AA, Kagan VE (2009) The role of nanotoxicology in realizing the 'helping without harm' paradigm of nanomedicine: lessons from studies of pulmonary effects of single-walled carbon nanotubes. J Intern Med 267:106–118
45. Shvedova AA, Kisin ER, Porter D, Schulte P, Kagan VE, Fadeel B, Castranova V (2009) Mechanisms of pulmonary toxicity and medical applications of carbon nanotubes: two faces of Janus? Pharmacol Ther 121:192–204
46. Oberdorster G (2009) Safety assessment for nanotechnology and nanomedicine: concepts of nanotoxicology. J Intern Med 267:89–105
47. Kagan VE, Bayir H, Shvedova AA (2005) Nanomedicine and nanotoxicology: two sides of the same coin. Nanomedicine 1:313–316
48. Chan VSW (2006) Nanomedicine: an unresolved regulatory issue. Regul Toxicol Pharmacol 46:218–224
49. Donaldson K, Aitken R, Tran L, Stone V, Duffin R, Forrest G, Alexander A (2006) Carbon nanotubes: a review of their properties in relation to pulmonary toxicology and workplace safety. Toxicol Sci 92:5–22
50. Tsuji JS, Maynard AD, Howard PC, James JT, Lam CW, Warheit DB, Santamaria AB (2006) Research strategies for safety evaluation of nanomaterials, Part IV: risk assessment of nanoparticles. Toxicol Sci 89:42–50
51. Moller P, Jacobsen NR, Folkmann JK, Danielsen PH, Mikkelsen L, Hemmingsen JG, Vesterdal LK, Forchhammer L, Wallin H, Loft S (2010) Role of oxidative damage in toxicity of particulates. Free Radic Res 44:1–46
52. Wang N, Huang DQ, Zhang J, Cheng JJ, Yu T, Zhang HQ, Bi SP (2008) Electrochemical studies on the effects of nanometer-sized tridecameric aluminum polycation on lactate dehydrogenase activity at the molecular level. J Phys Chem C 112:18034–18038
53. Yao KA, Huang DQ, Xu BL, Wang N, Wang YJ, Bi SP (2010) A sensitive electrochemical approach for monitoring the effects of nano-Al_2O_3 on LDH activity by differential pulse voltammetry. Analyst 135:116–120
54. McClendon S, Zhadin N, Callender R (2005) The approach to the Michaelis complex in lactate dehydrogenase: the substrate binding pathway. Biophys J 89:2024–2032
55. Gulotta M, Deng H, Deng H, Dyer RB, Callender RH (2002) Toward an understanding of the role of dynamics on enzymic catalysis in lactate dehydrogenase. Biochemistry 41:3353–3363
56. Kennedy AJ, Hull MS, Steevens JA, Dontsova KM, Chappell MA, Gunter JC, Weiss CA Jr (2008) Factors influencing the partitioning and toxicity of nanotubes in the aquatic environment. Environ Toxicol Chem 27:1932–1941
57. Moraru V, Lebovka N, Shevchenko D (2004) Structural transitions in aqueous suspensions of natural graphite. Colloids Surf A 242:181–187
58. Brant J, Lecoanet H, Hotze M, Wiesner M (2005) Comparison of electrokinetic properties of colloidal fullerenes (n-C60) formed using two procedures. Environ Sci Technol 39:6343–6351
59. Kato H, Mizuno K, Shimada M, Nakamura A, Takahashi K, Hata K, Kinugasa S (2009) Observations of bound Tween 80 surfactant molecules on single-walled carbon nanotubes in an aqueous solution. Carbon 47:3434–3440
60. Yu BZ, Yang JS, Li WX (2007) In vitro capability of multi-walled carbon nanotubes modified

with gonadotrophin releasing hormone on killing cancer cells. Carbon 45:1921–1927

61. Vitanov P, Harizanova A, Ivanova T, Dimitrova T (2009) Chemical deposition of Al2O3 thin films on Si substrates. Thin Solid Films 517:6327–6330
62. Di Sotto A, Chiaretti M, Carru GA, Bellucci S, Mazzanti G (2009) Multi-walled carbon nanotubes: lack of mutagenic activity in the bacterial reverse mutation assay. Toxicol Lett 184:192–197
63. Jiang J, Oberdorster G, Biswas P (2009) Characterization of size, surface charge, and agglomeration state of nanoparticle dispersions for toxicological studies. J Nanopart Res 11:77–89
64. Simon-Deckers A, Loo S, Mayne-L'hermite M, Herlin-Boime N, Menguy N, Reynaud C, Gouget B, Carriere M (2009) Size-, composition- and shape-dependent toxicological impact of metal oxide nanoparticles and carbon nanotubes toward bacteria. Environ Sci Technol 43:8423–8429
65. Braydich-Stolle LK, Schaeublin NM, Murdock RC, Jiang JK, Biswas P, Schlager JJ, Hussain SM (2009) Crystal structure mediates mode of cell death in TiO_2 nanotoxicity. J Nanopart Res 11:1361–1374
66. Karlsson HL, Cronholm P, Gustafsson J, Moeller L (2008) Copper oxide nanoparticles are highly toxic: a comparison between metal oxide nanoparticles and carbon nanotubes. Chem Res Toxicol 21:1726–1732
67. Shen M, Wang SH, Shi X, Chen XS, Huang QG, Petersen EJ, Pinto RA, Baker JR, Weber WJ (2009) Polyethyleneimine-mediated functionalization of multiwalled carbon nanotubes: synthesis, characterization, and in vitro toxicity assay. J Phys Chem C 113:3150–3156
68. Sayes CM, Reed KL, Subramoney S, Abrams L, Warheit DB (2009) Can in vitro assays substitute for in vivo studies in assessing the pulmonary hazards of fine and nanoscale materials? J Nanopart Res 11:421–431
69. Blaise C, Gagne F, Ferard JF, Eullaffroy P (2008) Ecotoxicity of selected nano-materials to aquatic organisms. Environ Toxicol 23:591–598
70. Zhu XS, Zhu L, Chen YS, Tian SY (2009) Acute toxicities of six manufactured nanomaterial suspensions to Daphnia magna. J Nanopart Res 11:67–75
71. Zheng HZ, Liu L, Lu YH, Long YJ, Wang LL, Ho KP, Wong KY (2010) Rapid determination of nanotoxicity using luminous bacteria. Anal Sci 26:125–128
72. Petrick JS, Ayala-Fierro F, Cullen WR, Carter DE, Vasken AH (2000) Monomethylarsonous acid (MMAIII) is more toxic than arsenite in chang human hepatocytes. Toxicol Appl Pharmacol 163:203–207
73. Olabarrieta I, L'Azou B, Yuric S, Cambar J, Cajaraville MP (2001) In vitro effects of cadmium on two different animal cell models. Toxicol In Vitro 15:511–517
74. Li HC, Zhang JS, Wang T, Luo WR, Zhou QF, Jiang GB (2008) Elemental selenium particles at nano-size (Nano-Se) are more toxic to Medaka (Oryzias latipes) as a consequence of hyper-accumulation of selenium: a comparison with sodium selenite. Aquat Toxicol 89:251–256
75. Cook SM, Aker WG, Rasulev BF, Hwang HM, Leszczynski J, Jenkins JJ, Shockley V (2010) Choosing safe dispersing media for C60 fullerenes by using cytotoxicity tests on the bacterium Escherichia coli. J Hazard Mater 176:367–373
76. Issa Y, Watts DC, Brunton PA, Waters CM, Duxbury AJ (2004) Resin composite monomers alter MTT and LDH activity of human gingival fibroblasts in vitro. Dent Mater 20: 12–20
77. Huang WH, Li YL, Wang H, Su MX, Jiang ZY, Ooi VEC, Chung HY (2009) Toxicological study of a Chinese herbal medicine, Wikstroemia indica. Nat Prod Commun 4:1227–1230
78. Rahban M, Divsalar A, Saboury AA, Golestani A (2010) Nanotoxicity and spectroscopy studies of silver nanoparticle: calf thymus DNA and K562 as targets. J Phys Chem C 114:5798–5803
79. Suttmann H, Retz M, Paulsen F, Harder J, Zwergel U, Kamradt J, Unteregger G, Stoeckle M, Lehmann J (2008) Antimicrobial peptides of the Cecropin-family show potent antitumor activity against bladder cancer cells. BMC Urol 8:1–7
80. L'Azou B, Jorly J, On D, Sellier E, Moisan F, Fleury-Feith J, Brochard P, Ohayon-Courtes C (2008) In vitro effects of nanoparticles on renal cells. Part Fibre Toxicol 5:1–14
81. Hu XK, Cook S, Wang P, Hwang HM (2009) In vitro evaluation of cytotoxicity of engineered metal oxide nanoparticles. Sci Total Environ 407:3070–3072
82. Khattak SF, Spatara M, Roberts L, Roberts SC (2006) Application of colorimetric assays to assess viability, growth and metabolism of hydrogel-encapsulated cells. Biotechnol Lett 28:1361–1370
83. Lantto TA, Colucci M, Zavadova V, Hiltunen R, Raasmaja A (2009) Cytotoxicity of curcumin, resveratrol and plant extracts from basil, juniper, laurel and parsley in SH-SY5Y and CV1-P cells. Food Chem 117:405–411

84. Lobo AO, Corat MAF, Antunes EF, Palma MBS, Pacheco-Soares C, Garcia EE, Corat EJ (2010) An evaluation of cell proliferation and adhesion on vertically-aligned multi-walled carbon nanotube films. Carbon 48:245–254
85. Krieg AF, Rosenblum LJ, Henry JB (1967) Lactate dehydrogenase isoenzymes a comparison of pyruvate-to-lactate and lactate-to-pyruvate assays. Clin Chem 13:196–203
86. Howell BF, McCune S, Schaffer R (1979) Lactate-to-pyruvate or pyruvate-to-lactate assay for lactate dehydrogenase: a re-examination. Clin Chem 25:269–272
87. Gutheil WG (1998) A sensitive equilibrium-based assay for d-lactate using d-lactate dehydrogenase: application to penicillin-binding protein/dd-carboxypeptidase activity assays. Anal Biochem 259:62–67
88. Larsen T (2005) Determination of lactate dehydrogenase (LDH) activity in milk by a fluorometric assay. J Dairy Res 72:209–216
89. Zewe V, Fromm HJ (1965) Kinetic studies of rabbit muscle lactate dehydrogenase. II. Mechanism of the reaction. Biochemistry 4:782–792
90. Klotzsch SG, Klotzsch HR, Haus M (1969) Inhibitor-contaminated NADH: its influence on dehydrogenases and dehydrogenase-coupled reactions. Clin Chem 15:1056–1061
91. Lovell SJ, Winzor DJ (1974) Effects of phosphate on the dissociation and enzymic stability of rabbit muscle lactate dehydrogenase. Biochemistry 13:3527–3531
92. Haid E, Lehmann P, Ziegenhorn J (1975) Molar absorptivities of beta-NADH and beta-NAD at 260 nm. Clin Chem 21:884–887
93. Tomaszek TA Jr, Moore ML, Strickler JE, Sanchez RL, Dixon JS, Metcalf BW, Hassell A, Dreyer GB, Brooks I, Debouck C, Meek TD (1992) Proteolysis of an active site peptide of lactate dehydrogenase by human immunodeficiency virus type 1 protease. Biochemistry 31:10153–10168
94. Arechabala B, Coiffard C, Rivalland P, Coiffard LJ, de Roeck-Holtzhauer Y (1999) Comparison of cytotoxicity of various surfactants tested on normal human fibroblast cultures using the neutral red test, MTT assay and LDH release. J Appl Toxicol 19:163–165
95. Zhu WL, Ma SP, Qu R, Kang DL (2006) Antidepressant-like effect of saponins extracted from Chaihu-jia-longgu-muli-tang and its possible mechanism. Life Sci 79: 749–756
96. Li S, Wang C, Wang MW, Li W, Matsumoto K, Tang YY (2007) Antidepressant like effects of piperine in chronic mild stress treated mice and its possible mechanisms. Life Sci 80: 1373–1381
97. Muller J, Huaux F, Fonseca A, Nagy JB, Moreau N, Delos M, Raymundo-Pinero E, Beguin F, Kirsch-Volders M, Fenoglio I, Fubini B, Lison D (2008) Structural defects play a major role in the acute lung toxicity of multiwall carbon nanotubes: toxicological aspects. Chem Res Toxicol 21:1698–1705
98. Yang H, Liu C, Yang DF, Zhang HH, Xi ZG (2009) Comparative study of cytotoxicity, oxidative stress and genotoxicity induced by four typical nanomaterials: the role of particle size, shape and composition. J Appl Toxicol 29:69–78
99. Zhao YF, Zhang N, Kong QZ (2006) Tetrazolium violet induces G0/G1 arrest and apoptosis in brain tumor cells. J Neurooncol 77:109–115
100. Vaucher RA, Teixeira ML, Brandelli A (2010) Investigation of the cytotoxicity of antimicrobial peptide P40 on eukaryotic cells. Curr Microbiol 60:1–5
101. Moran JH, Schnellmann RG (1996) A rapid beta-NADH-linked fluorescence assay for lactate dehydrogenase in cellular death. J Pharmacol Toxicol Methods 36:41–44
102. Liu YC, Gerber R, Wu J, Tsuruda T, McCarter JD (2008) High-throughput assays for sirtuin enzymes: a microfluidic mobility shift assay and a bioluminescence assay. Anal Biochem 378:53–59
103. Allain CC, Henson CP, Nadel MK, Knoblesdorff AJ (1973) Rapid single-step kinetic colorimetric assay for lactate dehydrogenase in serum. Clin Chem 19:223–227
104. Sepp A, Binns RM, Lechler RI (1996) Improved protocol for colorimetric detection of complement-mediated cytotoxicity based on the measurement of cytoplasmic lactate dehydrogenase activity. J Immunol Methods 196:175–180
105. Wolterbeek HT, van der Meer, Astrid JGM (2005) Optimization, application, and interpretation of lactate dehydrogenase measurements in microwell determination of cell number and toxicity. Assay Drug Dev Technol 3:675–682
106. Barillet S, Simon-Deckers A, Herlin-Boime N, Mayne-L'Hermite M, Reynaud C, Cassio D, Gouget B, Carriere M (2010) Toxicological consequences of TiO2, SiC nanoparticles and multi-walled carbon nanotubes exposure in several mammalian cell types: an in vitro study. J Nanopart Res 12:61–73
107. Lobo AO, Corat MAF, Antunes EF, Palma MBS, Pacheco-Soares C, Corat EJ (2009)

Cytotoxicity analysis of vertically aligned multi-walled carbon nanotubes by colorimetric assays. Synth Met 159:2165–2166
108. Yeh LK, Chen YH, Chiu CS, Hu FR, Young TH, Wang IJ (2009) The phenotype of bovine corneal epithelial cells on chitosan membrane. J Biomed Mater Res A 90A:18–26
109. Chen X, Liu J, Gu XS, Ding F (2008) Salidroside attenuates glutamate-induced apoptotic cell death in primary cultured hippocampal neurons of rats. Brain Res 1238:189–198
110. Lehmann J, Retz M, Sidhu SS, Suttmann H, Sell M, Paulsen F, Harder J, Unteregger G, Stoeckle M (2006) Antitumor activity of the antimicrobial peptide magainin II against bladder cancer cell lines. Eur Urol 50:141–147
111. De Juan BS, Von Briesen H, Gelperina SE, Kreuter J (2006) Cytotoxicity of doxorubicin bound to poly(butyl cyanoacrylate) nanoparticles in rat glioma cell lines using different assays. J Drug Target 14:614–622
112. Bembenek ME, Kuhn E, Mallender WD, Pullen L, Li P, Parsons T (2005) A fluorescence-based coupling reaction for monitoring the activity of recombinant human NAD synthetase. Assay Drug Dev Technol 3:533–541
113. Ivanova L, Uhlig S (2008) A bioassay for the simultaneous measurement of metabolic activity, membrane integrity, and lysosomal activity in cell cultures. Anal Biochem 379:16–19
114. Abe K, Matsuki N (2000) Measurement of cellular 3-(4,5-dimethylthiazol-2-yl)-2,5-diphenyltetrazolium bromide (MTT) reduction activity and lactate dehydrogenase release using MTT. Neurosci Res 38:325–329
115. Bartalits L, Nagy G, Pungor E (1984) Determination of enzyme activity in biological fluids by means of electrochemical oxidation of NADH at a modified glassy carbon electrode. Clin Chem 30:1780–1783
116. Tarmure C, Sandulescu R, Ionescu C (2000) Voltammetric determination of lactate dehydrogenase using a carbon paste electrode. J Pharm Biomed Anal 22:355–361
117. Eisenberg EJ, Cundy KC (1991) Amperometric high-performance liquid chromatographic detection of NADH at a base-activated glassy carbon electrode. Anal Chem 63:845–847
118. Santos-Alvarez NDL, Lobo-Castanon MJ, Miranda-Ordieres AJ, Tunon-Blanco P (2002) Amperometric determination of serum lactate dehydrogenase activity using an ADP-modified graphite electrode. Anal Chim Acta 457:275–284
119. Nowall WB, Kuhr WG (1995) Electrocatalytic surface for the oxidation of NADH and other anionic molecules of biological significance. Anal Chem 67:3583–3588
120. Di JW, Cheng JJ, Xu Q, Zheng H, Zhuang JY, Sun YB, Wang KY, Mo XY, Bi SP (2007) Direct electrochemistry of lactate dehydrogenase immobilized on silica sol–gel modified gold electrode and its application. Biosens Bioelectron 23:682–687

Chapter 9

Enzyme-Linked Immunosorbent Assay of IL-8 Production in Response to Silver Nanoparticles

Eun-Jeong Yang, Jiyoung Jang, Dae-Hyoun Lim, and In-Hong Choi

Abstract

Enzyme-linked immunosorbent assay (ELISA) for monitoring the effects of nanoparticles on immune cells is a conventional method of assessing the levels of cytokine that are released into the culture supernatant after the addition of nanoparticles to a macrophage culture. However, it has been found that the presence of nanoparticles can interfere with spectrophotometric analysis, used as an indicator test system; thus, it is necessary to thoroughly checked for the possibility of interference. In this chapter, the assessment method of cytokine production is covered in detail by utilizing the cytokine model produced by silver nanoparticles.

Key words: Cytokine, Nanoparticles, ELISA, Macrophages, Inflammation

1. Introduction

It is possible for nanoparticles to flow into cells, and even into nucleoplasm due to their potential to permeate into living organisms. Generally, nanoparticles penetrate into living organisms through the skin, respiratory organs, and digestive system; however, for medical purposes they are injected into the blood vessel. Primarily, the epithelial cells react to these nanomaterials, but once they flow into the living body, macrophages (immune cells) respond first.

Macrophages are major cells in the innate immunity (primary defense) that phagocytose foreign materials (in case of particles) or endocytose small particles and molecules. Furthermore, macrophages become activated after phagocytizing foreign materials and perform the function of initiating the adaptive immune responses by releasing a variety of cytokines. Currently, most composite nanomaterials are metal-, carbon-, and silica-based. Furthermore, antigenicity of nanomaterials is not as strong as against microbes, but the possibility to induce macrophage activation via other mechanisms, non-antigenic stimulation, is presented in this chapter.

Joshua Reineke (ed.), *Nanotoxicity: Methods and Protocols*, Methods in Molecular Biology, vol. 926,
DOI 10.1007/978-1-62703-002-1_9,

Some publications have reported that if nanoparticles flow into cells, they can trigger cytotoxicity by inducing mitochondrial perturbation and apoptosis; moreover, the immune cells produce active oxygen radicals that lead to the inflammatory cytokines. Thus, oxidative stress occurs if macrophages phagocytize nanomaterials, hence the activation of immune cells is induced either directly or indirectly, raising the possibility of immune system activation.

As for in-depth examples, carbon nanomaterials also trigger immune reactions, particularly multiwalled carbon nanotubes (MWCNT) set off more immune reaction (1) than single-walled carbon nanotubes (SWCNT) (2); and among metals and metal oxides, silver nanoparticles lead to the most robust immune reactions followed by gold nanoparticles and titanium dioxide (3, 4). Silica, polystyrene, and latex-based nanomaterials are also reported to trigger immunotoxicity (5, 6).

The simplest methods to measure the activation of macrophages (immune cells) are to assess the production of cytokines. This chapter covers the assessment of silver nanoparticle-based interleukin (IL)-8. IL-8 is a typical chemokine that mediates neutrophil migration. The enzyme-linked immunosorbent assay (ELISA) technique is explained herein.

ELISA was developed as a measuring technique for protein antigen and immunoglobulin in the 1970s (7). Since then, it has been widely used for assessing the presence of an antigen or antibody from various samples down to a single protein and after subjecting the specific antigen to an antibody reaction, measure another specific secondary antibody–antigen reaction. Since the secondary antibody is bound with enzyme, the amplification of a color reaction can be seen if substrate is added.

ELISA is widely used to measure cytokine levels and supplied by many companies due to its sensitive, specific, and reproducible nature. Sensitivity and specificity are determined by the quality of antibody. The drawback is that ELISA can only measure one type of antigen (cytokine) at a time. To make up for this insufficiency, cytokine protein array is applied to screen various types of cytokine (8–10). In addition to this drawback, the accurate quantitation is only possible at a density level within the standard curve, thus, it is imperative that samples with high-density cytokine are diluted.

In case of using ELISA and spectrophotometer in nanomaterial research, the subjected nanomaterials must be checked to determine whether or not they are hindering absorbance of the indicator system. As for the silver nanoparticles, the optical density might be influenced by nanomaterials seriously at high concentration and slightly at low concentration (Fig. 1). Therefore, the result should be assessed at 490 nm rather than the usual recommended range of 405–450 nm. Considering the background level, it is better to either calculate optical density or select the wavebands that are not subject to interferences.

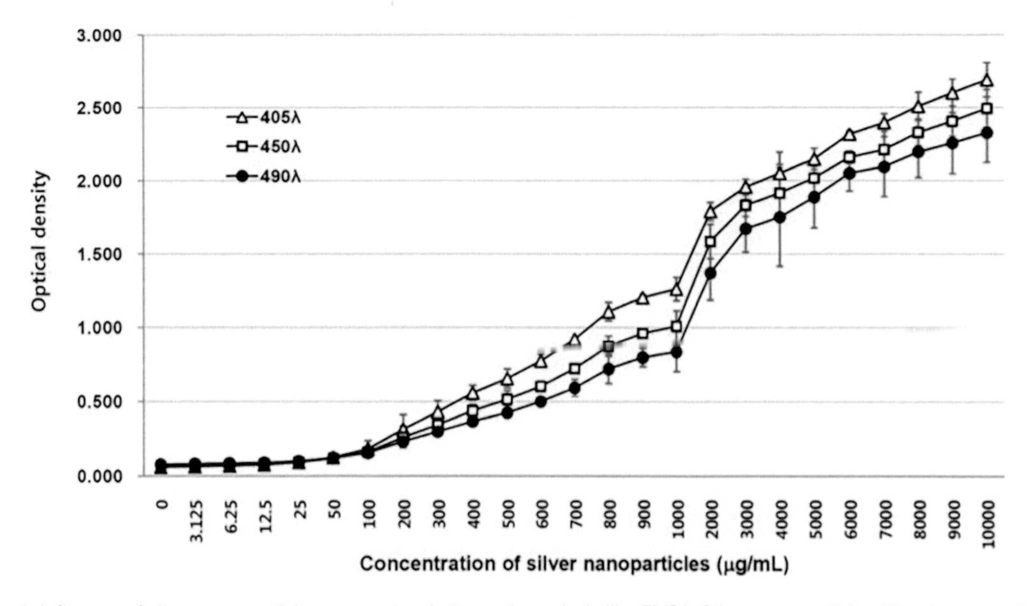

Fig. 1. Influence of silver nanoparticles on spectrophotoemrtic analysis like ELISA. Silver nanoparticles (5 nm) were serially diluted in cell culture medium and optical density was assessed at 405, 450, 490 nm, respectively (unpublished data).

With regard to silver nanoparticles exceeding 20 nm, no difficulties arise in the experiments conducted for measuring cytokines due to the fact that ELISA utilizes only a cell culture supernatant.

There are no major obstructions with the assessment since silver nanoparticles and cells are settled to the bottom, and culture supernatants are harvested after centrifugation. Selecting types of cytokine production for determining toxicity requires caution; that is, to explain the inflammatory cytokine response from tumor necrosis factor (TNF)-α or IL-6, both of which are not induced in the case of silver nanoparticles during early stage of exposure that show clear IL-8 production (Fig. 2), may be misleading.

2. Materials

1. Cell line: the human monocytic U937 cell line are obtained from ATCC (CRL1593.2™, Manassas, VA).
2. U937 cells are cultured in RPMI-1640 medium containing 10% fetal bovine serum (FBS). The medium also contains 2 mM L-glutamine, 100 U/mL penicillin, 100 μg/mL streptomycin. Cells are kept at a humidified 5% CO_2 incubator at 37°C.
3. T75 cm^2 and T25 cm^2 cell culture flasks, 24-well cell culture plates, and 96 multiwell plates are used for culture.
4. Polymyxin B.

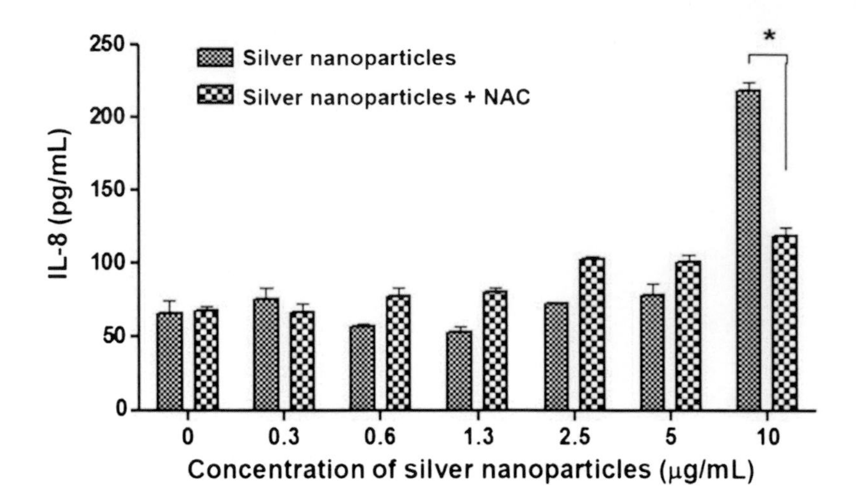

Fig. 2. Induction of IL-8 production by silver nanoparticles. Silver nanoparticles (5 nm) were added to macrophage cells and the culture supernatants were collected after 24 h. NAC was added to prevent the effect of reactive oxygen species. $^{*}p < 0.01$, *NAC* *N*-acetyl cysteine (unpublished data).

5. ELISA kit (BD OptEIM™, USA).
6. Coating buffer (pH 9.5) is composed of following elements: Dissolve 0.1 M sodium carbonate; 7.13 g $NaHCO_3$; and 1.59 g Na_2CO_3 in the deionized water 1 L. At this time, adjust pH level to 9.5 with 10 N NaOH and 10 N HCl. This solution can be prepared as required when conducting experiments, but must be stored at 2–8°C and must be used within 7 days.
7. Wash buffer consists of the following elements: phosphate-buffered saline (PBS) containing 0.05% Tween-20 (PBS: 80.0 g NaCl, 11.6 g Na_2HPO_4, 2.0 g KH_2PO_4, 2.0 g KCl in deionized water 10 L). The wash buffer can be made as required by the experiment, but it must be stored at 2–8°C and used within 3 days of preparation.

3. Methods

3.1. Cell Seeding

1. For adherent cells, initiate cell seeding with 10% FBS media on a multiwell plate at least 12 h prior to processing nanomaterials. The seeding density variation per well are dependent on the types of cells being utilized (see Note 1). Floating cells, i.e., U937 cells, are dispersed into $(0.5–1) \times 10^6$ cells/mL in RPMI-1640 with 10% FBS just before exposure to nanoparticles. Distribute 200 μL of cell suspension into each well of 24-well plate to make the final cell number, $(1–2) \times 10^5$ cells/well.
2. If a 6-well plate is used, set up a volume of 2 mL/well to fit 1×10^6 cells/well.

3. Media must be skimmed off the cells once they have stabilized in their plates after which the prepared samples (based on density) must be instilled with 10% FBS medium and then processed in a volume of 2 mL/well on a 6-well plate and 200 μL/well on a 24-well plate. In case of U937 cells, skimming off the medium is not required.
4. Add polymyxin B (1 μg/mL) to inactivate contaminated endotoxin (see Note 2).

3.2. Nanoparticle Exposure

1. LD_{50} should be determined by Cell Counting Kit-8 (DOJINDO Laboratories, Japan) and sublethal concentration is used for cytokine production (see Note 3).
2. Twofold working concentration of nanoparticles are prepared in RPMI-1640 medium without FBS.
3. Mix nanoparticles well. Vortexing for 1 min is enough for silver nanoparticles but silica nanoparticles require ultrasonication for 1 h.
4. Add 200 μL/well of nanoparticles (or same amount of medium containing cells) into a 24-well plate. Make sure that the final concentration of FBS is 5% (see Note 4).

3.3. Collection of Samples

1. After 24 h (or 18 h for the time convenience), it is necessary to separate the cells and nanomaterials from the supernatant by using the centrifugation. However, due to their extremely small size, the process of centrifugation of the nanomaterials is a difficult task. Therefore, if possible, the centrifugation process should be executed at the maximum speed for 3–5 min (see Note 5).
2. After centrifugation, transfer and store the isolated culture supernatant (300 μL) at −80°C in a deep freezer. For each sample, aliquote several tubes and discard the sample once it has been thawed (see Note 6).

3.4. ELISA Procedure

3.4.1. Capture Antibody Coating

1. After adding 100 μL each on a 96 multiwell plate, coat the capture antibody (diluted at the rate of 1:250) overnight in a 4°C humidified darkroom using the coating buffer.

3.4.2. Washing

1. Remove the coating solution by washing the samples 2–3 times with a wash buffer.

3.4.3. Block Plates with Assay Diluents

1. After washing 2–3 times, divide the assay diluents to 200 μL/well each, and run the reaction process for 1 h in dark conditions, at a room temperature.
2. Assay diluents are composed of 10% of FBS in PBS. Make assay diluents as needed when conducting the experiments and keep them at 2–8°C.

3.4.4. Add Standard or Sample

1. Divide the sample and standard to 100 μL/well each and incubate for 2 h in dark conditions at room temperature.
2. Wash the sample 3–5 times with wash buffer.
3. Process all standards and samples with the duplicates or triplicates.
4. Prepare the standards according to the following process:
   (a) After dispensing the well-blended deionized water (1 mL) to standard [Recombinant human IL-8, lyophilized (BD OptEIM™, USA)], divide the same quantity (~50 μL) into a microtube within 15 min at room temperature (see Note 7). It is possible to store the standard stock in a −80°C deep freezer for 6 months; however, it must be used within 8 h if stored at 2–8°C, and never left in room temperature.
   (b) Prepare the standards from the highest down to the lowest concentration after diluting 1/2 times with the assay diluents. A constant temperature of 2–8°C must be maintained when handling the standards, e.g., in case of IL-8, obtain a certain quantity from the stock standard and obtain 600 μL until it reaches 300 pg/mL. From this density level, secure 300 μL each and dilute 1/2 times with assay diluents, i.e., prepare the standards of IL-8 to 300, 150, 75, 37.5, 18.8, 9.4, 4.7 pg/mL and use assay diluents for zero standard. The highest density of the standards fluctuates depending on the type of cytokine; therefore, it must be verified in advance.
5. Prepare the samples according to the following process:
   (a) Dilute each sample to 1/2, 1/5, or 1/10 after dispensing the stock samples and, since the sample is processed with the duplicates or triplicates, prepare ample supply of 300–500 μL. The dilution ratio varies on the type of cells and processed samples of the cells, so it needs to be confirmed in advance.

3.4.5. Washing

1. Aspirate wells.
2. Wash the sample 5 times with wash buffer.
3. After last wash, invert plate and blot on absorbent paper to remove any residual buffer.

3.4.6. Add Detection Antibody

1. Add 100 μl of prepared working detector (detection antibody + streptavidin-HRP reagent) to each well.
2. Incubate for 1 h at a room temperature.

3.4.7. Final Washing

1. Wash the sample 5–7 times with wash buffer. This is the last step of washing process; therefore, it is important to

thoroughly soak each plate for 30–60 sec in a wash buffer at each washing step.

3.4.8. Adding Substrate

1. After dividing the substrate solution (SureBlue TMB Microwell Peroxidase Substrate, KPL, USA) to 100 μL/well, engage a reaction process for 30 min in dark conditions at room temperature.
2. When dividing the substrate solution, replace the pipette tip each time or refrain from touching the well. If the tip touches the fresh substrate solution, the discoloration process will be activated.

3.4.9. Stopping Reactions and Reading

1. If the color of substrate solution turns completely blue after 30 min, add the stop solution to 50 μL/well (1 M H_3PO_4 or 2 N H_2SO_4). If the stop solution is added, the color changes from blue to yellow, and the color becomes darker as the concentration of cytokine gets higher.
2. Analyze it with a spectrophotometer at 450 nm (or preferred wavelength) within 30 min. Before the reading, shake the plate thoroughly in order for the stop solution to blend completely inside the well.
3. Read the OD of control wells containing diluents and media and subtract the background value from the OD of sample or standard wells.

3.5. Analysis of Results

1. First, check the standard curve to confirm if the numerical value of R^2 is measured close to the value of 1.
2. It cannot be used if the numerical value of R^2 is measured under 0.90.

3.6. Limitation of the Procedure

1. If the expression rate of sample cytokine is higher than the maximum density of the standards, dilute the samples and process it again.
2. The interferences and hindrances cannot be eradicated completely due to the combination of soluble receptors and other proteins.
3. The suitability of assessment methods based on nanomaterials need to be determined before the application of ELISA. The colorless nanomaterials do not affect the test results; however, the colored nanomaterials such as reddish-brown Fe_3O_4 and yellow Ag require verification process to assess the impact of nanomaterials on assessment methods.
4. The read absorbance of ELISA is 450 nm, but if the value of OD is measured high due to the interferences from nanomaterials, then select within the least affected wavelength and apply.

4. Notes

1. Assess aggregation of nanoparticles by using dynamic light scattering (DLS) at least once a week or before adding nanoparticles into cell culture. Aggregation or the agglomeration process of silver nanoparticles and iron oxide nanoparticles can be easily observed in a culture medium with a low serum level after 24 h. It is assumed that the low serum density increases the ionic charge of culture solution which decreases the stability of particles.
2. Addition of polymyxin B is to inactivate endotoxin if present. Or estimation of LPS contamination is necessary to clarify the level of lipopolysaccharide.
3. In order to assess cytokine production, it is required to set it up at a lower density level than LD_{50}; therefore, it is necessary to check the viability or cytotoxictic concentration of the applicable nanoparticles first and then process it at lower density after determining LD_{50}.
4. It is imperative to stipulate the conditions of the test, because the density level or the addition of serum could affect the cytokine production of U937 cell.
5. Even after the centrifugation process, the small particles (e.g., 5 nm silver particle) do not completely settle to the bottom. Therefore, we should consider the fact that the extremely small nanoparticles, e.g., (5–10 nm) do not settle in centrifugation and could be included in the supernatant.
6. Check interference of test nanoparticles for optical density before starting the ELISA.
7. Depending on the type of proteins assessed, it is recommended to run a screening process in advance to find out how the nanomaterials induce the type of cytokine production by using cytokine protein array.

References

1. Ryman-Rasmussen JP, Cesta MF, Brody AR, Shipley-Phillips JK, Everitt JI, Tewksbury EW, Moss OR, Wong BA, Dodd DE, Andersen ME, Bonner JC (2009) Inhaled carbon nanotubes reach the subpleural tissue in mice. Nat Nanotechnol 4:747–751
2. Schipper ML, Nakayama-Ratchforod N, Davis CR, Kam NW, Chu P, Liu Z, Sun X, Dai H, Gambhir SS (2008) A pilot toxicology study of single-walled carbon nanotubes in a small sample of mice. Nat Nanotechnol 3:216–221
3. Braydich-Stolle L, Hussain S, Schlager JJ, Hofmann MC (2005) In vitro cytotoxicity of nanoparticles in mammalian germline stem cells. Toxicol Sci 88:412–419
4. AshaRani PV, Low Kah Mun G, Hande MP, Valiyaveettil S (2009) Cytotoxicity and genotoxicity of silver nanoparticles in human cells. ACS Nano 2009(3):279–290
5. Li X, Hu Y, Jin Z, Jiang H, Wen J (2009) Silica-induced TNF-alpha and TGF-beta1 expression in RAW264.7 cells are dependent on Src-ERK/AP-1 pathways. Toxicol Mech Methods 19:51–58
6. Manolova V, Flace A, Bauer M, Schwarz K, Saudan P, Bachmann MF (2008) Nanoparticles

target distinct dendritic cell populations according to their size. Eur J Immunol 38:1404–1413
7. Engvall E, Jonsson K, Perlmann P (1971) Enzyme-linked immunosorbent assay. II. Quantitative assay of protein antigen, immunoglobulin G, by means of enzyme-labelled antigen and antibody-coated tubes. Biochim Biophys Acta 251:427–434
8. Li Y, Schutte RJ, Abu-Shakra A, Reichert WM (2005) Protein array method for assessing in vitro biomaterial-induced cytokine expression. Biomaterials 26:1081–1085
9. Leng SX, McElhaney JE, Walston JD, Xie D, Fedarko NS, Kuchel GA (2008) ELISA and multiplex technologies for cytokine measurement in inflammation and aging research. J Gerontol A Biol Sci Med Sci 63:879–884
10. Huang RP, Huang R, Fan Y, Lin Y (2001) Simultaneous detection of multiple cytokines from conditioned media and patient's sera by an antibody-based protein array system. Anal Biochem 294:55–62

Chapter 10

Metabolomics Techniques in Nanotoxicology Studies

Laura K. Schnackenberg, Jinchun Sun, and Richard D. Beger

Abstract

The rapid growth in the development of nanoparticles for uses in a variety of applications including targeted drug delivery, cancer therapy, imaging, and as biological sensors has led to questions about potential toxicity of such particles to humans. High-throughput methods are necessary to evaluate the potential toxicity of nanoparticles. The omics technologies are particularly well suited to evaluate toxicity in both in vitro and in vivo systems. Metabolomics, specifically, can rapidly screen for biomarkers related to predefined pathways or processes in biofluids and tissues. Specifically, oxidative stress has been implicated as a potential mechanism of toxicity in nanoparticles and is generally difficult to measure by conventional methods. Furthermore, metabolomics can provide mechanistic insight into nanotoxicity. This chapter focuses on the application of both LC/MS and NMR-based metabolomics approaches to study the potential toxicity of nanoparticles.

Key words: Liquid chromatography/mass spectrometry (LC/MS), Metabolic profiling, Metabolomics, Nanoparticles, Nanotoxicology, Nuclear magnetic resonance spectroscopy (NMR)

1. Introduction

The field of nanosciences and related technologies is rapidly growing globally with the sale of products using nanotechnology estimated at one trillion dollars per year by 2015 (1–3). One major issue with the field of nanosciences is the risk of potential toxicity of particles at the nanometer scale. The field of toxicology plays a major role in the field of nanosciences in that methods can be employed to answer such questions as how nanomaterials affect an organism. Toxicology has long addressed the toxicity of chemical substances and drugs and the effects of particles on the micrometer and ultrafine scales, establishing itself as the ideal scientific discipline to assess the toxicity of nanoparticles (4, 5). Oberdörster (6) defined nanotoxicology as "the study of the adverse effects of engineered nanomaterials on living organisms and the ecosystems, including the prevention and amelioration of such adverse effects."

Joshua Reineke (ed.), *Nanotoxicity: Methods and Protocols*, Methods in Molecular Biology, vol. 926,
DOI 10.1007/978-1-62703-002-1_10, © Springer Science+Business Media, LLC 2012

To date, there have been no reports of toxicity or disease ascribed to nanomaterials in humans. Experimental evidence, however, indicates the likelihood for adverse responses due in part to the potential of nanomaterials to generate reactive oxygen species (ROS). Research on ultrafine particles, diesel exhaust particles, quartz, and asbestos has shown their potential to affect health adversely. Therefore, it is reasonable to assume that the novel physicochemical properties of nanomaterials may raise new toxicological concerns. Examples of human exposure include occupational encounters in manufacturing processes, use of sunscreens, toothpastes, cosmetic products containing nanoparticles, or from use of nanomaterials for drug therapy or imaging (3). There has also been an emphasis on the use of nanomaterials for both diagnostic and therapeutic purposes including targeted drug delivery, imaging, cancer therapy, and as biological sensors (7). However, the ability to specifically tune the physical and chemical properties of nanomaterials is also what makes them potentially toxic to cells and tissues (8). One potential for adverse effects is due in part to the large surface area-to-volume ratio of nanomaterials, facilitating reactions with biomolecules that could result in generation of ROS, conformational and membrane permeability changes, and mutations for example. Structure–activity relationships can be exploited to screen for nanoparticles toxicity by looking at interactions with cell membrane, cell uptake, and subcellular localization (9, 10). Nanoparticle uptake can also be examined using imaging techniques and fluorescence spectroscopy (11). Additionally, in a recent review, Xia et al. (3) described the importance of screening for ROS and oxidative stress in relation to the toxicity of nanomaterials. The generation of ROS can result in responses such as inflammation, apoptosis, necrosis, fibrosis, hypertrophy, metaplasia, and carcinogenesis (9).

Classical approaches to study toxicity do have some weakness whether studying drugs, chemicals, or nanoparticles. Therefore, high-throughput methods for screening of nanomaterials are essential for nanotoxicity assessment (3). Omics technologies (transcriptomics, proteomics, and metabolomics) can provide high-throughput screening information about endogenous markers in both in vitro and in vivo systems. Metabolomics, specifically, allows rapid screening for biomarkers of oxidative stress following introduction of nanomaterials into the system. In general, metabolic profiling involves investigating changes in the concentrations of endogenous metabolites within tissues or biofluids following a perturbation. The terms metabolomics and metabonomics have been applied to metabolic profiling where metabolomics refers to the measurable metabolite pool that exists within a cell or tissue under a particular set of environmental conditions (12). Metabonomics refers to "quantitative measurement of the dynamic multiparametric metabolic response of living systems to pathophysiological stimuli or genetic modification (13)." Metabolic profiling permits identification of biomarkers or patterns of biomarker changes related to nanotoxicity in biofluid samples such as urine and blood that can be collected with relative ease. The overall pattern of biomarkers is expected to differ based on the origin of the toxicity, specifically the organ that is damaged by a particular drug compound (14–17). Therefore, metabolomics can provide mechanistic insight into nanotoxicity.

2. Toxic Responses to Nanomaterials

Common responses that have been noted in both in vivo and in vitro systems following exposure to nanomaterials include an inflammatory response and induction of oxidative stress. Nanomaterials that contain transition metals may induce oxidative stress through interaction of reactive species generated during the inflammatory response. Fujita et al. (18) performed gene expression profiling on lung tissue from rats exposed to C(60) fullerenes through inhalation. The results showed that genes involved in an inflammatory response were upregulated. A separate study investigated the effects of zinc oxide (ZnO) particles on human umbilical vein and endothelial cells (19). A dose-dependent increase in oxidized glutathione levels was noted along with an increase in the expression of intercellular adhesion molecule-1 (ICAM-1), indicative of vascular endothelium inflammation. Semete et al. (20) reported the induction of cytokine production by nanoparticles engineered for oral administration. Deng et al. (21) demonstrated that negatively charged poly(acrylic acid)-conjugated gold nanoparticles can bind to fibrinogen and ultimately lead to the release of inflammatory cytokines.

3. Traditional Analytical Methods to Assess Toxicity of Nanoparticles

In vitro nanotoxicity experiments generally assess cell viability including proliferation, necrosis, apoptosis, or specific toxicity mechanisms such as DNA damage or oxidative stress using various assays (22). The most common assay to evaluate proliferation involves the cellular reduction of tetrazolium salts to produce formazan dyes. However, the mechanism of reduction of salts is not well understood. Numerous external factors such as media pH or additives to the cell culture can result in unreliable results. Thus, membrane integrity following exposure to nanoparticles is generally evaluated by monitoring the uptake of a dye such as Trypan Blue, Neutral Red, or propidium dye or by measuring the leakage of active enzymes into cell media via the LDH assay. Apoptosis assays include evaluation of morphological changes by light microscopy, the annexin-V assay, DNA laddering, the Comet assay, or the Tunel assay. The Comet assay, reported in multiple studies of nanoparticles toxicity (23–29), is most commonly used to assess DNA damage following exposure to a toxicant. Multiple studies of oxidative stress following nanoparticle exposure have been reported (26, 30–40). Measurements of oxidative stress using common methods are problematic due to high cost of instrumentation or in the case where fluorescent probe molecules are employed wherein high reactivity to a wide variety of reactive species is observed. Oxidative stress can also be determined by lipid and protein peroxidation or by the presence of DNA fragments. Other oxidative stress biomarkers include alterations in superoxide dismutase or glutathione production. Metabolomics allows detection of glutathione as well as other metabolites within the glutathione synthesis pathway, which may serve as better biomarkers of oxidative stress. Further, such markers can be applied to study both in vitro and in vivo systems.

While in vitro assays can be categorized as either viability or mechanistic studies, in vivo assays focus on blood serum chemistry and cell population, tissue morphology, or nanoparticles biodistribution. Biodistribution experiments can be useful in providing insight into localization, retention, and excretion of nanomaterials from the animal. However, such experiments generally rely on the use of a fluorescent tag, which may alter the biodistribution and clearance of nanomaterials. Another common method of assessing toxicity in vivo is through measuring serum chemistry parameters. However, although such measurements can detect many toxic agents, such biomarkers are not infallible. Finally, histological examination following killing of animals in a preclinical study is widely used to evaluate the toxicity of a compound or nanomaterial. However, histology results can be limited by subjectivity or introduction of artifacts in the preparation of the tissue for analysis. Further, the damage must be sufficient to be visualized by light microscopy, which might not always be the case. In vivo nanotoxicity studies have also been accomplished using newer methods such as microfluidics (41–44) and microelectrochemistry (45–47) to allow for dynamic measurements to originate directly from the animal. Metabolomics technologies also have the potential to provide more dynamic measurements of toxicity through sampling at multiple time intervals from the same subject. Further, the collection of biofluid samples are performed in a minimally invasive manner and do not rely upon implantation of a probe device, which is necessary for microfluidics and microelectrochemistry studies.

4. Metabolomics Methods

One of the major research focuses on metabolomics is the identification of early biomarkers of toxicity based on temporal studies of biofluids such as urine and serum. Figure 1 illustrates the basic protocol for a metabolomics analysis of samples generated from a nanotoxicity study through statistical analysis of spectral data to generate pathway information or to identify new toxicity biomarkers. Briefly, tissue or biofluid samples are collected and prepared for analysis following a specified dosing regimen, and then data are acquired on either one or more analytical platforms. Following data acquisition and processing, metabolites or differences in peak area are noted and used for statistical analyses to identify biomarkers of nanotoxicity and/or identify metabolic pathways that have been perturbed. Depending on the goal of a metabolomics investigation, biomarkers obtained from a metabolomics study can be used for predictive, prognostic, or diagnostic purposes. One advantage of using a temporal study with the same animal is that it is not necessary to understand the pharmacodynamics and pharmacokinetics of the toxin prior to the biomarker investigation. This is especially beneficial in the investigation of toxicity caused by nanoparticles since little is known about their kinetics or mechanisms of action. Further, temporal metabolic profiling studies may be carried out to determine whether the toxic insult to an animal causes temporary changes in the metabolic profile or sustained alterations that might potentiate susceptibility to a secondary insult. Since there is a wide concentration range and large chemical diversity in

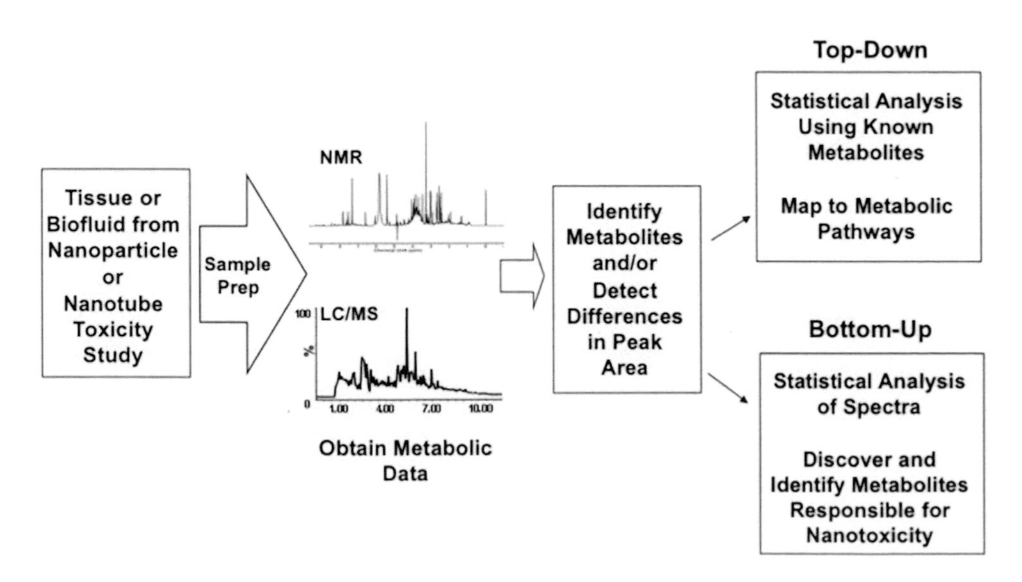

Fig. 1. General flow of a metabolomics study of nanotoxicity from sample collection and preparation through data analysis describing both the "top-down" and "bottom-up" approaches.

the metabolites in cells, tissues, and biofluids, many analytical techniques have been employed to detect the global changes induced by a toxin. The most commonly employed methods for metabolomics include liquid chromatography/mass spectrometry (LC/MS) techniques and nuclear magnetic resonance (NMR) spectroscopy. The advantages and disadvantages of the LC/MS and NMR platforms have been thoroughly reviewed (48–50). Other analytical techniques used in metabolomics are gas chromatography coupled to mass spectrometry (GC/MS), thin-layer chromatography coupled to mass spectrometry (TLC/MS), matrix-assisted laser desorption/ionization mass spectrometry (MALDI/MS), and Fourier transform mass spectrometry (FTMS).

4.1. Liquid-Chromatography/Mass Spectrometry

LC/MS offers greater sensitivity than NMR although there are several disadvantages to this technique including ion suppression or enhancement, lower reproducibility, not as quantitative by nature, and the ionization method will determine the types of analytes seen. The mass spectrum may consist of several thousand identifiable endogenous metabolites that may aid in the elucidation of mechanisms of toxicity. Recent advances in MS instrumentation and technology that have resulted in instruments with enhanced reproducibility and high mass measurement accuracy have increased the application of MS in metabolic profiling studies (51–56).

A major strength of MS methods is the greater sensitivity, which can detect metabolites at concentrations down to the picogram level. Since biofluid and tissue samples contain a wide range of molecules including acidic, basic, amphoteric, and neutral components, the ionization method and acquisition parameters will directly influence the types of metabolites that can be measured, making it possible that several experiments are required for a complete analysis. No single ionization technique in one analytical run is sufficient to detect every type of compound present in the complex biological matrices. Further issues with MS analyses include the effects of ion suppression or enhancement that can affect quantitation and reproducibility of the data.

Due to its high data dimensionality, MS can also be used as an imaging technique. Burnum et al. (57) described the use of matrix-assisted laser desorption/ionization (MALDI) MS for imaging of tissue samples. MALDI MS imaging can provide valuable information on peptides, proteins, lipids, metabolites, xenobiotics, and other endogenous compounds in complex tissue samples. Moreover, MALDI MS imaging includes spatial information that is lost when using techniques that require homogenization. The ability to detect not only global changes in metabolites in tissue slices without homogenization, but to provide a snapshot of how nanomaterial is distributed in tissues, makes MALDI MS imaging a potentially powerful tool for nanotoxicology.

4.2. Nuclear Magnetic Resonance Spectroscopy

NMR has the capacity to detect hundreds of metabolites in urine, serum, and tissues samples in a short amount of time. The use of NMR spectroscopy for metabolic profiling studies has grown due to the increased sensitivity offered by the availability of higher field strength instruments along with the introduction of the cryoprobe. NMR is highly advantageous for metabolomic studies owing to the fact that all the protons within a molecule have reproducible patterns and chemical shifts, and the data are quantitative by nature. Moreover, the data acquired using NMR are highly reproducible in both intra- and inter-laboratory experiments with less than 2 % variation between laboratories using spectrometers with different field strengths (58, 59). These advances in NMR technology have made it possible to measure hundreds of metabolites within several minutes. One weakness with NMR is that there is a great deal of overlap in the chemical shifts for many metabolites, and because of this additional NMR or MS studies are sometimes necessary to identify a metabolite of interest. Another limitation of NMR methods is its low sensitivity, which is unable to detect metabolites present at submicromolar concentrations.

Solid-state NMR studies using magic angle spinning (MAS) technology make it possible to measure intact tissue samples in a nondestructive manner, which is not achievable with MS. MAS NMR is a technique that can examine metabolic profiles in tissue samples. If additional biofluids are also analyzed using standard NMR spectroscopy, it may be possible to establish a direct link between the biofluid metabolic profile and metabolic profiles from tissue samples.

It is unrealistic to think that one single analytical technique, NMR or MS, can elucidate all of the metabolic responses to external stimuli such as exposure to a nanotoxin. However, the combination of both techniques on a sample set is able to provide a powerful means of elucidating changes in the metabolome due to nanotoxicity. These changes can be assessed in terms of the molecular pathways being perturbed and allow for the elucidation of the mechanism(s) at work following exposure to nanomaterials. NMR and MS are complimentary techniques that may have some overlap in identified metabolites, but may also identify unique metabolites on their own. The ability to link metabolites found using both techniques would increase one's confidence in biomarker identification, pathway analysis, and elucidation of toxic mechanism(s) of a particular material.

4.3. Analysis of Metabolomics Data

The analysis of metabolomics data generally follows one of two different protocols as shown in Fig. 1. One data analysis procedure, referred to as a "top-down" approach, uses public and in-house spectral databases to detect and provide quantitative or semiquantitative data for a specific set of metabolites. The second data analysis procedure analyzes the spectral data directly from several distinct treatment groups. The data are analyzed in such a way that potential new biomarkers or patterns can be found. Standards, spectral databases, and/or more analytical analyses are employed to determine the identity of a potential new biomarker or set of biomarkers. This type of data analysis is known as a "bottom-up" approach. In either case, the list of metabolites or spectral features can be further analyzed by unsupervised or supervised chemometric procedures.

In order to extract maximum information in complex spectroscopic data, a strong chemometrics support component is required. Chemometric analyses are necessary in order to develop statistical pattern recognition models, achieve optimal characterization of samples, and detect biomarkers from the diverse, highly dimensional "omics" datasets. "Chemometrics" is defined as the application of mathematical and statistical methods to chemistry (60, 61).

In the cases of NMR- and LC/MS-based metabolomics, many mathematical or statistical tools are involved to process spectra that include functions such as peak alignment, peak detection, and normalization. Efforts are continuously being made to use quality controls to aid in alignment of the NMR or LC/MS spectra collected from various samples in different analytical runs. Alignment of LC chromatograms is generally more problematic than NMR spectra due to the relatively lower reproducibility of the chromatography techniques including GC and LC. Peaks (chemical shifts for NMR and m/z for MS and retention times for GC or LC) can be detected and extracted from raw datasets. NMR peak signal intensities are determined by aligning the NMR spectra to a chemical shift standard and dividing the spectrum into discrete regions (referred to "bins" or "buckets") and integrating areas under the curve, which correspond to the metabolite concentrations in the samples. Generally, 0.04 ppm-wide bins are used, which will generate 200–250 "buckets" of data from a typical 10 ppm NMR spectrum. Similarly, peak areas of a particular m/z marker at certain retention times are integrated. Peak intensities must be normalized in each individual sample to take into account the effects of the sample dilution (for instance urine dilution) prior to any multivariate analysis. Total spectral intensity, creatinine levels for urine analyses, and other more sophisticated methods are currently used to normalize NMR-based and MS-based metabolomics data sets (62–66).

One simple and widely used pattern recognition technique is principal component analysis (PCA). PCA is an unsupervised clustering algorithm, which is capable of classifying sample groups based on their inherent similarity or dissimilarity corresponding to their biochemical composition. The major strength of the PCA technique is that no prior knowledge about the class of the samples is required. PCA can condense multivariate datasets into two types of plots known as the scores plot and the loadings plot. Figure 2a, b shows examples of PCA scores and loadings plots based on NMR

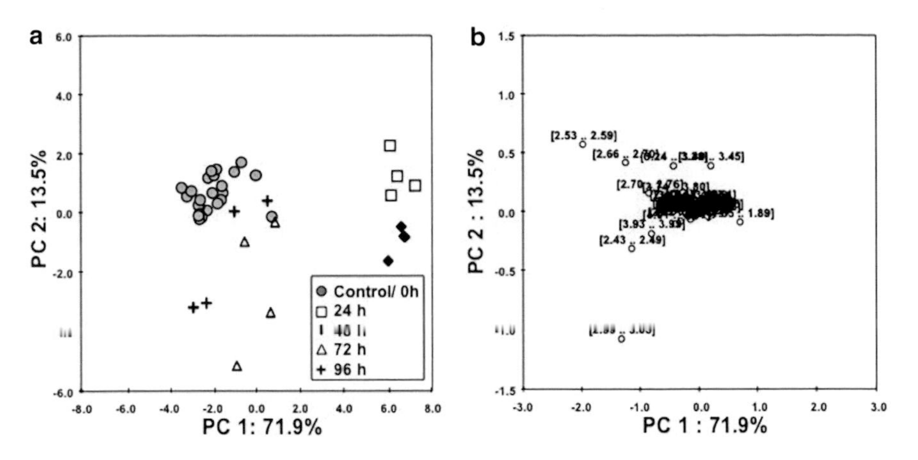

Fig. 2. Example PCA (**a**) scores and (**b**) loadings plot from a toxicity study. The scores plot linearly reflects the combinations of features detected in a raw spectrum. The loadings plot shows the weight of an individual feature, in this case an NMR spectral bin, which is used to interpret the pattern in the scores plot. For MS-based analyses, the features in a loadings plot would be related to a retention time and *m/z* pair.

data from a toxicity study. The scores plot (Fig. 2a), which linearly reflects the combinations of features detected in a raw spectrum can be used to establish models between sample groups. PCA can be based on the spectral features such as integrated bins from an NMR spectrum (Fig. 2b) or a LC or GC retention time and *m/z* pair. PCA can also be based on concentrations or peak area data from identified metabolites in a metabolomics approach. The loadings plot (Fig. 2b) shows the weight of an individual feature, which is used to interpret the pattern in the scores plot. A large number of spectra can be simultaneously compared with PCA, thus metabolic changes induced by nanotoxicity can be rapidly detected. Supervised methods, including partial least-square discriminant analysis (PLS-DA) (67, 68) and orthogonal signal correction (OSC) (69–71), are other statistical methods often used in metabolomic data mining. Both PCA and PLS-DA are very useful for pattern recognition after condensing variable dimensions through the combination of some or all features in sets of NMR and/or MS spectra.

PCA is a statistical technique that illustrates the largest changes in a data set, so it would not provide an accurate modeling system if the changes in biomarkers of interest are subtle compared to other changes detected during acquisition. In order to make useful comparisons of trajectories, defined as the systematic response to a physical perturbation over time, obtained from studies of different drugs as well as studies done in different species, it is necessary to scale and align the different trajectories. Keun et al. (72) introduced a procedure that permits direct comparison of trajectories, which is referred to as scaled-to-maximum, aligned, and reduced trajectories (SMART) analysis. The SMART analysis procedure first aligns the principal component trajectories by subtracting the pre-dose values from the remaining time points such that all trajectories start at the origin. Following pre-dose subtraction, the maximum differences between trajectories are used to scale them. Finally, the average trajectory for each dataset is calculated and compared. Not only does SMART analysis allow for comparison between data obtained in different studies and for different species or strains, but it also allows for a comparison of different dose levels of one drug. SMART analysis attempts to map one metabolic trajectory onto another metabolic trajectory to determine whether there are similar toxic mechanisms that occur between nanoparticles, dose levels, species, or strains. If there is a similar mechanism at work, the trajectories under comparison will coincide. SMART analysis was initially applied to data acquired using an NMR platform and made possible because of the linear response in NMR data. However, SMART analysis can be applied for trajectory analysis of LC/MS data or a set of known metabolites concentrations. This type of analysis has the potential to remove inter-laboratory, physiological, and phenotypical variations, to correlate dose–response data, and to correlate the toxic response in different species or strains of the same species.

4.4. Applications of Metabolomics to Study Nanoparticles

Silica (SiO_2) nanoparticles are widely used and can enter biological systems through multiple routes. They have been used in industrial manufacturing, disease labeling, drug delivery, cancer therapy, and biosensors (73). However, there have been reports of inflammatory responses, hepatotoxicity, and fibrosis following exposure (74–76). In a recent study, Lu et al. (73) applied a GC/MS-based metabonomics method to investigate the size- and dose–response following exposure to SiO_2 particles in mice. Histology and clinical chemistry results indicated a dose-dependent toxicity of SiO_2 particles 30 (SP30), 70 (SP70), and 300 (SP300) nm in diameter. GC/MS-based metabonomics analyses of liver tissue and serum showed similar changes in the metabolic response regardless of particle size and in some cases, prior to a traditional pathological change. Figure 3a, b shows the PCA scores plots from GC/MS data of liver and serum samples, respectively, for a control group, 10 mg/kg SP30, 40 mg/kg SP70 group, and 200 mg/kg SP300 group. All three dose groups clearly separated from control with the largest change in the SP70 group. Ten liver metabolites were found to be significantly changed in the same manner in each dose group and nine metabolites in the serum. PCA scores plots based on the significant metabolites for liver and serum are shown in Fig. 3c, d, respectively. The scores plots indicate that the selected metabolites for both sample type were used in separating the treated groups from the control group. Changes were noted in metabolites involved in energy, amino acid, lipid, and nucleotide metabolism. The significantly altered metabolites indicated impairment of the Krebs cycle and were also consistent with the occurrence of oxidative stress. The results demonstrated enhanced sensitivity of metabolomics methods over conventional histopathology and serum chemistry analysis.

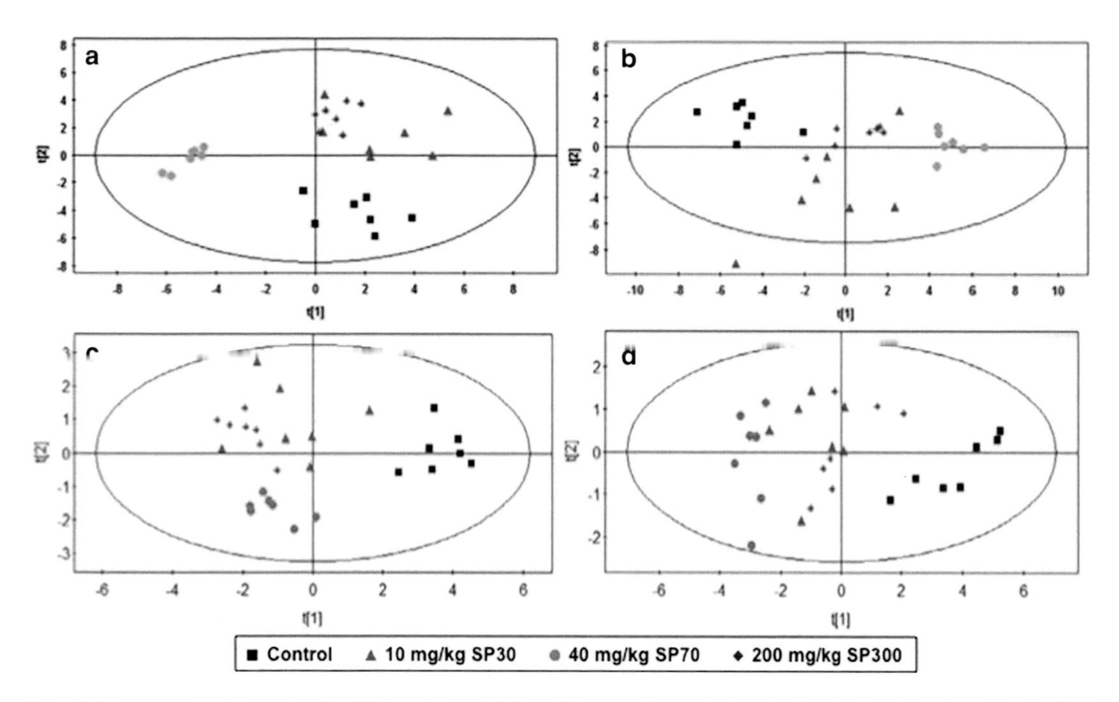

Fig. 3. PCA scores plots based on GC/MS data from (**a**) liver (**b**) serum for control and animals dosed with 10 mg/kg SP30, 40 mg/kg SP70, or 200 mg/kg SP300. (**c**) PCA scores plot based on ten significant metabolites from GC/MS analysis of liver samples. (**d**) PCA scores plot based on nine significant metabolites from GC/MS analysis of serum samples. Figures reproduced with permission from (73). IOP Publishing Ltd.

Tang et al. (77) evaluated urine samples using ^{1}H NMR to investigate the toxicological effects of titanium dioxide (TiO_2) in rats exposed intratracheally. Biochemical assays of the blood and histopathological analyses were also performed. Intratracheal instillation of 5 nm-TiO_2 particles altered the urinary metabolic profile with significant increases in lactate, taurine, acetate, and citrate noted. There was also a significant decrease in 2-oxoglutarate. All of the changes noted occurred within 8 h of exposure with the metabolic profiles showing a recovery response followed by a return toward normal levels at the endpoint of the experiment. The serum levels of alanine aminotransferase (ALT), blood urea nitrogen (BUN), and creatinine were consistent with the NMR analyses showing an increase within 8 h followed by recovery to normal values within 7 days.

5. Other Omics Methods

Recent reports have included gene expression analysis to investigate in vitro nanotoxicity (78, 79). Fowler and coworkers (80) investigated proteomic biomarkers in renal tubule cell cultures derived from male and female hamsters that had been exposed to gallium arsenide (GaAs) or indium arsenide (InAs) particles. While the particles used were not on the nanoscale (~5 mm in diameter), the results were relevant due to the phenomenon of agglomeration of nanoparticles into masses of similar size to the GaAs and InAs particles used in the study. Marked particle-specific changes in protein expression were noted in both males and females with an inhibition of protein synthesis in males following exposure to InAs.

Copper nanoparticles have a broad range of applications from industrial products to medical drugs and devices. Some studies have indicated that copper nanoparticles are acutely toxic to mice and may be a potential risk to human health. Yang et al. (81) report gene expression profile analysis of liver tissue from rats, which were orally administered with different doses of nano-copper or micro-copper for 5 days. A dose of 200 mg/kg nano-copper caused overt toxicity in male Wistar rats, while the same dose of micro-copper showed no overt pathology. Gene expression analysis showed 1,011 differentially expressed genes in the 200 mg/kg nano-copper dose group related to oxidoreductase activity, especially relevant in oxidative stress, energy metabolism, and signal transduction. Gene expression patterns induced by exposure to a subtoxic dose of nano-copper indicated cell stress or subtle cell injury that may predict overt toxicity at higher dose levels. A systems biology approach with gene expression, proteomics, and metabolomics may provide a means to better understand nanomaterial-induced toxicity.

6. Conclusions

While nanotoxicology is still an emerging field, metabolomics techniques reveal immense potential to evaluate the effects and toxicity of nanomaterials. A combination of NMR- and LC/MS-based metabolomics analyses can be used to evaluate the global changes within the metabolome of tissue and biofluid samples. Changes can also be traced through the application of chemometrics methods that can be used to extract potential biomarkers of nanoparticle-induced toxicity. Metabolomics methods can also be used to elucidate mechanisms of action with respect to nanoparticle-induced toxicity. Specifically, metabolomics can be applied to investigate specific pathways of interest including those related to oxidative stress, which has been reported following nanoparticle exposure and is generally difficult to measure using traditional methods. Furthermore, MALDI MS imaging can provide information about how nanomaterials are localized in particular tissues, which is also important for characterizing and understanding nanoparticle toxicity. In summary, the omics techniques can greatly impact the emerging field of nanotoxicology especially with regards to assessing the potential for adverse effects in humans.

Acknowledgments and Disclaimer

The views presented in this chapter do not necessarily reflect those of the U.S. Food and Drug Administration.

References

1. Maynard AD, Aitken RJ, Butz T, Colvin V, Donaldson K, Oberdorster G, Philbert MA, Ryan J, Seaton A, Stone V, Tinkle SS, Tran L, Walker NJ, Warheit DB (2006) Safe handling of nanotechnology. Nature 444:267–269
2. Meng H, Xia T, George S, Nel AE (2009) A predictive toxicological paradigm for the safety assessment of nanomaterials. ACS Nano 3:1620–1627
3. Xia T, Li N, Nel AE (2009) Potential health impact of nanoparticles. Annu Rev Public Health 30:137–150
4. Borm PJ (2002) Particle toxicology: from coal mining to nanotechnology. Inhal Toxicol 14:311–324
5. Borm PJ, Kreyling W (2004) Toxicological hazards of inhaled nanoparticles–potential implications for drug delivery. J Nanosci Nanotechnol 4:521–531
6. Oberdorster G (2010) Safety assessment for nanotechnology and nanomedicine: concepts of nanotoxicology. J Intern Med 267:89–105
7. Schrand AM, Rahman MF, Hussain SM, Schlager JJ, Smith DA, Syed AF (2010) Metal-based nanoparticles and their toxicity assessment. Wiley Interdiscip Rev Nanomed Nanobiotechnol 2:544–568
8. Kagan VE, Bayir H, Shvedova AA (2005) Nanomedicine and nanotoxicology: two sides of the same coin. Nanomedicine 1:313–316
9. Nel A, Xia T, Madler L, Li N (2006) Toxic potential of materials at the nanolevel. Science 311:622–627
10. Oberdorster G, Oberdorster E, Oberdorster J (2005) Nanotoxicology: an emerging discipline evolving from studies of ultrafine particles. Environ Health Perspect 113:823–839
11. Chithrani BD, Chan WC (2007) Elucidating the mechanism of cellular uptake and removal of protein-coated gold nanoparticles of different sizes and shapes. Nano Lett 7:1542–1550
12. Fiehn O (2002) Metabolomics–the link between genotypes and phenotypes. Plant Mol Biol 48:155–171
13. Nicholson JK, Lindon JC, Holmes E (1999) 'Metabonomics': understanding the metabolic responses of living systems to pathophysiological stimuli via multivariate statistical analysis of biological NMR spectroscopic data. Xenobiotica 29:1181–1189
14. Ebbels TM, Keun HC, Beckonert OP, Bollard ME, Lindon JC, Holmes E, Nicholson JK (2007) Prediction and classification of drug toxicity using probabilistic modeling of temporal metabolic data: the consortium on metabonomic toxicology screening approach. J Proteome Res 6:4407–4422
15. Lindon JC, Keun H, Ebbels TMD, Pearce JMT, Holmes E, Nicholson JK (2005) The consortium for metabonomic toxicology (COMET): aims, activities and achievements. Pharmacogenomics 6:691–699
16. Nicholson J, Keun H, Ebbels T (2007) COMET and the challenge of drug safety screening. J Proteome Res 6:4098–4099
17. Robertson DG, Reily MD, Sigler RE, Wells DF, Paterson DA, Braden TK (2000) Metabonomics: evaluation of nuclear magnetic resonance (NMR) and pattern recognition technology for rapid *in vivo* screening of liver and kidney toxicants. Toxicol Sci 57:326–337
18. Fujita K, Morimoto Y, Endoh S, Uchida K, Fukui H, Ogami A, Tanaka I, Horie M, Yoshida Y, Iwahashi H, Nakanishi J (2010) Identification of potential biomarkers from gene expression profiles in rat lungs intratracheally instilled with C(60) fullerenes. Toxicology 274:34–41
19. Tsou TC, Yeh SC, Tsai FY, Lin HJ, Cheng TJ, Chao HR, Tai LA (2010) Zinc oxide particles induce inflammatory responses in vascular endothelial cells via NF-kappaB signaling. J Hazard Mater 183:182–188
20. Semete B, Booysen LI, Kalombo L, Venter JD, Katata L, Ramalapa B, Verschoor JA, Swai H (2010) In vivo uptake and acute immune response to orally administered chitosan and PEG coated PLGA nanoparticles. Toxicol Appl Pharmacol 249:158–165
21. Deng ZJ, Liang M, Monteiro M, Toth I, Minchin RF (2010) Nanoparticle-induced unfolding of fibrinogen promotes Mac-1 receptor activation and inflammation. Nat Nanotechnol 6:39–44
22. Marquis BJ, Love SA, Braun KL, Haynes CL (2009) Analytical methods to assess nanoparticle toxicity. Analyst 134:425–439

23. Barnes CA, Elsaesser A, Arkusz J, Smok A, Palus J, Lesniak A, Salvati A, Hanrahan JP, Jong WH, Dziubaltowska E, Stepnik M, Rydzynski K, McKerr G, Lynch I, Dawson KA, Howard CV (2008) Reproducible comet assay of amorphous silica nanoparticles detects no genotoxicity. Nano Lett 8:3069–3074
24. Colognato R, Bonelli A, Ponti J, Farina M, Bergamaschi E, Sabbioni E, Migliore L (2008) Comparative genotoxicity of cobalt nanoparticles and ions on human peripheral leukocytes in vitro. Mutagenesis 23:377–382
25. Jacobsen NR, Pojana G, White P, Moller P, Cohn CA, Korsholm KS, Vogel U, Marcomini A, Loft S, Wallin H (2008) Genotoxicity, cytotoxicity, and reactive oxygen species induced by single-walled carbon nanotubes and C(60) fullerenes in the FE1-Mutatrade markMouse lung epithelial cells. Environ Mol Mutagen 49:476–487
26. Kang SJ, Kim BM, Lee YJ, Chung HW (2008) Titanium dioxide nanoparticles trigger p53-mediated damage response in peripheral blood lymphocytes. Environ Mol Mutagen 49: 399–405
27. Karlsson HL, Cronholm P, Gustafsson J, Moller L (2008) Copper oxide nanoparticles are highly toxic: a comparison between metal oxide nanoparticles and carbon nanotubes. Chem Res Toxicol 21:1726–1732
28. Kisin ER, Murray AR, Keane MJ, Shi XC, Schwegler-Berry D, Gorelik O, Arepalli S, Castranova V, Wallace WE, Kagan VE, Shvedova AA (2007) Single-walled carbon nanotubes: geno- and cytotoxic effects in lung fibroblast V79 cells. J Toxicol Environ Health A 70:2071–2079
29. Mroz P, Pawlak A, Satti M, Lee H, Wharton T, Gali H, Sarna T, Hamblin MR (2007) Functionalized fullerenes mediate photodynamic killing of cancer cells: type I versus Type II photochemical mechanism. Free Radic Biol Med 43:711–719
30. Choi O, Hu Z (2008) Size dependent and reactive oxygen species related nanosilver toxicity to nitrifying bacteria. Environ Sci Technol 42:4583–4588
31. Clarke SJ, Hollmann CA, Zhang Z, Suffern D, Bradforth SE, Dimitrijevic NM, Minarik WG, Nadeau JL (2006) Photophysics of dopamine-modified quantum dots and effects on biological systems. Nat Mater 5:409–417
32. Isakovic A, Markovic Z, Todorovic-Markovic B, Nikolic N, Vranjes-Djuric S, Mirkovic M, Dramicanin M, Harhaji L, Raicevic N, Nikolic Z, Trajkovic V (2006) Distinct cytotoxic mechanisms of pristine versus hydroxylated fullerene. Toxicol Sci 91:173–183
33. Lee KJ, Nallathamby PD, Browning LM, Osgood CJ, Xu XH (2007) In vivo imaging of transport and biocompatibility of single silver nanoparticles in early development of zebrafish embryos. ACS Nano 1:133–143
34. Lin W, Huang YW, Zhou XD, Ma Y (2006) In vitro toxicity of silica nanoparticles in human lung cancer cells. Toxicol Appl Pharmacol 217:252–259
35. Long TC, Saleh N, Tilton RD, Lowry GV, Veronesi B (2006) Titanium dioxide (P25) produces reactive oxygen species in immortalized brain microglia (BV2): implications for nanoparticle neurotoxicity. Environ Sci Technol 40:4346–4352
36. Pulskamp K, Diabate S, Krug HF (2007) Carbon nanotubes show no sign of acute toxicity but induce intracellular reactive oxygen species in dependence on contaminants. Toxicol Lett 168:58–74
37. Sayes CM, Gobin AM, Ausman KD, Mendez J, West JL, Colvin VL (2005) Nano-C60 cytotoxicity is due to lipid peroxidation. Biomaterials 26:7587–7595
38. Schrand AM, Braydich-Stolle LK, Schlager JJ, Dai LM, Hussain SM (2008) Can silver nanoparticles be useful as potential biological labels? Nanotechnology 19:235104, 13 pp
39. Shukla R, Bansal V, Chaudhary M, Basu A, Bhonde RR, Sastry M (2005) Biocompatibility of gold nanoparticles and their endocytotic fate inside the cellular compartment: a microscopic overview. Langmuir 21:10644–10654
40. Wick P, Manser P, Limbach LK, Dettlaff-Weglikowska U, Krumeich F, Roth S, Stark WJ, Bruinink A (2007) The degree and kind of agglomeration affect carbon nanotube cytotoxicity. Toxicol Lett 168:121–131
41. Cellar NA, Burns ST, Meiners JC, Chen H, Kennedy RT (2005) Microfluidic chip for low-flow push-pull perfusion sampling in vivo with on-line analysis of amino acids. Anal Chem 77:7067–7073
42. Hogan BL, Lunte SM, Stobaugh JF, Lunte CE (1994) On-line coupling of in vivo microdialysis sampling with capillary electrophoresis. Anal Chem 66:596–602
43. Sandlin ZD, Shou M, Shackman JG, Kennedy RT (2005) Microfluidic electrophoresis chip coupled to microdialysis for in vivo monitoring of amino acid neurotransmitters. Anal Chem 77:7702–7708
44. Zhou SY, Zuo H, Stobaugh JF, Lunte CE, Lunte SM (1995) Continuous in vivo monitoring of amino acid neurotransmitters by microdialysis sampling with on-line derivatization and capillary electrophoresis separation. Anal Chem 67:594–599

45. Aragona BJ, Cleaveland NA, Stuber GD, Day JJ, Carelli RM, Wightman RM (2008) Preferential enhancement of dopamine transmission within the nucleus accumbens shell by cocaine is attributable to a direct increase in phasic dopamine release events. J Neurosci 28: 8821–8831
46. Cheer JF, Aragona BJ, Heien ML, Seipel AT, Carelli RM, Wightman RM (2007) Coordinated accumbal dopamine release and neural activity drive goal-directed behavior. Neuron 54: 237–244
47. Ewing AG, Bigelow JC, Wightman RM (1983) Direct in vivo monitoring of dopamine released from two striatal compartments in the rat. Science 221:169–171
48. Dunn WB, Ellis DI (2005) Metabolomics: current analytical platforms and methodologies. Trends Anal Chem 24:285–294
49. Lenz EM, Wilson ID (2007) Analytical strategies in metabonomics. J Proteome Res 6:443–458
50. Robertson DG (2005) Metabonomics in toxicology: a review. Toxicol Sci 85:809–822
51. Plumb RS, Granger JH, Stumpf CL, Johnson KA, Smith BW, Gaulitz S, Wilson ID, Castro-Perez J (2005) A rapid screening approach to metabonomics using UPLC and oa-TOF mass spectrometry: application to age, gender and diurnal variation in normal/Zucker obese rats and black, white and nude mice. Analyst 130: 844–849
52. Plumb RS, Johnson KA, Rainville P, Shockcor JP, Williams R, Granger JH, Wilson ID (2006) The detection of phenotypic differences in the metabolic plasma profile of three strains of Zucker rats at 20 weeks of age using ultra-performance liquid chromatography/orthogonal acceleration time-of-flight mass spectrometry. Rapid Commun Mass Spectrom 20:2800–2806
53. Plumb RS, Jones MD, Rainville PD, Nicholson JK (2008) A rapid simple approach to screening pharmaceutical products using ultra-performance LC coupled to time-of-flight mass spectrometry and pattern recognition. J Chromatogr Sci 46:193–198
54. Plumb RS, Rainville PD, Potts WB 3rd, Johnson KA, Gika E, Wilson ID (2009) Application of ultra performance liquid chromatography-mass spectrometry to profiling rat and dog bile. J Proteome Res 8:2495–2500
55. Wilson ID, Nicholson JK, Castro-Perez J, Granger JH, Johnson KA, Smith BW, Plumb RS (2005) High resolution "ultra performance" liquid chromatography coupled to oa-TOF mass spectrometry as a tool for differential metabolic pathway profiling in functional genomic studies. J Proteome Res 4:591–598
56. Yang J, Xu G, Zheng Y, Kong H, Wang C, Zhao X, Pang T (2005) Strategy for metabonomics research based on high-performance liquid chromatography and liquid chromatography coupled with tandem mass spectrometry. J Chromatogr A 1084:214–221
57. Burnum KE, Frappier SL, Caprioli RM (2008) Matrix-assisted laser desorption/ionization imaging mass spectrometry for the investigation of proteins and peptides. Annu Rev Anal Chem (Palo Alto Calif) 1:689–705
58. Dumas M-E, Maibaum EC, Teague C, Ueshima H, Zhou B, Lindon JC, Nicholson JK, Stamler J, Elliott P, Chang Q, Holmes E (2006) Assessment of analytical reproducibility of 1 H NMR spectroscopy based metabonomics for large-scale epidemiological research: the INTERMAP study. Anal Chem 78: 2199–2208
59. Keun HC, Ebbels TMD, Antti H, Bollard ME, Beckonert O, Schlotterbeck G, Senn H, Niederhauser U, Holmes E, Lindon JC, Nicholson JK (2002) Analytical reproducibility in 1 H NMR-based metabonomic urinalysis. Chem Res Toxicol 15:1380–1386
60. Deming SN (1986) Chemometrics: an overview. Clin Chem 32:1702–1706
61. Lavine B, Workman J (2008) Chemometrics. Anal Chem 80:4519–4531
62. Bollard ME, Stanley EG, Lindon JC, Nicholson JK, Holmes E (2005) NMR-based metabonomic approaches for evaluating physiological influences on biofluid composition. NMR Biomed 18:143–162
63. Crockford DJ, Keun HC, Smith LM, Holmes E, Nicholson JK (2005) Curve-fitting method for direcct quantitation of compounds in complex biological mixtures using 1 H NMR: application in metabonomic toxicology studies. Anal Chem 77:4556–4562
64. Schnackenberg LK, Sun J, Espandiari P, Holland RD, Hanig J, Beger RD (2007) Metabonomics evaluations of age-related changes in the urinary compositions of male Sprague Dawley rats and effects of data normalization methods on statistical and quantitative analysis. BMC Bioinformatics 8(Suppl 7):S3
65. Sysi-Aho M, Katajamaa M, Yetukuri L, Oresic M (2007) Normalization method for metabolomics data using optimal selection of multiple internal standards. BMC Bioinformatics 8:93
66. Warrack BM, Hnatyshyn S, Ott KH, Reily MD, Sanders M, Zhang H, Drexler DM (2009) Normalization strategies for metabonomic analysis of urine samples. J Chromatogr B Analyt Technol Biomed Life Sci 877:547–552

67. Bertram HC, Malmendal A, Petersen BO, Madsen JC, Pedersen H, Nielsen NC, Hoppe C, Molgaard C, Michaelsen KF, Duus JO (2007) Effect of magnetic field strength on NMR-based metabonomic human urine data. Comparative study of 250, 400, 500, and 800 MHz. Anal Chem 79:7110–7115
68. Chen P, Liu J (2007) Metabonomics and diabetes mellitus. Adv Ther 24:1036–1045
69. Pan Z, Gu H, Talaty N, Chen H, Shanaiah N, Hainline BE, Cooks RG, Raftery D (2007) Principal component analysis of urine metabolites detected by NMR and DESI-MS in patients with inborn errors of metabolism. Anal Bioanal Chem 387:539–549
70. Wagner S, Scholz K, Donegan M, Burton L, Wingate J, Volkel W (2006) Metabonomics and biomarker discovery: LC-MS metabolic profiling and constant neutral loss scanning combined with multivariate data analysis for mercapturic acid analysis. Anal Chem 78:1296–1305
71. Wold S, Sjostrom M (1998) Chemometrics, present and future success. Chemometr Intell Lab Syst 44:3–14
72. Keun HC, Ebbels TMD, Bollard ME, Beckonert O, Antti H, Holmes E, Lindon JC, Nicholson JK (2004) Geometric trajectory analysis of metabolic responses to toxicity can define treatment specific profiles. Chem Res Toxicol 17:578–587
73. Lu X, Tian Y, Zhao Q, Jin T, Xiao S, Fan X (2011) Integrated metabonomics analysis of the size-response relationship of silica nanoparticles-induced toxicity in mice. Nanotechnology 22:055101, 16 pp
74. Cao Q, Zhang S, Dong C, Song W (2007) Pulmonary responses to fine particles: differences between the spontaneously hypertensive rats and wistar kyoto rats. Toxicol Lett 171: 126–137
75. Cho WS, Choi M, Han BS, Cho M, Oh J, Park K, Kim SJ, Kim SH, Jeong J (2007) Inflammatory mediators induced by intratracheal instillation of ultrafine amorphous silica particles. Toxicol Lett 175:24–33
76. Nishimori H, Kondoh M, Isoda K, Tsunoda S, Tsutsumi Y, Yagi K (2009) Silica nanoparticles as hepatotoxicants. Eur J Pharm Biopharm 72: 496–501
77. Tang M, Zhang T, Xue Y, Wang S, Huang M, Yang Y, Lu M, Lei H, Kong L, Yuepu P (2010) Dose dependent in vivo metabolic characteristics of titanium dioxide nanoparticles. J Nanosci Nanotechnol 10:8575–8583
78. Hauck TS, Ghazani AA, Chan WC (2008) Assessing the effect of surface chemistry on gold nanorod uptake, toxicity, and gene expression in mammalian cells. Small 4:153–159
79. Zhang T, Stilwell JL, Gerion D, Ding L, Elboudwarej O, Cooke PA, Gray JW, Alivisatos AP, Chen FF (2006) Cellular effect of high doses of silica-coated quantum dot profiled with high throughput gene expression analysis and high content cellomics measurements. Nano Lett 6:800–808
80. Fowler BA, Conner EA, Yamauchi H (2008) Proteomic and metabolomic biomarkers for III-V semiconductors: and prospects for application to nano-materials. Toxicol Appl Pharmacol 233:110–115
81. Yang B, Wang Q, Lei R, Wu C, Shi C, Yuan Y, Wang Y, Luo Y, Hu Z, Ma H, Liao M (2010) Systems toxicology used in nanotoxicology: mechanistic insights into the hepatotoxicity of nano-copper particles from toxicogenomics. J Nanosci Nanotechnol 10:8527–8537

Chapter 11

Nanoparticle Uptake Measured by Flow Cytometry

Yuko Ibuki and Tatsushi Toyooka

Abstract

The uptake of nanoparticles by cells is an important factor to assess nanotoxicity. In general, the nanoparticles taken up by the cells have been identified by transmission electron microscope, inductively coupled plasma mass spectrometry, etc.; however, the methods required an immense amount of time and effort. Flow cytometry (FCM) has been used and developed in the fields of biochemistry and clinical hematology, and has advantages to analyze thousands of cells in seconds. We recently clarified that the side-scatter(ed) light of FCM could be used as a guide to measure uptake potential of nanoparticles. Here, we describe the protocol for screening of the uptake potential of nanoparticles using FCM.

Key words: Nanoparticle, Flow cytometry, Side-scatter light, GFP

1. Introduction

Flow cytometry (FCM) is an analytic technique using the principles of light scattering, light excitation, and emission of fluorochrome molecules to generate specific multiparameter data from particles and cells in the size range of 1–40 μm diameter (1, 2). This was developed in the 1970s and rapidly became an essential instrument for the clinical examination of hemogram and biochemical experiments for determination of cell cycle (3), apoptosis (4, 5), expression of some antigens on cell membrane (6), etc. In the flow chamber in FCM, as particles or cells flow in single file past the intersection of the light beam (mainly 488 nm laser), light is scattered in various directions. If there are fluorochromes with particles and cells, they become excited and result in fluorescent emission. These data are gathered and multiparametric analysis is performed.

Light that is scattered in the forward direction, typically up to 20° offset from the laser beam's axis, is called the forward-scatter(ed) (FS) light. The FS intensity roughly equates to the cell's and particle's size. Light measured approximately at a 90° angle to

Joshua Reineke (ed.), *Nanotoxicity: Methods and Protocols*, Methods in Molecular Biology, vol. 926,
DOI 10.1007/978-1-62703-002-1_11, © Springer Science+Business Media, LLC 2012

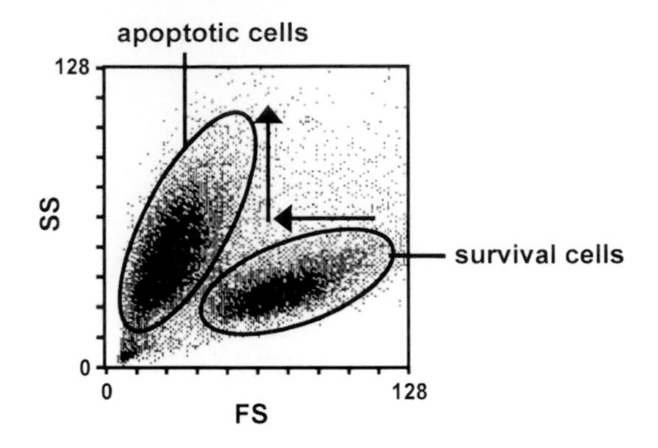

Fig. 1. Analysis of survival/apoptosis by flow cytometric light scatter. FS and SS of the cells going into apoptosis were analyzed using FCM (instrument: EPICS-XL).

the excitation line is called side-scatter(ed) (SS) light. The SS intensity provides information about the granular content within a cell and a particle. Both FS and SS are unique for every cell and particle, and a combination of the two has been used to differentiate different cell types in a heterogeneous sample (1, 2). Figure 1 shows the dot plot of a typical pattern of cell survival/death (apoptosis). It is well established that cells undergoing apoptosis have decreased cell size (cell shrinking) and increased granularity (nuclear spheral follicles (apoptotic body) are produced) (7). Therefore, in apoptotic cells, FS intensity decreases and SS intensity increases (arrows in Fig. 1). By using FS and SS, we developed the method for determination of cellular uptake potential of nanoparticles (8). When particles are taken up by the cells, SS intensity consequently increases without change of FS intensity. This evaluation of uptake potential of nanoparticles using SS in FCM is suitable for initial screening of nanotoxicity because particular treatments (staining, labeling, etc.) apart from preparation of single cell suspension are not required. In addition, statistically valid information about cell populations is quickly obtained since large numbers of cells are analyzed in a short period of time.

To get more information about nanotoxicity, it is possible to add signals of fluorescence by staining of intracellular molecules with fluorescent agents. Propidium iodide (PI) intercalates into the major groove of double-stranded DNA and produces a highly fluorescent adduct that can be excited at 488 nm with a broad emission centered around 610 nm. PI is used to distinguish between living cells and dead cells by its characteristics passing through cell membrane of dead and damaged cells (9). Clarification of relationship between uptake of particles and PI-staining would suggest definitive information of toxicity induced by uptake of nanoparticles. Fluorescence analysis of proteins expressed by treatment with nanoparticles also gives useful information. The cells transfected with plasmid encoding green fluorescence protein (GFP) fusion protein which expression increases after uptake of nanoparticles could realize high-throughput analysis of change of specific proteins.

The greatest benefit of FCM analysis is the possibility to perform multiparametric analysis by gathering the following information: particle uptake, cell death, relative protein expression, etc. In this chapter, we introduce the method to evaluate uptake of nanoparticles using FS and SS in FCM. In addition, multiparametric analysis with PI staining (cell death) and GFP-tagged protein is shown.

2. Materials

2.1. Nanoparticles

Titanium dioxide (TiO_2) particles (listed size; 5 nm) and zinc oxide (ZnO) particles (listed size; <100 nm) (Sigma-Aldrich, St. Louis, MO) (see Note 1).

2.2. Cell Culture and Cell Staining

Prepare all solutions using ultrapure water (prepared by purifying deionized water to attain a sensitivity of 18 MΩ cm at 25°C) and analytic grade reagents.

1. Cell culture medium: Ham's F-12 medium supplemented with 10% fetal bovine serum (FBS), penicillin (100 U/ml), and streptomycin (0.1 mg/ml). Dulbecco's modified Eagle medium (DMEM) supplemented with 10% FBS, penicillin (100 U/ml), and streptomycin (0.1 mg/ml) (see Note 2). Store at 4°C until use.
2. G-418: G-418 disulfate salt (Sigma-Aldrich, St. Louis, MO, USA) in water (30 mg/ml). Store at −20°C until use.
3. Phosphate-buffered saline (PBS): 137 mM NaCl, 2.7 mM KCl, 10 mM Na_2HPO_4, 1.76 mM KH_2PO_4, pH 7.4 (see Note 3). Store at 4°C until use.
4. EDTA-PBS solution: 0.02% ethylenediaminetetraacetic acid (EDTA) in PBS. Sterilize by autoclave. Store at 4°C until use.
5. Trypsin/EDTA-PBS solution: 0.125% Trypsin and 0.01% EDTA in PBS (see Note 4). Store at −20°C until use.
6. PI solution: PI in water (1 mg/ml). Store at 4°C until use.
7. Polystyrene cell culture dish (see Note 5).

2.3. Cell Lines

Many kinds of cells would be suitable to analyze cellular uptake potential of nanoparticles. In this chapter, we introduce protocol using three cell lines as follows:

1. Chinese hamster ovary (CHO)-K1 cells (Japanese Collection of Research Bioresources) (see Note 6).
2. CHO-p21 cells (see Note 7).
3. HaCaT cells (see Note 8).

3. Methods

3.1. Subculture and Preparation of Cells for Experiments

1. Plate equivalent number of cells in dishes (100 mm dishes for subculture) with suitable medium (10 ml for 100 mm dish) and culture at 37°C in 5% CO_2. Maintain the cells with medium-change every 3 or 4 days. Check the cells under microscope every day. Passage the cells before confluent.
2. Wash cells with PBS and add 600 μl of Trypsin/EDTA-PBS solution. Incubate for 3–5 min at 37°C in 5% CO_2 (see Note 9).
3. Add 2 ml culture medium and suspend the cells. Passage a part of the cells to a new dish for subculture (dilution: 1/5–1/20) or 35 mm dishes for experiments.
4. Culture the cells for a few days. Prepare 10^5–10^7 cells per each dish at the day of experiment. Treatment with nanoparticles should be performed in exponentially growing cells (see Note 10).
5. Adjust culture medium for 1 ml/35 mm dish before the start of experiments.

3.2. Preparation and Treatment of Nanoparticles

1. Weigh nanoparticles (<15 mg) into a 1.5 ml plastic microtube (length 37 mm) under the static electricity removed condition.
2. Add culture medium and suspend the nanoparticles at an initial concentration of 1 or 10 mg/ml.
3. Sonicate the tube for 1 min in a bath-type sonicator (ultrasonic wave frequency: 20 kHz, ultrasonic wave output power: 320 W, Bioruptor; Cosmo Bio, Japan) (see Note 11).
4. Add the particles to cells in culture dishes immediately after the sonication at the concentration ranging from 1 to 1,000 μg/ml, and incubate for predetermined time at 37°C in 5% CO_2 (see Note 12).
5. Detach the cells from culture dishes to prepare single cell suspension for FCM. Wash the cells gently, but thoroughly, for at least three times with PBS (in HaCaT cells, wash with EDTA-PBS) to remove particles which are not taken up. Add 200 μl of Trypsin/EDTA-PBS solution. Incubate for 3–5 min at 37°C in 5% CO_2. In the case of HaCaT cells, add 100 μl of EDTA-PBS at first, then further add 100 μl of Trypsin/EDTA-PBS and incubate for at least 10 min at 37°C in 5% CO_2 (see Note 13).
6. Add 1 ml of culture medium containing 10% FBS to the culture dishes and recover the cells by gentle pipetting (see Note 14).
7. Transfer the cells into FCM tubes assigned to instrument and add PI solution at a final concentration of 10 μg/ml. Stand on ice until analysis. Each sample should contain 10^5–10^7 cells/ml.

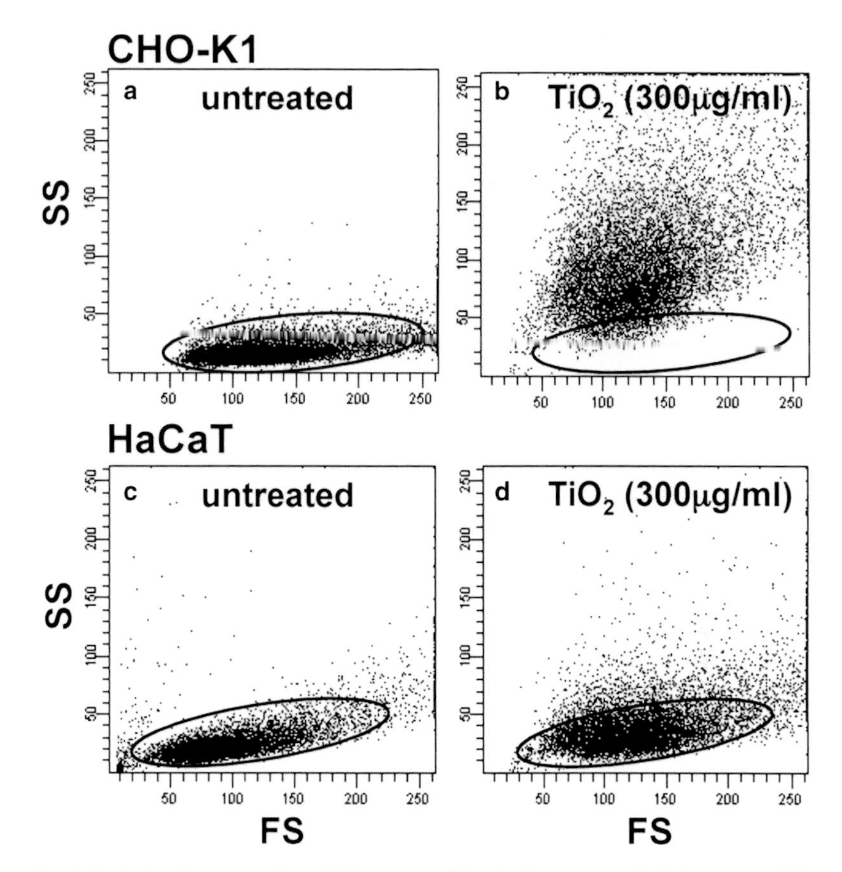

Fig. 2. Analysis of incorporation of TiO_2 nanoparticles by flow cytometric light scatter. CHO-K1 (**a**, **b**) and HaCaT (**c**, **d**) cells were treated with TiO_2 nanoparticles (300 μg/ml) for 1 h, trypsinized and suspended in the culture medium. FS and SS were analyzed using FCM (instrument: FACS CANT II).

3.3. Acquisition of FS and SS by FCM

1. Start-up FCM (see Note 15) according to instrument's protocol. About 5–10 min is needed for initializing and warming-up.
2. Verify instrument optical alignment and fluidics using polystyrene beads impregnated with the appropriate fluorochrome according to instrument's protocol (see Note 16).
3. Set parameters of FS and SS and make a dot plot sheet (horizontal axis: FS, vertical axis: SS) according to instrument's software. Both FS and SS are represented as linear scale.
4. Set threshold in FS to omit noise (see Note 17).
5. Flow untreated cells (control), prepared in Subheading 3.2, at the speed of hundreds of cells/s (see Note 18). Each dot represents an individual cell that has passed through the intersection of laser beam (see Fig. 2).
6. Adjust sensitivity of FS. The dots should be plotted in the middle of the sheet (see Fig. 2a, c).
7. Adjust sensitivity of SS to be plotted in the bottom of the sheet (see Fig. 2a, c) (see Note 19).

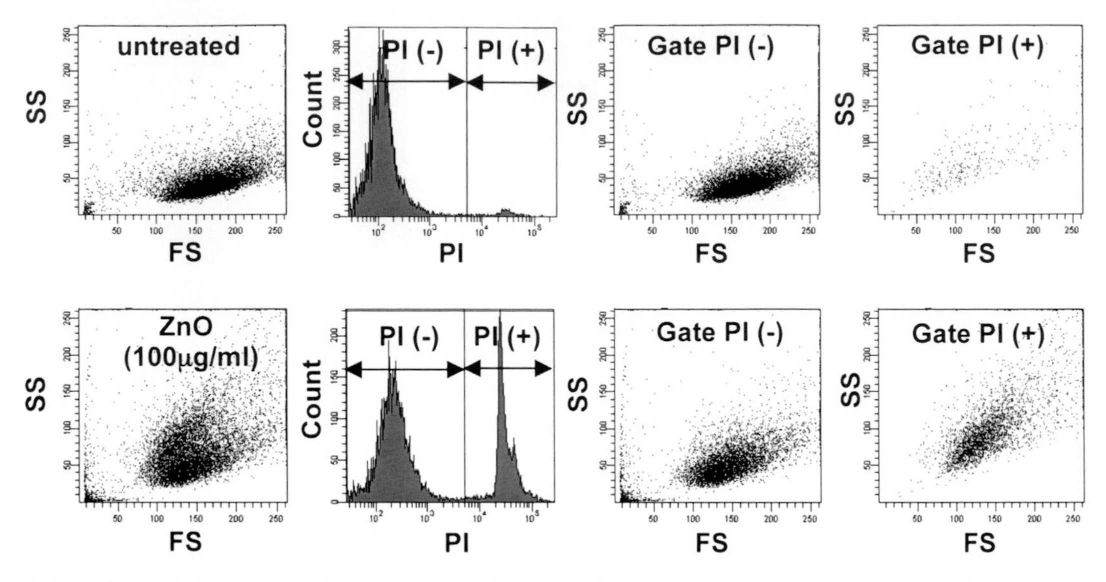

Fig. 3. Analysis of toxicity dependent on incorporation of nanoparticles—gating of PI-stained cells. CHO-K1 cells were treated with ZnO nanoparticles (100 μg/ml) for 6 h, trypsinized and suspended in the culture medium containing PI (10 μg/ml). FS, SS, and fluorescence from PI were analyzed using FCM (instrument: FACS CANT II). PI-unstained cells [PI(−)] and stained cells [PI(+)] were gated and the intensity of SS was analyzed.

8. Flow cells treated with nanoparticles (sample treated with the highest concentration) and adjust sensitivity to locate the dots to the suitable position (see Fig. 2b, d) (see Note 20).
9. Analyze all samples without changing the sensitivities. Acquire 5,000–20,000 cells per sample at a flow rate of about 1,000 cells/s.

3.4. Gating of PI-Stained Cells

1. FS and SS settings are same as Subheading 3.3.
2. Set parameter of fluorescent emission to detect the relative fluorescence from PI-staining (see Note 21) and make a histogram sheet (horizontal axis: FL log (orange or red), vertical axis: number of events [cell number)] according to manufacture's soft wear. The relative fluorescence from PI is represented as log scale.
3. Adjust sensitivity for detecting fluorescent emission of PI.
4. Analyze all samples without changing the sensitivities. Acquire 5,000–20,000 cells per sample at a flow rate of about 1,000 cells/s.
5. Gate PI-unstaining cells [PI(−)] or PI-staining cells [PI(+)] (see Fig. 3) to analyze the cytotoxicity of nanoparticles (see Note 22).

3.5. Analysis of GFP-Protein

1. FS and SS settings are same as Subheading 3.3.
2. Set parameter of fluorescent emission to detect the relative fluorescence from GFP (see Note 23) and make a histogram

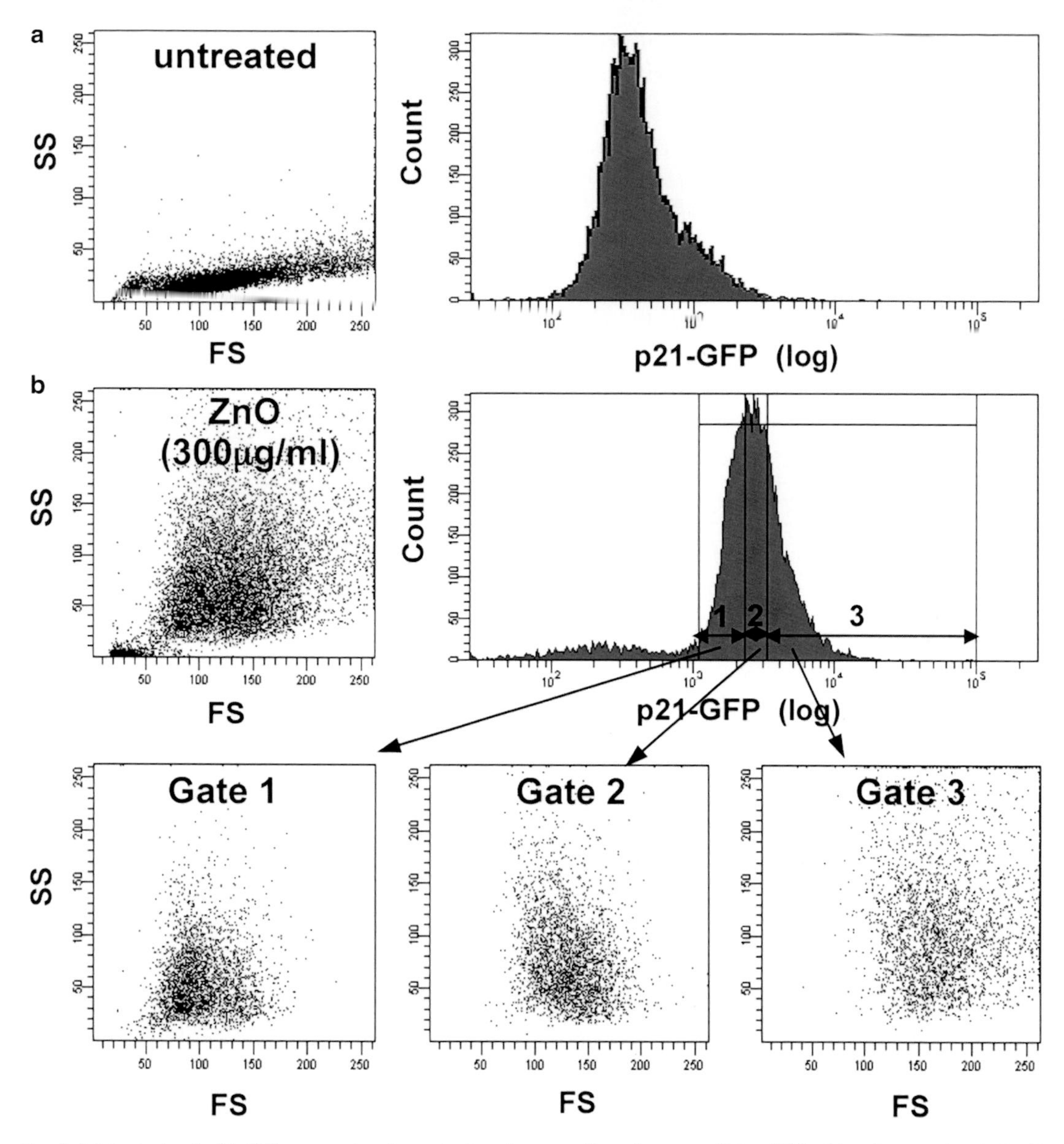

Fig. 4. (**a**) Analysis of p21-GFP expression dependent on incorporation of nanoparticles. CHO-p21 cells were treated with ZnO nanoparticles (300 μg/ml) for 4 h, trypsinized and suspended in the culture medium. FS, SS, and fluorescence from GFP were analyzed using FCM (instrument: FACS CANT II). Uptake of nanoparticles (SS intensity) according to expression of p21-GFP was analyzed by gating in histogram of p21-GFP. Untreated, (**b**) Treated with ZnO nanoparticles (300 μg/ml).

(horizontal axis: FL log (green), vertical axis: number of events [cell number)] according to instrument's software. Relative fluorescence from GFP is represented as log scale.

3. Flow untreated cells suspended in Subheading 3.2 at the speed of hundreds of cells/s.
4. Adjust sensitivity for detecting fluorescent emission of GFP (see Fig. 4a) (see Note 24).

5. Flow cells treated with nanoparticles (see Note 25) and adjust sensitivity to locate the histogram to the suitable position (see Fig. 4b).
6. Analyze all samples without changing the sensitivities. Acquire 5,000–20,000 cells per sample at a flow rate of about 1,000 cells/s.
7. Gate three different area of GFP intensity (see Fig. 4b).
8. Analyze the relationships between uptake of nanoparticles and expression of GFP-tagged protein (see Note 26).

4. Notes

1. FS and SS in FCM are applicable to analysis of cellular uptake potential for all types of nanoparticles.
2. Culture mediums, FBS, and antibiotics can be obtained from Gibco®; Invitrogen (Carlsbad, CA, USA), Sigma-Aldrich (St. Louis, MO), etc.
3. Simple method for preparing PBS: add about 800 ml distilled water to a 1 L graduated cylinder or glass beaker. Weigh 8 g NaCl, 0.2 g KCl, 1.44 g Na_2HPO_4, 0.24 g KH_2PO_4, and transfer the cylinder. Mix and adjust the pH to 7.4 with HCl and NaOH. Make up to 1 L with water. Sterilize by autoclave.
4. Dispense and mix 5.85 ml PBS (see Subheading 2.2, item 3), 6.5 ml EDTA-PBS (see Subheading 2.2, item 4), and 0.65 ml Trypsin solution (2.5%) in 15 ml polypropylene conical tube under aseptic conditions.
5. Select proper size culture dishes suitable for your experiments. Generally, 35 or 60 mm dishes were manageable. In this chapter, we introduce the case to use 35 mm dishes for experiments.
6. CHO-K1 cell line is a subclone of the parental CHO cell line, which was derived from the ovary of an adult Chinese hamster. The cells are maintained in Ham's F-12 (see Subheading 2.2, item 1).
7. CHO-p21 cells are CHO-K1 cells which stably expressed GFP-tagged p21 established by our group and Dr. Manabu Koike (Nat. Inst. Radiol. Sci. Japan). The cells are maintained in Ham's F-12 containing G418 (300 μg ml).
8. HaCaT cells are an immortalized human keratinocytes provided kindly by Dr. Fusening (German Cancer Research Center, Germany). The cells are maintained in DMEM.
9. In the case of HaCaT cells, add 2 ml of EDTA-PBS and stand for at least 10 min at 37°C in 5% CO_2. Discard the EDTA-PBS, then treat with 600 ml of Trypsin/EDTA-PBS for 3–5 min.

10. Cell density in culture dish may affect the uptake potential of nanoparticles. It would be better to determine proper cell density to obtain reproducible results.
11. It would be better to confirm the size distribution of nanoparticles in culture medium after sonication. Particles might aggregate and the dispersion size might be larger than listed size. We use Zetasizer Nano (Nano-ZS, Malvern Instruments, UK) to measure the size distribution (10).
12. The uptake potential is different in cell types. Fibroblasts like CHO-K1 are easier to uptake nanoparticles than keratinocytes like HaCaT (10). If you analyze only uptake potential of nanoparticles in general fibroblasts, incubation for 1 h would be enough.
13. Treatment with Trypsin/EDTA-PBS solution for excessively long times should be avoided to prevent cell aggregation.
14. Do not recover the cells with PBS. It causes aggregation of cells.
15. The standard FCM (an air-cooled argon gas laser emitting a monochromatic beam of light fixed at 488 nm at 15 mW of power) is enough to analyze uptake of nanoparticles. For example, EPICS-XL, Cytomics FC 500, Quanta SC (Beckman Coulter, Inc. Brea, CA), FACSCalibur™, FACS CANT™ II (Becton Dickinson, Franklin Lakes, NJ), etc. The instruments are capable of detecting some parameters: FS, SS, and at least three fluorescent emissions (green, yellow–orange and red) utilizing filter reflection and passage.
16. Polystyrene fluorescent microspheres are used as standard beads. According to each instrument, the beads are sold. The fluorescence emission of the dye is contained within the fluorospheres ranges from about 500 to 700 nm when excited at 488 nm.
17. By setting threshold in FS, small-sized cells and particles cannot be detected. Bacteria contaminated in cultured medium and nanoparticles untaken in cells are deleted.
18. To set the sensitivity during flowing, above a certain flow rate is required, whereas total flow volume should be kept to the minimum to prevent the excess loss of sample.
19. As SS intensity increases after uptake of nanoparticles, the location of the dots in untreated cells is better to be in the bottom of the dot plot sheet (circles in Fig. 2). The location of FS doesn't move unless excess particles are taken up by the cells.
20. SS intensity increases according to the amounts of nanoparticles taken up by the cells. The uptake is different between cell lines. HaCaT cells shows lower uptake than CHO-K1 cells (Fig. 2c, d).

21. PI, a DNA specific dye, can be excited by 488 nm and emit in around 610 nm, which can be measured in either the orange or red channel.
22. An important characteristic of FCM data analysis is to selectively visualize the cells of interest while eliminating results from unwanted particles, e.g. dead cells and debris. This procedure is called gating. In this analysis of uptake of nanoparticles, there are two patterns of gating using fluorescent emission of PI according to the purpose. One is to delete damaged and dead cells. In the process of trypsinization of cells, a part of the cells is sometimes damaged. To omit the damaged and dead cells and analyze the uptake in living cells, dead cells were stained by PI and deleted using gating of non-staining cells. Another is to analyze toxicity of nanoparticles from relationships between cell death and amounts of uptake of particles. By gating PI-positive cells, we can clarify whether more nanoparticles were taken up by the dead cells.
23. GFP can be excited by 488 nm and emit in around 510 nm, which can be measured in green fluorescence.
24. Slide the histogram of untreated cells to the left side in the sheet.
25. The p21-GFP increases at 2–4 h after exposure to nanoparticles (This might be due to inhibition of proteasomal degradation).
26. If the treatment with nanoparticles increases expression of p21-GFP, the cells having higher SS intensity would show higher GFP intensity.

References

1. Ormerod MG (2008) Flow cytometry—a basic introduction. http://flowbook.denovosoftware.com/. Accessed Sept 2010.
2. Radcliff G, Jaroszeski MJ (1998) Basics of flow cytometry. Methods Mol Biol 91:1–24
3. Darzynkiewicz Z, Smolewski P, Bedner E (2001) Use of flow and laser scanning cytometry to study mechanisms regulating cell cycle and controlling cell death. Clin Lab Med 21:857–873
4. Bedner E, Li X, Gorczyca W, Melamed MR, Darzynkiewicz Z (1999) Analysis of apoptosis by laser scanning cytometry. Cytometry 35: 181–195
5. Riccardi C, Nicoletti I (2006) Analysis of apoptosis by propidium iodide staining and flow cytometry. Nat Protoc 1:1458–1461
6. Sklar LA, Carter MB, Edwards BS (2007) Flow cytometry for drug discovery, receptor pharmacology and high-throughput screening. Curr Opin Pharmacol 7:527–534
7. Donner KJ, Becker KM, Hissong BD, Ahmed SA (1999) Comparison of multiple assays for kinetic detection of apoptosis in thymocytes exposed to dexamethasone or diethylstilbesterol. Cytometry 35:80–90
8. Suzuki H, Toyooka T, Ibuki Y (2007) Simple and easy method to evaluate uptake potential of nanoparticles in mammalian cells using a flow cytometric light scatter analysis. Environ Sci Technol 41:3018–3024
9. al-Rubeai M, Emery AN, Chalder S (1991) Flow cytometric study of cultured mammalian cells. J Biotechnol 19:67–81
10. Toyooka T, Amano T, Suzuki H, Ibuki Y (2009) DNA can sediment TiO_2 particles and decrease the uptake potential by mammalian cells. Sci Total Environ 407:2143–2150

Chapter 12

Determining Biological Activity of Nanoparticles as Measured by Flow Cytometry

Jennifer F. Nyland

Abstract

Flow cytometry is a powerful tool to evaluate cellular responses at the single cell level. Applicability to evaluating biological activity of in vitro and in vivo exposure to compounds is limited only by the number of fluorochrome emission spectra a particular instrument can detect and the availability of antibodies specific for a particular cellular protein. Here, I describe the general method considerations and provide an example experimental design for utilizing flow cytometry to evaluate the biological activity of nanoparticles on primary murine immune cells.

Key words: Flow cytometry, Nanoparticles, Antibody, Fluorochrome

1. Introduction

Fluorescence-assisted cell sorting (FACS) or flow cytometry (FCM) has grown to become a powerful tool to analyze individual cellular responses to various environmental situations. Originating in the field of microscopy, FCM utilizes light and light scatter as well as fluorescence emission patterns to identify the physical and chemical characteristics of a single cell (or other things roughly the size of a single cell).

What can you measure? A virtual smorgasbord of parameters, including intrinsic and extrinsic structural and functional parameters. In Table 1, I have presented some of the parameters for which FCM can be used, but understand that this list is certainly not comprehensive.

But what is biological activity? I define it as induced changes in the expression levels of cellular proteins, indicative of signal induction, leading to cellular activation, differentiation, or apoptosis. This is a very broad definition. As such, the method presented here is very general. When designing FCM experiments to test the biological

Joshua Reineke (ed.), *Nanotoxicity: Methods and Protocols*, Methods in Molecular Biology, vol. 926,
DOI 10.1007/978-1-62703-002-1_12,

Table 1
Some parameters measurable with flow cytometry

	Structural	Functional
Intrinsic	Cell size and shape Cytoplasmic granularity Pigments, porphyrins, etc.	Redox state
Extrinsic	DNA content and base ratio	Surface and intracellular receptors
	Nucleic acid sequence Chromatin structure RNA content Total protein Surface and intracellular antigens	Surface charge Membrane integrity Membrane organization Endocytosis Cell generation number
	Surface sugars Lipids	Enzyme activity Oxidative metabolism DNA synthesis and degradation Gene expression Cytoplasmic pH Membrane potential

activity of nanoparticles, one must consider the type of nanoparticles (including composition, size, and functional groups), route of exposure (if conducting in vivo exposure experiments), and the types of cells to be analyzed by FCM. A brief word on nanoparticles "dosing" and the source of the cells to be analyzed for biological effects: careful consideration should be given to the concentration or dose of nanoparticles, whether the exposure is in vitro or in vivo, and of course the timing and route of exposure.

1.1. FCM Instruments

There are many manufacturers of FCM instruments with wide and varying capabilities. It is important, therefore, to understand what your particular instrument is capable of detecting and what it cannot. Some FCM instruments are able to sort specific subsets of cells. Some can only detect a handful of fluorochromes while others can detect many. Some fluorochromes overlap in their emission spectra and some are quite nicely separated. Once you are familiar with the capabilities of your FCM instrument (the laser and optical filter combinations), you can easily design experiments to utilize its full capability.

There are many other references available which detail the methodology behind the FCM technology. This is not the focus of this chapter and I would instead refer you to *Practical Flow Cytometry* (1) as an excellent reference.

1.2. Fluorochromes

There are also a wide range of fluorochromes from which to choose; select the fluorochromes that your instrument can detect and those whose emission spectra do not overlap. Where possible, balance pairing between the intensity of fluorochromes and the likely level of parameter of interest on the cell. That is, use more intensely fluorescing fluorochromes to detect lower frequency parameters and less intense fluorochromes for higher frequency parameters. For example, allophycocyanin (APC) (an intense fluorochrome with a relatively narrow emission peak) used to detect a low frequency surface protein, such as CD25, in conjugation with another lower intensity fluorochrome and higher frequency protein, such as CD3, will help to balance the analysis of multicolor flow cytometry.

1.3. Evaluating Changes in Cellular Activation and Maturation States

With the plethora of antibodies to cellular proteins available, particularly to mouse proteins, evaluation of the effects of exposures to nanoparticles on cellular activation and maturation states is simply a matter of choosing the combinations of antibodies and fluorochromes to match the desired parameter. Multicolor FCM can allow for identification of cells of a particular subset, for example murine B cells with anti-CD19 antibody, and within that subset levels of expression of co-stimulatory molecules such as CD80 and CD86 can provide an approximation of activation. Additional staining for CD27 on these cells could demonstrate differentiation or maturation into plasma cells (antibody-forming B cells).

Typically stains to evaluate cellular activation and maturation states focus on extracellular staining. This means that the cells are kept cold and intact on ice while the fluorochrome-labeled antibodies are added. Care is taken to avoid permeabilization of the cellular membrane.

A more general method for examining activation in response to stimuli is evaluation of intracellular Ca^{2+} concentration. This is accomplished through the use of intracellular dyes such as Indo-1 pentaacetoxymethyl ester (Indo-1 AM) or Fluo-4-acetoxymethyl ester (Fluo-4 AM) which are high affinity calcium indicators. The latest generation of flow cytometers allows for addition of test compounds, such as drugs or nanoparticles, over the time course of cellular acquisition by the instrument. This permits a dynamic reading of calcium concentration over time (2). Measurements of intracellular calcium concentration can provide an indication of second messenger response and occur within nanoseconds. However, this method does not allow for determination of which signaling pathways are activated nor which proteins are changed following stimulus.

1.4. Evaluating Changes in Intracellular Signals

It is also possible to utilize FCM to evaluate intracellular protein levels. In this instance, extracellular stains are fixed and the cellular membrane permeabilized. There are a number of commercially available fix/permeabilization buffers, but an easy buffer to use is

0.3% saponin (w/v) in phosphate-buffered saline (PBS). Dilute all intracellular staining antibodies and wash cells in 0.3% saponin/PBS to maintain the membrane permeability before analysis. Be sure to resuspend the cells in PBS or appropriate cell fixation solution (such as *p*-formaldehyde) before analysis on the instrument.

A new application of an old fixation buffer (zinc salt-based fixation) typically used for histochemical analyses was recently proposed (3). In this method, the use of zinc-based fixation, rather than formaldehyde-based fixation, enables the user to stain cells for both extracellular and intracellular proteins and preserve DNA for subsequent PCR analyses.

1.5. Direct or Indirect Effects

Some of the parameters discussed above allow for clear distinctions between direct and indirect effects of nanoparticles to cells. If you can see the nanoparticles binding directly to the cell or receptor, for example, then that is a direct effect. If, however, you see changes in surface protein expression, is that because the nanoparticles bound to a promotion element in the DNA to turn on expression of that protein or is it because the nanoparticles "stressed" the cell and caused activation of stress-response pathways leading to upregulation of that surface protein? To muddle the situation further, what happens if the nanoparticle activates one set of cells in your model system which then release signaling proteins and activate the cells you are examining by FCM? Unless you are able to visualize the nanoparticles in specific cells, FCM may not be the best method to evaluate the effects observed.

I have been fortunate to work with some talented chemical engineers who have created nanoparticles with specific dyes covalently bound within the matrix of the nanoparticles (4). These siloxane (silicate–silicone matrix) nanoparticles can theoretically incorporate any dye with a Si. Through careful selection of dyes which do not overlap with FCM fluorochrome emission spectra, we were able to visualize cells with (Fig. 1b) and without (Fig. 1a) nanoparticles. Then, gating on the cells with nanoparticles, further biological effects could be evaluated. In Fig. 2, I show an FCM dot plot for cells stained with two markers to evaluate viability and cells undergoing apoptosis. These are only cells which were positive for the nanoparticles in Fig. 1b.

Other investigators have taken advantage of the light scatter capacity of nanoparticles for detection with FCM (5). In this case, TiO_2 nanoparticles at concentrations as low as 5–10 nanoparticles per cell could be detected by FCM due to the substantial light scattering and reflection of the particles. Multicolor FCM following nanoparticle treatment allows for evaluation of the direct effects of nanoparticles on protein expression on and within cells.

Thus, depending upon the properties of the nanoparticles and the capabilities of the FCM instrument, very complex evaluations of biological activity of nanoparticles can be completed.

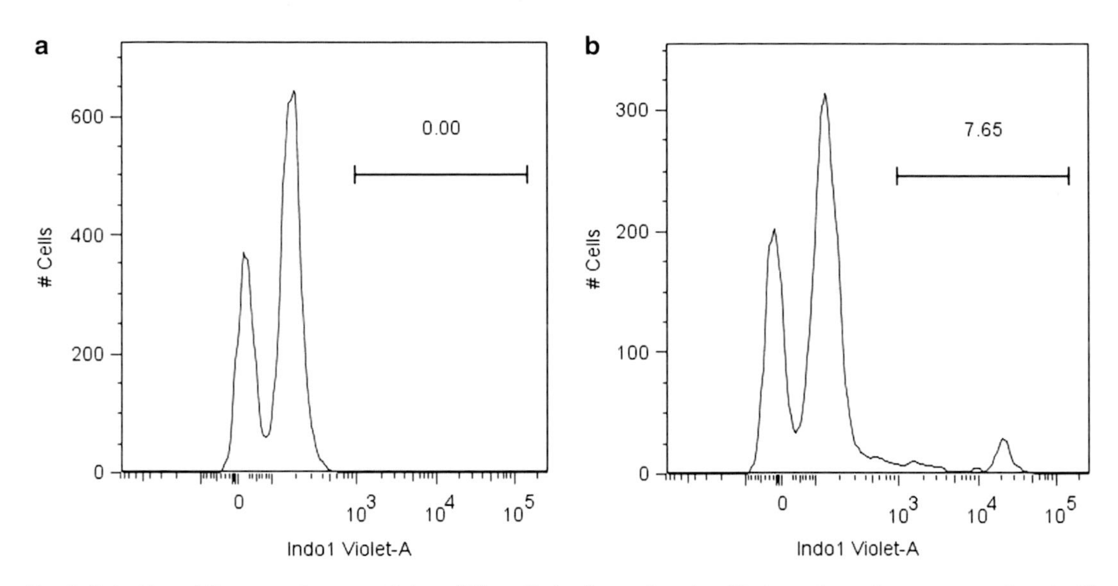

Fig. 1. Detection of fluorescent nanoparticles within cells by flow cytometry. Murine splenocytes were cocultured with (**a**) media alone or (**b**) dansylamide-siloxane nanoparticles for 48 h in phenol red-free culture media. Cells positive for nanoparticles fluoresce at >10^3 in the indo-violet channel (*x*-axis), as marked by gate on histogram.

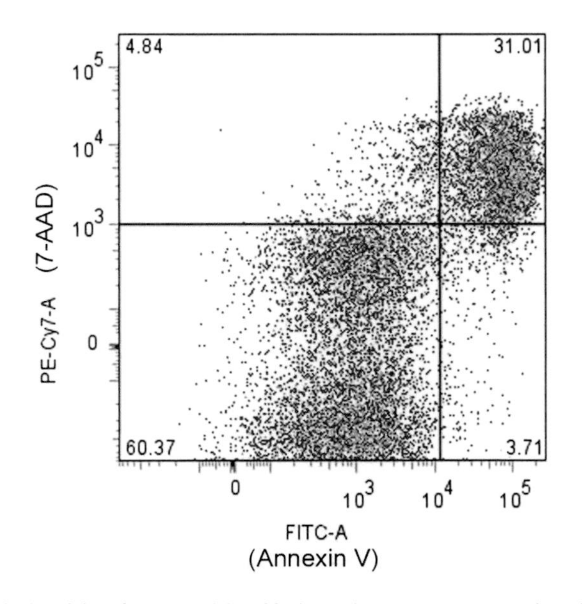

Fig. 2. Assessment of biological activity of nanoparticles. Murine splenocytes were cocultured with dansylamide-siloxane nanoparticles for 48 h. Cells were stained with fluorescein isothiocyanate (FITC)-conjugated annexin V and phycoerythrin (PE)-Cy^7-conjugated 7-aminoactinomycin D (7-AAD). Cells positive for nanoparticles were evaluated for annexin V (*x*-axis) and 7-AAD (*y*-axis) staining to assess apoptosis and viability (Color figure online).

All these will influence the combinations of parameters that can and should be measured. Since the combinations are seemingly endless, and extremely dependent on the FCM instrument, types of cells, and the likely biological changes induced in those cells, recognize that this methods chapter represents only one possible example which will need to be adjusted and tailored to your specific experimental question and needs.

2. Materials

2.1. Cell Source, Isolation, and Culture

1. There are many sources of cells available including cell lines from various species. In this chapter, I provide a protocol using a single cell suspension of immune cells derived from mouse spleen.
2. 50 ml conical tubes.
3. 5 cm petri dishes.
4. Glass slides with frosted ends, wrapped in paper and sterilized in autoclave.
5. 24-Well culture plates (CellBIND, Corning, Lowell, MA).
6. 0.4 μm cell strainer (BD Falcon, BD Biosciences, Bedford, MA).
7. Tris–ammonium chloride (TAC) buffer: 8.3 g ammonium chloride (NH_4Cl), 100 ml 0.1M Tris–HCl, bring to 1 L with ultra pure water. Adjust pH to 7.2 with 0.1 M NaOH or 0.1 M HCl and filter sterilize.
8. Supplemented RPMI: 500 ml RPMI 1640 medium without phenol red, 50 ml heat-inactivated FBS, 5 ml 1 M HEPES buffer, 5 ml 5,000 IU penicillin–5,000 μg/ml streptomycin, 5 ml 200 mM l-glutamine.
9. lipopolysaccharide (LPS): 400 ng/ml LPS diluted with sterile PBS.

2.2. FCM Buffers

1. FCM cell staining buffer: *1% BSA/PBS*: 10 g bovine serum albumin (BSA), 1 L PBS. Store at 4°C. Prepare fresh, keep cold, and use immediately.
2. Permeabilization buffer: *0.3% saponin/PBS*: 300 mg saponin, 100 ml PBS. Store protected from light at 4°C; expires after 1 month (see Note 1).
3. FCM cell fixation buffer: *2% p-formaldehyde*: 2 g paraformaldehyde, 100 ml PBS. Adjust pH to 7.25 with 0.1 M NaOH or 0.1 M HCl. Store protected from light at 4°C; expires after 1 month (see Note 2).

2.3. Cell staining and FCM Acquisition

1. 1.5 ml snap-cap tubes.
2. V-bottom 96-well plates (Greiner Bio-One, Monroe, NC).

3. Methods

3.1. Isolation of Cells for Experiments

1. Remove spleen from mouse and place in 3 ml sterile PBS in 50 ml conical tube.
2. Transfer spleen and PBS to 5 cm petri dish.
3. Gently smash tissue between the frosted ends of the slides.
4. Filter single cell suspension back into the 50 ml conical tube through the 0.4 μm filter.
5. Wash petri dish with additional PBS to collect all cells (2× 5 ml) and filter into 50 ml conical tube.
6. Centrifuge to pellet the cells (200 × *g*, 10 min, 4°C) (see Note 3).
7. Lyse red blood cells: resuspend the cell pellet in 5 ml TAC buffer and incubate on ice 10 min. Stop reaction with FCM staining buffer (30 ml).
8. Centrifuge to pellet the immune cells as in step 6.
9. Resuspend cell pellet in 1 ml supplemented RPMI and count cells with hemocytometer.

3.2. Cell Culture and In Vitro Nanoparticles Treatment

1. Plate equivalent numbers of cells (1×10^6) per well in 24-well plates.
2. Add 100 μl diluted nanoparticles (see Note 4).
3. For one half of the cultured wells, add 50 μl of 400 ng/ml LPS.
4. Bring each well to a final volume of 1 ml with supplemented RPMI.
5. Culture the cells for 2 days at 37°C in 5% CO_2, gently agitating cultures once every 4 h.
6. Harvest cells with gentle pipetting and transfer into 1.5 ml tubes.
7. Centrifuge to collect cell pellet and resuspend in 200 μl FCM staining buffer (see Note 5).

3.3. Cell Staining for FCM Acquisition

3.3.1. Surface Staining

1. Transfer cells to v-bottom 96-well plate with one cell culture aliquot per well (see Note 6).
2. Centrifuge the entire plate (200 × *g*, 10 min, 4°C) to pellet cells.
3. Decant by gently inverting the entire plate.
4. Add 50 μl diluted fluorochrome-conjugate antibody specific for your cell surface protein of interest. Antibody is diluted in FCM buffer to achieve the appropriate intensity of signal on your particular FCM instrument and for the particular type of

cell and protein of interest combination. Mix each cell culture aliquot by pipetting.

5. Incubate 30 min on ice protected from light (see Note 7).
6. Wash each cell culture aliquot with 150 μl FCM buffer.
7. Centrifuge and decant.
8. Repeat steps 4–7 as necessary for additional cell surface markers, up to as many colors as your FCM instrument can separate.
9. If you are interested in intracellular components, proceed on to Subheading 3.3.2, otherwise go directly to step 7 of Subheading 3.3.2.

3.3.2. Intracellular Staining

1. Add 200 μl permeabilization buffer per well and incubate 60 min on ice.
2. Centrifuge and decant.
3. Add 50 μl diluted fluorochrome-conjugate antibody specific for your cell surface protein of interest. Antibody is diluted in FCM buffer. Mix each cell culture aliquot by pipetting.
4. Incubate 30 min on ice protected from light.
5. Wash each cell culture aliquot with 150 μl FCM buffer.
6. Centrifuge and decant.
7. Resuspend the cell pellet in 200 μl FCM cell fixation buffer and transfer to the appropriate tube for analysis on the FCM instrument.

4. Notes

1. This is the recommended permeabilization buffer for intracellular staining. I find it better to use this saponin permeabilization buffer rather than 0.2% Tween-20/PBS as I prefer to keep all my cells, stains, and buffers at 4°C throughout the preparation of the cells for FCM analysis (Tween-20/PBS works best at room temperature for permeabilization). This is particularly important if dual-conjugated fluorochromes are to be used (e.g., APC-Cy^7) which can decouple under certain circumstances.
2. To dissolve *p*-formaldehyde, heat the solution to 70°C (do not heat above 70°C) in a chemical fume hood with stirring. Allow the solution to cool to room temperature before adjusting the pH. Check the pH before use. An alternative fixation buffer (as mentioned in the introduction) is the zinc salt-based fixation buffer: 0.5 g zinc acetate, 0.5 g zinc chloride, 50 mg calcium acetate, 100 ml 0.1 M Tris–HCl pH 7.8. For this method I utilize the *p*-formaldehyde.

3. Once you have isolated your cells in a single cell suspension, carry out all procedures on ice or at 4°C.
4. My typical practice for in vitro assessment is to conduct each individual experiment in triplicate, at a minimum, for each condition. Additionally, I utilize serial dilutions, either twofold or tenfold for the nanoparticles in order to observe a range of biological effects. There continues to be some debate regarding dosing with nanoparticles. In an effort to avoid ambiguity, I choose to use concentrations and provide size and shape information in my manuscript methods. This method necessitates knowing the number of nanoparticles per unit volume, which we have typically calculated with multiple field counts under a TEM (4).
5. Once you have harvested your cells from the in vitro cultures, carry out all procedures on ice or at 4°C.
6. Alternatively, if you are interested in more protein combinations than your FCM instrument can visualize simultaneously, you can divide the ~1×10^6 cells per culture well into multiple aliquots. Each aliquot would then be added to a separate well in the v-bottom 96-well plate and different antibody stains added to the different aliquots. I choose to use v-bottom 96-well plates for antibody staining as this streamlines the process for me.
7. From this point on, that is once you have added your first fluorochrome to the cells, protect the cells from light. This can be as simple as covering your ice bucket with aluminum foil.

References

1. Shapiro HM (2003) Practical flow cytometry, 4th edn. Wiley, Hoboken
2. Vines A, McBean GJ, Blanco-Fernández A (2010) A flow-cytometric method for continuous measurement of intracellular $Ca2^+$ concentration. Cytometry A 77A:1091–1097
3. Jensen UB, Owens DM, Pedersen S, Christensen R (2010) Zinc fixation preserves flow cytometry scatter and fluorescence parameters and allows simultaneous analysis of DNA content and synthesis, and intracellular and surface epitopes. Cytometry A 77A:798–804
4. Nyland JF, Bai JJ, Katz HE, Silbergeld EK (2009) In vitro interactions between splenocytes and dansylamide dye-embedded nanoparticles detected by flow cytometry. Nanomedicine 5:298–304
5. Zucker RM, Massaro EJ, Sanders KM, Degn LL, Boyes WK (2010) Detection of TiO_2 nanoparticles in cells by flow cytometry. Cytometry A 77A:677–685

Chapter 13

Whole Cell Impedance Biosensoring Devices

Evangelia Hondroulis and Chen-Zhong Li

Abstract

Nanotechnology is rapidly growing and has great potential in various fields such as biomedical engineering, drug delivery, environmental health, pharmaceutical industries and even electronics and communication technologies. However, with this rapid development, these new nanoscale materials (including nanotubes, nanowires, nanowhiskers, fullerenes or buckyballs, and quantum dots) might have unintended human health and environmental hazards. Testing for toxicological parameters is a necessary first step toward ensuring the compatibility of nanomaterials for medical applications and for the safety of the environment. Here, we describe an array formatted electrical impedance sensing (EIS) system that is capable of measuring nanotoxicity in real time.

Key words: Nanotoxicity, Biosensor, Cell impedance, MEMS, Chips, Ecotoxicity, Kinetic analysis, Nanoparticles, Devices

1. Introduction

Biosensors are becoming a valuable tool for detecting toxic chemical compounds in industrial products, chemical substances, environmental samples (e.g., air, soil, and water), or biological systems (e.g., bacteria, virus, or tissue components). Most current biosensors are used to detect enzymes, DNA/RNA, and immunological components (1). Biosensors that incorporate whole cells can have an advantage over these previous biosensors as they are able to provide information about the total physiological effect of a toxin toward the whole cell.

Traditional biological methods for measurement of cellular activity and proliferation are used in nanotoxicity studies. These methods include, for example, mitochondrial reduction of tetrazolium salts into an insoluble dye (MTT test) and enzyme lactate dehydrogenase (LDH) release tests. These methods are used as markers for cell viability and consist of procedures that are time consuming (24–96 h) and provide only a general sense of toxicity

Joshua Reineke (ed.), *Nanotoxicity: Methods and Protocols*, Methods in Molecular Biology, vol. 926,
DOI 10.1007/978-1-62703-002-1_13,

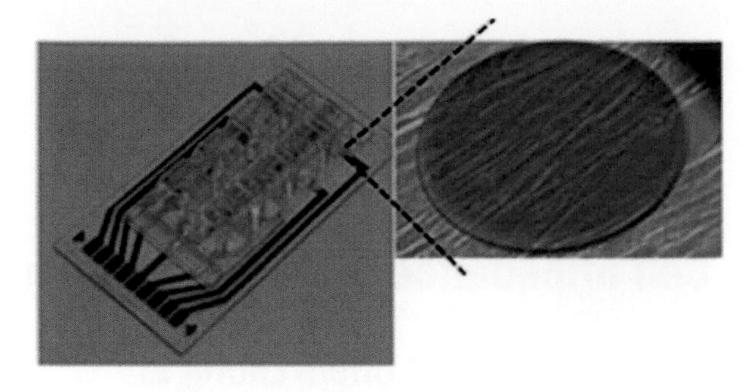

Fig. 1. Design of the EIS chip showing the array of eight detecting gold electrodes with culture wells places on *top* and a zoom in on the cells attached on an electrode (Color figure online).

as they show results only at a final time point (2). As a result, the kinetic model (absorption, distribution, metabolism, and excretion) of nanoparticle uptake cannot be observed using these conventional methods. Following biological exposure, the particles may transport across cell membranes, especially into mitochondria, causing internal damage that may affect the cell behavior and over time, leading to cell death (3).

In this method, we use an array formatted electrical impedance sensing (EIS) system that is capable of revealing the kinetic effects of gold nanoparticles (AuNPs 10, 100 nm), silver nanoparticles (AgNPs 10 nm,100 nm), single wall carbon nanotubes (SWCNTs cut, uncut), and cadmium oxide (CdO) when exposed to human lung fibroblasts (CCL-153™).

The EIS chip consists of an array of eight detecting gold electrodes (250 μm in diameter) each on the bottom of individual tissue culture wells of volume 9 mm × 9 mm × 10 mm to measure the resistance produced by growing cell monolayers over the electrodes and can detect changes in resistance to AC current flow (approximately 1 μA) that may occur with changes in the cell layer after nanoparticle exposure (Fig. 1). A gold counter electrode (7 mm × 46 mm) is linked to each individual detecting electrode. The impedance of the circuit with a resistance (R) and a capacitor (C) in series can be measured for each well by applying an alternating potential to the two types of electrodes present in the EIS chip through a 1 MW resistor (Fig. 2). The EIS offers compact structure, easy use, and capability to measure multiple samples simultaneously in real time, a critical feature in monitoring cytotoxicity.

2. Materials

2.1. Nanomaterial Preparation

The nanomaterials in this study—AuNPs, AgNPs, and SWCNTs—were chosen for the reason that they have attracted substantial research efforts for potential biomedical and energy applications (4–8) (see Note 1). CdO, which is extremely toxic, is employed to

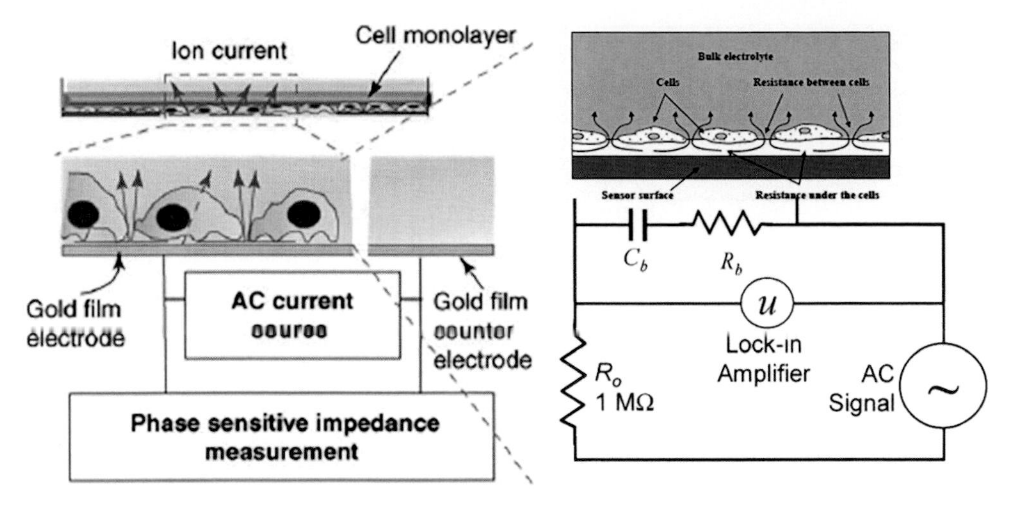

Fig. 2. Schematic of EIS chip and its underlying mechanism.

demonstrate the ability of the EIS to measure kinetic effects of the nanoparticles before and after exposure to the cells. Cadmium is a toxic material that has been shown to cause lysosomal damage and DNA breakage in mammalian cells and disrupt mitochondrial function and promote apoptosis (9).

2.1.1. Synthesis of AuNPs

AuNPs have been reported to exhibit inert properties, as previous studies have shown that AuNPs with a size less than 100 nm did not induce any adverse effects in human cells (10, 11). Due to these inert properties, AuNPs are often employed in research for various applications such as gene and drug delivery transfection vectors, DNA-binding agents, and in various imaging systems (12).

1. Hydrogen tetrachloraurate ($HAuCl_4$) (40 ml, 1.0 mM).
2. Trisodium citrate ($Na_3C_6H_5O_7$) (4 ml, 1%).

2.1.2. Synthesis of AgNPs

Silver has been shown to exhibit a strong toxicity to a wide range of microorganisms and is widely used in antibacterial solutions (13), and when eukaryotic cells are exposed to AgNPs an apoptotic effect is noticed (14).

1. Sodium borohydride ($NaBH_4$) (60 ml, 2 mM).
2. Silver nitrate ($AgNO_3$) (4 ml, 1.0 mM).

2.1.3. Preparation of the Single-Walled Carbon Nanotubes

Rapid advancements in single-walled carbon nanotubes (SWCNTs) research call for a clearer picture of their cytotoxicity as the effects of SWCNT surface chemistry, surface area, and aggregation on the cell cycle is not well established (9).

1. SWCNTs.
2. Sulfuric acid (H_2SO_4).
3. Nitric acid (HNO_3).

2.1.4. Characterization of Nanomaterials

1. Multimode Nanoscope IIIa system (Veeco Instruments, Santa Barbara, CA).
2. 200 keV transmission electron microscope (Hitachi HF-2000 FEG).

2.1.5. Preparation of Cell Culture Medium Suspension of Nanomaterials

1. F-12K medium (ATCC, Rockville, MD).
2. Fetal bovine serum (Invitrogen, Carlsbad, CA).
3. L-glutamine solution 100× (Invitrogen, Carlsbad, CA).
4. AuNPs, AgNPs, SWCNTs.

2.2. EIS Chip Fabrication

1. Chromium.
2. "4 × 4" clear glass microscope slide.
3. Positive photoresist AZ1518 (AZ Electronic Materials).
4. LOR resists coatings 3B (MicroChem, Newton, MA).
5. AZ400K developer.
6. Ti thin film.
7. rf-magnetron sputtering system (AJA).
8. MicroChem's Remover PG.
9. *SU*-8.

2.3. Cell Culture

The main risk associated with the use and manufacture of nanomaterials is human exposure. One of the more common impacts of the use of nanomaterials is the emission of hazardous air pollutants which contain particulate matter on the order of 1–100 nm in size. Any material in the respirable size range, less than 100 nm in diameter, may have toxic effects on the lung fibroblasts after inhalation (15). In particular, nanoparticles with sizes less than 20 nm affect the alveolar region of the lung, and therefore human lung fibroblasts (CCL-153™) are chosen for this method (16). More recently, the use of nanomaterials in biomedical sciences has placed nanomaterials directly in contact with biological materials, and thus it is necessary to observe their interaction closely.

1. Suggested cell line: CCL-153™ human lung fibroblasts (ATCC, Rockville, MD).
2. F-12K medium (ATCC, Rockville, MD).
3. Fetal bovine serum (Invitrogen, Carlsbad, CA).
4. Trypsin EDTA (Mediatech, Herndon, VA).
5. L-glutamine solution 100× (Invitrogen, Carlsbad, CA).
6. Pen–Strep solution 100× (Invitrogen, Carlsbad, CA).
7. T25 cell culture flask (Corning Inc., Corning, NY).
8. Bright-Line Hemacytometer (Hausser Scientific, Horsham, PA).
9. Shell Lab CO_2 Series Incubator (Sheldon Mfg. Inc., Cornelius, OR).

2.4. Real Time Measurements of Nanoparticle Cytotoxicity

1. Electric cell substrate impedance sensing (ECIS) 1600RE (Applied Biophysics, Troy, NY).
2. Nanomaterials: AuNPs, AgNps, Cadmium oxide, SWCNTs.
3. Confluent 25 cm^2 flask of CCL-153™.
4. 0.25% Trypsin, 0.03% EDTA solution to prepare cell suspensions.
5. Cell culture media: F-12K medium containing 10% fetal bovine serum, 0.3 mg/ml L-glutamine, 100 U/ml penicillin, and 100 mg/ml streptomycin.

3. Methods

3.1. Nanomaterial Synthesis

The AuNPs and AgNPs are fabricated at different sizes (10 and 100 nm) to further understand the size effects of the particles and the underlying mechanisms that govern the uptake of nanoparticles into cells. Most cell types undergo phagocytosis or endocytosis to eliminate large particles that interact with the cell membrane; however, the smaller sized particles may bypass the natural mechanical barriers of the cells which may result in severe tissue damage (17). Once inside the cells, the nanoparticles may interfere with the functions of the cell's organelles and other biomolecular structures leading to damage and even death of the cell. The exact mechanisms of this process are still not fully understood (10, 11). However, with the EIS, the effects of the cellular uptake of the nanoparticles on the proliferation of the cells can be demonstrated.

3.1.1. Synthesis of AuNPs

1. Add 40 ml of 1.0 mM $HAuCL_4$ to a 250 ml Erlenmeyer flask.
2. Place a stir bar in the Erlenmeyer flask and place the flask on a hot plate.
3. Bring the solution to a boil while continuously stirring.
4. Add 4 ml of 1% $Na_3C_6H_5O_7$ drop wise into the boiling $HAuCL_4$ solution.
5. After 3 min, the color of the solution will start to change from light purple to a deep burgundy (approximately 10 min). Samples taken at the beginning of the color change will yield 100 nm AuNps and samples taken after 10 min of stirring yield 10 nm AuNPs.

3.1.2. Synthesis of AgNPs

1. Add 60 ml of 2 mM $NaBH_4$ to a 250 ml Erlenmeyer flask.
2. Place the flask in an ice-bath, add a stirbar to the flask, and place the flask on a stir plate.
3. Add 4 ml of 1.0 mM $AgNO_3$ drop wise into the flask and continue to stir.

4. After 3 min, the color of the solution will start to turn yellow. Samples taken at the beginning of the color change will yield 100 nm AgNPs and samples taken after 10 min of stirring yield 10 nm AgNPs.

3.1.3. Preparation of the SWCNTs

The length of the SWCNTs is an important factor to consider when testing for cytotoxicity. Previous works have shown that ultra-short SWCNTs can be used as reinforcing agents to enhance the mechanical properties of certain polymers in various biomedical applications (18). However, the shorter SWCNTs may affect the cells in a different manner than their longer counterparts as they may be able to enter and damage the cells more easily.

1. Place the SWCNTs into a 3:1 mixture of 95% H_2SO_4—60% HNO_3.
2. Sonicate the solution for 5 h.
3. Dry the sample in an oven for 24 h.

3.1.4. Characterization of Nanomaterials

The nanoparticles in this method, AuNPs (10, 100 nm), AgNPs (10, 100 nm), and SWCNTs (cut, uncut), all should be characterized using TEM and AFM. Nanomaterial characterization is necessary as specific properties of the materials depend on their size and purity. Phase imaging and force spectroscopy are performed by using a Multimode Nanoscope IIIa system from Veeco Instruments (Santa Barbara, CA, USA) in air with a relative humidity 40–50% at room temperature. A 200 keV transmission electron microscope (Hitachi HF-2000 FEG) is used to analyze the structure and composition of the nanomaterials.

1. Disperse the samples in ethanol and sonicate for 15 min. Place the suspension onto a copper grid covered with a carbon thin film for analysis.

3.2. EIS Chip Fabrication

Photolithographic technology was employed for the gold patterning of the microelectrode array on the substrate (Fig. 3). A mask defines the area of the chip that will be exposed to UV light. The mask is placed on top of the photoresist during the exposure process so that only the photoresist under the transparent part of the mask gets exposed.

3.2.1. Substrate Cleaning and Dehydration

1. Clean Microscope slide with acetone and rinse with DI water.
2. Bake at 200°C for 5 min on a contact hotplate.

3.2.2. Sacrifice Layer Coating

1. Spin coat LOR 3B at 2,500 rpm for 45 s.
2. Bake on a hotplate at 150°C for 10 min.
3. Spin coat the AZ1518 positive photoresist on top of the LOR layer at 3,500 rpm for 45 s.
4. Bake in an oven at 110°C for 6 min.

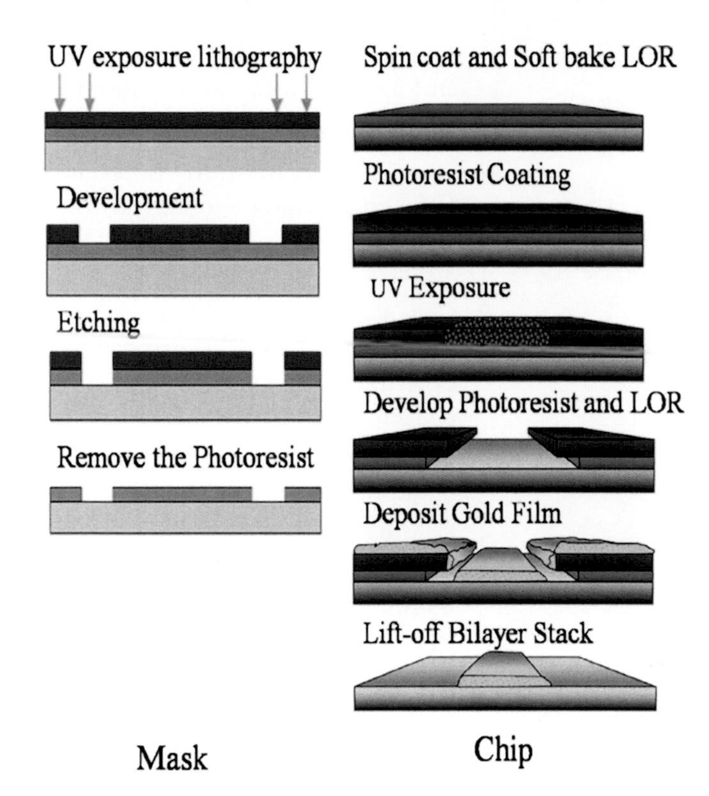

Fig. 3. Flow diagram of fabrication of the mask (*left*) and the array chip (*right*).

3.2.3. Exposure and Development

1. Expose photoresist through a mask under 12.5 mW/cm^2 for 15 s (total dose is 187.5 mW/cm^2).
2. Submerge slide in AZ400K developer for 60 s for development.

3.2.4. Deposition

1. Deposit 10 nm of Ti thin film with a growth rate of 3 nm/min at room temperature by a rf-magnetron sputtering system (AJA) using a Ti target in Ar ambient (10 sccm flow rate) at 75 W of rf power, 5 mTorr pressure, and room temperature.
2. Deposit a 150 nm (7 min) Au thin film on top of the Ti thin film in Ar ambient (10 sccm flow rate) at 75 W of rf power, 5 mTorr pressure and room temperature.

3.2.5. Lift-Off

1. Submerge the slide in MicroChem's Remover PG to dissolve the sacrificial layer to form the desired pattern.

3.2.6. Protection Layer Coating

1. Spin coat SU-8 onto the slide at 2,000 rpm for 45 s.
2. Bake on a hotplate at 95°C for 13 min.

3.2.7. Final Exposure and Development

1. Expose the photoresist through the mask under 12.5 mW/cm² for 24 s (total dose is 300 mW/cm²).
2. Bake on a hotplate for 16 min at 95°C.
3. Dip the substrate in *SU*-8 developer for 4 min and rinse with IPA and DI water.

3.3. Cell Culture

1. Sterilize all glassware for handling cell cultures and media.
2. Cells should be seeded in a 25 cm² flask and placed in an incubator (37°C, 5% CO_2 atmosphere) for at least 24 h prior to experimentation so that the cells reach confluency with a final concentration of 10^5 cells/ml.
3. Remove the flask of cells from the incubator and place in hood or appropriate place for cell inoculation.
4. Remove medium and rinse flask with 2 ml of Trypsin–EDTA solution.
5. Remove the solution and add an additional 1–2 ml of Trypsin–EDTA solution. Allow the flask to sit at room temperature (or at 37°C) for 2–10 min until the cells detach.
6. Add 3 ml of fresh culture medium to the flask, aspirate, and dispense into a 15 ml centrifuge tube.
7. Centrifuge the cell suspension at 100 × g for 5 min and remove the supernatant.
8. Resuspend the cells with 10 ml of cell culture media.
9. Remove 0.2 ml of the cell suspension and fill both sides of the hemocytometer chamber.
10. Count viable cells in each of the four corner squares bordered by triple lines, omitting cells lying on these lines. This is repeated for the second side of the chamber.
11. Calculate the viable cell concentration per ml using the following formula:

    $C_1 = t \times 1/4 \times 10^4$

    t = total viable cell count of four corner squares

    $1/4$ = correction to give mean cells per corner square

    10^4 = conversion factor for counting chamber

    C_1 = initial cell concentration per ml
12. Make the appropriate dilutions to the cells in the centrifuge tube if necessary.
13. Store centrifuge in incubator.

3.4. Real Time Measurements of Nanoparticle Cytotoxicity

The embodiment of this procedure is that the cells are placed in each well, drift downward and attach themselves onto the electrode surface over time. The current flowing though each electrode is now impeded by the cell monolayer, increasing the impedance and resistance measurements. From Ohm's law, $V=I\ R$, it is possible to monitor the cell attachment and proliferation from the increase in impedance or resistance measurements. Nanotoxicity measurements can be made by monitoring the change in resistance of each electrode. An extremely toxic material would cause the cells to die and detach from the electrode at a rapid rate which, in turn, would decrease the resistance reading of the system (see Note 2). On the other hand, a nontoxic material would not affect the attachment of the cells; hence the resistance measurements would be similar to the measurements of the cells by themselves (control) (see Note 3).

1. Place 200 ml of warmed media in the eight wells of the ECIS array.
2. Insert array into the slot at the base of the array holder.
3. Turn the black knob on the array holder to lower the clamp until the pins are in contact with the gold pads on the array and tighten until finger tight.
4. On the PC desktop, double click the *ECIS 1600R* icon to open the software.
5. From the main menu, choose the Acquire Data tab.
6. Select 8W1E, then select wells 1–8, and click on the *Electrode Check* button.
7. Record the electrode values.
   (a) Values should be in the range of 1,600–2,000.
8. Remove the array from the holder and place in hood or appropriate area for cell inoculation.
9. Place the centrifuge tube of cells in hood or appropriate area for cell inoculation.
10. Remove array lid from the array and aspirate the media from each well.
11. Add 400 ml of cell suspension into the appropriate wells.
    (a) It is suggested that at least one well be filled with only media to obtain a proper baseline reading for the media being used.
12. Place the inoculated array in the holder and lower the clamp to make contact.
13. In the 1600R software, choose New Experiment from the Aquire Data menu.
14. Name the file and write any specific conditions in the Comments page.
15. Set up the parameters (e.g., time between points and duration of the run) according to the specific materials to be tested.
16. Click on Start Run

4. Notes

1. Results will vary among nanomaterials tested.
2. In this method we report CdO as one of the materials to test due to its known toxic effects. Results should show steady resistance readings when the CdO is added to the cells at the beginning of the run indicating disruption in cell attachment. Another test that can be performed with CdO is to allow the cells to grow on the electrode to confluency over a period of 24 h before adding the cadmium oxide. This setup would provide the kinetics of the interaction of the toxin with the cells.
3. The cytotoxic measurements of AuNPs (10, 100 nm) toward CCL-153™ should show little difference in resistance values to those observed for the control (cells only), whereas AgNPs and SWCNTs should exhibit some noticeable changes in readings depending on the preparation and characteristics of the materials.

Acknowledgments

The work is currently supported by NIH 1 R15 ES021079-01 of Department of Defense/US Army Medical Research and Material Command, Wallace H. Coulter Foundation, and NSF MRI 0821582. We would like to thank the Advanced Material Engineering Research Institute (AMERI) at FIU for allowing us to use the MEMS facilities.

References

1. Luong JHT, Male KB, Glennon JD (2008) Biosensor technology: technology push versus market pull. Biotechnol Adv 26(5):492–500
2. Lewinski N, Colvin V, Drezek R (2008) Cytotoxicity of nanoparticles. Small 4(1):26–49
3. Wilhelm C, Gazeau F, Roger J, Pons JN, Bacri JC (2002) Interaction of anionic superparamagnetic nanoparticles with cells: kinetic analyses of membrane adsorption and subsequent internalization. Langmuir 18(21): 8148–8155
4. Banerjee R, Katsenovich Y, Lagos L, Senn M, Naja M, Balsamo V, Pannell KH, Li C-Z (2010) Functional magnetic nanoshells integrated nanosensor for trace analysis of environmental uranium contamination. Electrochim Acta. 17(27): 3120–3141
5. Liu C, Alwarappan S, Li C-Z (2010) Design and characterization of novel membraneless enzymatic biofuel cell based on graphene nanosheets. Biosens Bioelectron 7:1829–1833
6. Alwarappan S, Li C-Z (2010) Simultaneous detection of dopamine, ascorbic acid and uric acid at electrochemically activated carbon nanotube biosensors. Nanomedicine 6:52–57
7. Alwarappan S, Prabhulkar S, Durygin A, Li C-Z (2009) The effect of electrochemical pretreatment on the sensing performance of single walled carbon nanotubes. J Nanosci Nanotechnol 9: 2991–2996
8. Li C-Z, Choi W-B, Chuang C-H (2008) Enhancement of photocurrents by finite-sized SWNT based thin films. Electrochim Acta 54:821–828

9. Tian F, Cui D, Schwarz H, Estrada GG, Kobayashi H (2006) Cytotoxicity of single-wall carbon nanotubes on human fibroblasts. Toxicol In Vitro 20:1202–1212
10. Brayner R (2008) The toxicological impact of nanoparticles. Nano Today 3:48–55
11. Connor EE, Mwamuka J, Gole A, Murphy CJ, Wyatt MD (2005) Gold nanoparticles are taken up by human cells but do not cause acute cytotoxicity. Small 1:325–327
12. Ghosh PS, Kim CK, Han G, Forbes NS, Rotello VM (2008) Efficient gene delivery vectors by tuning the surface charge density of amino acid-functionalized gold nanoparticles. ACS Nano 2:2213–2218
13. Elechiguerra JL, Burt JL, Morones JR, Camacho-Bragado A, Gao X, Lara HH, Yacaman MJ (2005) Interaction of silver nanoparticles with HIV-1. J Nanobiotechnology 3:6–16
14. Arora S, Jain J, Rajwade JM, Paknikar KM (2009) Interactions of silver nanoparticles with primary mouse fibroblasts and liver cells. Toxicol Appl Pharmacol 236:310–318
15. Mossman BT, Borm PJ, Castranova V, Costa DL, Donaldson K, Kleeberger SR (2007) Mechanisms of action of inhaled fibers, particles and nanoparticles in lung and cardiovascular diseases. Part Fibre Toxicol 4:1–4
16. Elder A, Vidyasagar S, DeLouise L (2009) Physicochemical factors that affect metal and metal oxide nanoparticle passage across epithelial barriers. Wiley Interdiscip Rev Nanomed Nanobiotechnol 1:434–450
17. Barnes PJ, Shapiro SD, Pauwels RA (2003) Chronic obstructive pulmonary disease: molecular and cellular mechanisms. Eur Respir J 22:672–688
18. Xinfeng S, Balaji S, Quynh PP, Patrick PS, Jared LH, Lon JW, James MT, Robert MR, Antonios GM (2008) In vitro cytotoxicity of single-walled carbon nanotube/biodegradable polymer nanocomposites. J Biomed Mater Res A 86:813–823

Chapter 14

Free Energy Calculation of Permeant–Membrane Interactions Using Molecular Dynamics Simulations

Paolo Elvati and Angela Violi

Abstract

Nanotoxicology, the science concerned with the safe use of nanotechnology and nanostructure design for biological applications, is a field of research that has recently received great attention, as a result of the rapid growth in nanotechnology. Many nanostructures are of a scale and chemical composition similar to many biomolecular environments, and recent papers have reported evident toxicity of selected nanoparticles. Molecular simulations can help develop a mechanistic understanding of how structural properties affect bioactivity. In this chapter, we describe how to compute the free energy of interactions between cellular membranes and benzene, the main constituent of some toxic carbonaceous particles, with well-tempered metadynamics. This algorithm reconstructs the free energy surface and accelerates rare events in a coarse-grained representation of the system.

Key words: Free energy calculation, Cellular membranes, Well-tempered metadynamics, Nanotoxicity, Molecular dynamics, Lipid bilayers

1. Introduction

Nanoparticles exhibit unique physicochemical properties different from fine particles of the same composition, and they have been exploited in a wide number of novel applications and products (1–3). Some of the most promising applications include structural engineering, electronics, optics, consumer products, alternative energy, water remediation, and medicinal uses as therapeutic, diagnostic, and drug delivery devices.

The unique physicochemical properties of nanomaterials are attributable to their size, chemical composition, solubility, shape, surface structure, and aggregation. Although so impressive, the novel properties raise concerns about adverse effects on biological systems (4–13).

Joshua Reineke (ed.), *Nanotoxicity: Methods and Protocols*, Methods in Molecular Biology, vol. 926,
DOI 10.1007/978-1-62703-002-1_14,

To gain insight into the nanoscale phenomena associated with nanoparticles interacting with biological systems, computer simulations such as molecular dynamics (MD) have become an appealing and powerful tool, owing to advances in computing power. MD simulations of biomolecules treat the molecules as classical particles interacting through a potential energy function. The evolution of the system over time is determined via numerical integration of Hamilton's equations of motion, typically discretized into steps on the order of femtoseconds. The information produced by such MD simulations is an atomic-resolution model of structural transitions and conformational equilibria in the system of interest (14–18).

The purpose of this chapter is to illustrate the capability of molecular dynamics to assess the interactions of permeants with biological systems. Specifically, we are interested in the nano–bio interface that comprises the thermodynamic exchanges between benzene, a carcinogen to humans, and biological components. The free energy of interaction between benzene and cellular membranes provides information on the ability of nanostructures to pass through membranes designed to act as barriers. This possibility can create many opportunities for toxic effects to occur.

There exist a variety of techniques to determine the free energy landscapes of a system (19). Among them, metadynamics (20) has been employed to study various systems, like protein folding, chemical reactions, and phase transition (see (21) and reference therein). Metadynamics, first suggested by Laio and Parrinello in 2002 (20), is used to improve the sampling of a system and reconstruct the free energy landscape. The algorithm assumes that the system can be described by a few collective variables s_α, which discern the states (initial, final, and intermediates) of the system, as well as the slow events that are relevant to the process of interest.

During a molecular dynamics simulation, the metadynamics algorithm allows the system to visit new states by periodically adding a Gaussian-shaped repulsive potential to the previously visited free energy locations. The sum of these Gaussian is the bias potential, $V(\{s_\alpha\},t)$:

$$V(\{s_\alpha\},t) = \sum_i^N w_0 \exp\left(-\sum_\alpha^d \frac{[s_\alpha(i\tau) - s_\alpha]^2}{2\sigma_\alpha^2}\right)$$

where w_0 and σ are the height and width of the Gaussians, τ is the time between the deposition of two consecutive Gaussians, N is the number of added Gaussians, and d is the number of collective variables (CVs).

As the simulation proceeds, more Gaussians are added at a constant rate, until the system explores the full energy landscape. The repulsive potential makes it possible for the system to cross the lowest energy barrier between local basins or, when all the accessible minima are explored, to diffuse freely between states.

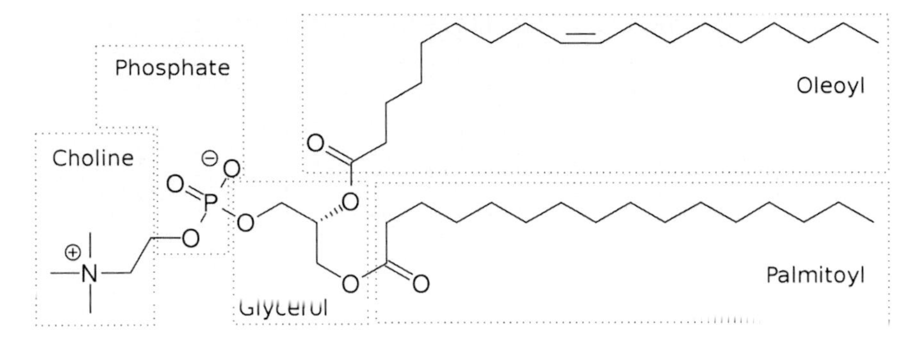

Fig. 1. Structural formula of a POPC molecule.

One great advantage of metadynamics is the fact that it does not require an initial estimate of the energy landscape to explore.

In this chapter we use a recent refinement of this technique, namely well-tempered metadynamics (22). The difference between metadynamics and well-tempered metadynamics lies in the gradual reduction of the heights of the deposited Gaussians. Mathematically the height of the Gaussians is expressed as:

$$w(t) - w_0 \exp\left(-\frac{V(\{s_\alpha(t)\},t)}{k_B \Delta T} \right)$$

where k_B is the Boltzmann constant, T is the temperature of the system, and $T + \Delta T$ is the fictitious temperature of the CVs. This modification allows the bias potential to converge, thus solving one of the major problems of the original formulation of the metadynamics algorithm. The final free energy $F(\{s_\alpha\})$ can be then recovered by using the fact that:

$$\lim_{t \to \infty} V(\{s_\alpha\},t) = -\frac{\Delta T}{T + \Delta T} F(\{s_\alpha\})$$

The reader is referred to the cited literature for more comprehensive details.

As a system of interest for this study, we studied the interactions of benzene with a lipid bilayer composed of palmitoyl-oleoyl-phosphatidylcholine (POPC, see Fig. 1). Lipids together with membrane proteins constitute the skeleton of biological membranes that show high compositional complexity and versatile functional capabilities (23).

Among pollutants, benzene has been a well-studied chemical. Over the last years, a great volume of research has been reported on the exposure to benzene associated with increased risk of developing hematopoietic disorders. Benzene can cause, among other diseases, cancers of the lung, liver, stomach, esophagus, intestine, nasopharynx; acute myelogenous leukemia, acute lymphocytic leukemia, acute erythroleukemia, myelomonocytic, promyelocytic, undifferentiated and hairy cell leukemias; chronic myelogenous or lymphocytic leukemias; Hodgkin's lymphoma, non-Hodgkin's lymphoma (24–26), and multiple myeloma (27–29). The recent "Benzene 2009" symposium of international experts has focused on the health effects and mechanisms of toxicity of benzene, highlighting new aspects of exposure assessment as well as exploration of the range of diseases caused by benzene in humans.

The methods described below for this system are broken into four subsections: membrane input files, permeant input files, system input files, and free energy calculation.

2. Materials

2.1. Softwares

Free energy calculations, associated with the passive transport of small molecules through a cellular membrane, can be performed with a variety of techniques, and for each of them, different combinations of software and scripting can be used. Here we propose a selection of cross-platforms and freely available softwares that can be used to reproduce the example. For each of them we report the version used for the calculations. Please refer to the documentation of each software for installation and general usage instructions.

- *Avogadro*. This is an open-source molecular builder and visualization tool. It can be downloaded from http://avogadro.openmolecules.net/ (Version 1.00).
- *NAMD* (30). This is a parallel molecular dynamics code. It was developed by the Theoretical and Computational Biophysics Group in the Beckman Institute for Advanced Science and Technology at the University of Illinois at Urbana-Champaign. It can be downloaded from http://www.ks.uiuc.edu/Research/namd (free registration required) (Version 2.7b4).
- *PLUMED* (31). This is a plugin for free energy calculation of molecular systems, which works with NAMD (and other MD codes). It can be downloaded (free registration required) from http://merlino.mi.infn.it/~plumed/PLUMED/Home.html (Version 1.2) (see Note 1).
- *VMD* (32). This is a molecular visualization program for displaying, post-processing analysis of molecular systems as well as NAMD simulation setup. It can be downloaded (free registration required) from http://www.ks.uiuc.edu/Research/vmd (Version 1.8.7).

For the simulations reported in this chapter, we used the CHARMM force field (FF). The file containing the parameters for CHARMM27 (8, 33) (toppar_c35b2_c36a2.tgz) can be downloaded from http://mackerell.umaryland.edu/CHARMM_ff_params.html, while general CHARMM FF (34) (version 2b5) can be found at http://dogmans.umaryland.edu/~kenno/cgenff/download.html.

3. Methods

The system of interest, benzene interaction with a POPC lipid bilayer, was simulated at 310 K (human body temperature), using a Langevin thermostat and, when needed, at 1 atm using the Nose-Hoover Langevin piston method (35, 36). The equation of motion was integrated using a timestep of 1 fs for all the interactions except the electrostatics ones, which were evaluated every 2 fs (37). Van der Waals potentials were smoothed to zero between 10 and 12 Å; electrostatic interactions were modeled with the Particle Mesh Ewald method (38) (1 Å grid spacing and cubic interpolation) and cubic periodic boundary condition were applied.

Below we report the steps necessary to build the configurations of the system (benzene and lipid bilayer) and to run the simulations.

3.1. Membrane Input Files

1. The CHARMM-GUI (39) website contains an excellent tool to prepare the initial membrane configuration files. By visiting the "Membrane generator" section of the website (http://www.charmm-gui.org/?doc=input/membrane) and by following the step-by-step instructions shown there, the initial configuration and topology file for the lipid bilayer can be generated. In this example we use a membrane composed by 72 POPC molecules (36 in each leaflet) separated by 40 Å of water. Since we do not use the CHARMM code, we can stop at the step 3 of the online procedure and save the three files containing the initial positions ("step5_assembly.pdb"), topology ("step5_assembly.xplor.psf"), and box information ("step5_assembly.str"). The last file is only needed because it contains the size of the box that has to be specified in the NAMD input file (e.g., *cellBasisVector1*) (see Note 2).
2. The next step is to run at least 500 steps of energy minimization of the membrane in water, followed by a relaxation run in the NPzAT ensemble. The relaxation should be long enough to allow equilibration of the system. For the system under study, 1 ns is likely enough to show that the length of the *z*-axis of the periodic box is oscillating around its equilibrium value. For sake of clarity, we refer to the pdb file, which describes the final relaxed membrane configuration (see Fig. 2), and the psf file as "mem.pdb" and "mem.psf," respectively.

3.2. Permeant Input Files

1. For small molecules, it is easy to manually prepare the initial pdb file. Use the Avogadro's "Draw tool" to draw an approximate shape of benzene with alternating double bonds, then optimize the structure and save it as "benz.pdb."
2. In order to generate the psf file, we need to modify "benz.pdb" so that the atom names (characters 13–16) and the residue name (characters 18–20) correspond to the one listed in the topology file. For benzene, we can locate the relevant

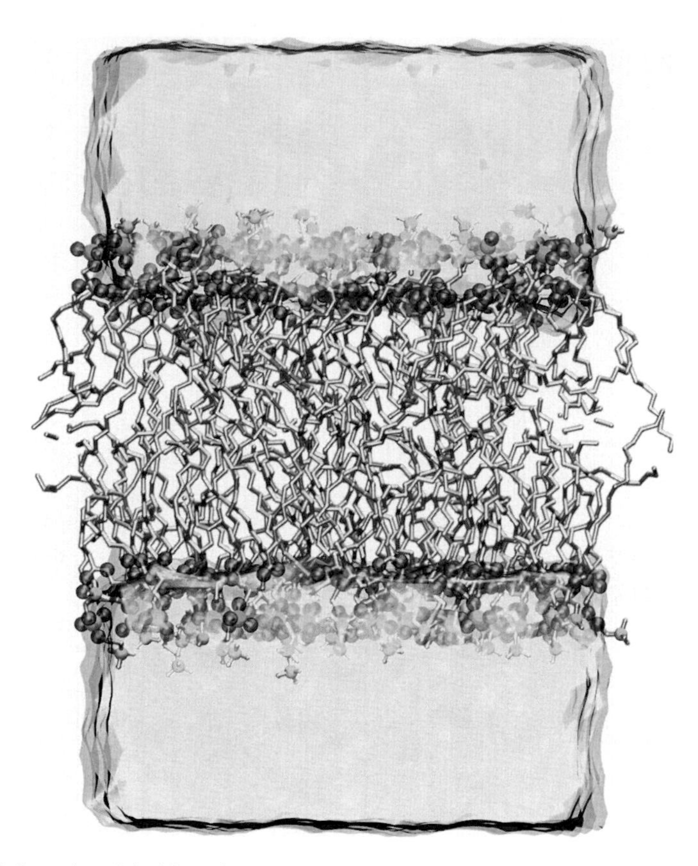

Fig. 2. Snapshot of the bilayer/water system. Three regions are clearly distinguishable, the water phase (*semi-transparent boxes*), the interface regions where the heads of the lipids (POPC oxygens are shown as *dark spheres*) coexist with water and the hydrophobic region composed by the aliphatic legs of POPC. The graphics were generated using VMD.

names by searching the string "RESID BENZ" in the file "top_all36_cgenff.rtf." After the change, VMD can be used to check that the right names (Mouse->Label->Atoms) and connectivity are used.

3. We then use psfgen, a tool in the NAMD package (see Chapter 4 of the NAMD manual for more information), to build the benzene's psf file. This task can be done by writing a script (see "buildpsf" for an example) and then running it (by executing "psfgen buildpsf") (see Note 3).

   buildpsf:

   ```
   topology top_all36_cgenff.rtf
   segment BENZ {pdb benz.pdb}
   coordpdb benz.pdb BENZ
   ### Write the PDB and PSF files
   writepdb out.pdb
   writepsf out.psf
   exit
   ```

3.3. System Input Files

1. Before merging the pdb and psf files of the permeant ("out.pdb" and "out.psf") with the ones of the membrane ("mem.pdb" and "mem.psf"), we will change the relative position of the benzene with respect to the membrane, in order to control the initial conditions. Open both pdb files with VMD and drag the benzene fragment (Mouse->Move->Molecule) in the water phase at about 10–15 Å from the surface of the membrane. To have a clearer idea of the distances involved, the ruler grid (Extensions->Visualizations->Ruler) can be activated. Once the benzene is at the desired position, "benz.pdb" has to be updated with the new coordinates (see Note 4).
2. A single pdb and psf file can be prepared by running another script that uses psfgen, like the following:

```
merge:
readpsf mem.psf
coordpdb mem.pdb
readpsf out.psf
coordpdb out.pdb
writepsf sys.psf
writepdb sys.pdb
exit
```

3. Due to the possible bad contacts that are present in the newly constructed system, a energy minimization run (500 steps) followed by another relaxation is needed: 500 ps simulation in the NPzAT ensemble is likely enough to equilibrate the system used in this example (see Note 5).

3.4. Free Energy Calculation

1. The system is now set up for the free energy calculation with well-tempered metadynamics. In order to do that, we need first to define a collective variable in the plumed input file (*meta.inp*). The CV is a function of the microscopic coordinates of the system that includes all the slow modes of the process under study and at the same time, it discriminates among the relevant states. In the case of the passive permeation of a membrane, a possible choice for the CV is the z projection of the distance between the center of mass of the membrane and the center of mass of the benzene molecule. In order to define these two groups, we need to list the atoms that compose the groups. We select all the atoms composing the benzene (e.g., LOOP 19915 19926 1), and some heavy atoms, e.g., the phosphorous, the nitrogen, and the carbon atoms at the end of each aliphatic chain, for the membrane (see Note 6). Once the groups that are needed to define the CV and the value of the Gaussian width (SIGMA) are provided, it is easy to write the actual collective variable, e.g., DISTANCE LIST <MEM> <BENZ> DIR Z SIGMA 0.25 (see Note 7).

2. Other well-tempered metadynamics parameters that need to be specified are the initial energy deposition rate and the effective sampling temperature of the CV space. In order for the system to have enough time to equilibrate after the addition of a new Gaussian, an initial deposition rate in the range of 0.2–0.4 kcal/mol/ps is suggested. The choice of the initial rate is not critical and it does not affect the final result but only the initial transient "filling" period. For our systems, benzene and POPC bilayer, we add a new Gaussian every 500 fs (e.g., W_STRIDE 500, since the MD timestep is of 1 fs), with an initial height of 0.2 kcal/mol (HEIGHT 0.2). The fictitious CV temperature determines the rate of decay of the Gaussian height and therefore the maximum free energy barrier that can be crossed with well-tempered metadynamics. Therefore, the choice of the bias factor is extremely important. Since we expect the barrier to be less than 6 kcal/mol (see Note 8), a bias factor of 10 should be appropriate (see Note 9).
3. Finally, to reduce the computational cost and avoid interactions with the periodic image, a wall can be put at 35 Å from the center of the membrane (UWALL CV 1 KAPPA 60 LIMIT 35) (See Note 10). The final PLUMED input file (meta.inp) will look like:

```
### General options
    PRINT W_STRIDE 125
    HILLS W_STRIDE 500 HEIGHT 0.2
    WELLTEMPERED SIMTEMP 310 BIASFACTOR 10
### definition of groups
# Selected atom to represent the center of mass of membrane
    MEM->
    LOOP 1 9648 134 # N atoms
    LOOP 20 9648 134 # P1 atoms
    LOOP 88 9648 134 # C218 atoms
    LOOP 131 9648 134 # C316 atoms
    MEM<-
    # All the atoms of benzene
    BENZ->
    LOOP 19915 19926 1 # all the benzene atoms
    BENZ<-
### definition of CVs
    DISTANCE LIST <MEM> <BENZ> DIR Z SIGMA 0.25
    UWALL CV 1 KAPPA 60 LIMIT 35
    ENDMETA
```

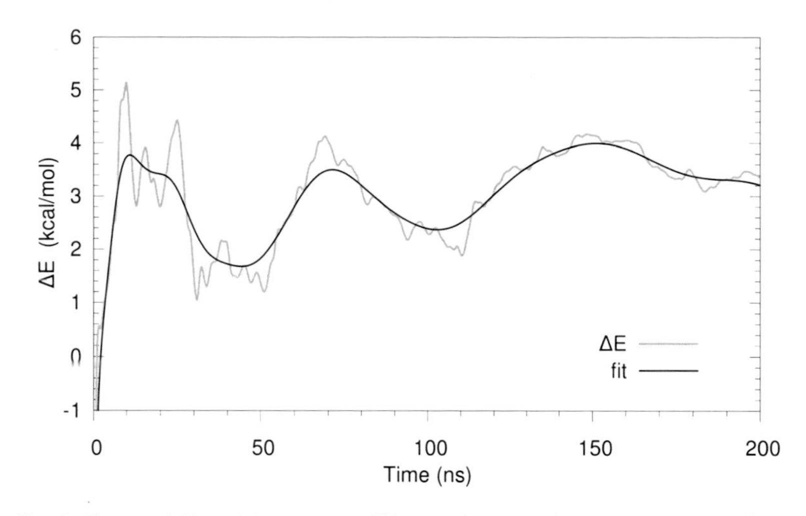

Fig. 3. Time evolution of free energy difference between the states corresponding to benzene inside (CV <5) and outside (CV >30) the bilayer.

4. The input file of NAMD does not need any special modification except for the addition of the lines "*plumed on*" and "*plumedfile meta.inp*."
5. The simulation will take several nanoseconds to converge. Convergence can be monitored by checking several parameters. First, it is important that the system has explored all the values of interest of the CV several times. This can be checked by plotting the first (time) and second (CV value) columns of the COLVAR file. When this condition is met, we can monitor the evolution of the difference in free energy between two states (e.g., the benzene in water and inside the membrane). If the states are properly selected, after an initial phase in which the basins are filled, the free energy difference will show a dumped oscillation around the final value (see Fig. 3 or (15)). From this point, the simulation can be continued to reduce the error (which goes to zero for infinite time) in the final free energy surface (see Note 11).
6. The final free energy (see Fig. 4) shows that the benzene molecule is more stable in the membrane than in water by about 3.7 kcal/mol. However, the minimum in the free energy is not located in the center of the membrane but around 8 Å (with a difference of about 0.7 kcal/mol), which corresponds to the interface between the hydrophobic and the hydrophilic regions of the membrane (see Note 12). This result is likely due to the reduced entropy of the aliphatic chains of the lipids when the benzene molecule is inside the hydrophobic region. However, since the thermal energy is slightly more than 0.6 kcal/mol, benzene can present near both interface regions, therefore potentially taking part in a reaction that happens on both sides of the cellular membranes.

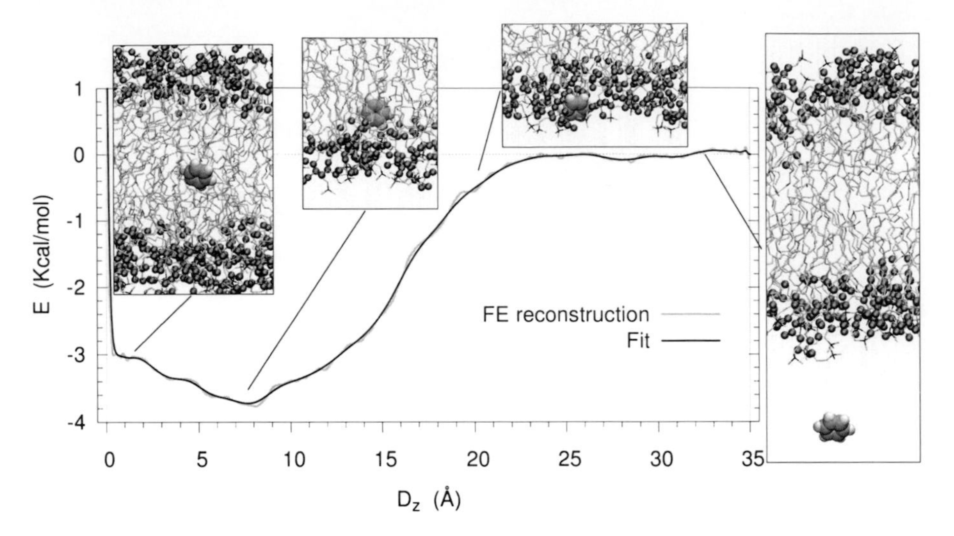

Fig. 4. Free energy profile of benzene insertion in POPC reconstructed using the well-tempered metadynamics method. The selected snapshots of the system illustrate the positions of benzene with respect to the center of the membrane.

4. Notes

1. Recent versions of the NAMD code have their own implementation of collective variable-based free energy calculations; however, the PLUMED plugin provides a convenient framework for introducing new collective variables. Moreover, almost the same input file can be used with many different codes, for both classical and ab initio MD. For example, it would be particularly straightforward to repeat this example by employing LAMMPS (40) instead of NAMD. The plumed input file is the same, except for some unit conversions (since the values in PLUMED are in the same unit as the underlying MD code) and the conversion of the CHARMM topology files. The latter problem can be solved by using the ch2lmp tools present in LAMMPS.
2. The size of the system, in particular the amount of water present in the system, can be changed to reduce the computational cost of the simulations. However, the minimum value of z is limited by the cutoff radius of the nonbonded interaction and from the size of the permeant since periodic boundary conditions are applied. The size of the box should never decrease to the point that the permeant in the water phase is simultaneously within the cutoff distance of one leaflet and the periodic image of the other.
3. For the script to work, both the topology file (*top_all36_cgenff.rtf*) and the parameter file (*par_all36_cgenff.prm*) are required to be in the same directory.
4. The initial position of the permeant can be chosen differently and in case multiple walkers are employed (see Note 10), is

preferable to start from different configurations. However, the complexity of correcting unphysical configuration (e.g., an aliphatic chain passing through the aromatic ring) makes positioning the benzene in the water a simpler solution. These considerations are general, since creating a cavity in a solvent poses no problem as long as some solvent molecules can be removed from the system.

5. When merging the psf and pdb files of the permeants with the ones of the membrane sometimes some molecule of water need to be removed, since some of them can be trapped in an unlikely configuration by the merging procedure (e.g., water inside a fullerene). For that purpose, the psfgen command *delatom* can be used. More information and examples can be found on NAMD webpages.
6. For the purpose of the center of mass calculation, the whole membrane was approximate with some selected heavy atoms. The reason is that the frequent computation of the center of mass of all the atoms of the lipid bilayer would sensibly slow down the simulation and strongly reduce the possibility to run in parallel since this part of the PLUMED code is serial. Different selections are also possible as long as the computed center of mass has a smooth trajectory that is similar to the one of the whole membrane.
7. The choice of sigma is related to the system and depends on the choice of CVs. It determines the resolution and the error in the reconstructed free energy profile (41). A wider Gaussian will help fill the free energy wells faster but will also increasingly affect the configurations that are further away from the point of the Gaussian drop in the space of the CV. This situation can lead to the loss of local details and in extreme cases to inaccurate free energy surfaces.
8. Estimating the free energy barrier is required to minimize the computational cost of this simulation. The experimental value (42) of the free energy of the transfer of benzene from water to hexadecane shows that the hydrophobic phase is favored by ~ 2.93 kcal/mol, while MD simulations of the insertion of benzene from water in DOPC (42) and DPPC (43) membranes reported 5.7 and 1.2 kcal/mol, respectively.
9. A bias factor of 10 for a simulation at 310 K means that barriers of the order of $10 \times 310 \times 0.0019859 = \sim 6.0$ kcal/mol (0.0019859 is the ideal gas constant in kcal/mol/K) can be crossed during a well-tempered metadynamics simulation. Bigger bias factor can be chosen without affecting the result; however, a longer time to reach the convergence would be needed. It is important to note that bias factor can be changed on the fly.

10. Free energy simulations can be quite demanding from the computational point of view. To reduce the time needed for the computation (although not the total computational time), multiple walkers can be used: with this option multiple simulations run in parallel and concur to reconstruct the same free energy surface by simultaneously adding bias in different places of the CV space (for details see PLUMED manual).
11. Another way to check the convergence is to plot the heights of the hills vs. time (columns 1 and 4 of the HILLS file). This provides an indication of the simulation status. As a general guideline, if the deposition rate is consistently lower than 0.05 kcal/mol/ps the simulation can be stopped. This procedure, however, should be used as a quick test not as a substitute of the plot of the free energy difference between different basins.
12. The reconstructed free energy may present some more or less marked oscillations near walls, in this case both at 35 Å (added in the input file) and 0 Å (deriving from the definition of distance, which is always greater or equal to 0). These numerical artifacts disappear with the convergence of the free energy surface and by choosing a smaller energy deposition rate.

Acknowledgments

This work is funded by a National Science Foundation grant NSF-CAREER (CBET 0644639). Computational time Research was supported in part by the National Science Foundation through TeraGrid resources. The authors are grateful to Dr. M. Bonomi for his invaluable feedback and useful suggestions.

References

1. Nel AE, Madler L, Velegol D, Xia T, Hoek EMV, Somasundaran P, Klaessig F, Castranova V, Thompson M (2009) Understanding biophysicochemical interactions at the nano-bio interface. Nat Mater 8:543–557
2. Geraci CL, Castranova V (2010) Challenges in assessing nanomaterial toxicology: a personal perspective. Wiley Interdiscip Rev Nanomed Nanobiotechnol 2:569–577
3. Fubini B, Ghiazza M, Fenoglio I (2010) Physico-chemical features of engineered nanoparticles relevant to their toxicity. Nanotoxicology 4:347–363
4. Oberdorster G (2009) Safety assessment for nanotechnology and nanomedicine: concepts of nanotoxicology. J Intern Med 267:89–105
5. Nel A, Xia T, Madler L, Li N (2006) Toxic potential of materials at the nanolevel. Science 311:622–627
6. Mahmood M, Dervishi E, Xu Y, Li ZR, Al-Muhsen MA, Ali N, Whitlow M, Biris AS (2009) Cytotoxicity and biological effects of functional nanomaterials delivered to various cell lines. In: Shastri VP, Lendlein A, Liu L, Mikos A, Mitragotri S (eds) Advances in material design for regenerative medicine, drug delivery and targeting/imaging. Cambridge University Press, pp 79–85
7. Oberdorster G, Stone V, Donaldson K (2007) Toxicology of nanoparticles: a historical perspective. Nanotoxicology 1:2–25

8. Jiang JK, Oberdorster G, Biswas P (2009) Characterization of size, surface charge, and agglomeration state of nanoparticle dispersions for toxicological studies. J Nanopartic Res 11: 77–89
9. Oberdorster G, Elder A, Rinderknecht A (2009) Nanoparticles and the brain: cause for concern? J Nanosci Nanotechnol 9:4996–5007
10. Rushton EK, Jiang J, Leonard SS, Eberly S, Castranova V, Biswas P, Elder A, Han XL, Gelein R, Finkelstein J, Oberdorster G (2010) Concept of assessing nanoparticle hazards considering nanoparticle dosemetric and chemical/biological response metrics. J Toxicol Environ Health Part A Curr Issues 73:445–461
11. Vanwinkle BA, Bentley KLD, Malecki JM, Gunter KK, Evans IM, Elder A, Finkelstein JN, Oberdorster G, Gunter TE (2009) Nanoparticle (NP) uptake by type I alveolar epithelial cells and their oxidant stress response. Nanotoxicology 3:307–318
12. Myojo T, Ogami A, Oyabu T, Morimoto Y, Hirohashi M, Murakami M, Nishi K, Kadoya C, Tanaka I (2010) Risk assessment of airborne fine particles and nanoparticles. Adv Powder Technol 21:507–512
13. Canady RA (2010) The uncertainty of nanotoxicology: report of a society for risk analysis workshop. Risk Anal 30:1663–1670
14. Monticelli L, Salonen E, Ke PC, Vattulainen I (2009) Effects of carbon nanoparticles on lipid membranes: a molecular simulation perspective. Soft Matter 5:4433–4445
15. Freddolino PL, Harrison CB, Liu YX, Schulten K (2010) Challenges in protein-folding simulations. Nat Phys 6:751–758
16. Fiedler SL, Violi A (2010) Simulation of nanoparticle permeation through a lipid membrane. Biophys J 99:144–152
17. Choe S, Chang R, Jeon J, Violi A (2008) Molecular dynamics simulation study of a pulmonary surfactant film interacting with a carbonaceous nanoparticle. Biophys J 95: 4102–4114
18. Chang R, Violi A (2006) Insights into the effect of combustion-generated carbon nanoparticles on biological membranes: A computer simulation study. J Phys Chem B 110:5073–5083
19. Chipot C, Pohorille A (2007) Free energy calculations. Springer, Berlin
20. Laio A, Parrinello M (2002) Escaping free-energy minima. Proc Natl Acad Sci U S A 99:12562–12566
21. Laio A, Gervasio FL (2008) Metadynamics: a method to simulate rare events and reconstruct the free energy in biophysics, chemistry and material science. Rep Prog Phys 71:126601
22. Barducci A, Bonomi M, Parrinello M (2010) Linking well-tempered metadynamics simulations with experiments. Biophys J 98:L44–L46
23. Singer SJ, Nicolson GL (1972) Fluid mosaic model of structure of cell-membranes. Science 175:720–731
24. Savitz DA, Andrews KW (1997) Review of epidemiologic evidence on benzene and lymphatic and hematopoietic cancers. Am J Ind Med 31: 287–295
25. Andrews KW, Savitz DA, Kupper LL, Millikan RC, Loomis DP, Aldrich TE (1997) Occupational sunlight exposure and malignant melanoma. Am J Epidemiol 145:162
26. Smith MT, Jones RM, Smith AH (2007) Benzene exposure and risk of non-Hodgkin lymphoma. Cancer Epidemiol Biomarkers Prev 16:385–391
27. Goldstein BD (1990) Is exposure to benzene a cause of human multiple-myeloma. Ann N Y Acad Sci 609:225–234
28. Mehlman MA (1991) Dangerous and cancer-causing properties of products and chemicals in the refining and petrochemical industry. 4. Human health and environmental hazards resulting from oil and oil products. J Clean Technol Environ Sci 1:103–121
29. Infante PF (2006) Benzene exposure and multiple myeloma—a detailed meta-analysis of benzene cohort studies. In: Mehlman MA, Soffritti M, Landrigan P, Bingham E, Belpoggi F (eds) Living in a chemical world: framing the future in light of the past. Wiley, John & Sons, pp 90–109
30. James CP, Rosemary B, Wei W, James G, Emad T, Elizabeth V, Christophe C, Robert DS, Laxmikant K, Klaus S (2005) Scalable molecular dynamics with NAMD. J Comput Chem 26:1781–1802
31. Bonomi M, Branduardi D, Bussi G, Camilloni C, Provasi D, Raiteri P, Donadio D, Marinelli F, Pietrucci F, Broglia RA, Parrinello M (2009) PLUMED: a portable plugin for free-energy calculations with molecular dynamics. Comput Phys Commun 180:1961–1972
32. William H, Andrew D, Klaus S (1996) VMD: visual molecular dynamics. J Mol Graph 14:33–38
33. Feller SE, MacKerell AD (2000) An improved empirical potential energy function for molecular simulations of phospholipids. J Phys Chem B 104:7510–7515
34. Vanommeslaeghe K, Hatcher E, Acharya C, Kundu S, Zhong S, Shim J, Darian E, Guvench O, Lopes P, Vorobyov I, Jr ADM (2010) CHARMM general force field: A force field for

drug-like molecules compatible with the CHARMM all-atom additive biological force fields. J Comput Chem 31:671–690

35. Feller SE, Zhang Y, Pastor RW, Brooks BR (1995) Constant pressure molecular dynamics simulation: the Langevin piston method. J Chem Phys 103:4613–4621
36. Martyna GJ, Tobias DJ, Klein ML (1994) Constant pressure molecular dynamics algorithms. J Chem Phys 101:4177–4189
37. Tuckerman M, Berne BJ, Martyna GJ (1992) Reversible multiple time scale molecular dynamics. J Chem Phys 97:1990–2001
38. Darden T, York D, Pedersen L (1993) Particle mesh Ewald: an N log(N) method for Ewald sums in large systems. J Chem Phys 98:10089–10092
39. Sunhwan J, Taehoon K, Vidyashankara GI, Wonpil I (2008) CHARMM-GUI: a web-based graphical user interface for CHARMM. J Comput Chem 29:1859–1865
40. Plimpton S (1995) Fast parallel algorithms for short-range molecular dynamics. J Comput Phys 117:1–19
41. Laio A, Rodriguez-Fortea A, Gervasio FL, Ceccarelli M, Parrinello M (2005) Assessing the accuracy of metadynamics. J Phys Chem B 109:6714–6721
42. Zhu T, Li J, Hawkins GD, Cramer CJ, Truhlar DG (1998) Density functional solvation model based on CM2 atomic charges. J Chem Phys 109:9117
43. Bemporad D, Essex JW, Luttmann C (2004) Permeation of small molecules through a lipid bilayer: a computer simulation study. J Phys Chem B 108:4875–4884

Chapter 15

Screening of Fullerene Toxicity by Hemolysis Assay

Federica Tramer, Tatiana Da Ros, and Sabina Passamonti

Abstract

Fullerene is a compound formed during carbon burst that has been produced synthetically starting from the 1990s. The spherical shape and the characteristic carbon bonds of this allotrope (C_{60}) have made it a suitable molecule for many applications. During the last decade, the low aqueous solubility of this molecule has been improved by chemical functionalization allowing the use of fullerene derivatives in biological fluids. The characterization of the toxicity potential of fullerenes is therefore of growing interest for any biomedical application. Intravenous injection is one of the possible routes of their administrations and therefore red blood cells are among the first targets of fullerene cytotoxicity. Human red blood cells are easily available and separated from plasma. Membrane disruption by toxic compounds is easily detected in red blood cells as release of hemoglobin in the cell medium, which can be assayed spectrophotometrically at $\lambda = 415$ nm. Due to the high molar extinction coefficient of hemoglobin, the assay can be performed on a small amount of both red blood cells and the test compounds, which might be available only in small quantities. So, the hemolysis assay is a simple screening test, whose results can guide further investigations on cytotoxicity in more complex experimental models.

Key words: Fullerene derivatives, Toxicity, Hemolysis assay, Spectrophotometry

1. Introduction

Nano-sized materials and nano-scaled processes are very important in biological and materials science, being actively introduced in biomedicine as diagnostic and therapeutic. An important area of research in modern nanomaterials science involves carbon-based materials, among which fullerenes take the priority. Fullerenes are spherical molecules built up of fused pentagons and hexagons where all double bonds are conjugated. The smallest stable, and also the most abundant fullerene, obtained by usual preparation methods is C_{60}, the third carbon allotrope. Unfortunately, the fullerene molecule is completely insoluble in aqueous media and aggregates very easily (1). Among several strategies to increase their hydrophilicity, chemical functionalization with amino acid, carboxylic acid,

Joshua Reineke (ed.), *Nanotoxicity: Methods and Protocols*, Methods in Molecular Biology, vol. 926, DOI 10.1007/978-1-62703-002-1_15, © Springer Science+Business Media, LLC 2012 polyhydroxyl group, or amphiphilic polymers is one of the most promising (2–4). Fullerenes molecules allow derivatizations by addition reaction, with a remarkable regioselectivity.

Manipulation of surface chemistry and molecular makeup has created a wide family of fullerenes, which exhibit drastically different behaviors (5), allowing the investigation of their action in different biomedical fields. Some fullerene derivatives can act as antioxidant and neuroprotective agents, based on their capability to react with oxygen and nitric oxide radical species or by blocking glutamate receptors (6–14). Other molecules behave as inhibitors of different enzymes (15–21), but it has also been demonstrated an antibacterial activity probably due to cell membrane disruption (22). Klumpp et al. reported the use of poly-fulleropyrrolidinium salts as drug delivery (23). Metal atoms can be entrapped inside the cavity of fullerene spheres allowing the formation of a nanomaterial family called endohedral metallofullerenes. Its most appealing applications are in the nuclear medicine field as fullerene-based contrast agents and radiotracers (24). All these expected applications involve different routes of administration as oral, dermal, pulmonary, or injection routes.

It is, however, still poorly known if some characteristics, such as composition, crystal, or aggregate size, shape, water solubility, and surface modifications/functionalization of nanoparticles, could affect normal physiological functions of the cell and cause cytotoxicity (25).

In some applications (e.g., drug delivery or contrast agents), fullerenes might be administered intravenously and red blood cells could, therefore, represent the first targets of cytotoxic action. One of the most direct methods to study the cytotoxicity of molecules is to investigate their hemolytic properties.

The hemolytic assay has many advantages: red blood cells are easily obtained, they don't need to be cultured and hemoglobin release upon membrane disruption is easily assessed spectrophotometrically ($\lambda = 415$ nm). Moreover, due to the high value of hemoglobin's molar extinction coefficient, a very small number of cells consumed. Currently, there are some papers reporting the cytotoxicity induced by fullerenes both in in vivo and in vitro experiments (18, 25), but only few report their effect on red blood cells (26, 27).

2. Materials

Unless otherwise specified, reagents were purchased from Sigma-Aldrich, Milan, Italy. Fullerene was purchased from BuckyUSA and dipeptide Boc-Lys(Boc)-ProOH from Bachem. Solutions were prepared with ultrapure water of resistivity = 18.2 MΩ cm (Milli-Ro and Milli Q, Millipore Co., Bedford, MA). This standard is referred to as "water" in this text.

2.1. Blood Collection

1. Anticoagulant solution (ACD solution): 70 m*M* citric acid trisodium salt dihydrate, 38.01 m*M* citric acid monohydrate, and 123.63 m*M* D-(+)-glucose monohydrate. Weigh 1.036 g of citric acid trisodium salt dehydrate, 0.4 g of citric acid monohydrate, 1.225 g of D-(+)-glucose monohydrate and make up to 50 mL with water. Filter the solution through 0.22 μm Nalgene filter and store in aliquots at −20°C.
2. 5 *M* NaCl solution: weigh 146.1 g of NaCl and make up to 500 mL with water.
3. 0.5 *M* Ethylenediaminetetraacetic acid (EDTA) solution: weigh 14.612 g of EDTA powder and make up to 80 mL. Mix and adjust pH with NaOH (see Note 1). Make up to 100 mL with water. Store at 4°C.
4. 0.2 *M* Phosphate buffer: prepare 600 mL of 0.2 *M* Na_2HPO_4 and 200 mL of 0.2 *M* of NaH_2PO_4. Add 17.035 g of Na_2HPO_4 and make up to 600 mL with water. In a different cylinder, add 6.24 g of $NaH_2PO_4 \cdot 2H_2O$ and make up to 200 mL with water. Adjust the pH of the Na_2HPO_4 solution to pH 7.4 by adding the NaH_2PO_4 solution dropwise. Store at 4°C.
5. Phosphate-buffered saline (PBS) 10×: combine 500 mL of 0.2 *M* phosphate buffer with 278 mL of 5 *M* NaCl solution and make it up to 1 L with water in a cylinder. Store at room temperature (see Note 2).
6. PBS 1×: add 100 mL of PBS 10× and make it up to 1 L with water. Store at 4°C.
7. Phosphate-buffered saline EDTA (PBSE): combine 100 mL of PBS 10× with 2 mL of 0.5 *M* EDTA solution and make it up to 1 L with water. Store at 4°C.

2.2. Preparation of a Standard Suspension of Red Blood Cells

1. PBS 10× prepared as above (Subheading 2.1, step 5).
2. PBS 1× prepared as above (Subheading 2.1, step 6).
3. 1% (v/v) Triton X-100 solution: add 0.15 mL of Triton X-100 and 14.85 mL of water into a 15 mL Falcon, mix and store at 4°C.
4. 10% (v/v) DMSO solution: combine 500 μL of DMSO and 500 mL of PBS 10× and make up to 5 mL with water into a 15 mL Falcon.

3. Methods

3.1. Fullerene Derivatives Preparation

Figure 1 shows the fullerene molecules mentioned in this study.

3.1.1. Preparation of Compound A

1. Dissolve 1 g of fullerene (C_{60}) in a round bottom flask with 500 mL of toluene.
2. Add 249 mg of *N*-methylglycine (CH_3-NH-CH_2-COOH) and 210 mg of formaldehyde (HCOH).
3. Heat the solution at 120°C and stir under reflux for 2 h, by means of the stirring and heating plate equipped with a paraffin oil bath.
4. Pour the crude solution on the top of a glass chromatographic column filled with silica gel (63–200 μm) already wet with toluene.
5. Elute the column with toluene/ethyl acetate (7:3).
6. Recover the product by eluting it from the column.
7. Evaporate the solvent by means of a rotavapor.
8. Dissolve the residue in dichloromethane (5 ml) and filter the solution on a cotton filter.
9. Concentrate the solution up to 1 ml of solvent, by gently fluxing nitrogen stream.
10. Precipitate the compound by adding 9 ml of methanol.
11. Centrifuge at 3,000 × *g* for 10 min.
12. Remove the supernatant with a pipette.
13. Dry the compound under vacuum with a vacuum pump.
14. Dissolve the compound in 5 mL of dichloromethane in a screw-topped vial.
15. Add 10 mL of methyl iodide (CH_3I), close the stopper, and stir for 24 h at 80°C by means of the stirring and heating plate equipped with a paraffin oil bath.
16. Transfer the solution into a round bottom flask.
17. Evaporate the solvent under reduced pressure by rotavapor.
18. Remove the solid from the flask.
19. Dry the compound (A) under vacuum.

3.1.2. Preparation of Compound B

1. Dissolve 1 g of fullerene (C_{60}) in a round bottom flask with 500 mL of toluene.
2. Add 856 mg of the amino acid 1 (Boc-HN-CH_2CH_2-O-CH_2CH_2-O-CH_2CH_2-NH-CH_2-COOH) and 210 mg of formaldehyde (HCOH).

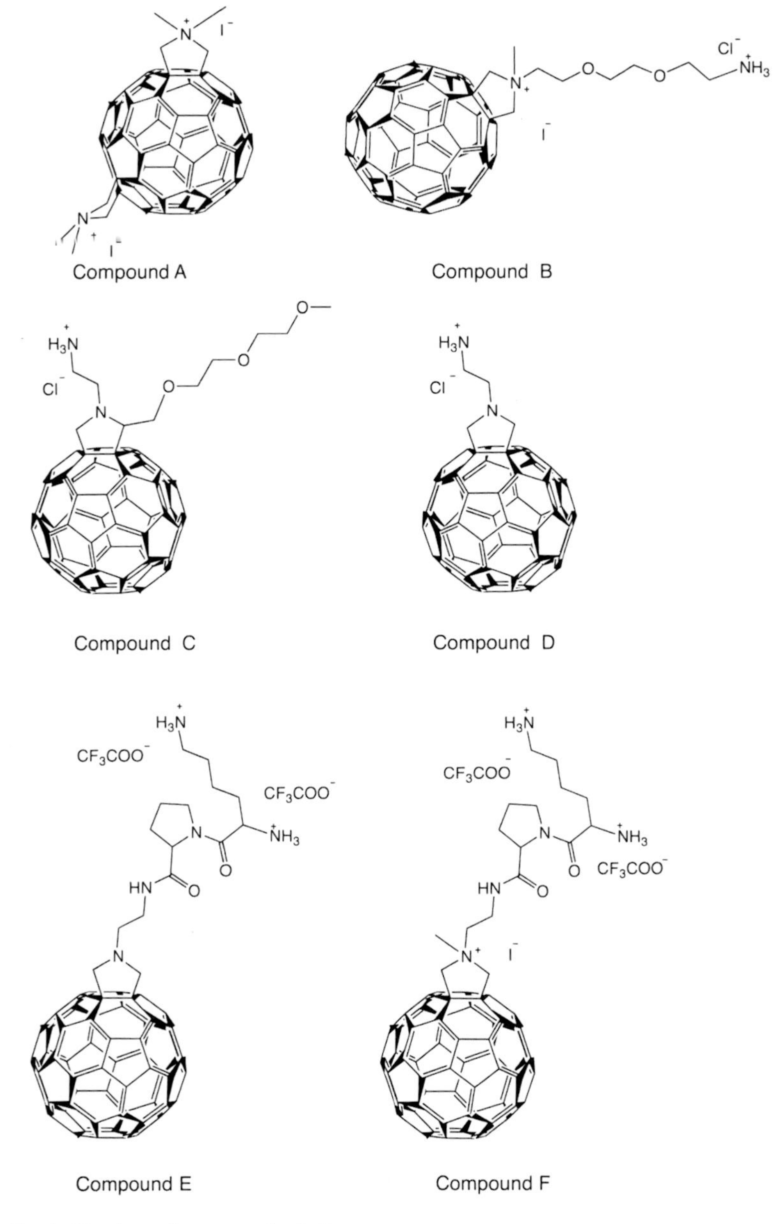

Fig. 1. Structure of compounds *A–F*.

3. Heat the solution at 120°C and stir under reflux for 2 h, by means of the stirring and heating plate equipped with a paraffin oil bath.
4. Pour the crude solution on the top of a glass chromatographic column filled with silica gel (63–200 μm) already wet with toluene.
5. Elute the column with toluene/ethyl acetate (9:1).
6. Recover the product by eluting it from the column.
7. Evaporate the solvent by means of a rotavapor.

8. Dissolve the residue in dichloromethane (5 ml) and filter the solution on a cotton filter.
9. Concentrate the solution up to 1 ml of solvent by gently fluxing nitrogen stream.
10. Precipitate the compound by adding 9 ml of methanol.
11. Centrifuge at 3,000 × *g* for 10 min.
12. Remove the supernatant with a pipette.
13. Dry the compound under vacuum.
14. Dissolve 50 mg of the compound in 5 mL of dichloromethane in a screw-topped vial.
15. Add 10 mL of methyl iodide (CH_3I), close the stopper, and stir for 24 h at 80°C by means of the stirring and heating plate equipped with a paraffin oil bath.
16. Transfer the solution into a round bottom flask.
17. Evaporate the solvent under reduced pressure by rotavapor.
18. In a round bottom flask, dissolve the product in 100 mL of methanol.
19. Bubble gaseous HCl through the suspension for 20 min, cooling the solution in an ice bath.
20. Stir the obtained solution at room temperature for 2 h.
21. Evaporate the solvent under reduced pressure by rotavapor.
22. Dissolve the product in 5 mL of methanol and transfer the solution into a centrifuge tube.
23. Concentrate the solution up to 1 ml of solvent by gently fluxing nitrogen stream.
24. Precipitate the compound by adding 9 ml of diethyl ether.
25. Centrifuge at 3,000 × *g* for 10 min.
26. Remove the supernatant with a pipette.
27. Dry the compound (B) under vacuum with a vacuum pump.

3.1.3. Preparation of Compounds C and D

1. Dissolve 1 g of fullerene (C_{60}) in a round bottom flask with 500 mL of toluene.
2. Heat the solution at 120°C and stir under reflux for 2 h, by means of the stirring and heating plate equipped with a paraffin oil bath.
3. Pour the crude solution on the top of a glass chromatographic column filled with silica gel (63–200 μm) already wet with toluene.
4. Elute the column with toluene/ethyl acetate (9:1).
5. Recover the product by eluting it from the column.
6. Evaporate the solvent by means of a rotavapor.

7. Dissolve the residue in dichloromethane (5 ml) and filter the solution on a cotton filter.
8. Concentrate the solution up to 1 ml of solvent by gently fluxing nitrogen stream.
9. Precipitate the compound by adding 9 ml of methanol.
10. Centrifuge at 3,000 × *g* for 10 min.
11. Remove the supernatant with a pipette.
12. Dry the compound under vacuum.
13. Dissolve the product in 100 mL of methanol in a round bottom flask.
14. Bubble gaseous HCl through the suspension for 20 min, cooling the solution in an ice bath.
15. Stir the obtained solution at room temperature for 2 h.
16. Evaporate the solvent under reduced pressure by rotavapor.
17. Dissolve the product in 5 mL of methanol and transfer the solution into a centrifuge tube.
18. Concentrate the solution up to 1 ml of solvent by gently fluxing nitrogen stream.
19. Precipitate the compound by adding 9 ml of diethyl ether.
20. Centrifuge at 3,000 × *g* for 10 min.
21. Remove the supernatant with a pipette.
22. Dry the compounds (C) or (D) under vacuum with a vacuum pump.

3.1.4. Synthesis of Compound E

1. Dissolve 45 mg of compound D in 25 mL of dichloromethane (CH_2Cl_2) (solution 1).
2. Add 14 mg of triethylamine ($N(CH_2CH_3)_3$) to the solution 1 and stir at room temperature (solution 2).
3. In another tube dissolve 20 mg of Boc-Lys(Boc)-ProOH in 25 mL of CH_2Cl_2 (solution 3).
4. Add 14 mg of 1-ethyl-3-(3-dimethylaminopropyl) carbodiimide (EDCI) and 9 mg of hydroxybenzotriazole (HOBt) (solution 4).
5. Stir at room temperature for 45 min.
6. Mix the solutions 2 and 4 and let the reaction occur for 2 h at room temperature.
7. Evaporate the solvent under reduced pressure by means of rotavapor.
8. The crude is dissolved in 10 mL of toluene.
9. Pour the crude solution on the top of a glass chromatographic column filled with silica gel (63–200 μm) already wet with toluene.

10. Elute the column with toluene/isopropanol 99/1.
11. Recover the product by eluting it from the column.
12. Evaporate the solvent by means of a rotavapor.
13. Dissolve the residue in dichloromethane (5 ml) and filter the solution on a cotton filter.
14. Concentrate the solution up to 1 ml of solvent by gently fluxing nitrogen stream.
15. Precipitate the compound by adding 9 ml of methanol.
16. Centrifuge at 3,000 × *g* for 10 min.
17. Remove the supernatant with a pipette.
18. Dry the compound under vacuum.
19. Dissolve 18 mg of the coupling product in 10 mL of CH_2Cl_2 and add 5 mL of trifluoroacetic acid (TFA).
20. Stir at room temperature for 18 h.
21. Evaporate the solvent and the acid under vacuum.
22. Precipitate the compound from toluene.
23. Dry the compound (E) under vacuum.

3.1.5. Synthesis of Compound F

1. Dissolve 45 mg of compound D in 25 mL of CH_2Cl_2 (solution 1).
2. Add 14 mg of $N(CH_2CH_3)_3$ and stir at room temperature (solution 2).
3. Dissolve 20 mg of Boc-Lys(Boc)-ProOH in 25 mL of CH_2Cl_2 (solution 3).
4. Add to solution, 14 mg of EDCI and 9 mg of HOBt (solution 4).
5. Stir at room temperature for 45 min.
6. Mix solutions 2 and 4 and let the reaction occur for 2 h at room temperature.
7. Evaporate the solvent under reduced pressure by means of rotavapor.
8. The crude is dissolved in 10 mL of toluene.
9. Pour the crude solution on the top of a glass chromatographic column filled with silica gel (63–200 μm) already wet with toluene.
10. Elute the column with toluene/isopropanol 99/1.
11. Recover the product by eluting it from the column.
12. Evaporate the solvent by means of a rotavapor.
13. Dissolve the residue in dichloromethane (5 ml) and filter the solution on a cotton filter.
14. Concentrate the solution up to 1 ml of solvent by gently fluxing nitrogen stream.

15. Precipitate the compound by adding 9 ml of methanol.
16. Centrifuge at 3,000 × *g* for 10 min.
17. Remove the supernatant with a pipette.
18. Dry the compound under vacuum.
19. Dissolve 15 mg of the coupling product in 10 mL of CH_2Cl_2 and add 3 mL of methyl iodine in a seal vial and stir at 80°C for 18 h.
20. Evaporate the solvent under vacuum.
21. Precipitated the compound from toluene by centrifugation at 3,000 × *g* for 10 min.
22. Dry the compound under vacuum.
23. Dissolve 18 mg of the coupling product in 10 mL of CH_2Cl_2 and add 5 mL of TFA.
24. Stir at room temperature for 18 h.
25. Evaporate the solvent and the acid under vacuum.
26. Dry the compound (F) under vacuum.

3.1.6. Fullerene Stock Solutions

1. Dissolve fullerenes in DMSO at 1 m*M*.

3.2. Determination of Fullerene Derivatives Absorbance at λ = 415 nm (Blanks)

Fullerene dissolved in PBS 1× absorb at λ = 415 nm. They therefore interfere with the spectrophotometric analysis of hemoglobin. In order to obtain a reliable blank value to subtract to hemolysis samples, it is convenient to create a calibration curve of fullerenes dissolved in PBS. In this way, fullerenes available in small quantities will not be wasted in running blank samples.

1. Dilute 2 mL of fullerenes dissolved in DMSO as follows: 2 mL in PBS 10× and 16 mL of water in order to have a 100 μM stock solution (PBS:DMSO 9:1, v/v) (see Note 3).
2. Choose 6 different fullerene concentrations in the range 0.5–80 μ*M*.
3. Prepare 6 different 1.5 mL eppendorf tubes for each fullerene solution and fill them according to the following scheme (volumes are expressed in μL):

Final [Fullerene] (μ*M*)	Fullerene solution	PBS 1×	PBS 1×
0.5	5	995	500
2	20	980	500
10	100	900	500
30	300	700	500
50	500	500	500
70	700	300	500

4. Keep the eppendorf tubes in ice and prepare the samples as described in Subheading 3.3 (see Note 4).
5. Centrifuge the blank samples at 9,000 × *g* for 5 min (see Note 5).
6. Set the spectrophotometer (λ = 415 nm) to zero with PBS 1×.
7. Prepare as many cuvettes as the number of samples and add 400 μL of water in each one.
8. Add 400 μL of each supernatant of the blank samples to the cuvettes, previously prepared (see above, step 7).
9. Analyze spectrophotometrically (λ = 415 nm).
10. Plot A_{415} data versus [fullerene], as shown in Fig. 2 (see Note 6).
11. The A_{415} data, whether obtained experimentally or by regression analysis, are named A_B. Since 12 fullerene concentrations will be tested by the hemolysis assay, 12 values of A_B will be used, i.e., A_{B1}–A_{B12}.

3.3. Blood Collection

1. Use a sterile syringe to collect a suitable volume (5–10 mL) of blood by venipuncture of a human donor (see Note 7).
2. Immediately remove the needle from the syringe, gently push down the plunger, and transfer the blood into a 50 mL Falcon tube.
3. Quickly add ACD solution in a volume equal to 10% (v/v) of the blood volume and gently mix.

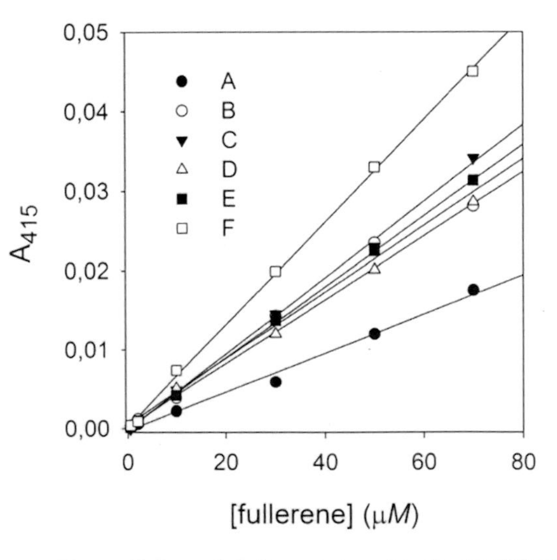

Fig. 2. Dependence of A_{415} on [fullerene]. Fullerenes were dissolved in PBS:DMSO (9:1, v/v) in optical cuvettes with 1 cm of light path. The *lines* are the best fitting curves using the equation $y = a + bx$, where $y = A_{415}$, x = [fullerenes], a = intercept on the y-axis, b = extinction coefficient (μM^{-1} cm^{-1}). The following ε were calculated (M^{-1} cm^{-1}): $\varepsilon_A = 246 \pm 11$, $\varepsilon_B = 419 \pm 24$, $\varepsilon_C = 481 \pm 9$, $\varepsilon_D = 403 \pm 8$, $\varepsilon_F = 646 \pm 9$. In all cases, $r^2 > 0.99$.

4. Centrifuge the blood at 600 × *g* for 10 min, in order to separate cells from plasma.
5. Remove the supernatant by gentle aspiration using a vacuum pump (see Note 8).
6. Resuspend the cell pellet in 1 volume of cold PBSE and centrifuge again as above at 10°C.
7. Repeat steps 4 and 5 using 4 volumes of PBSE.
8. Resuspend cells using 4 volumes of PBS 1× and repeat steps 4 and 5 twice.
9. Keep the red blood cell pellet (2–4 mL).

3.4. Preparation of a Standard Suspension of Red Blood Cells

In order to work with uniform red blood cells suspensions, these are assessed for their hemoglobin content. The latter is measured spectrophotometrically at $\lambda = 415$ nm and the red blood cell suspension is adequately diluted before being lysed.

1. Add 4 volumes of cold PBS 1× to the red blood cell pellet, so as to obtain an approximate hematocrit of 20% (suspension A).
2. Put 100 μL of suspension A into an eppendorf tube and add 2 volumes of PBS 1× (suspension B).
3. Lyse the suspension by adding Triton solution as follows (see Note 9):

   Cuvette **1** (blank): 640 μL PBS 1× + 160 μL 1% Triton solution (see Note 10);

   Cuvette **2** (sample): 540 μL PBS 1× + 160 μL 1% Triton solution + 100 μL (suspension B) (see Note 11).
4. Measure A_{415} of blank and set the spectrophotometer to zero.
5. Measure A_{415} of suspension B. It should be 1.1–1.8.
6. Adjust the hematocrit of suspension A in order to obtain $A_{415} = 1.1–1.8$ at the end of the procedure (steps 2–5) (see Note 12).

3.5. Assay of Hemolysis

Toxic compounds can damage red blood cells to different extents. To evaluate the percent of hemolysis, the 100% hemolysis must be assessed by using disruptors of the phospholipidic bilayer. One of these is the detergent Triton X-100.

3.5.1. Preparation of Complete (100%) Hemolysis Sample

1. Add to an eppendorf tube: 755 μL of water, 100 μL of PBS 10×, 100 μL of 1% Triton X-100 solution, 45 μL of suspension A.
2. Add 500 μL of cold PBS 1× and keep in ice.
3. These samples will be processed along the others, as described below (Subheading 3.5.2, steps 9–11).

3.5.2. Hemolysis Samples

Solutions of fullerenes (100 μ*M*) and 10% (v/v) DMSO solution must be at room temperature, PBS 1× must be kept in ice. Run at least triplicate tests.

1. Prepare 12 eppendorf tubes for each fullerene solution.
2. Prepare 12 eppendorf tubes for a fullerene-free solution (see Note 13).
3. Prepare eppendorf tubes as shown below (volumes are expressed in μL) (see Note 14)

Sample	Final [Fullerene] (μ*M*)	Fullerene solution or 10% DMSO sol.	PBS 1×
1	0.5	5	950
2	1	10	945
3	2	20	935
4	4	40	915
5	6	60	895
6	10	100	855
7	20	200	755
8	30	300	655
9	40	400	555
10	50	500	455
11	60	600	355
12	80	800	155

4. Incubate at 37°C for 5 min.
5. Add 45 μL of suspension A in each eppendorf tube, timing every 10 s and then vortex.
6. Incubate at 37°C for 30 min.
7. In the meantime, prepare as many cuvettes as the number of samples and put in 400 μL water.
8. Stop the lysis reaction by adding 500 μL of cold PBS 1× to each eppendorf tube, by timing 10 s between the samples, and immediately put them in ice.
9. Centrifuge all the samples (including 100% hemolysis sample, see above Subheading 3.5.1, step 3) at 9,000×*g* for 5 min.
10. Set the spectrophotometer (λ = 415 nm) to zero with PBS 1×.
11. Withdraw 400 μL of each supernatant resulting from step 9 and add it to the cuvettes previously prepared at step 7.
12. Carefully mix and analyze A_{415} (see Note 15).
13. The A_{415} value obtained with the 100% hemolysis sample (Subheading 3.5.1, step 3) is named A_{H100}.

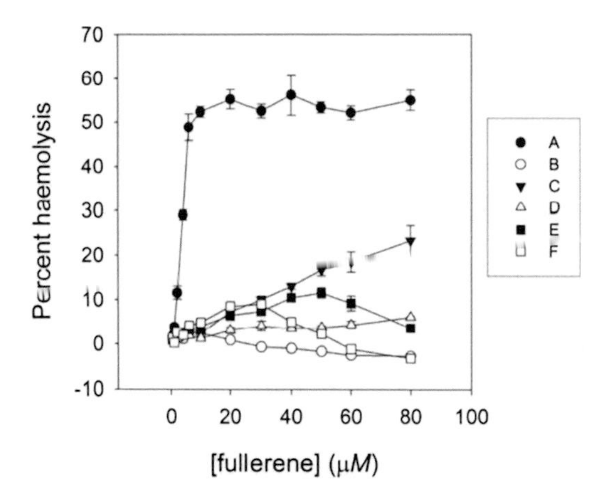

Fig. 3. Dependence of percent of hemolysis on [fullerene]. Values are means ± sem ($n = 3$). The chemical structure of compounds *A–F* are shown in Fig. 1.

14. The hemolysis A_{415} value obtained with the fullerene-free solution, containing only 10% DMSO, is named A_D.
15. The hemolysis A_{415} value obtained with the fullerene solutions (1–12, see above scheme) are named A_{F1}–A_{F12}.

3.6. Data Analysis

1. The net values of hemolysis are calculated as follows, from the measurements obtained at step 11 of Subheading 3.2 and steps 14–15 of Subheading 3.5.2:

$$X_1 = A_{F1} - A_D - A_{B1};\ X_2 = A_{F2} - A_D - A_{B2};\ \ldots X_{12} = A_{F12} - A_D - A_{B12}$$

2. The percent values of hemolysis are calculated as follows, from the measurements obtained at step 13 of Subheading 3.5.2, step 1 of Subheading 3.6:

$$\text{Percent hemolysis} = 100 \cdot X_{1-12} / A_{H100}$$

3. Plot a graph of percent hemolysis versus [fullerene] (Fig. 3).

4. Notes

1. Concentrated NaOH (6 *N*) can be used at first to narrow the gap from the starting pH of the EDTA solution to the required pH. From then on, it would be better to use a series of NaOH dilutions (e.g., 2, 1, 0.5, 0.1 *N*) to avoid a sudden shift of pH above the required value.
2. Very small differences in the ionic strength of the solution cause cell lysis.
3. The final concentration of DMSO is 10%.
4. It is possible to determine the absorbance of fullerene solutions the day before.

5. This step is done in order to treat blanks in the same way as hemolysis samples.
6. Linear regression analysis of the data enables to calculate values of A_{415} of any [fullerene], provided it is interpolated with the experimental data.
7. It is possible to obtain a "buffy coat" from the local transfusion bank. This is the upper, lighter portion of the blood clot forming after density gradient centrifugation. It is particularly rich in white blood cells and platelets. These cells will be discarded with the supernatants resulting from the centrifugations, whereas the red blood cells will be recovered as a sediment. If you are using the buffy coat, put 5 mL of this in a 50 mL Falcon tubes and add 5 mL of PBSE, centrifuge at $500 \times g$ for 20 min, resuspend again in 1 volume of PBS 1×, and then proceed as indicated at steps 7–9 of Subheading 3.3.
8. If the supernatant is red due to hemoglobin release, collect the cell pellet on the bottom of the Falcon tube by a Pasteur pipette and transfer it to a new tube.
9. 1% Triton X-100 solution is needed in order to cause complete cells lysis.
10. Use plastic cuvettes with 1 cm path.
11. Suspension A is 24-fold diluted in cuvette 2.
12. If $1.3 \leq A_{415} \leq 1.8$, this means that suspension A is around 24% (v/v), so dilute it in order to bring it to 20% (v/v). Gradually dilute suspension A also if $A_{415} \geq 2$. It often happens that suspension A needs to be diluted more than ten times. If $A_{415} \leq 1.1$, this means that suspension A is too diluted, so centrifuge it again and resuspend cells in less volume, then measure A_{415} as at step 3.
13. This test is needed to assess hemolysis due to the fullerene-free solution.
14. Fullerene derivatives could have different solubility in water solutions, so prior to use mix them very well.
15. When possible, use a microplate reader with a 96-well plate. Add to each well 200 μL of samples and read all samples (including blanks) at the wavelength closest to 415 nm (e.g. 405 nm).

References

1. Prato M (1997) [60]Fullerene chemistry for materials science applications. J Mater Chem 7:1097–1109
2. Beuerle F, Hirsch A (2009) Synthesis and orthogonal functionalization of [60]fullerene e, e, e-trisadducts with two spherically defined addend zones. Chemistry 15:7434–7446
3. Bianco A, Da Ros T, Prato M, Toniolo C (2001) Fullerene-based amino acids and peptides. J Pept Sci 4:208–219
4. Jagadeesan D, Eswaramoorthy M (2010) Functionalized carbon nanomaterials derived from carbohydrates. Chem Asian J 5: 232–243

5. Nakamura E, Isobe H (2003) Functionalized fullerenes in water. The first 10 years of their chemistry, biology, and nanoscience. Acc Chem Res 36:807–815
6. Witte P, Beuerle F, Hartnagel U, Lebovitz R, Savouchkina A, Sali S, Guldi D, Chronakis N, Hirsch A (2007) Water solubility, antioxidant activity and cytochrome C binding of four families of exohedral adducts of C60 and C70. Org Biomol Chem 5:3599–3613
7. Gharbi N, Pressac M, Hadchouel M, Szwarc H, Wilson SR, Moussa F (2005) [60]Fullerene is a powerful antioxidant in vivo with no acute or subacute toxicity. Nano Lett 5:2578–2585
8. Sun T, Xu Z (2006) Radical scavenging activities of alpha-alanine C60 adduct. Bioorg Med Chem Lett 16:3731–3734
9. Enes RF, Tomé AC, Cavaleiro JA, Amorati R, Fumo MG, Pedulli GF, Valgimigli L (2006) Synthesis and antioxidant activity of [60]fullerene-BHT conjugates. Chemistry 12: 4646–4653
10. Hu Z, Guan W, Wang W, Huang L, Xing H, Zhu Z (2007) Synthesis of β-alanine C60 derivative and its protective effect on hydrogen peroxide-induced apoptosis in rat pheochromocytoma cells. Cell Biol Int 31:798–804
11. Yang J, Alemany LB, Driver J, Hartgerink JD, Barron AR (2007) Fullerene-derivatized amino acids: Synthesis, characterization, antioxidant properties, and solid-phase peptide synthesis. Chem Eur J 13:2530–2545
12. Dugan L, Gabrielsen J, Yu S, Lin T, Choi D (1996) Buckminsterfullerenol free radical scavengers reduce excitotoxic and apoptotic death of cultured cortical neurons. Neurobiol Dis 3:129–135
13. Dugan LL, Turetsky DM, Du C, Lobner D, Wheeler M, Almli CR, Shen CKF, Luh TY, Choi DW, Lin TS (1997) Carboxyfullerenes as neuroprotective agents. Proc Natl Acad Sci USA 94:9434–9439
14. Jin H, Chen WQ, Tang XW, Chiang LY, Yang CY, Schloss JV, Wu JY (2000) Polyhydroxylated C(60), fullerenols, as glutamate receptor antagonists and neuroprotective agents. J Neurosci Res 62:600–607
15. Yang X, Chen Z, Meng X, Li B, Tan X (2007) Inhibition of DNA restrictive endonucleases and Taq DNA polymerase by trimalonic acid C60. Chin Sci Bull 52:1802–1806
16. Iwata N, Mukai T, Yamakoshi Y, Hara S, Yanase T, Shoji M, Endo T, Miyata N (1998) Effects of C_{60}, a fullerene, on the activities of glutathione S-transferase and glutathione-related enzymes in rodent and human livers. Fullerenes Nanotubes Carbon Nanostruct 6:213–226
17. Lai YL, Chiang LY (1997) Water-soluble fullerene derivatives attenuate exsanguination-induced bronchoconstriction of guinea-pigs. J Auton Pharmacol 17:229–235
18. Nielsen GD, Roursgaard M, Jensen KA, Poulsen SS, Larsen ST (2008) *In vivo* biology and toxicology of fullerenes and their derivatives. Basic Clin Pharmacol Toxicol 103:197–208
19. Yang X, Meng X, Li B, Chen Z, Zhao D, Tan X, Yu Q (2008) Inhibition of in vitro amplification of targeted DNA fragment and activity of exonuclease I by a fullerene-oligonucleotide conjugate. Biologicals 36:223–226
20. Wang Z, Zhao J, Li F, Gao D, Xing B (2009) Adsorption and inhibition of acetylcholinesterase by different nanoparticles. Chemosphere 77:67–73
21. Innocenti A, Durdagi S, Doostdar N, Strom TA, Barron AR, Supuran CT (2010) Nanoscale enzyme inhibitors: fullerenes inhibit carbonic anhydrase by occluding the active site entrance. Bioorg Med Chem 18:2822–2828
22. Bosi S, Da Ros T, Castellano S, Banfi E, Prato M (2000) Antimycobacterial activity of ionic fullerene derivatives. Bioorg Med Chem Lett 10:1043–1045
23. Klumpp C, Lacerda L, Chaloin O, Da Ros T, Kostarelos K, Prato M, Bianco A (2007) Multifunctionalised cationic fullerene adducts for gene transfer: design, synthesis and DNA complexation. Chem Commun (Camb) 36:3762–3764
24. Bosi S, Da Ros T, Spallato G, Prato M (2003) Fullerene derivatives: an attractive tool for biological applications. Eur J Med Chem 38: 913–923
25. Johnston HJ, Hutchison GR, Frans M, Christensen FM, Aschberger K, Stone V (2010) The biological mechanisms and physicochemical characteristics responsible for driving fullerene toxicity. Toxicol Sci 114:162–182
26. Trpkovic A, Todorovic-Markovic B, Kleut D, Misirkic M, Janjetovic K, Vucicevic L, Pantovic A, Jovanovic S, Dramicanin M, Markovic Z, Trajkovic V (2010) Oxidative stress-mediated hemolytic activity of solvent exchange-prepared fullerene (C60) nanoparticles. Nanotechnology 21:375102
27. Bosi S, Feruglio L, Da Ros T, Spalluto G, Gregoretti B, Terdoslavich M, Decorti G, Passamonti S, Moro S, Prato M (2004) Haemolytic effects of water-soluble fullerene derivatives. J Med Chem 47:6711–6715

Chapter 16

Assessment of In Vitro Skin Irritation Potential of Nanoparticles: RHE Model

P. Balakrishna Murthy, A. Sairam Kishore, and P. Surekha

Abstract

The skin irritation test is designed for the prediction of acute skin irritation of nanoparticles by measurement of its cytotoxic effect, as reflected in the MTT assay, on the Reconstructed Human Epidermis (RHE) model. RHE tissues are commercially available.

Key words: In vitro, Reconstructed human epidermis, MTT, Skin irritation, Nanoparticles, Nanotoxicity

1. Introduction

Nanoparticles are generally defined as engineered structures with at least one dimension less than 100 nm. Application fields range from medical imaging, new drug delivery technologies to various industrial products.

Due to the expanding use of nanoparticles, the risk of human exposure rapidly increases and reliable toxicity test systems are urgently needed (1). If all new nanoparticles were to be tested in animals, taking into consideration manipulations in composition, size, formulation, contaminants, and routes of exposure, then hundreds of thousands of animals would be required to fully assess the potential hazard of these materials. There is currently a need to develop and validate in vitro assays for assessing the potential toxicity of the ever-expanding range of nanoparticles (2). The in vivo rabbit skin irritation test is the most criticized one in consideration of pain and distress to the animal. In view of 3R (reduction, refinement, replacement) the RHE model is proposed as an alternative model for skin irritation with nanomaterials (3).

Reconstructed tissues such as RHE are biological models widely used in safety or efficacy prescreening tests. Moreover, these

Joshua Reineke (ed.), *Nanotoxicity: Methods and Protocols*, Methods in Molecular Biology, vol. 926, DOI 10.1007/978-1-62703-002-1_16, © Springer Science+Business Media, LLC 2012 models have been validated for regulatory purposes as replacement to animal testing for skin corrosivity and skin irritation (4). Theses assays usually rely on evaluating cytotoxicity and cytokine release.

So far, the published results demonstrate that these models are pretty similar to normal human skin. The in-depth characterization of reconstructed skin models is still ongoing. In addition, reconstructed human skin models present clear advantages (structure and biology of the epidermis, commercial availability as an in vitro screening tool) as compared to rodent skin for human risk assessment.

Each test substance (test material, negative and positive controls) is topically applied concurrently on three tissue replicates for 42 min at room temperature (RT, comprised between 18 and 24°C). Exposure to the test substance is followed by rinsing with phosphate buffer saline (PBS) and mechanically dried. Epidermis are then transferred to fresh medium and incubated at 37°C for 42 additional hours. Cell viability is assessed by incubating the tissues for 3 h with 0.3 mL MTT solution (1 mg/mL). The formazan crystals are extracted using 1.5 mL isopropanol for 2 h at RT and quantified by spectrophotometry at 570 nm wavelength. Sodium dodecyl sulfate (SDS 5%), and PBS-treated epidermis are used as positive and negative controls, respectively. For each treated tissue, the cell viability is expressed as the percentage of the mean negative control tissues. Values less than 50% is qualified the test substance as irritant.

2. Materials

2.1. RHE Kit (RHE Is a Reconstructed Epidermis Made by Skin Ethic Company, France.) (see Note 1)

1. Epidermal units (the area of epidermal surface and number of units will be supplied as per order) (see Note 2).
2. Maintenance medium (see Note 3).
3. Assay medium.
4. Quality controls of the kit.
5. Temperature indicator (see Note 4).

2.2. Reagents

1. Growth culture medium (SGM).
2. Phosphate-buffered saline (PBS).
3. SDS solution: The 5% SDS solution must be made in weight/volume. First weigh the SDS then add distilled water to the necessary volume to reach the final concentration of 5% w/v.
4. MTT solution: First make a stock solution. Dissolve MTT powder (3-(4,5-dimethylthiazol-2-yl)-2,5-diphenyltetrazolium bromide) to a final concentration of 5 mg/mL in PBS. Always protect the solution from light (see Note 5). Filter solution

with a 0.22 μm filter. Divide the MTT solution into 1 mL aliquots in sterile, dark 1.5 mL microtubes. Stock solution can be stored for up to 1 year at −20°C. On the day of testing, thaw the stock solution (5 mg/mL) and dilute it with pre-warmed maintenance medium at room temperature to 1 mg/mL.

5. Isopropanol.

3. Methods

One sterile 12-well plate can be used for one test material (see Note 6). For a given test material, the same plate needs to be used for all steps of the protocol. Prior to the transfer of epidermis from their transport packaging, the 12-well plates need to be labeled. Mark all plate lids with the code number of the test material (three wells per test material), or negative control (three wells) or positive control (three wells).

3.1. Preparation and Pre-incubation (Day 0)

The following steps need to be conducted in sterile conditions:

1. Fill an appropriate number of 12-well plates with 1 mL growth culture medium (SGM) (Fig. 1).
2. Remove the adhesive tape from the agarose plate containing epidermal tissues. Open the tissue kit and remove the absorbent paper.
3. Use sterile forceps to take off tissues from the agarose, clean the bottom of the insert on sterile absorbent paper or gauze to remove eventual remaining agarose pieces (Fig. 2) (see Note 7).
4. Check visually that no agarose is remaining and transfer the tissue on fresh medium by first sloping the insert before complete insert setting.

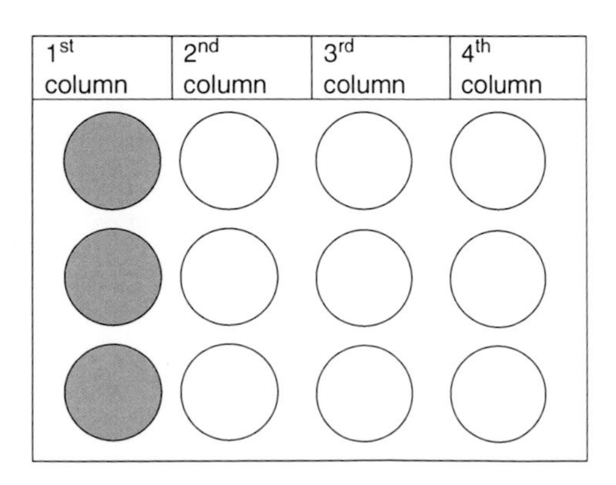

Fig. 1. Twelve-well plates filled with growth medium represented in *pink* in the first column (Color figures online).

Fig. 2. Removal of agarose from the bottom of the RHE unit.

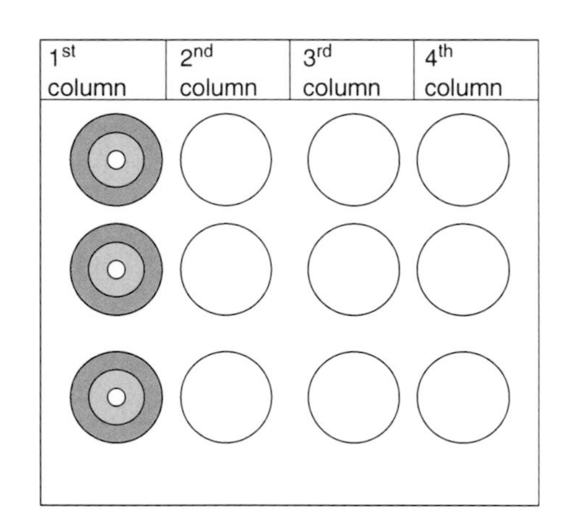

Fig. 3. The epidermal units placed in the growth medium filled wells for pre-incubation. *Yellow* represents epidermal unit, *pink* represents growth medium (Color figures online).

5. Check the absence of air bubbles by watching underneath the 6-well plate.
6. Place the RHE tissues at 37°C, 5% CO_2 until test substance application.
7. Label inserts by replicate treatment order (to be followed in the next step): Ex: Rep 1; rep 2; rep 3 (Fig. 3) (see Note 8).
8. Fill appropriate numbers of 12-well plates with 300 μL growth culture medium and proceed to pre-incubation for at least 2 h prior to test substance application (see Note 9).
9. Place the RHE tissues at 37°C, 5% CO_2 until test substance application.

3.2. Application of Test Substances and Rinsing (Day 1)

1. Pre-warm the maintenance culture medium at room temperature.
2. Fill the second column (three wells) of a 12-well assay plate with growth culture medium (300 μL per well, adjustable

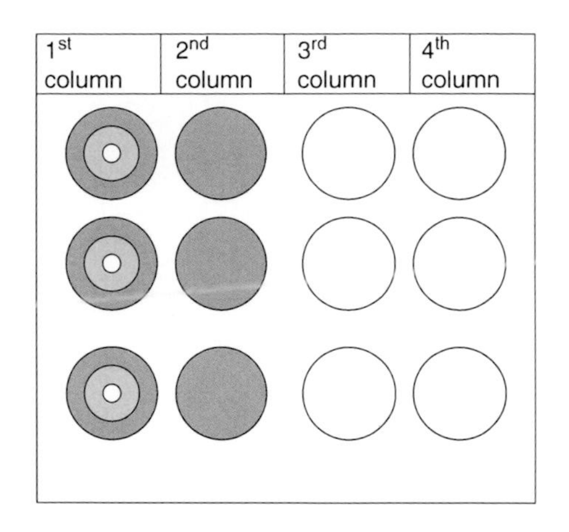

Fig. 4. Before application of test material, to the pre-incubated epidermal units, the second column is filled with growth medium represented in *pink* (Color figures online).

Fig. 5. The epidermal units were removed from the first column with forceps.

multistep pipette). Verify the plate lid labeling (name of test material (three wells per test material), or negative control (three wells) or positive control (three wells)) (Fig. 4).

3. Use sterile forceps to transfer tissues by first sloping the insert before complete insert setting at the air–liquid interface (Figs. 5 and 6).
4. Check for the absence of air bubbles by watching underneath the 12-well plate.
5. Check the presence of all materials/equipment necessary for test substance application, washing, drying, and post-incubation steps.

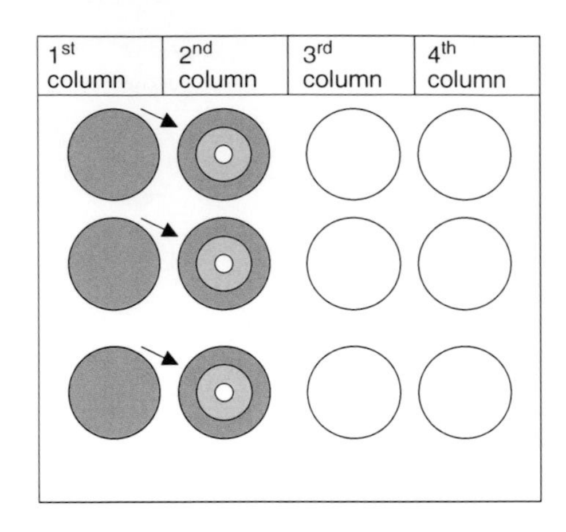

Fig. 6. The pre-incubated epidermal units are transferred from first column to second column. *Arrows* represents the transfer of tissues from first column to second column.

3.2.1. Topical Applications: 42 min Treatment

It is strongly recommended to perform these steps under sterile conditions.

1. Three tissues per test substance should be used (three replicates). The application order is important since it will be the same for washing.
2. Record the exact timings.
3. Due to the application timing of 42 min, the application and rinsing phases should be performed in a minimum of two steps for testing the internal controls (NC and PC) and the test substances.

3.2.2. For Liquid Test Materials (Nanoparticle Suspension) (see Note 10)

1. Apply 16 μL ± 0.5 μL (i.e. 32 μL/cm^2) on the top of each epidermis, using a positive displacement pipette.
2. Gently spread it on the epidermis surface (three per test substance: replicate 1, replicate 2, and replicate 3). Ensure to cover the entire surface of the epidermal units.
3. Keep the plate (lid on) containing the treated RHE tissues for 42 min exposure (±1 min) in the ventilated cabinet sterile conditions at room temperature.

3.2.3. For Solid Test Materials (see Note 10)

1. Apply 10 μl distilled water using a positive displacement pipette to the epidermal surface in order to improve further contact between the powder and the epidermis. Gently spread with the pipette.
2. Apply 16 mg ± 2 mg (32 mg/cm^2) of the powder to the epidermis surface.
3. Gently spread if necessary on the epidermal surface with a curved spatula.

4. Three tissues per test substance should be used. The application order is important since it will be the same for rinsing.
5. Keep the plate (lid on) containing the treated RHE tissues for 42 min exposure (±1 min) in the ventilated cabinet sterile conditions at room temperature.

3.3. End of the Treatment and Removal of the Test Material

It is strongly recommended to work in laminar flow hood to prevent contamination. Strictly respect the application order. In order to prevent pollution the lids should be put on the plates continuously during the rinsing and drying steps.

3.3.1. After 42 min Exposure (±1 min)

1. Remove the treated units using forceps, and rinse thoroughly with 25 ml sterile PBS by filling and emptying the tissue inserts to remove all residual test material from the epidermal surface. Rinse thoroughly with 25 ml of sterile PBS by filling and emptying the tissue inserts (Fig. 7) (see Note 11).
2. Place the units on an absorbent paper, remove the remaining PBS from the epidermal surface by gently taping, and sweep the surface with a cotton bud if necessary without damaging the epidermis (Figs. 8 and 9).
3. Transfer the blotted tissue units in the new maintenance medium prefilled wells (third column) (Fig. 10).

3.3.2. Posttreatment Incubation: 42 h

1. Incubate the rinsed epidermis at 37°C, 5% CO_2, 95% humidified atmosphere for 42 h (± 60 min). Record incubations start time.
2. Label appropriate numbers of polypropylene tubes with caps (three tubes per tissue unit): test material name, replicate number, and date.
3. At the end of the 42 h ± 1 h incubation period, remove the plates from the incubator, and shake the plates containing the

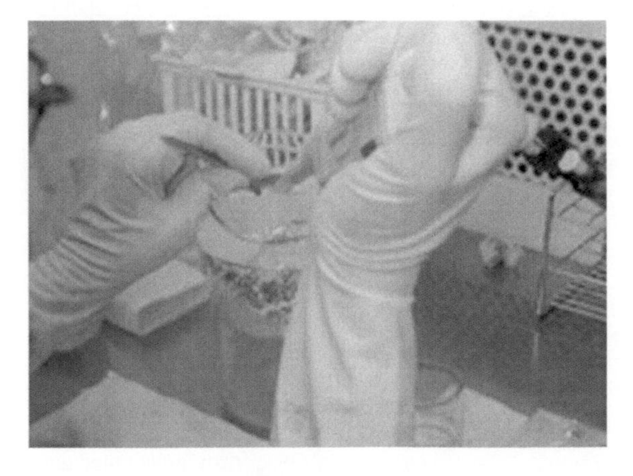

Fig. 7. The epidermal units are washed with PBS by filling and emptying using pipette.

Fig. 8. The epidermal units are dried using absorbent paper.

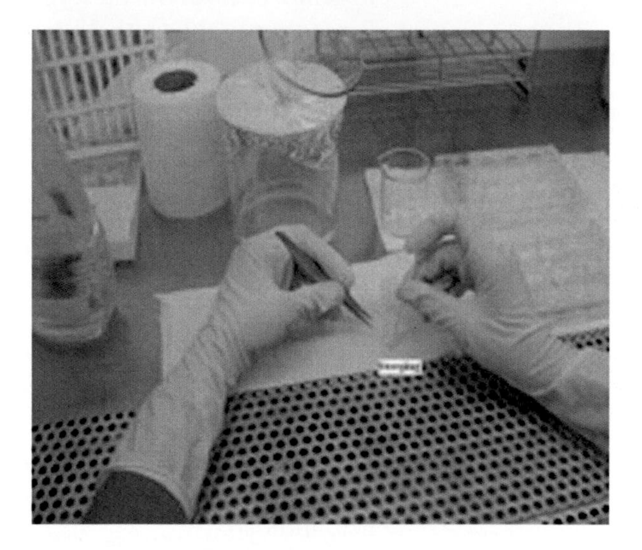

Fig. 9. Sweeping of excess PBS with cotton bud.

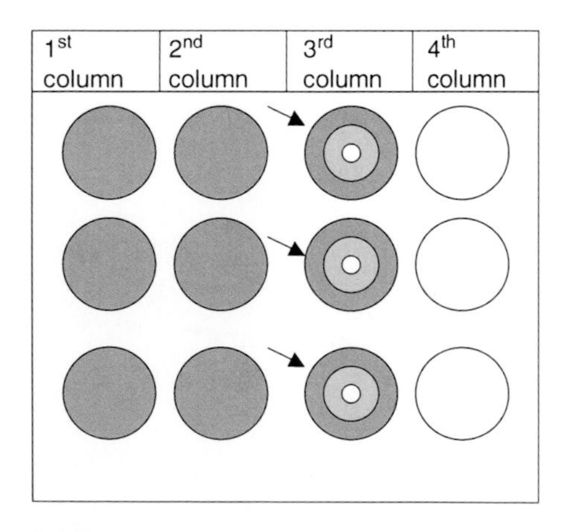

Fig. 10. The washed epidermal tissues were transferred from second column to third column containing fresh growth medium. *Arrows* represents the transfer of tissues from second to third column.

treated epidermis (lids on) on a plate shaker for 2 min, medium speed (300 rpm/min). This step helps to homogenize the released mediators in the medium before sampling.

4. Transfer 3 × 500 μL of incubation medium from each tissue to the pre-labeled tubes. Store frozen at –20°C until analyses (up to 1 year).
5. In addition, freeze 15 mL of the growth culture medium that was used as diluent. The inflammatory cytokines (IL-1α) can be quantified using ELISA (see Notes 12–14).

3.4. MTT Test After the 42 h Incubation Period

1. Tissue viability is assessed by MTT reduction measurement, after the 42 h (±60 min) incubation at 37°C, 5% CO_2, 95% humidified atmosphere.
2. Fill the wells of the 12-well plate (fourth column) with 300 μL per well with MTT working solution prepared in assay medium.
3. Transfer the tissue units to the MTT working solution filled wells. Before transferring, sweep excess maintenance medium on the unit bottom with absorbent paper before setting down on the well (fourth column). Replace the lid on the plate (Fig. 11).
4. Incubate for 3 h (±5 min) at 37°C, 5% CO_2, and 95% humidified atmosphere. Record start time of incubation. (Viable cells metabolize MTT; the blue color of the epidermis is due to the intra cellular formation of formazan crystals.) (Fig. 12).
5. At the end of the incubation, record observations.
6. Label with appropriate numbers on new 24-well plate(s).
7. Fill the plate(s) with 800 μL isopropanol at the end of the 3 h (±5 min) incubation in MTT solution.

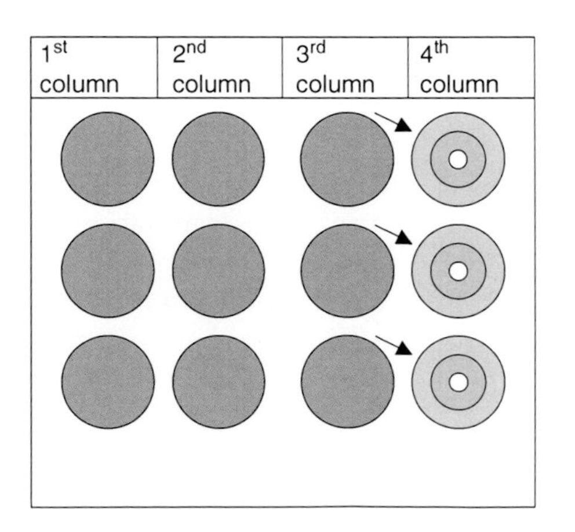

Fig. 11. After 42 h of post-incubation the epidermal units were transferred from third column to fourth column containing MTT working solution. *Light yellow* represents MTT solution. *Arrows* represents the transfer of tissues from third to fourth column (Color figures online).

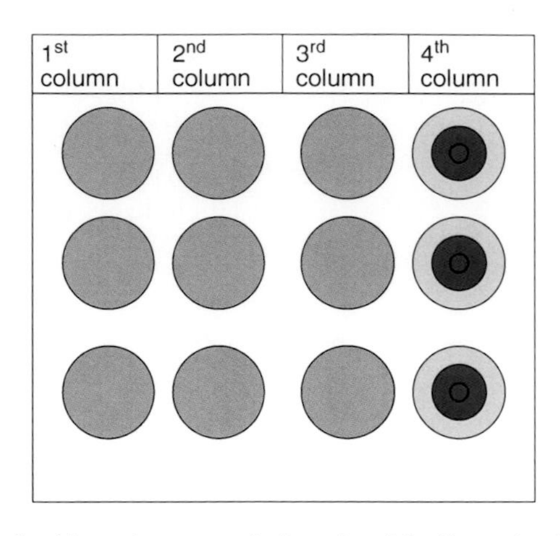

Fig. 12. After 3 h, the *blue color* represents the color of the tissue due to formazan crystals. *Yellow* represents MTT solution (Color figures online).

8. Record observations and comments after the 3 h MTT incubation.
9. Use sterile forceps to transfer treated tissues.
10. Dry the insert bottom of the treated tissue on sterile absorbent paper or gauze.
11. Transfer the tissues in isopropanol solution. Add 700 μL isopropanol solution on the top of each tissue. Ensure that tissue is completely covered by the isopropanol solution (see Note 15).
12. Consciously protect plate(s) from evaporation by stretching three parafilm layers over the plate and adding the lid on the plate.
13. Incubate for 2 h (±5 min) at room temperature with gentle agitation (about 150 rpm) for formazan extraction, or over night for about 16–18 h at +4°C protected from light.
14. Document isopropanol extraction start time.
15. The plates containing the tissues need to be sealed carefully to prevent evaporation and cross contaminations.

3.5. Absorbance/Optical Density Measurements

1. At the end of the 2 h (±5 min) or overnight formazan extraction incubation time, open the plate and remove the three parafilm layers.
2. Isopropanol solution is used as blank (six replicates).
3. Maintain the insert with forceps and pierce tissue and polycarbonate filter with a tip in order to get the whole extraction solution in the corresponding well (see Note 16).
4. Homogenize the extraction solution by pipetting three times up and down to complete formazan crystals solubilization.

PLATE1

	1	2	3	4	5	6	7	8	9	10	11	12
A	EMPTY	blank	blank	blank	blank	blank	blank	EMPTY	EMPTY	EMPTY	EMPTY	EMPTY
B	EMPTY	NC	NC	NC	NC	NC	NC	NC	NC	NC	EMPTY	EMPTY
C	EMPTY	PC	PC	PC	PC	PC	PC	PC	PC	PC	EMPTY	EMPTY
D	EMPTY	C1	C1	C1	C1	C1	C1	C1	C1	C1	EMPTY	EMPTY
E	EMPTY	C2	C2	C2	C2	C2	C2	C2	C2	C2	EMPTY	EMPTY
F	EMPTY	C3	C3	C3	C3	C3	C3	C3	C3	C3	EMPTY	EMPTY
G	EMPTY	C4	C4	C4	C4	C4	C4	C4	C4	C4	EMPTY	EMPTY
H	EMPTY	EMPTY	EMPTY	EMPTY	EMPTY	EMPTY	EMPTY	EMPTY	EMPTY	EMPTY	EMPTY	EMPTY
		TISSUE1			TISSUE2			TISSUE3				

Fig. 13. Flow chart for 96-well plate usage for optical density reading. *PC* Positive control, *NC* Negative control, *C1* Nanoparticle 1, *C2* Nanoparticle 2, *C3* Nanoparticle 3, *C4* Nanoparticle 4.

5. Transfer per tissue 3 × 200 μL extraction solution per well (three wells per tissue, i.e., three replicates per tissue) into a 96-well plate (see Note 17) (Fig. 13).
6. Read the optical densities (OD) using a 96-well plate spectrophotometer ideally at 570 nm wavelength (eventually between 540 and 600 nm).
7. Do not use the "empty" wells for readings. Record the results.

3.6. Acceptance Criteria

1. Negative control (NC) acceptance criteria: The NC data meet the acceptance criteria if the mean OD value of the three tissues is ≥1.2 at 570 nm according to the historical database. The standard deviation value is considered valid if it is ≤18%, according to the Performance Standards (5).
2. Positive control (PC) acceptance criteria: The PC data meet the acceptance criteria if the mean viability, expressed as % of the NC, is <40% and the standard deviation value is ≤18%.

3.7. Killed Epidermis Preparation (for MTT-Interacting Substances)

1. Transfer living epidermis in 24-well plate and place them at −20°C (or −80°C) for at least 48 h (three tissues/MTT-interacting test substance).
2. Thaw killed tissues, before use, on 300 μL of maintenance medium for 1 h (±5 min) at room temperature.
3. Further use of killed tissues is similar to living tissues.

3.8. Data Calculations Steps (5)

1. Blanks—Calculate the OD mean from the six replicates for each plates OD blank.
2. Negative PBS-treated controls: Subtract blanks mean value from individual tissues OD. Corrected OD mean for the three tissues corresponds to 100% viability. ODNC = ODNCraw − ODblank.
3. Calculate the OD mean per tissue (three replicates). The mean OD for all tissues corresponds to 100% viability = mean ODNC.
4. Positive control (SDS 5%): Subtract blanks mean value from individual tissues OD. ODPC = ODPCraw − ODblank.

5. Calculate the OD mean for each individual tissue (three replicates).
6. Calculate the viability per tissue: %PC = (ODPC/mean ODNC) × 100.
7. Calculate the mean viability for all tissues: Mean PC = Σ %PC/ number of tissues.
8. Tested compound: Subtract blanks mean value from individual tissues: OD ODTT = ODTTraw − ODblank.
9. Calculate the OD mean for each individual tissue (three replicates).
10. Calculate the viability per tissue: %TT = (ODTT/mean ODNC) × 100.
11. Calculate the mean viability for all tissues: Mean TT = Σ %TT/ number of tissues.
12. Standard deviations are calculated on OD and % viabilities.

3.8.1. Data Calculations for MTT-Interacting Substances

1. Test substances that interfere with MTT can produce nonspecific reduction of the MTT. It is necessary to evaluate the OD due to nonspecific reduction and to subtract it before calculations of viability %.
2. Nonspecific MTT reduction calculation (NSMTT).
   (a) ODKu: untreated killed tissues OD.
   (b) ODKT: test substance-treated killed tissues OD.
   (c) NSMTT = ((ODKT − ODKU)/ODNC) × 100.
   (d) If NSMTT is >30% relative to the negative control: additional steps must be undertaken if possible or the test substance must be considered as incompatible with the test.
   (e) Treated tissue True MTT metabolic conversion (TODTT).
   (f) ODTT: test substance-treated viable tissues.
   (g) TODTT = (ODTT − (ODKT − ODKU)).
   (h) Relative viability % = (TODTT/ODNC) × 100.

3.8.2. Data Calculations for Coloring Test Substances Able to Stain Tissues

For chemicals detected as able to color the tissues, it is necessary to evaluate the nonspecific OD due to the residual chemical staining (unrelated to any mitochondrial activity) and to subtract it before calculations of the "true" viability %.

1. Nonspecific staining % (NSS %).
   (a) ODCT: coloring test substance-treated tissue (incubated in maintenance medium before isopropanol extraction).
   (b) ODPBS: control PBS-treated tissue (incubated in maintenance medium before isopropanol extraction).

   (c) NSS% = (ODCT/ODPBS) × 100.
   (d) If NSS % is >30% relative to the negative control: additional steps must be undertaken if possible or the test substance must be considered as incompatible with the test.
2. True MTT metabolic conversion (TODDT).
   (a) ODCT: coloring test substance treated tissues (MTT incubation).
   (b) TODCT: true MTT metabolic conversion for coloring test substance-treated tissue.
   (c) TODCT = ODTT − ODDT.
   (d) Relative viability % = (TODCT/ODNC) × 100.
   (e) Data calculations for coloring test substances which are also MTT interacting test substances.
   (f) Calculate corresponding NSMTT and NSS.
   (g) If (NSMTT % + NSS %) is > 30% relative to the negative control: additional steps must be undertaken if possible or the test substance must be considered as incompatible with the test.
3. True MTT metabolic conversion for dye test substances which are also MTT-interacting test substances (TODDTT).
   (a) ODCT: coloring test substance-treated tissues (MTT incubation).
   (b) TODCT: true MTT metabolic conversion for coloring test substance-treated tissue.
   (c) TODCTT = (ODTT − (TODCT + TODTT)).
   (d) Relative viability % = (TODCTT/ODNC) × 100.

3.9. Data Interpretation: Prediction Model

According to EU classification, the irritancy potential of test substances is predicted for distinguishing between R38 skin irritating and no-label (non-skin irritating) test substances (6). In the present study, the irritancy potential of test substances is predicted by mean tissue viability of tissues exposed to the test substance. The test substance is considered to be an irritant to skin (R38), if the mean relative viability after 42 min exposure and 42 h post-incubation is less or equal (≤) to 50% of the negative control (see Tables 1, 2, and 3).

Table 1
The prediction model (PM) for classification of test material

Criteria for in vitro interpretation	Classification
Mean tissue viability is ≤50%	R38, irritant (I)
Mean tissue viability is >50%	No label, nonirritant (NI)

Table 2
Flow chart for various test substances (standard/MTT interacting/colored/colored + MTT interacting)

Condition 1	Condition 2	Condition 3
All test substances (standard, MTT interacting, coloring test substances)	Coloring test substances	MTT-interacting test substances
Use of *living RHE tissues* + PBS negative control tissues + SDS positive control tissues	Use of *living RHE tissues*	Use of *killed RHE tissues* + Untreated negative control Killed tissues
42 min exposure + 42 h post-incubation 3 h MTT incubation	3 h medium incubation	3 h MTT incubation
2 h/overnight isopropanol extraction OD = specific OD + nonspecific OD	OD = nonspecific staining (NSS)	OD = nonspecific MTT reduction (NSMTT)

Table 3
Case by case test conditions guidance

	Medium coloration	Tissue staining	MTT interaction	Test conditions
Case 1	–	–	–	1
Case 2	+	–	–	1
Case 3	–	+	–	1 + 2
Case 4	+	+	–	1 + 2
Case 5	–	–	+	1 + 3
Case 6	+	–	+	1 + 3
Case 7	–	+	+	1 + 2 + 3
Case 8	+	+	+	1 + 2 + 3

4. Notes

1. Soon after receipt of the kit, the following aspects to be checked: (1) date of shipping on the package and (2) color of the nutritive agar and its pH. If the color is orange it is acceptable, but if the color is yellow or violet it is not acceptable. If any abnormalities are noted, the supplier needs to be contacted

immediately. Leave the RHE kit in packaging at room temperature in a microbiological safety cabinet until use.

2. Gloves are to be worn while handling the units. The epidermal units can be kept at room temperature (37°C).
3. The maintenance and growth medium need to be stored at 4°C and the shelf life of epidermal units and the medium are 10 days. The maintenance and assay medium supplied with the kit need to be pre-warmed at room temperature prior to the experiment.
4. Temperature indicators are to verify whether the kit has exposed to a temperature above 40°C. Inspect the color of temperature indicator. If the color is pale gray it is acceptable, but if the color is dark gray it is not acceptable.
5. MTT solution is light sensitive. Protect it from light using silver paper or appropriate material. Work in ventilated cabinets to prevent accidental contact wear protective gloves, and if necessary a mask and/or safety glasses.
6. Act quickly as the tissue cultures dry out rapidly when not in contact with medium.
7. Make sure that the medium is never above the epidermal units.
8. Based on the day of tissue receipt, the time of incubation and volume of growth medium needs to be employed. If the tissues are received on day 17 or earlier, the above procedure is applicable. If the tissues are received on day 18 or later, the procedure is the same except the amount of growth medium and incubation time.
9. Use one plate per test substance to prevent any adjacent effects of test substance fumes.
10. Chemicals are applied undiluted.
11. The washing and rinsing of tissues need to be done very carefully, since the nanoparticles may interfere with the medium and readings.
12. While estimating the cytokines, the optical density due to the nanoparticle suspensions needs to be measured in wells without tissue; the Optical Density of Nano Particle suspension Without Tissue (OD_{NPWOT}). The accurate blank correction has to be done by subtracting the optical density of nanoparticle suspension without tissue.
13. The IL-1 alpha measurements can be done only for tissues with a viability >50% after treatment.
14. For thawing of frozen incubation medium do not use water bath because it may lead to cytokine damage.

15. Be careful to prevent isopropanol evaporation in 12- or 96-well plates.
16. After use, the epidermis, the material, and all media in contact with it should be decontaminated (e.g., by using a 10% solution of bleach or appropriate containers) prior to disposal.
17. It is recommended to fill not more than 42 wells per plate and to make the readings in the same run (Fig. 13).

Acknowledgments

We thank SkinEthic Laboratories, France for their RHE kit and procedure. We also thank Prof. P.V. Subbarao, Bigtec, Bangalore for critically reading the book chapter.

References

1. Kroll A, Pillukat MH, Hahn D, Schnekenburger J (2009) Current *in vitro* methods in nanoparticle risk assessment: limitations and challenges. Eur J Pharm Biopharm 72:370–377
2. Stone V, Johnston H, Schins RPF (2009) Development of *in vitro* systems for nanotoxicology: methodological considerations. Crit Rev Toxicol 39(7):613–626
3. Sairam Kishore A, Surekha P, Balakrishn Murthy P (2009) Assessment of the dermal and ocular irritation potential of multi-walled carbon nanotubes by using *in vitro* and *in vivo* methods. Toxicol Lett 191:268–274
4. OECD guideline for testing of chemicals number 439 (2010) In vitro skin irritation: reconstructed human epidermis test method. Organization for Economic Cooperation and Development, Paris, France
5. Skin Ethic skin irritation test-42bis using the reconstructed human epidermis (RHE) model. Skin Ethic Laboratories, Version-2, 2009
6. OECD guideline for testing of chemicals number 404 (2002) Acute dermal irritation/corrosion, 6 pp. Organization for Economic Cooperation and Development, Paris, France

Chapter 17

In Vivo Methods of Nanotoxicology

Khaled Greish, Giridhar Thiagarajan, and Hamidreza Ghandehari

Abstract

The new field of nanotoxicology is steadily emerging in parallel with rapid advances made in nanotechnology to evaluate biological impact of intended and non-intended nanomaterial exposure over time as their human applications constantly increase. Over the last decade nanotoxicology methods have mostly relied on in vitro cell-based characterizations that do not account for the complexity of in vivo systems with respect to biodistribution, metabolism, hematology, immunology, and neurological ramifications. Comprehensive in vivo studies addressing the toxicity of nanoscale materials are scarce mainly because the field is still nascent. Efforts in standardizing methodology to study the in vivo safety of these materials are currently undertaken by various government agencies and research organizations. Here, we discuss the need for in vivo nanotoxicity studies, outline some of the important methods, and comment on practical considerations in carrying out such studies.

Key words: In vivo, Nanotechnology, Biocompatibility, Nanotoxicology, Safety

1. Introduction

Compared to the exponential number of publications appearing about nanoparticles, only 3% of the scientific effort is devoted to study their critical biological effects (Fig. 1). Nanomaterials can have unique and hazardous effects on biological systems. For example, protein interactions that are widely ascribed to bare nanoparticles can translate into serious clotting consequences in vivo when they initiate clotting cascades (1, 2). Similar interactions of nanoparticles with plasma membranes can result in significant hemolysis of red blood corpuscles (RBCs) in circulation. Currently, there are multiple efforts in the United States and the rest of the world to stipulate safety studies of nanoscale constructs for biomedical purposes. The National Nanotechnology Initiative (NNI)

Khaled Greish and Giridhar Thiagarajan contributed equally to this work.

Joshua Reineke (ed.), *Nanotoxicity: Methods and Protocols*, Methods in Molecular Biology, vol. 926,
DOI 10.1007/978-1-62703-002-1_17, © Springer Science+Business Media, LLC 2012

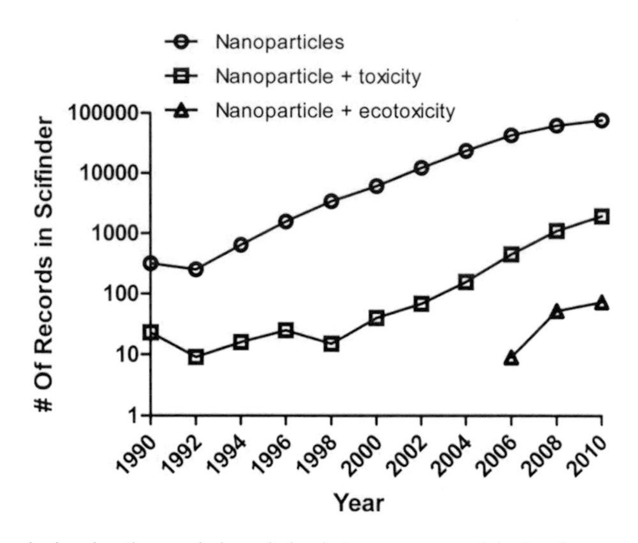

Fig. 1. Graph showing the gap in knowledge between nanoparticle development and their safety and ecological impact with respect to the number of publications.

in the USA was created to coordinate research and development of new nanoscale science and engineering as well as to understand the biological impact of this burgeoning field on the environment and in living organisms (3). Efforts in this area involve identifying potential occupational risk, biomedical exposure, and environmental exposure (Fig. 2). As an example, comprehensive assays and tests for physicochemical and in vitro characterization of nanoparticles have been elaborated by the Nanotechnology Characterization Laboratory (NCL) at the National Cancer Institute (NCI) to guide the research community on the safety of nanomaterials for human use (4). This chapter elaborates on the possible methods to evaluate in vivo nanotoxicity and highlights important considerations while planning such studies.

1.1. Need for In Vivo Nanotoxicity Studies

Cultured cells have been extensively utilized to test the safety of nanoconstructs (5, 6) and recently in silico computational models have also been used to predict toxicity. These methods delve into isolated aspects of the biological system whereas the in vivo machinery is much more complex with interdependent pathways that are just impossible to capture in a single in vitro experiment. Toxicity assays performed in animal models are probably the closest and most representative of the human condition. The following section briefly discusses merits and limitations of in vitro studies and emphasizes the need for in vivo evaluation of nanotoxicity.

1.1.1. In Vitro Assays for Nanotoxicity

In vitro toxicity methods were first developed to examine the detailed mechanisms underlying toxicity of certain chemicals as observed in vivo. In vitro assays usually involve measurement of cell viability and a battery of other expanding tests such as assessment of plasma integrity, integrity of mitochondria, biomolecule synthesis,

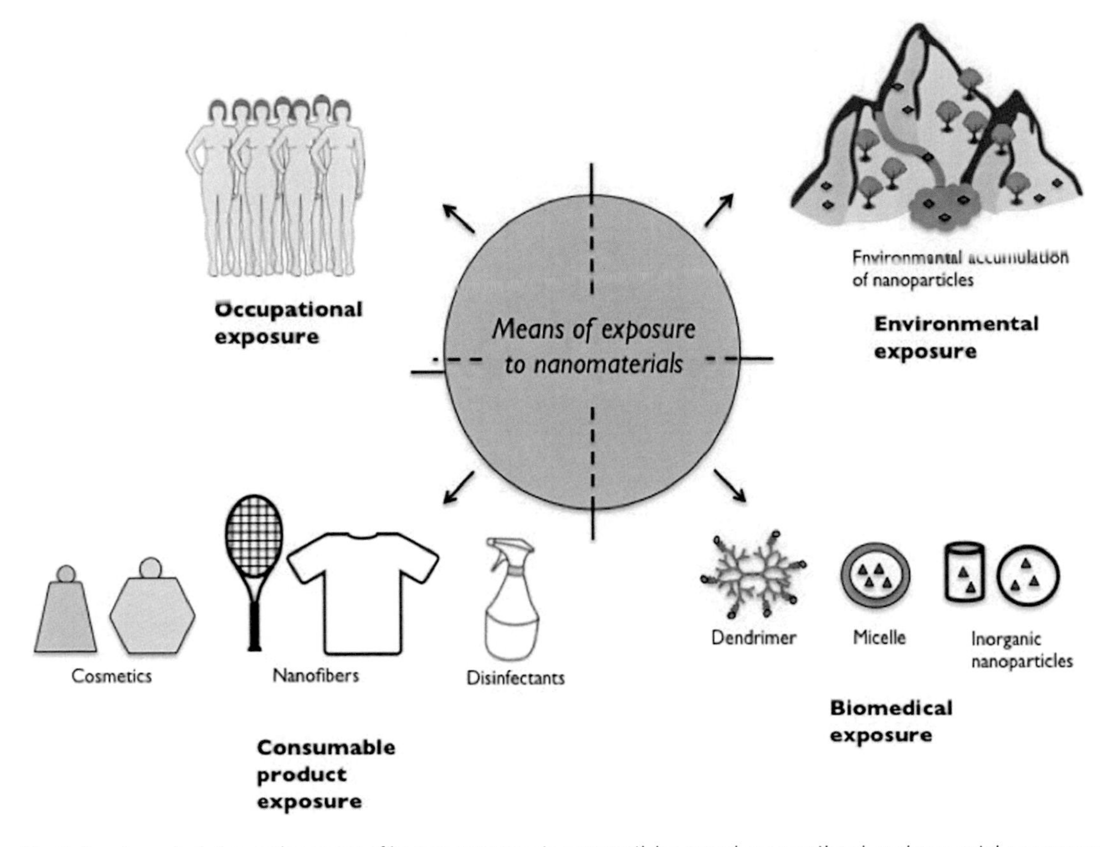

Fig. 2. A cartoon depicting major routes of human exposure to nanoparticles namely occupational, environmental, consumer products, and biomedical exposure.

cell division, and redox potentials (5). Both quantitative chemical assays and qualitative imaging systems have been developed to address these end points. Recently much advancement to the existing two-dimensional cell systems has been developed to serve as an intermediate between cultured cells and animal models. One such model is the cell sheet with intact extracellular matrices which have been developed as an alternative to cultured cells that have the ability to mimic not only the phenotype of the cell type under study but also some of their physico-mechanical properties (7). Recently, researchers have also developed a three-dimensional model for studying foreign body responses in vitro consisting of fibroblasts and macrophages mimicking the biological system (8). Although these models are improvements over the 2D cultured systems, they leave much to be desired as compared to in vivo models. Cell lines used for in vitro toxicity assays do not represent the cell population exposed to the nanomaterial under an in vivo setting due to various factors such as route of exposure, absorption, distribution, metabolism, or excretion. In addition, cells in vitro are usually under a controlled environment that does not allow many complex interactions that exist in vivo such as signaling between cell types through

Table 1
Advantages and disadvantages of in vitro and in vivo toxicity assays

	Advantages	Disadvantages
In vitro toxicity assays	Ethical issues for use of animals for experimentation are not a concern	In vitro environment lacks entire set of stimuli (e.g., cytokines, growth factors, etc.)
	Cost effective	Cultured cells do not represent in vivo phenotype
	Homogenous and controlled environment	Monolayered cultures do not represent in vivo responses
	Eliminate secondary effects such as inflammation	Cultured cell experiments do not replicate chronic exposure
	Easy to delineate mechanisms of toxicity	Lack of standard methods for cell-line maintenance and protocols
In vivo toxicity assays	Captures complex cellular and tissue interactions	Species to species variations
	Can understand the influence of route of exposure on toxicity	Higher costs of maintenance, time, and labor intensive
	Absorption, metabolism, biodistribution, excretion effects can be studied	Ethical restrictions and animal welfare concerns
	Chronic exposure effects can be studied	Expertise and special training requirements necessary to handle animals

cytokines, hormones, chemokines, etc. Furthermore absence of vascular system implies complex hematological interactions, and other end points including systemic inflammation, ischemia, or hyperpermeability cannot be measured. However, acknowledging their limitations, in vitro assays pose as an attractive alternative to in vivo nanotoxicity assays due to relative simplicity and feasibility. The advantages and limitations inherent to in vitro toxicity assays are listed in Table 1.

1.1.2. Predictive Mathematical Models for In Vivo Nanotoxicity

Use of in silico models is an interesting approach to address toxicity of new materials through integrated computational systems accounting for multiple variables associated with the material/biological interactions. The term can be confused with toxicity in silico, which actually describes reaction of nanomaterials with biological molecules such as proteins or DNA bound to silicon microarrays. Mathematical modeling is mainly used to assess safety, especially the pharmacokinetic properties (i.e. absorption, distribution, metabolism, excretion) based on structurally similar compounds. These modeling systems have been adapted to predict nanotoxicity. However, in this type of evaluation well-defined biological, toxicological, or pharmacological endpoints observed with similar nanoparticles should be available in order to predict potential nanotoxicity. In the case of newly engineered nanoparticles there is obviously little toxicological or pharmacological data available and only after sufficient data

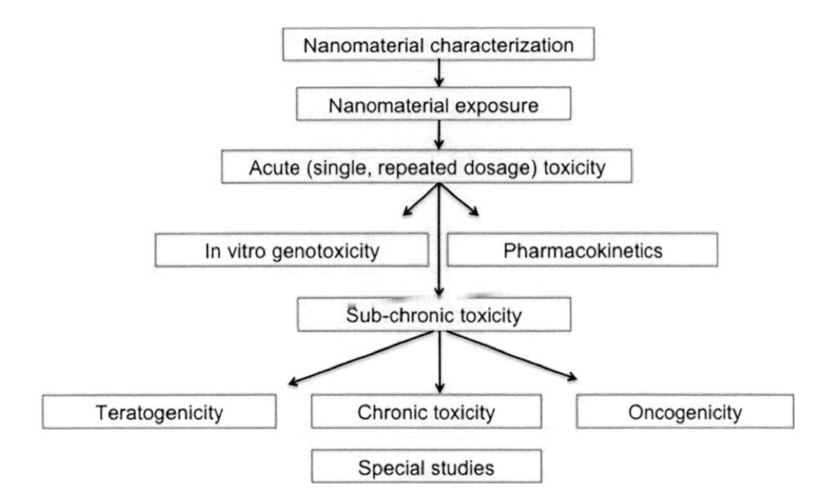

Fig. 3. Flowchart showing the array of toxicological assays available related to nanomaterials safety.

is generated can the potential cell or tissue damage be predicted based on chemical structures, functional groups, shape, density, or size of a particular class of nanomaterials (9) (e.g., metal oxide nanoparticles vs. mesoporous silica nanoparticles). Validating the results of the mathematical model by testing actual in vivo toxicity will always be required to ascertain the accuracy of such systems. Given the limitations of in vitro and in silico methods, consistent and accurate in vivo evaluation is essential.

2. In Vivo Nanotoxicity Methods

A classical scheme for toxicity assessment is represented in Fig. 3. While detailed toxicity assessments are described in toxicology textbooks (10–12), we will attempt to summarize the principles of these methodologies in the following section and relate them to the assessment of nanomaterials. Toxicity assessments are classified into acute, subacute, sub-chronic, and chronic studies. Absorption, distribution, metabolism, excretion, and pharmacokinetics (ADME/PK), carcinogenic and teratogenic studies are also required for complete evaluation of nanomaterial toxicity.

2.1. Characterization of Nanoparticles and Their Purity

Thorough and systematic physicochemical characterization of nanomaterials is essential prior to studying their toxic effects in vivo. One of the main reasons is to ensure the accuracy and reproducibility of the observed effects of toxicity with the studied nanoparticles. Toxicity of nanoparticles is significantly influenced by their physicochemical properties such as size, shape, surface charge, charge density, composition, density of structure, presence of pores, and surface activating sites. Hence documenting the characteristics of the nanoparticle under evaluation for in vivo toxicity becomes crucial in order to correlate the observed biological effects in vivo. Knowledge of nanoparticle properties is indispensible for understanding their biological behavior in a complex in vivo environment. Importantly characterization data is important to facilitate comparison of toxicity results of a given nanoparticle with other nanoconstructs. Instrumentation techniques have developed considerably in the past decade to sufficiently characterize the physicochemical properties of nanoparticles including dynamic light scattering techniques for size determination, zeta potential measurements for relative particle charge, electron (scanning and transmission) microscopy to qualitatively analyze size and shape, nuclear magnetic resonance to accurately determine chemical structures, X-ray photoelectron spectroscopy to measure chemical composition, chemical and electronic states, mass spectrometry for mass, and elemental analysis and spectroscopy techniques that measure absorption, emission, or scattering of either wavelength or frequency. An important note of caution while characterizing nanomaterials is to evaluate these properties under physiologically relevant conditions. For an in-depth perspective about nanomaterial characterization, the readers are directed to references 2, and 13–15.

It is critical to ensure the purity of nanomaterials while assessing toxicity so that any nonspecific toxicity attributed to material impurities or biological contaminations can be excluded. Chromatographic techniques such as high pressure liquid chromatography or size exclusion chromatography can indicate presence of impurities in the sample. Some of the other techniques mentioned above such as mass spectrometry and spectral analyses also provide valuable information regarding contaminations or impurities.

Endotoxins or lipopolysaccharides (LPS) are small bacterially derived hydrophobic molecules which can easily contaminate nanomaterials. The presence of LPS can considerably influence in vivo toxicity due to their pronounced biological effects. Intravenous injection of an endotoxin dose as low as 1 ng/ml was found to produce pyrogenic reactions and shock through the release of tumor necrosis factor (TNF), interleukin (IL)-6 and IL-1 from activated monocytes and macrophages in rabbits (16). The commonly used FDA-approved techniques for endotoxin detection are the rabbit pyrogen test and limulus amoebocyte lysate (LAL) assay (17).

2.2. Acute Studies

The objectives of acute toxicity studies are usually to identify the maximum tolerated dose (MTD) and no observable effect level (NOEL) of nanoparticle dosage. While dose in classical toxicological studies are measured in terms of mg of test material administered per kg of animal weight, nanotoxicity studies can substantially deviate from this classical requirement. Substantial surface area per mass of nanosized materials is common, which can significantly influence the material properties and consequently its biological safety behavior. Hence surface area of nanoparticles plays an important role. In addition, particles can settle, diffuse, and aggregate according to their size, density, and surface properties. The definition of dosage for nanoparticles is more dynamic, complicated, and less comparable across particles as compared to conventional small molecular weight toxins (18). Nanotoxicology should be conducted based on both the classical system of toxicology referring to the mg/kg exposure in order to be able to compare with other toxins, and dosage based on surface area to clarify the effect of nanoscale reactivity on toxicity. We will refer to the classical dose of mg/kg in this chapter, as a consensus has not been arrived for dose definition of nanomaterials.

In addition to dose finding evaluation, acute studies focus on organ-specific toxic effects and elucidate their mechanisms of toxicity. Such studies typically span 14 days after a single dose or repeated dose administration. At least two species are required to conclude results from these studies, preferably one rodent and one non-rodent species. In the case of single dose administration, a minimum of five animals/sex/dose is needed with rodents or one animal/sex/dose in the case of primates. The following parameters need to be monitored during the study:

- *Response to administered dose*: Within few minutes after administration of nanomaterials, hematological, cardiac, and neuronal responses can occur. It is important to keep the animal under close observation for at least 30 min following administration to observe these changes.
- *Weight change*: This simple and feasible parameter is one of the most sensitive indicators of overall health of the animal. Fluctuation in weight change can reflect adverse response to the materials, which is otherwise difficult to detect by other tests. While this parameter is very sensitive, it is also nonspecific. Once dramatic weight change (>10%) is detected, further investigation is required to determine the cause of such toxicity and the target organs or tissues involved.
- *Clinical observation*: Function of all organ systems should be examined to evaluate clinical changes. Cardiovascular system function can be evaluated by the presence of cyanosis of the tail, mouth, or footpads. Vasodilatation can be assessed by redness of the skin and vasoconstriction by coldness of the body. Respiratory effects manifest as dyspnea. Locomotive system can be studied by posture, coordination of movements, or tremors. Ocular signs can be observed by the presence of lacrimation or ptosis. Gastrointestinal function can be monitored by food consumption and the quality of dropping (e.g., watery stool). Fur and skin should be examined for ruffling. The clinical examination frequency should be at least

once per day, and once an adverse effect is noticed observation frequency should be increased to twice per day. More stringent clinical observation can involve behavioral changes although they are less common in toxicity studies. Imaging procedures for specific organ toxicity such as ultrasound, X-ray, computed tomography (CT), and magnetic resonance imaging (MRI) can also be used.

- *Mortality*: Utmost care should be exercised to ensure humane and ethical use of test animals. The animals should not be left to die in agony and once severe side effects are noted during one observation point the animal should be humanely euthanized before the next observation point (within the same day) if they have not recovered.
- *Clinical pathology*: One of the main objectives of acute studies is to identify the target organs with toxicity. Blood samples acquired within few minutes after the animal is euthanized allow investigation of specific organ damage. During blood collection animals need to be killed by CO_2 asphyxiation. Use of cervical dislocation can result in low yield of blood. In comparison to cardiac puncture, collection of blood from inferior vena cava is more efficient. Blood samples should be aspired using a large needle (i.e. 26 G) to avoid hemolysis. Whole blood can be used for analyzing complete blood count that could indicate bone marrow function and also nonspecific immunological interactions. For biochemical assays of parameters related to different organs, plasma should be separated within 30 min of blood collection to ensure integrity of the measured enzyme levels. Once plasma is collected it can be used directly or frozen in −80°C until further analysis. Plasma can be used to check for liver function by measurement of aspartate aminotransferase (AST), alanine aminotransferase (ALT), total bilirubin, total protein, and albumin. Kidney function can also be assessed by blood urea nitrogen and creatinine levels in plasma as well as by measuring the electrolyte levels. Similarly, cardiac function is assessed by measuring levels of lactate dehydrogenase (LDH) and creatine phosphokinase (CPK). Exocrine pancreatic function can be measured by amylase levels. Integrity of endocrine system can be evaluated by measuring circulating hormones such as insulin, thyroid hormones, mineralocorticoids, and glucocorticoids.
- *Gross necropsy*: Valuable information related to specific toxicity of nanomaterials can be collected from necropsy of test animals. For example, necropsy revealed severe intestinal bleeding as shown in Fig. 4 after administration of G7 amine terminated PAMAM dendrimers intravenously in CD-1 mice (19). After gross examination of each major organ, they should be dissected in total and their weight measured. Damage should be

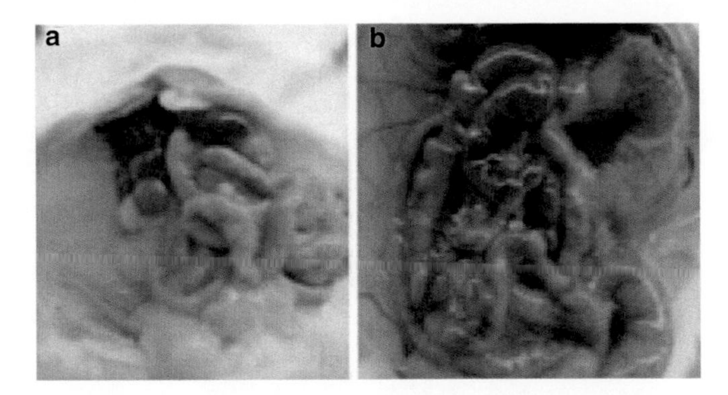

Fig. 4. Images showing severe intestinal bleeding observed with intravenous administration of amine terminated Generation 7 PAMAM dendrimers at doses of 10 mg/kg in CD-1 mice. (**a**) Normal intestine of control mice; (**b**) Bleeding in the intestine of CD-1 mice after i.v. administration of amine terminated G7 dendrimers. Adapted from (19) (Color figure online).

minimized to the organ's integrity during dissection so as to have reliable gross pathological observation as well as to report organ weight accurately. Harvested organs can then be fixed in 10% formalin for 1–2 h, washed, and fresh 10% formalin added again to remove tissue debris. Even though the visual quality of tissues decreases with time when stored in formalin, the tissue samples can be preserved up to 6 months in that state. Once tissue is ready for staining it needs to be transferred to 70% ethanol and then embedded in paraffin, thin-sectioned, and mounted on glass microscopic slides using standard histopathological techniques. Well-preserved tissue samples are essential to elucidate mechanistic studies. Figure 5 shows histology sections of the lung and liver from mice treated with single wall carbon nanotubes (SWCNT) in order to investigate for signs of toxicity (20).

2.3. Subacute Studies

Subacute toxicity studies are usually the extension of acute toxicity evaluations. They are tailored to detect adverse effects that can develop beyond the 2 weeks time period of acute toxicity studies. It is known that toxicological effects run parallel to the area under a concentration vs. time curve. For that reason subacute studies can be of special importance in nanotoxicology of materials with long circulatory half-life such as silica, gold, silver nanoparticles, and carbon nanotubes. Subacute toxicities have basically the same design as the acute study with an extended time period of observation, which typically lasts 4–5 weeks. These studies can also involve more animal numbers than acute studies (i.e., ten animals/sex/dose) for rodents or three animals/sex/dose in the case of primates. Larger number of animals is needed to detect small deviations from control values to account for recovery mechanisms that can ensue within this time frame.

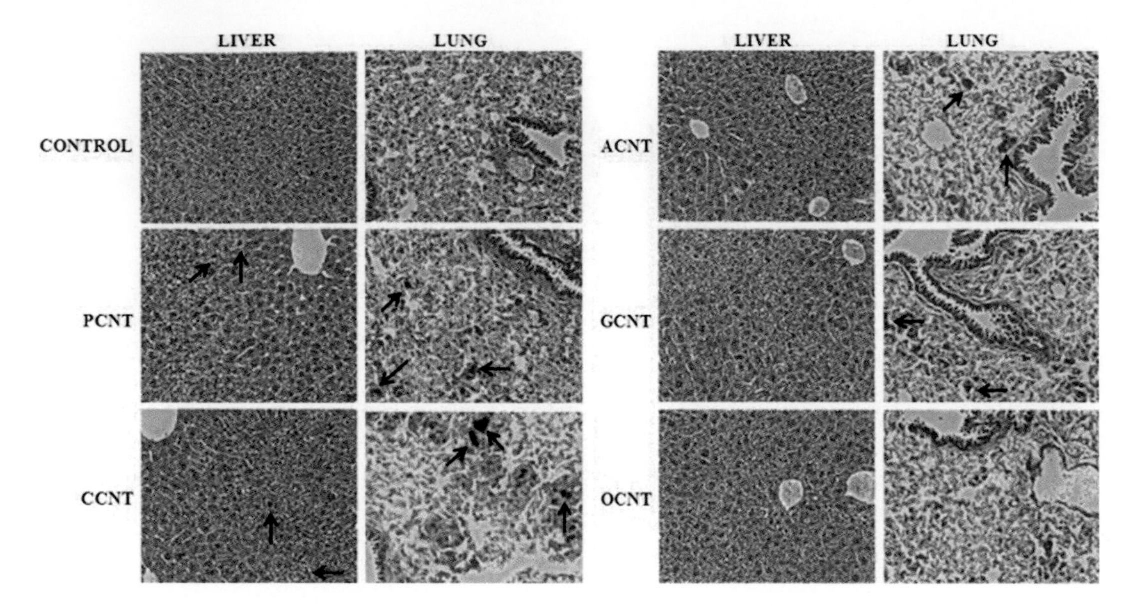

Fig. 5. Representative histopathological images of liver and lung tissues of control mice and various single-walled carbon nanotubes (SWCNT) exposed mice 7 days post-intravenous exposure. *Arrows* point to the nanotube localization. Reproduced with permission from (20) (Color figure online).

2.4. Absorption, Distribution, Metabolism, Excretion, and Pharmacokinetics (ADME/PK)

ADME/PK studies are of special importance to evaluate the occupational health impact due to the steadily increasing production of nanomaterials amongst other routes of exposure (Fig. 2). In addition, these studies can aid in assessing the potential hazard of the nanomaterials building up slowly in our ecosystem or even in the general population. An integral part of ADME/PK studies are to administer nanomaterials in a fashion similar to the expected or intended exposure route. Metabolism, biodistribution, and subsequent toxicity are to a large extent influenced by the first pass effect. It is thus not practical to conclude results from a toxicity study involving silica or carbon nanotubes injected intraperitoneally while natural exposure usually occurs via the pulmonary or dermal route (21, 22). Nanoscale materials have two unique pharmacokinetic patterns that distinguish them from classical small molecular weight drugs. First, nanoparticles tend to have longer blood circulation half lives (23), and second, they generally localize in the liver and spleen. Both liver and spleen are essential components of the reticuloendothelial system (RES) with professional phagocytes. Understanding pharmacokinetics of nanoparticles is an integral part of assessing their toxicity as it can facilitate an in-depth understanding of nanomaterial concentration in specific organs, which is a prelude to phase I clinical study in human. Based on unique biodistribution pattern of certain nanomaterials, it is appropriate to include longer observation time points for plasma and organ assay which can extend up to 2 months in some cases (24). Special care should be given to quantification of nanomaterials in isolated RES organs such as liver, spleen, bone marrow, lymph nodes, and intestinal

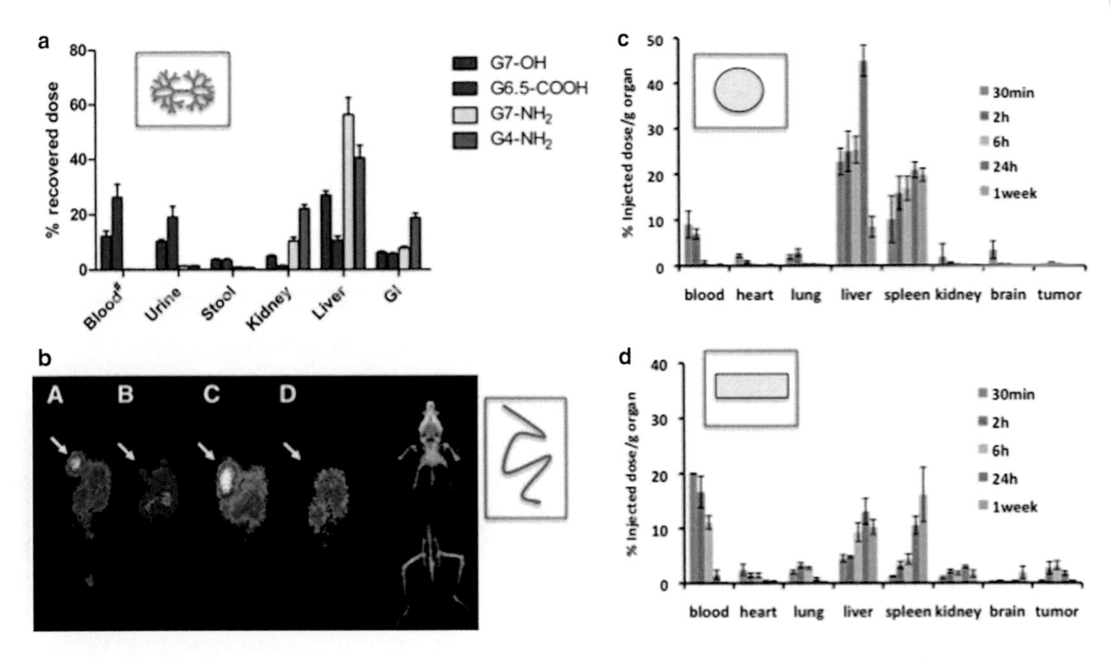

Fig. 6. Biodistribution of various nanoconstructs in animal models. *Panel (**a**) Organ localization of poly(amido amine) (PAMAM) dendrimers of different sizes (generations) and surface groups radiolabeled with ^{125}I, 8 h after intravenous administration in CD-1 mice (19); (**b**) Scintigraphy images of SCID mice with DU145 xenografts 24 h after i.v. administration of various *N*-(2-hydropropyl)methacrylamide copolymers labeled with ^{99}Tc (29): (**A**) HPMA copolymer–RGD4C, (**B**) HPMA copolymer–RGE4C, (**C**) RGD4C–DPK, (**D**) RGE4C–DPK. RGD—Targeting ligand binding to integrins and RGE—scramble sequence for RGD. *Arrows* indicate localization of nanoparticle in the tumor; (**c**) biodistribution of gold nanospheres in non-metastatic orthotropic ovarian tumor-bearing mice at various time points analyzed by ICP-MS (30); (**d**) biodistribution of gold nanorods in non-metastatic orthotropic ovarian tumor-bearing mice at various time points analyzed by ICP-MS (30). *Nanoconstructs can be polymeric (*branched*, *Panel a*, or *linear*, *Panel b*), or inorganic (*Spheres*, *Panel c*, or *Rods*, *Panel d*). Architectural depictions of nanoconstructs in the insets are not in scale (Color figure online).

peyer's patches. The most popular method for tracking nanoparticle uptake in vivo involves radiolabeling either by gamma emitters such as ^{125}Iodine or beta emitters such as ^{14}Carbon (25–28). Gamma emitters are usually preferred since quantification is direct and involves fewer steps as compared to beta emitters. In the latter case tissue homogenization is required to release the radioactive particles, which then chemically interact with a second indicator to release a color that is proportional to the amount of radioactivity. Figure 6a) shows the biodistribution profile of PAMAM dendrimers with differential surface charge and size labeled with ^{125}I at 8 h after intravenous administration in CD-1 mice (19). Radioactive labeling also has the advantage of follow-up in live animal over a period of time by gamma cameras or simply by a gamma counter. These factors can greatly improve the reliability of data and the power of statistical analysis as a single animal can represent multiple time points. In addition, it can significantly lower the number of animals needed to conduct the study. Figure 6b) shows radio scintigraphy of ^{99}Tc labeled nanoscale *N*-(hydroxypropyl)methacrylamide copolymers imaged in mice (29).

With tracking of nanoconstructs using radiolabels, it is essential to evaluate the stability of the association between the radiolabel and the nanoparticle so that the biodistribution results represent the localization of the nanoparticle and not the detached radiolabel. The use of fluorescence labeling to track nanoparticles on the other hand should be critically evaluated to exclude natural tissue fluorescence and bleaching of these dyes. Another attractive detection method to overcome these problems is by inductively coupled plasma mass spectroscopy (ICP-MS). ICP-MS combines ease of sample introduction and quick analysis with ICP technology combined with accuracy and lower detection limits of a mass spectrometer. ICP-MS is efficient in detection of metal-based nanoparticles (NP) such as gold (30), silver, and quantum dots (31). Figure 6c and d shows the biodistribution profile of gold nanorods and spheres in tumor-bearing mice at various time points analyzed by ICP-MS (30). ADME/PK studies also involve the typical rule of two species and it can span with time points from a few minutes to few weeks. In addition, histology slices of each organ can be prepared and examined using autoradiography. The radioactive intensities from various sections of the tissue are then correlated with the total biodistribution and rate of particle clearance. The most popular animal models used in these studies are mice, rats, rabbits, and guinea pigs.

2.5. Sub-chronic and Chronic Studies

Many of the nanosized materials are nonbiodegradable, consisting of metal oxides such as gold, carbon, and silica nanoparticles. In these systems the acute and subacute observations are insufficient to assess safety, as the expected residence time is much longer. A typical observation period for sub-chronic studies is 13 weeks and 18–30 months for chronic studies. In both conditions repeated exposure of the nanomaterial is investigated. Sub-chronic studies are usually performed to address three objectives. First, these studies define the toxicity of repeated dose (usually identified from the acute toxicity studies). Second, they are essential to support initiation of Phase I clinical trials and ideally they should address the dosing strategy in human applications. The third objective is to provide sufficient information for chronic studies. Sub-chronic and chronic studies are time and resource consuming and are performed to satisfy requirements of approval for human use by regulatory agencies. In addition to the parameters listed above for acute toxicity studies, sub-chronic and chronic studies should include the following:

- *Ophthalmological examinations*: Examination should be performed both prior to and at the completion of the study by an experienced veterinary ophthalmologist.
- *Cardiovascular function*: Usually involves measuring blood pressure, heart rate, and electrocardiography (ECG). These examinations should be performed prior to study and periodically during the course of the study.

- *Neurotoxicology*: A battery of clinical signs and symptoms are usually involved in the identification of specific end points. Observations such as frequent alteration of body posture, seizure, and poor response to stimuli are usually indicative of neurological complications and require further investigation.
- *Immunotoxicology*: Usually involves frequent measurement of complete blood count to identify the leukocytic and lymphocytic cell population in blood. Both elevation and decrease in cell numbers can indicate immunological involvement. Further information about the type of abnormal leukocyte levels can elucidate potential mechanism of immunological reaction. Under such circumstances histopathology of lymphoid tissues (lymph nodes, thymus, and spleen) is required. Globulin production and albumin to globulin ratio in plasma are usually required as well. In both sub-chronic and chronic studies, a larger number of animals compared to acute studies are necessary (10–15 animals/sex/dose for rodents and 2–3 animals/sex/dose in the case of primates).

2.6. Genotoxicity and Carcinogenic Studies

These studies aim at identifying three interrelated outcomes namely mutagenic, teratogenic, and carcinogenic potential of nanomaterials. In vitro methods involving bacterial and mammalian cells are accepted for screening purposes. The most widely used tests are Ames test, comet test, and micronuclei test. In Ames test, histidine-dependent *Salmonella typhimurium* are challenged with the test compound, and mutations can be detected by the bacteria's ability to survive in histidine depleted environment (32). The second in vitro assay is the comet test in which cells exposed to the material are lysed and nuclear material run on a gel electrophoresis to separate DNA based on its molecular weight in which the presence of a comet shape indicates DNA damage (33). The third common in vitro assay is related to scoring the micronuclei (MNi) in treated cells. MNi presence indicates chromosomal fragmentation or damage during the anaphase of damaged cells (34). Even though in vitro assays are usually accepted in providing strong evidence related to genotoxicity, there is a need for in vivo validation of the in vitro data. There are two widely accepted in vivo genotoxicity tests, namely metaphase chromosomal analysis and bone marrow micronucleus test. The mouse bone marrow micronucleus test was developed by Schmid in 1975 as an alternative to the cytogenic studies on mammalian bone marrow cells (35). The micronucleus test detects damage to chromosomes or the mitotic apparatus (resulting in micronucleus formation) in the bone marrow cells of the animals exposed to a test compound. MNi are formed as a consequence of acentric fragments of chromosomes or entire chromosomes lagging behind at the anaphase stage of cell division due to chromosome breakage or mitotic spindle apparatus damage (36). Modification of the test can involve the use of flow cytometry to sort abnormal peripheral RBCs instead of bone marrow cells (36). In the classic form of micronucleus test, CD-1 mice are usually employed and can be randomly allocated to test genotoxicity of a nanomaterial. A positive control group exposed to known genotoxic material, i.e., mitomycin C at 12 mg/kg should be included, as well as a negative control group with just the vehicle injection. In this test, animals are usually euthanized after 2–7 days postinjection. Both femurs then need to be dissected, cleared of muscle tissue, and proximal epiphysis cut with sterile scissor. The bone marrow then can be flushed with a syringe, and cells collected into 2 ml tubes filled with fetal calf serum (FCS). Collected cells can be centrifuged and washed with FCS twice, then fixed in methanol, stained with Geimsa stain, and dried. The stained smear can then be examined under light microscope and micronucleated cells counted. Teratogenic studies are not routinely required as part of toxicity studies. When teratogenic effect is suspected, studies usually follow the International Conference on Harmonisation of Technical Requirements for Registration of Pharmaceuticals for Human Use (ICH) guidelines (37). ICH guidelines involve segments standardizing the test procedures for evaluation of fertility and reproductive performance, embryo-fetal development, perinatal, and postnatal analysis on both the maternal and newborn cases. In vivo carcinogenic studies typically involve long-term repeated exposure of the nanomaterials and follow-up of animals for the development of tumors. This study is simple with one end point measurement (tumor development). The study usually involves large number of animals (>50 animal) per species and usually extends up to 30 months.

3. Comments and Notes

1. Proper characterization of the nanomaterials prior to administration is a crucial factor in assessing in vivo toxicity. The physicochemical attributes of nanomaterials such as purity, charge, charge density, size, shape, density, and surface area can play an important role in causing adverse toxicity as discussed in Section 2.1.
2. The time period of toxicity studies should be carefully decided in relation to both the intended use of the nanomaterial and the expected time frame of exposure based on the interaction between nanomaterials and the biological system. It is an important factor in deciding between an acute toxicity study which usually spans 2 weeks and chronic study that can take up to 2 years.
3. Route of administration should mimic the actual exposure of the nanomaterial; possible administration routes include nasal

inhalation, ingestion, and exposure through skin as well as intravenous administration.

4. Dose and dosing regimen can be designed within the range of expected exposure. A dose regimen of single or multiple exposures should consider the intended exposure or the potential application of the nanomaterial. A simple conversion factor of animal dose to human dose can be found elsewhere (38). The acquired data for use should delineate MTD and NOEL. Controversy regarding MTD definition exists; however, a simple and reliable definition is as follows: "the maximum dose associated with less than 10% loss of animal weight" (39). In case of nanomedicines intended for therapeutic use, the therapeutic index should be acquired which is the ratio of the drug that produces toxicity to the dose that produces the desired clinical effect.
5. In choosing the dose range a single escalation dose or multiple escalation doses can be applied. In either case the dose that is expected to produce no observable effect as well as the dose that is expected to produce adverse effects should be chosen. Dose escalation can include log scale or nominal scale spanning at least 3–5 doses. Parallel dosing can be designed to reduce the duration of these studies. However, serial dosing will allow economical use of resources so that higher doses are given only after a lower dose is tolerated.
6. Expected mechanism(s) of toxicity with nanomaterials can guide the thorough investigation of specific organs, cells, or tissues that can be affected by a particular nanomaterial. For example nanoparticles have been shown to be taken up by macrophages to a much greater extent than epithelial cells (40). It is thus appropriate to consider evaluation of macrophage population in the liver and spleen when testing safety of nanoparticles.
7. Target population expected to be exposed to a particular nanoparticle should be replicated in test animals for these toxicity studies. For example, it is inappropriate to test the safety profile of a nanomedicine system in male mice designed for treating ovarian cancer.
8. Chronic studies are necessary, as previous experience with asbestos has shown the effect of time in the development of granulomatous changes and mesothelioma.
9. In choosing the rodent model, rats are usually preferred to mice due to similar hemodynamics to human.
10. The requirement of two species (one rodent and one nonrodent) excludes the species-specific toxicity as well as the confirmation of similar pattern of toxicity.

11. Regulatory agencies now require the use of MTD as the goal rather than lethal dosage (LD) for human use of animals. LD can be acquired through mathematical modeling of data.
12. In evaluating animal weight a normalized weight with respect to day 0 is more preferable than absolute weights in each group as it facilitates similar basis for all tested groups regardless of the small weight differences found on day 0.
13. Ascertaining the health of animals and exclusion of colony infection are essential. Usually a set of three animals should be dedicated to test for infectious pathogens (e.g., mouse parvoviruses).
14. In reporting the organ weight the organ index (gm/body weight) is more valuable than the absolute gram weight of a specific organ. This percentage value can account for individual animal toxicity, which could be overlooked while averaging organ weights of many animals.
15. In H&E staining examination of different organ-specific endpoint(s) should be set and semiquantitative systems evaluated for this end point(s) should be employed to give a better understanding of the results.
16. Upon tissue harvesting at necropsy it is always helpful to snap freeze in liquid nitrogen organs that will not be used immediately. This can preserve the tissue for protein, DNA, and RNA separations to evaluate further mechanistic studies and may not be accounted for during organ collection.
17. It is critical to include late time points in PK studies of nanomaterials, e.g., weeks to account for the prolonged circulation time of nanomaterials.
18. In PK studies, accounting for the injected dose can serve as an internal quality control measure. Pharmacokinetic studies should, whenever possible, utilize preparations that are representative of those intended for toxicity testing and clinical use. They should also employ a route of administration that is relevant to the anticipated clinical studies. Patterns of absorption may be influenced by formulation, concentration, site of exposure, and volume.
19. For in vivo genotoxicity micronucleus test time points at which animals need to be killed can largely depend on plasma half-life of the test nanomaterials. Two or three half lives after administration can ensure enough exposure to the bone marrow cells in order to assess genetic damage.
20. As an alternative to in vivo methods, hematotoxicity can also be analyzed ex vivo. Figure 7 shows images of hemolysis after treatment with generation 7 amine terminated dendrimers with human blood. Necessary approvals need to be obtained before such studies are done.

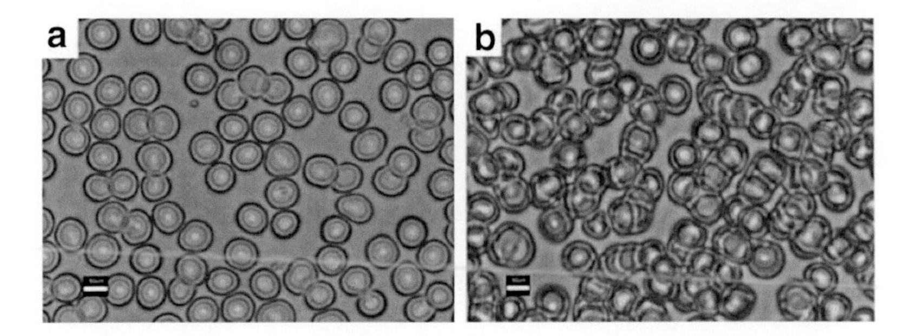

Fig. 7. Inverted microscopic images of RBCs in human blood showing hemolysis after treatment with G7 amine terminated PAMAM dendrimers at dose of 10 mg/kg. (**a**) Control, (**b**) G7 amine terminated dendrimer treatment.

21. For in vivo genotoxicity assessment it is usually preferable to use young animals (6 weeks of age) where the bone marrow is more active and more vulnerable to damage.
22. Reproductive toxicity studies that evaluate the effect of nanoparticles in reproduction and offspring should also be evaluated as a part of the chronic toxicity studies.

Acknowledgments

The authors wish to thank Dr. Alexander Malugin for valuable discussions on in vitro nanotoxicity assays. Financial support was provided by the National Institutes of Health (R01DE019050) and the Utah Science Technology and Research (USTAR) initiative.

References

1. Borm PJ, Kreyling W (2004) Toxicological hazards of inhaled nanoparticles–potential implications for drug delivery. J Nanosci Nanotechnol 4:521–531
2. Oberdorster G, Oberdorster E, Oberdorster J (2005) Nanotoxicology: an emerging discipline evolving from studies of ultrafine particles. Environ Health Perspect 113:823–839
3. Roco MC (2006) Nanotechnology's future. Sci Am 295:39
4. Assay cascades—Nanotechnology Characterization Laboratory-Assay Cascades – http://ncl.gove/assay_cascade.asp, 2010
5. Gormley AJ, Ghandehari H (2009) Evaluation of toxicity of nanostructures in biological systems. Wiley, Chichester
6. Kroll A, Pillukat MH, Hahn D, Schnekenburger J (2009) Current in vitro methods in nanoparticle risk assessment: limitations and challenges. Eur J Pharm Biopharm 72:370–377
7. Yang S, Leong KF, Du Z, Chua CK (2001) The design of scaffolds for use in tissue engineering. Part I. Traditional factors. Tissue Eng 7:679–689
8. Holt DJ, Chamberlain LM, Grainger DW (2010) Cell-cell signaling in co-cultures of macrophages and fibroblasts. Biomaterials 31:9382–9394
9. Sayes C, Ivanov I (2010) Comparative study of predictive computational models for nanoparticle-induced cytotoxicity. Risk Anal 30: 1723–1734
10. Derelanko MJ, and Hollinger MA, Handbook of Toxicology, CRC Press, Boca Raton
11. Hodgson E (2004) Introduction to toxicology. Wiley, Hoboken

12. Stine KE, Brown TM (2006) Principles of toxicology, 2nd edn. CRC Press, Boca Raton
13. Oberdorster G, Maynard A, Donaldson K, Castranova V, Fitzpatrick J, Ausman K, Carter J, Karn B, Kreyling W, Lai D, Olin S, Monteiro-Riviere N, Warheit D, Yang H (2005) Principles for characterizing the potential human health effects from exposure to nanomaterials: elements of a screening strategy. Part Fibre Toxicol 2:8
14. Powers KW, Brown SC, Krishna VB, Wasdo SC, Moudgil BM, Roberts SM (2006) Research strategies for safety evaluation of nanomaterials. Part VI. Characterization of nanoscale particles for toxicological evaluation. Toxicol Sci 90:296–303
15. Murdock RC, Braydich-Stolle L, Schrand AM, Schlager JJ, Hussain SM (2008) Characterization of nanomaterial dispersion in solution prior to in vitro exposure using dynamic light scattering technique. Toxicol Sci 101:239–253
16. Fiske MJ, Fredenburg RA, VanDerMeid KR, McMichael JC, Arumugham R (2001) Method for reducing endotoxin in Moraxella catarrhalis UspA2 protein preparations. J Chromatogr B Biomed Sci Appl 753:269–278
17. Magalhaes PO, Lopes AM, Mazzola PG, Rangel-Yagui C, Penna TC, Pessoa A Jr (2007) Methods of endotoxin removal from biological preparations: a review. J Pharm Pharm Sci 10:388–404
18. Teeguarden JG, Hinderliter PM, Orr G, Thrall BD, Pounds JG (2007) Particokinetics in vitro: dosimetry considerations for in vitro nanoparticle toxicity assessments. Toxicol Sci 95: 300–312
19. Griesh K, Thiagarajan G, Herd H, Price R, Bauer H, Hubbard D, Burckle A, Sadekar S, Yu T, Anwar A, Ray A, Ghandehari H (2010) Size and surface charge significantly influence toxicity of silica and dendritic Nanotoxicology. 2011 Jul 28. [Epub ahead of print]
20. Dowell P, Robinson K, Greish K, Ghandehari H, Nan A (2010) In vivo evaluation of the effect of physicochemical properties of functionalized single-walled carbon nanotubes on acute toxicity and biodistribution (in press)
21. Hudson SP, Padera RF, Langer R, Kohane DS (2008) The biocompatibility of mesoporous silicates. Biomaterials 29:4045–4055
22. Poland CA, Duffin R, Kinloch I, Maynard A, Wallace WA, Seaton A, Stone V, Brown S, Macnee W, Donaldson K (2008) Carbon nanotubes introduced into the abdominal cavity of mice show asbestos-like pathogenicity in a pilot study. Nat Nanotechnol 3:423–428
23. Gref R, Luck M, Quellec P, Marchand M, Dellacherie E, Harnisch S, Blunk T, Muller RH (2000) 'Stealth' corona-core nanoparticles surface modified by polyethylene glycol (PEG): influences of the corona (PEG chain length and surface density) and of the core composition on phagocytic uptake and plasma protein adsorption. Colloids Surf B Biointerfaces 18: 301–313
24. Balasubramanian SK, Jittiwat J, Manikandan J, Ong CN, Yu LE, Ong WY (2010) Biodistribution of gold nanoparticles and gene expression changes in the liver and spleen after intravenous administration in rats. Biomaterials 31:2034–2042
25. Leu D, Manthey B, Kreuter J, Speiser P, DeLuca PP (1984) Distribution and elimination of coated polymethyl (2-14C)methacrylate nanoparticles after intravenous injection in rats. J Pharm Sci 73:1433–1437
26. Troster SD, Kreuter J (1992) Influence of the surface properties of low contact angle surfactants on the body distribution of 14C-poly(methyl methacrylate) nanoparticles. J Microencapsul 9:19–28
27. Gref R, Minamitake Y, Peracchia MT, Trubetskoy V, Torchilin V, Langer R (1994) Biodegradable long-circulating polymeric nanospheres. Science 263:1600–1603
28. Bazile D, Prud'homme C, Bassoullet MT, Marlard M, Spenlehauer G, Veillard M (1995) Stealth Me.PEG-PLA nanoparticles avoid uptake by the mononuclear phagocytes system. J Pharm Sci 84:493–498
29. Line BR, Mitra A, Nan A, Ghandehari H (2005) Targeting tumor angiogenesis: comparison of peptide and polymer-peptide conjugates. J Nucl Med 46:1552–1560
30. Arnida, Janat-Amsbury MM, Ray A, Peterson CM, Ghandehari H (2011) Geometry and surface characteristics of gold nanoparticles influence their biodistribution and uptake by macrophages. Eur J Pharm Biopharm 77:417–423
31. Yang RS, Chang LW, Wu JP, Tsai MH, Wang HJ, Kuo YC, Yeh TK, Yang CS, Lin P (2007) Persistent tissue kinetics and redistribution of nanoparticles, quantum dot 705, in mice: ICP-MS quantitative assessment. Environ Health Perspect 115:1339–1343
32. Mortelmans K, Zeiger E (2000) The Ames Salmonella/microsome mutagenicity assay. Mutat Res 455:29–60
33. Singh NP, McCoy MT, Tice RR, Schneider EL (1988) A simple technique for quantitation of low levels of DNA damage in individual cells. Exp Cell Res 175:184–191

34. Fenech M (2007) Cytokinesis-block micronucleus cytome assay. Nat Protoc 2:1084–1104
35. Schmid W (1975) The micronucleus test. Mutat Res 31:9–15
36. Sarto F, Finotto S, Giacomelli L, Mazzotti D, Tomanin R, Levis AG (1987) The micronucleus assay in exfoliated cells of the human buccal mucosa. Mutagenesis 2:11–17
37. The International Conference on Harmonisation of Technical Requirement of Registration of Pharmaceuticals for Human Use (ICH)
38. Reagan-Shaw S, Nihal M, Ahmad N (2008) Dose translation from animal to human studies revisited. FASEB J 22:659–661
39. Saffiotti U, Page NP (1977) Releasing carcinogenesis test results: timing and extent of reporting. Med Pediatr Oncol 3:159–167
40. Lai Y, Chiang P-C, Blom J, Li N, Shevlin K, Brayman T, Hu Y, Selbo J, Hu L (2008) Comparison of in vitro nanoparticles uptake in various cell lines and in vivo pulmonary cellular transport in intratracheally dosed rat model. Nanoscale Res Lett 3:321–329

Chapter 18

The Luminescent Bacteria Test to Determine the Acute Toxicity of Nanoparticle Suspensions

Ana Garcia, Sonia Recillas, Antoni Sánchez, and Xavier Font

Abstract

Luminescent bacteria, *Vibrio fischeri*, is used in an ecotoxicological test to determine toxicity of water samples. In comparison to other ecotoxicological tests, the use of luminescent bacteria reports final toxicity values in a short time (minutes). Luminescent toxicity test can be used to determine toxicity of nanoparticles suspensions.

Key words: Toxicity test, Nanoparticles, *Vibrio fischeri*

1. Introduction

In recent years there has been an exponential use of nanoparticles or nanodevices in water treatment and others fields such as in catalysis and medicine due to the high performance obtained. This means that an increasing amount of nanoparticles will be released into the environment through production processes or its use (1, 2). Therefore it is necessary to evaluate, in a short period of time, the impact of these nanomaterials in the environment and human health.

Toxicity of a water sample or a water-soluble substance can be measured by different methods. Methods using fish (zebrafish, rainbow trout) (3) or the little crustacean *Daphnia magna* (4) are standardized and accepted. Also algae (5), anaerobic bacteria (6), or plant seeds (6) are widely used. The use of the luminescent bacteria *Vibrio fischeri* as a target organism for toxicity test is also used in water and wastewater toxicity determination (7).

Vibrio fischeri is a prokaryotic, luminescent, gram negative, saprophytic marine bacteria. McElroy (8) reported that the natural bioluminescence of these bacteria is related to flavin mononucleotide (FMN), a long chain aldehyde, oxygen, and the enzyme

Joshua Reineke (ed.), *Nanotoxicity: Methods and Protocols*, Methods in Molecular Biology, vol. 926,
DOI 10.1007/978-1-62703-002-1_18,

luciferase. Bacterial luciferase seems to be coupled to cellular respiration via NADH and FMN. Toxic substances influence this sensitive metabolism and inhibit the bioluminescence (9).

The light emitted reduction is directly related to the relative toxicity of the sample. To report the toxicity value the term effective concentration 50% (EC_{50}) is used. EC_{50} corresponds to the toxic concentration that provokes a 50% reduction in the light emission, and it is obtained from plotting the percentage of luminescence reduction against concentration after 5 and 15 min incubation time. In general, it can be assumed that values of EC_{50} (%) ≤ 25 correspond to highly toxic substances; 25–50 to moderately toxic substances; 51–75 to toxic substances; >75 to slightly toxic substances; and >100 to nontoxic substances (10).

The main advantage of the bioluminescent test, in comparison with other ecotoxicological tests, is that the time needed to obtain the results is very short (15–30 min) and it is an easy test to perform. Moreover, *Vibrio fischeri* toxicity test used for the determination of toxicity in water is widely accepted and standardized in the ISO 11348-3 standard normative. However, due to the marine origin of *Vibrio fischeri*, water samples must be osmotically adjusted as necessary.

The broadly used luminescent bacteria toxicity assay is the Microtox test, commercialized by Strategic Diagnostics Incorporated that bought in 2001 by Azur Environmental (formerly Microbics Corporation). However other commercial tests can be found (BioTox, LUMIStox, ToxAlert).

2. Materials

1. *Test bacteria*: Lyophilized bioluminescent bacteria (*Vibrio fischeri* NRRL B-11177) should be obtained commercially and conserved at −20 to −25°C until 1 year.
2. *Reconstitution solution*: This solution can be commercially obtained (see Note 1).
3. *Diluent solution*: Sodium chloride (2%) solution (see Note 2).
4. *Apparatus*: A luminometer is used to measure the bioluminescence. The luminometer cell temperature must be set at 15 ± 1°C (see Note 3).

3. Methods

Before conducting the assay nanoparticle suspension pH must be checked, and it should be adjusted in the range of 6.5–8.5, ideally 7.0 ± 0.2 (11). It is important to check the presence of precipitates or nanoparticles agglomeration due to pH change. Sodium citrate (0.5 M) or citric acid (0.5 M) could be used to adjust pH.

3.1. Experimental Methods

The method described allows measuring the acute toxicity from a sample concentration of 100–0%.

1. Place 1,000 μL of reconstitution solution in a cuvette at 5°C.
2. Place 1,000 μL of diluents in 5 cuvettes at 15°C.
3. Adjust sample osmolarity by mixing 10 mL of sample with 0.2 mg of NaCl. Ensure that no precipitate is formed. Otherwise this should indicate nanoparticle destabilization.
4. Make 1:2 serial dilutions according to Fig. 1. The test should be performed in triplicate, so the scheme diagram presented in Fig. 1 must be repeated three times, starting from the same initial osmotically adjusted sample.
5. Wait for 5 min to homogenize temperatures.
6. Take lyophilized bacteria vial from the freezer. Reconstitute the bacteria by quickly pouring 1,000 μL of reconstitution solution (step 1) to the lyophilized bacteria vial (see Note 4). Mix the vial by swirling the vial three or four times and pour the mixture back into the cuvette to maintain it at 5°C.

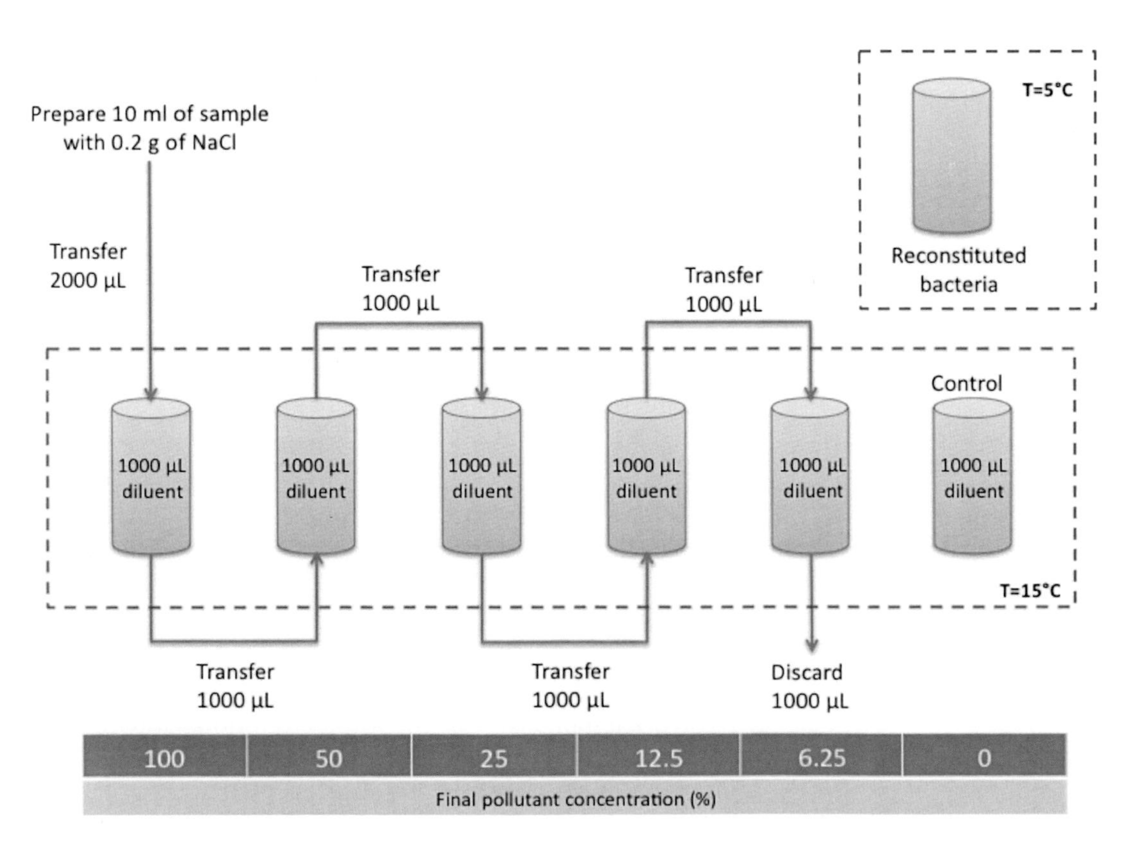

Fig. 1. Dilution scheme for the whole effluent toxicity test.

7. Transfer 10 μL of reconstituted bacteria to each sample cuvette prepared in step 4 and mix by shaking (see Note 5).
8. Wait for the contact time (5, 15, or 30 min) and read all cuvettes in the same order as reconstituted bacteria were added in step 7. *Vibrio fischeri* light emission is measured at 490 nm.
9. Note the light emission values (I) at the given time (t) for the samples (s) and the control (c). (see Note 6).

3.2. Calculation

Gamma value (G) is the ratio of the control (c) light output at a given time (t) to the light output for a given sample (s) at the same test time, minus 1 (Eq. 1).

$$G_t = \frac{I_{tc}}{I_{ts}} - 1. \quad (1)$$

The effect at a given time for a given sample concentration (s) can be calculated according to Eq. 2.

$$\%\text{Effect}_t = \left[\frac{G_t}{1+G_t}\right] \times 100. \quad (2)$$

Finally, the correlation sample concentration/effect for a given contact time can be described according to Eq. 3 that corresponds to a straight line and EC_{50} can be calculated.

$$\log(C_t) = b \times \log(G_t) + \log(a). \quad (3)$$

In those cases in which nonlinear regression must be used to correlate the obtained values, the Probit statistical normal distribution can be applied.

4. Notes

1. According to Azur Environmental (12) Reconstitution solution is a 0.01% sodium chloride solution. However UNE-EN-ISO 11348-3 methodology use distilled water to perform the reconstitution (11).
2. All the dilutions performed during the assay should be maintained at 15 ± 1°C, while reconstituted bacteria should be maintained at 4 ± 3°C (11).
3. Commercial luminometers especially designed to perform bioluminescent test are equipped with different wells to maintain the correct temperature of the sample dilutions and controls (15°C) and for the reconstituted bacteria (5°C). Commercial equipment is also distributed with proper software to control time and perform final calculations.

4. Once reconstituted it should be used within 1–2 h. If the reconstituted bacteria must be used after 2 h of its reconstitution, its sensitivity should be monitored periodically with a standard such as phenol (EC_{50} = 13–26 mg/L) (12).

5. Final results will be reported relating the time of contact between the bacteria and the sample, normally 5, 15, or 30 min. It is very important to use a chronometer to establish the initial time for each cuvette. Normally, 10–20 s must be waited, which will be enough time to perform the light level reading for each cuvette (step 8).

6. Since nanoparticle suspensions are normally obtained by the use of a stabilizer (hexamethylene tetramine, tetramethyl ammonium hydroxide, sodium citrate, or others) to calculate the real toxicity of the nanoparticle, the stabilizer toxicity must also be determined by using the same methodology previously described.

References

1. Moore MN (2006) So nanoparticles present ecotoxicological risks for the health of the aquatic environment? Environ Int 32:967–976
2. Brar SK, Verma M, Tyagi RD, Surampalli RY (2010) Engineered nanoparticles in wastewater sludge – evidence and impacts. Waste Manag 30:504–520
3. Sharma VK (2009) Aggregation and toxicity of titanium dioxide nanoparticles in aquatic environment – a review. J Environ Sci Health A Tox Hazard Subst Environ Eng 44:1485–1495
4. Persoone G, Baudo R, Cotman M, Blaise C, Thompson KC, Moreira-Santos M, Vollat B, Torokne A, Han T (2009) Review on the acute Daphnia magna toxicity test – evaluation of the sensitivity and the precision of assays performed with organisms from laboratory cultures or hatched from dormant eggs. Knowl Manag Aquat Ecosyst 393:01
5. Ramakrishnan B, Megharaj M, Venkateswarlu K, Naidu R, Sethunathan N (2010) The impacts of environmental pollutants on microalgae and cyanobacteria. Crit Rev Environ Sci Technol 40:699–821
6. Barrena R, Casals E, Colón J, Font X, Sánchez A, Puntes V (2009) Evaluation of the ecotoxicity of model nanoparticles. Chemosphere 75: 850–857
7. Kahru A, Dubourguier HC, Blinova I, Ivask A, Kasemets K (2008) Biotests and biosensors for ecotoxicology of metal oxide nanoparticles: a minireview. Sensors 8:5153–5170
8. McElroy WD (1961) Bioluminescence. In: Gunsalus IC, Stanier RY (eds) The bacteria, vol 2. Academic, New York
9. Johnson BT (2005) Microtox® acute toxicity test. In: Blaise C, Férard J-F (eds) Small-scale freshwater toxicity investigations, vol 1. Springer, The Netherlands, pp 69–105
10. Anand KV, Chinnu MK, Kumar RM, Mohan R, Jayavel R (2009) Formation of zinc sulfide nanoparticles in HMTA matrix. Appl Surf Sci 255:8879–8882
11. UNE-EN-ISO 11348–3, Water quality. Determination of the inhibitory effect of water samples on the light emission of Vibrio fischeri (Luminiscent bacteria test). Part 3: Method using freeze-dried bacteria
12. Azur Environmental (1999) Instruction manual of the Microtox Analyzer model 500

Chapter 19

The Primacy of Physicochemical Characterization of Nanomaterials for Reliable Toxicity Assessment: A Review of the Zebrafish Nanotoxicology Model

John P. Bohnsack, Shoeleh Assemi, Jan D. Miller, and Darin Y. Furgeson

Abstract

Engineered nanomaterials (ENMs) have become increasingly prevalent in the past two decades in academic, medical, commercial, and industrial settings. The unique properties imbued with nanoparticles, as the physiochemical properties change from the bulk material to the surface atoms, present unique and often challenging characteristics that larger macromolecules do not possess. While nanoparticle characteristics are indeed exciting for unique chemistries, surface properties, and diverse applications, reports of toxicity and environmental impacts have tempered this enthusiasm and given cause for an exponential increase for concomitant nanotoxicology assessment. Currently, nanotoxicology is a steadily growing with new literature and studies being published more frequently than ever before; however, the literature reveals clear, inconsistent trends in nanotoxicological assessment. At the heart of this issue are several key problems including the lack of validated testing protocols and models, further compounded by inadequate physicochemical characterization of the nanomaterials in question and the seminal feedback loop of chemistry to biology back to chemistry. Zebrafish (*Danio rerio*) are emerging as a strong nanotoxicity model of choice for ease of use, optical transparency, cost, and high degree of genomic homology to humans. This review attempts to amass all contemporary nanotoxicology studies done with the zebrafish and present as much relevant information on physicochemical characteristics as possible. While this report is primarily a physicochemical summary of nanotoxicity studies, we wish to strongly emphasize that for the proper evolution of nanotoxicology, there must be a strong marriage between the physical and biological sciences. More often than not, nanotoxicology studies are reported by groups dominated by one discipline or the other. Regardless of the starting point, nanotoxicology must be seen as an iterative process between chemistry and biology. It is our sincere hope that the future will introduce a paradigm shift in the approach to nanotoxicology with multidisciplinary groups for data analysis to produce predictive and correlative models for the end goal of rapid preclinical development of new therapeutics into the clinic or insertion into environmental protection.

Key words: Zebrafish, Nanotoxicology, Physicochemical characteristics, Review, Nanoparticles, Characterization

Joshua Reineke (ed.), *Nanotoxicity: Methods and Protocols*, Methods in Molecular Biology, vol. 926,
DOI 10.1007/978-1-62703-002-1_19,

1. Introduction

The nanotechnology revolution is well underway with a cornucopia of novel and exciting nanomaterials for multiple uses. Concern of these *engineered nanomaterials* (ENMs), possessing at least one dimension of ≤100 nm, is the dramatic research gap of vital assessments of associated "nanotoxicity." These seminal studies are not trivial and are of vital concern as a single physicochemical characteristic, such as particle size, zeta potential, or morphology, may have profound effects upon an ENM's impact in vitro, in vivo, and globally upon the environment itself.

Nanotoxicology is an interdisciplinary field that must bridge both the complex physical and chemical properties of nanoparticles and their interaction with biological systems. Unfortunately, the nanotoxicity literature is dominated by biological studies which at the most rarely boast expertise in physical science and at the least under appreciate the ramifications of physicochemical properties on nanotoxic stresses in living organisms. The intent of this review is to emphasize that nanotoxicology, in all its shapes and forms, must begin with (a) a rigorous assessment of the physical chemistry behind the system being studied; (b) next, biological identification of biological stresses, phenotypic and/or genotypic; and (c) a return to the chemistry for reformulation to alleviate these stresses, or confirm the physical property contributing to these stresses, in order to rapidly advance preclinical development of ENMs or utility in the environment. This review is to provide a broad background of current zebrafish nanotoxicity studies, as a representative model, with various ENMs with and without physicochemical alteration.

ENMs and their synthesis is a growing trillion-dollar industry with novel uses being discovered and investigated daily (1). Already, ENMs are used or being developed, for the medical, pharmaceutical, chemical, information and communication technology, energy, automotive, aerospace, cosmetics, textile, and agricultural industries (2). This splurge of growth is in part explained due to the differing properties between an ENM and the bulk material, where surface-active electrons at the atomic scale promote seemingly unexpected properties compared to the bulk material. This anomaly is commonly termed the "quantum effect" (3, 4). A representative list of physicochemical characteristics affected by quantum effects includes: homo-/heterogeneity composition, surface functionalization, surface charge, shape, particle size distribution, purity, density, and concentration. Furthermore, modification of these properties may allow for specialized pharmacokinetics (toxicokinetics) including ab(d)sorption, biodistribution, metabolism, excretion, bioreactivity, immunoreactivity, and biotolerance (Table 1).

Table 1
A selective list of biological considerations pertinent to nanotoxicology

Biological issue	Example study
Bioavailability	(14)
Bioaccumulation	(252)
Biodistribution	(31)
Necrosis/apoptosis e.g. Acridine Orange, TUNEL	(41)
ROS—gene expression e.g. HSP-70, GSH-pi, GCLc, ferritin	(129)

Consequently, these characteristics imbue different nanotoxicological responses than their bulk counterparts (5) and as such must be carefully and efficiently studied, so the full scope of their effects and dangers can be assessed.

Nanotoxicology's modest beginnings were clearly identified in 2005 evolving from the study of ultrafine particles (6), already in the research limelight due to asbestos (7), coal combustion (8), and diesel fuels (9). As the ENM industry continued to grow, more nanoparticles (ENMs) entered the environment (10). Since then, research has demonstrated toxicity changes in situ (11). In response to these intriguing, yet alarming reports, the number of publications concerning nanotoxicity has exponentially increased from 70 in 2003 to 1,020 in 2009 (12), but other research found using models based on private and public firms using ENMs and current cost of toxicity testing that a precautionary $1.18B is necessary to adequately screen ENMs for their toxic properties (13). Without a massive surge of nanotoxicology research, then the lack of regulations in the status quo (1) will remain unchanged, as the current hodgepodge does not adequately demonstrate the current threat to human safety. Numerous academic, industrial, governmental, and private agencies/consortiums are vigorously addressing this critical issue with substantial scientific funding including: BAG, BMU Nanodialog, Greenpeace, IANH (International Alliance for NanoEHS Harmonisation), IEC, IRGC (International Risk Governance Council), ISO, NIST, NNI, OECD (Organisation for Economic Co-operation and Development), Royal Academy, SCENHIR (Scientific Committee on Emerging and Newly Identified Health Risks), and the Swiss "Action Plan on Synthetic Nanomaterials." Sadly, the United States is not a world leader in nanotoxicology research despite its dominance in producing innovative and novel ENMs.

Even with ignoring budget deficiencies, nanotoxicology faces a lurking challenge. The lack of sufficient characterization of nanoparticles and their intrinsic biological reactivity, biodistribution, and bioavailability creates deficits in studying integrity and usefulness, as correlation cannot be effectively established. For example, a recent study of fluorescent SiO_2 nanoparticles was unable to fully ascertain safety because of the lack of information on bioavailability (14), and others have found discrepancies between *manufacturer* reported characteristics and *empirical* characteristics (15, 16). Assessment of risk requires knowledge of exposure and dosing (17, 18), since certain dosing measurements may not accurately reflect the number of nanoparticles or their surface area in solution or the proper concentration dose (19, 20). Further complicating dosing metrics is the variability within nanotoxicity studies over frequency of dosing, and time point of initial dose, both of which have shown to have co-relatable effects on toxicity (21). A tangle of different methods for measuring the dose, ensuring consistent dispersion, toxicity evaluation, and dose–response and modeling (22) culminates in a chaotic jumble that allows for few precise correlations. Stuck in the middle of this problem is the bridge between in vitro cell culture and in vivo studies, as results may differ due to poorly understood and complex biological interactions (23, 24). The call for valid and reproducible in vivo nanotoxicity models has already been issued (25), and the range of studies using in vivo models has increased, but without addressing the aforementioned problems.

Zebrafish (*Danio rerio*) has emerged as one of the most prevalent and promising models for assessing nanotoxicity. The zebrafish is already a highly used and respected model for drug screening and environmental toxicology assays (26). A seminal characteristic of the zebrafish is its close homology (approx. 83.1%) (27) to the human genome (28). It also possesses comparable tissue types (kidney, spleen, liver, blood–brain barrier—lack of lungs and two chambered heart), immunogenic responses (26), and physiological responses to common xeno-substances such as oxidative stress and foreign bodies (29). Additionally, the zebrafish's quasi-transparent embryos and larvae allow for biological and optical clarity (30), i.e. quick viewing of nanoparticle aggregates within the embryos or chorion space (described in more detail below), and assessment of damage (31). Additionally, this property allows for all stages of zebrafish development to be observed and divergence from the well-outlined development path quickly identified (30). In opposition to mammalian models, their small size (3–4 cm adults and 4 mm early larvae, categorized in relevant literature from hatching to approximately 5 dpf) decreased housing and husbandry costs, equivalent longevity to mammalian models and generation time, higher embryo yields, (200–300 eggs per the breeding cycle of 5–7 days), and noninvasive visualization of organ systems makes the zebrafish a comparably low cost and effective model (26, 28–30, 32).

The zebrafish has two important considerations for nanotoxicity assays. The first is the presence of a selective barrier called the chorion that surrounds the embryo during development. While the chorion allows for materials to pass into the embryo via passive diffusion as well as numerous chorion pore channels reported as 0.5–0.7 μm in diameter (31), it has been shown that it can selectively block particles that are small enough to pass through it (14). Studies utilizing the zebrafish model should test for uptake of nanoparticles in both the chorion space and the embryo. The second consideration is the choice of media such as eggwater, a mixture of salts and minerals. Certain nanoparticles have been shown to aggregate, disperse (16), or change in other ways, i.e. change in the zeta potential (surface charge density) of particles and of the egg water (33), suggesting that these external components (barriers) must be considered in tandem to the behaviors of ENMs themselves.

2. Exposure of Engineered Nanomaterials

ENMs are already present in the environment with growing numbers every day. Sites for exposure include factories (34), dust storms, forest fires, volcanoes, oceans, diesel and engine exhaust, indoor pollution, building demolition sites, groundwater, consumer products, and food contamination (35). Methods for exposing zebrafish to NPs include oral, waterborne, and microinjection exposures. Each of these methods has their distinct advantages and disadvantages based on the NPs used and the hypothesis to be proven. ENMs may be taken up by the zebrafish through four primary sites of exposure: chorion (31), gills, skin (32), and digestive tract (36) through three primary routes of exposure. For example, recent studies have found that SiO_2 cannot pass through the chorion (14), suggesting that microinjection is a better choice. Other studies have found that CuNPs are preferentially taken up through the gills and not through the digestive tract (37, 38). Microinjection is useful because it bypasses the protective chorion or skin, and directly exposes NPs to the developing embryo or fish but consequentially is not indicative of natural environmental exposure. Oral and waterborne exposures are the same, as both require suspending fish in an aqueous solution, but differ when the fish are exposed, i.e. embryo or adult and where the particles accumulate in fish, i.e. skin, gills, and/or digestive tract. This more closely mirrors naturally occurring exposure, but suspending particles may require chemicals that change toxicity (39), or NPs may not enter the fish (16).

3. Chemical Contributions to Consider in Nanomaterial Toxicity

There are a number of considerations that weigh into ENMs toxicity (Table 2). Previous studies have already found that steps in the formulation process such as synthesis scheme (40), catalysts (40), solvents (40), washing (41), and attached surfactants (39) can contribute to toxicity in either a beneficial or detrimental aspect to both in vitro and in vivo systems. Additionally, there have been documented cases in both zebrafish and other studies that demonstrate discrepancies between manufacturer characterization and experimental characterization (16). Actual ENMs physicochemical properties must also be considered when evaluating toxicity as a vast number of studies have found differential toxicity based entirely on slight modification of properties such as size (42), surface charge (42), surface functionalization (43), shape (44), pH (43), ionic strength (45), and endotoxin attachment (46). This list is not exhaustive, and numerous other physicochemical properties may in some way contribute to toxicity in synergistic or novel

Table 2
A selective list of physicochemical characteristics and formulation considerations pertinent to evaluating nanotoxicology

Physicochemical characteristic	Example study
Homo-/heterogeneity	(43)
Synthesis scheme Reproducibility Co-solvents/excipients	(40) (40)
Purity	(126)
Size distribution	(42)
Shape	(44)
Density	(154)
Surface Functionalization	(43)
Charge/Charge density	(42)
Surfactants Ligand exchange Attachment method Endotoxins	(39) (21) (137) (46)
pH	(40)
Ionic Strength	(45)

mechanisms. Furthermore, physicochemical properties may not be an indicator of toxicity across different ENMs species (47), and as such additional research is necessary to discover the unique contributions each physicochemical property may imbue within each specific physiognomy and/or chemical composition for ENMs.

4. Physicochemical Characterization Techniques

The extraordinary properties of ENMs are primarily attributed to their nanoscale structure, size, and shape. Any surface modification either imparted prior to application or acquired in the environment, can significantly alter properties such as size, porosity, and surface charge, which in turn can change the distribution and interactions of nanoparticles in the desired environment.

Physicochemical characterization of nanoparticles for properties such as size distribution, agglomeration state, shape, crystal structure, chemical composition, surface area, surface chemistry, surface charge, and porosity have been strongly recommended (48). A brief review of several popular nanoparticle characterization methods with possible applications in zebrafish nanotoxicology is given below:

4.1. Mass Spectrometry

In mass spectrometry (MS) the components of a sample are rapidly converted into ions and are resolved based on their mass-to-charge ratio, providing qualitative and quantitative information about the atomic and molecular composition of inorganic and organic materials. A mass spectrometer has three essential components. An ion source ionizes the sample (usually to cations). A mass analyzer then separates and sorts the ions according to their mass and charge, and finally, the separated ions are measured by a detector. Different MS techniques utilize different methods for each of the ionization, mass analysis, and detection steps (49). Liquid samples are often ionized using electro-spray ionization. Biological samples can be analyzed using matrix-assisted laser desorption/ionization (MALDI). MALDI is a soft ionization procedure in which the sample is mixed with excess matrix and is dried on a MALDI plate to protect the molecule from being destroyed by direct laser beam (and to facilitate vaporization and ionization). Surface ions of solid samples can be analyzed using secondary ion mass spectrometry (SIMS), where the surface is bombarded with high-energy ions (usually oxygen or cesium), resulting in sputtering of secondary ions from the sample surface. Oxygen and cesium are more likely to ionize the atoms emitted from the surface, and thus be detected. Elemental analysis can be performed by depth or by area.

A sector type mass analyzer, which is a magnetic or electric field that affects the velocity of the ions according to their m/z ratio can also be used. In a time-of-flight (TOF) mass analyzer ions are accelerated using an electric field of known strength and the time for the ions to reach the detector is measured. The mass-to-charge ratio of an ion is proportional to the square of its drift time (Eq. 1):

$$\frac{m}{z} = \frac{2t^2 K}{L^2}, \tag{1}$$

where t is the drift time, L is the drift length, m is the mass, K is the kinetic energy of the ion, and z is the ion charge.

In quadrupole type mass analyzers the ion beam passes through four metal rods. AC and DC potentials are applied to the quadrupole so that only ions with the desired m/z ratio can pass through the analyzer at a time. MS techniques have been routinely used for qualitative and quantitative analysis of macromolecules and nanoparticles. Navin et al. (50) used MALDI-TOF to characterize colloidal platinum nanoparticles capped with PVP through experiments with capped and non-capped nanoparticles, assuming that the 1.85 nm particles were capped by only one PVP and had a spherical geometry. Size was comparable to those measured by TEM and XRD.

Although MS techniques have been used routinely for nanoparticle detection and identification, the technique has not been fully explored in zebrafish studies. One interesting area is the study of zebrafish embryo proteomics. Link et al. (51) used MALDI-TOF MS to identify the proteins in the early zebrafish embryos. After separating the yolk proteins, the embryo proteins were separated using 2D gel electrophoresis. Protein spots were digested and analyzed by MALDI-TOF and the peptide mass fingerprints were searched against databases. Hanisch et al. (52) demonstrated that proteomic signatures are important markers in zebrafish toxicity assessment. The embryonal proteome was altered after 48 h of exposure to model toxins, while the organisms seemed visibly intact.

Isaacson (20) used LC/MS with electro-spray ionization for quantitative determination of fullerene uptake by zebrafish embryos with a detection limit of 0.02 mg/L, allowing them to obtain in vivo LD_{50} data.

4.2. Inductively Coupled Plasma-Mass Spectrometry

Inductively coupled plasma-mass spectrometry (ICP-MS) is a highly sensitive mass spectrometry technique, which utilizes an inductively coupled plasma as the ion source. ICP-MS can detect metals and some nonmetals at concentrations below one part per trillion and is capable of monitoring isotopic speciation for the ions of choice. ICP-MS spectrometers can accept solid as well as liquid samples. Solid samples are introduced into the ICP by way of a laser ablation system and aqueous samples are introduced by way of a nebulizer, which aspirates the sample with high velocity argon, forming a fine mist to be vaporized in the plasma torch. Chromatographic techniques such as HPLC and FFF (see 4.4) can be coupled with ICP-MS to provide information on metal ion concentration along the distribution.

Boyle et al. (53) analyzed the effect of metals on the zebrafish reproductive system using ICP-MS and HPLC–ICP-MS (for Arsenic). Zebrafish fed with polychaete *Nereis diversicolor* collected from a metal-impacted estuary for 68 days showed reduced reproductive output, characterized by 47% reduction in cumulative egg production, 30% reduction in cumulative number of spawns, reduction in the average number of eggs produced per spawn and hatch rate. The research showed that although the *N. diversicolor* was contaminated with several metals (Ag, As, Cd, Cu, Fe, and Zn), only inorganic As was accumulated in the reproductive tissues of the zebrafish. Other studies have investigated the toxicity of Ni (44) and U (54) in zebrafish embryos and zebrafish brain as well as bioaccumulation and uptake of Cu and Cu nanoparticles in zebrafish (37).

4.3. Static and Dynamic Light Scattering

Static light scattering instrumentation exploits the scattering property of particles when illuminated by a laser light. The intensity of the scattered light depends on several factors including the polarizability of the particle and the ratio of the light wavelength to the particle diameter. Light scattering can be modeled only if the particle size is much smaller or comparable with the wavelength of the incident light.

If the particle diameter is much smaller than the wavelength of the incident beam, ($r<\lambda/20$), the intensity of the scattered light (I) can be calculated from the Rayleigh model (Eq. 2):

$$I = I_0 \frac{(1+\cos^2\theta)8\pi^4 r^6}{R^2\lambda^4}\left(\frac{n^2-1}{n^2+1}\right)^2, \qquad (2)$$

where λ is the wavelength, I_0 is the intensity of the incident beam, R is the distance from the light source to the particle, θ is the scattering angle, n is the refractive index of the particle, and r is the radius of the particle.

If the polarizability of the sample is known, the absolute size or molecular weight can be measured using light scattering, assuming a spherical geometry and no multiple or backscattering. For non-spherical particles multi-angle light scattering can provide information about size and shape.

In dynamic light scattering (DLS), also known as Photon Correlation Spectroscopy (PCS), the particles are illuminated with laser and time-dependent fluctuation and the scattering intensity is recorded. The decay in the autocorrelation of the intensity trace is used to calculate the diffusion coefficient through numerical fits of known distributions.

Static and dynamic light scattering techniques are excellent methods for characterization of homogeneous, monodispersed, spherical particles, which preferably have the same polarizability and refractive index. Ideally, and in the absence of multiple scattering, the ensemble of those identical particles should produce a pattern similar to that of a single particle, but proportional to the number of particles in the ensemble. However, in addition to different scattering patterns, the presence of larger particles can completely change the average size, since the intensity of the scattered light has a power relation with the radius ($r \alpha I^6$). Particle size distributions can be calculated from DLS either assuming some standard form such as log-normal or without any such assumptions. One advantage of light scattering is that it can measure the size in a very short time. Combination of LS with separation techniques eliminates the problems posed by sample broadness. Tadjiki et al. (55) used SdFFF-LS to detect and quantify unlabeled 70-nm silica nanoparticles in rat lung tissue homogenate and human endothelial cell lysate. The isolated nanoparticles were quantified using the linear relationship between the particle number and the area under the fractograms with a detection limit of 1.6 μg or 3.8×10^9 nanoparticles.

Light scattering has been primarily used for nanoparticle size characterization in zebrafish nanotoxicity studies. Johnston et al. (36) used DLS to study aggregation of TiO_2, CeO_2 and ZnO nanoparticles that were used in exposure studies to zebrafish and rainbow trout. DLS measured a hydrodynamic size of 200 nm for the TiO_2 in deionized water, regardless of the TiO_2 concentration (50–2,000 mg/L). TEM images showed a smaller size for the TiO_2 nanoparticles. DLS measurements showed aggregation of TiO_2 particles in reverse osmosis and tap water.

In a study of toxicity of nanocopper to zebrafish, Griffitt et al. (37) used DLS to measure the size of nanocopper particles over the course of the exposure and reported an increase in size of nanocopper particles from 0.5 to 1 μm (samples agitated prior to measurement). SEM of the same sample dispersed in water also showed a wide disparity in size, while TEM images showed a size of 50–60 nm (Fig. 1). A particle size distribution of 0.015–0.25 μm was obtained by centrifugal particle sizing.

4.4. Field-Flow Fractionation

Field-flow fractionation (FFF) is a chromatography-like technique in which the partition of sample species is achieved in a thin, open channel, and the particle size distribution can be calculated directly from the first principles (56, 57). In FFF an external field is applied

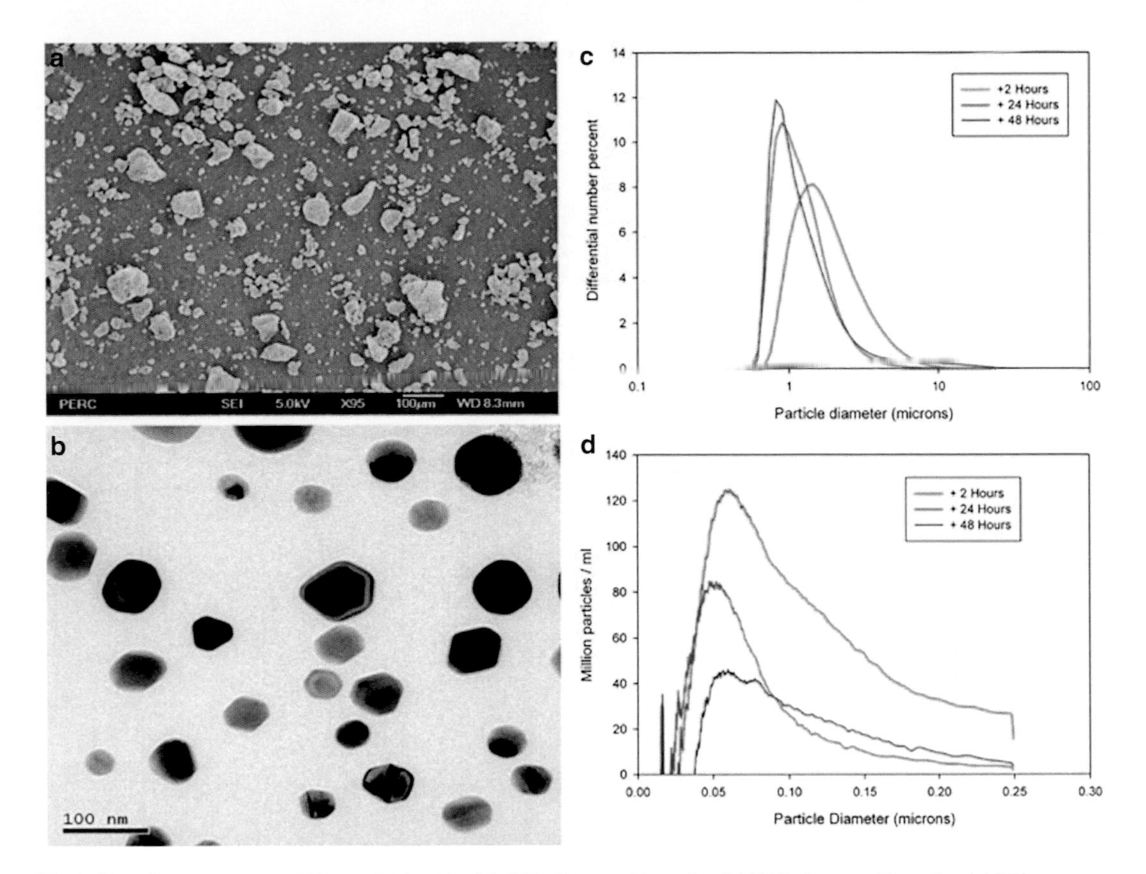

Fig. 1. Size of nanocopper particle as obtained by (**a**) SEM, dispersed in water, (**b**) TEM, dispersed in water, (**c**) DLS, suspension agitated prior to measurement, and (**d**) centrifugal particle sizing, samples were not agitated prior to measurement (from ref. 37) (Color figure online).

at the right angle to the direction of carrier flow. Separation takes place in a thin ribbon-shaped channel where the carrier flow in this channel is laminar. In the presence of the applied field, particles are forced to migrate toward the accumulation wall where they form a concentration gradient that causes diffusional migration against the direction of the field-induced flux. Particles reach equilibrium as these two motions (i.e. the field and diffusion based fluxes) balance each other.

At equilibrium each particle component forms an equilibrium cloud whose average thickness l depends on the magnitude of the diffusion coefficient and the strength of the interaction of the field with the particles. Lower flow velocities are found near the walls of the channel than near the center because of frictional drag. Larger particles that possess less diffusional motion and higher interaction with the applied field will be caught up in the slower moving streams near the channel wall and elute later than the smaller particles (Fig. 2).

Different fields can be applied which result in calculation of different parameters. Flow FFF (FlFFF where the applied field is

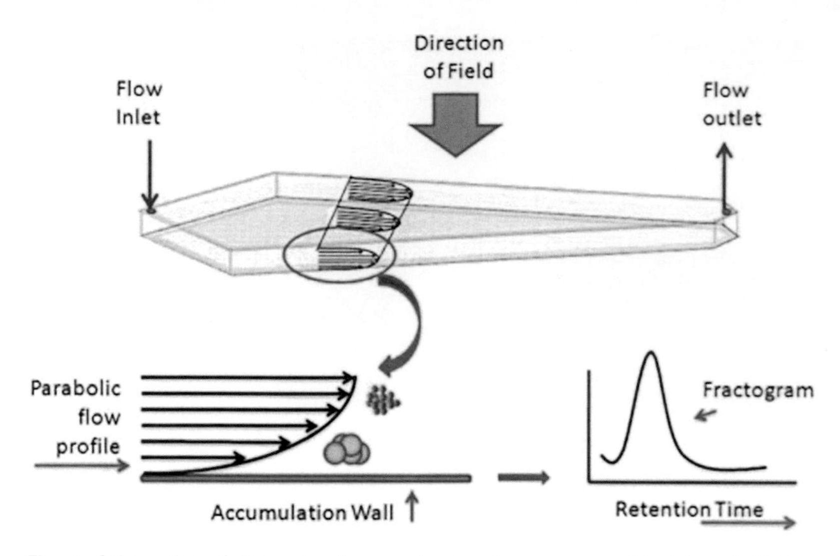

Fig. 2. Schematics of the separation mechanism in an asymmetrical flow field-flow fractionation channel (from ref. 58).

a flow of the carrier) and sedimentation FFF (SdFFF where the applied field is centrifugal force) are commonly used for size characterization of nanoparticles and macromolecules.

In asymmetrical FlFFF the hydrodynamic diameter (d_h) can be calculated directly from the retention time (t_r) (Eq. 3).

$$d_{\mathrm{h}} = \frac{2kTt_{\mathrm{r}}}{\pi\eta w^2 \ln\left(1 + a\dfrac{V_{\mathrm{c}}}{\dot{V}}\right)}, \tag{3}$$

In SdFFF the channel has been placed in a centrifuge basket. The applied field is the centrifugal force and the measured parameter is buoyant mass. In SdFFF the equivalent spherical diameter (d) of a well-retained particle can be calculated directly from the retention time (t_r) (Eqs. 4 and 5):

$$d = \left(\frac{36kTt_{\mathrm{r}}}{\pi wG\Delta\rho t^0}\right)^{1/3}, \tag{4}$$

$$t^0 = \frac{V^0}{\dot{V}}, \tag{5}$$

where k is the Boltzmann constant. T is the temperature (K), w is the channel thickness (m), η is viscosity and V_c is the flow rate of the carrier fluid, G is the centrifugal acceleration (ms^{-2}), Δρ is the density difference between the particle and the carrier fluid (kg/m^3). t^0 is the void time (s), the time for a non-retained component to exit the channel, V^0 is the channel void volume (m^3), and V is the volumetric carrier flow rate (m^3/s).

An advantage of FFF is the absence of packing material inside the channel. Therefore sample size distribution can be measured in a variety of solution conditions. Fractions can be taken along the size distribution and examined by auxiliary methods (58). FFF-LS eliminates the multiple scattering problems for the LS detectors (59, 60). FFF-ICPMS can provide size and elemental distribution (59). In a study of toxicity of nanosilver to aquatic organisms *Daphnia magna, Pimephales promelas*, and *Pseudokirchneriella*, Kennedy et al. (61) used FFF, DLS, and TEM to characterize nanosilver particles. For rod-shaped, heterogeneous samples both FFF and DLS gave larger hydrodynamic diameter than those deduced from TEM images. For spherical nanosilver, FFF results were closer to those obtained by TEM. For suspensions with large particle size distributions general agreement was observed between TEM, DLS, and FFF results. The difference was attributed to the tendency of DLS to skew the effective particle diameter towards the larger size in a polydispersed system, or the possibility of cutting off the FFF run prematurely and thus not observing the larger size distribution.

4.5. Electron Microscopy

Transmission electron microscope generates highly resolved images of the sample using the wave properties of moving electrons. In a TEM, electrons emitted from an electron gun pass through the sample. The transmitted beam is projected onto a fluorescent viewing screen, which will be excited as a result of the impact by the electrons to produce a visible magnified image of the sample. TEM offers detailed information on the size of inorganic particles down to a few nanometers (62), but obtaining an accurate statistical representation can be very tedious. Besides, sample preparation often involves drying, which might result in aggregation of the sample. Since the sample preparation involves drying, images need to be obtained in deionized water rather than the sample matrix. Inference of the three-dimensional structures from two-dimensional images can be very difficult (62) but can be accomplished in some cases with stereological transformation (63). With the advent of high-resolution TEM, imaging sub-nanometer particles has been accomplished. TEM can be used as stand-alone or as an auxiliary method in combination with other characterization techniques such as LS, MS, and FFF. Biological samples can be examined by cryogenic TEM, where the sample is frozen and sections are obtained. Both conductive and nonconductive samples can be analyzed by TEM.

The scanning electron microscope (SEM) scans the sample with a high-energy beam of electrons in a raster scan pattern. An image is obtained as a result of the interaction of the electrons in the beam (primary ions) and the atoms of the sample. High-resolution images can be obtained by using secondary ion (ions emitted from the sample) detectors. Backscattered electron imaging is useful in distinguishing one material from another, since the yield of the collected backscattered electrons increases monotonically with the sample's atomic number. In addition, SEM can be equipped with energy dispersive X-ray for compositional analysis of the sample. Conductive samples yield better resolution. However, imaging of nonconductive samples is also possible. Samples need to be dried, which makes imaging in the matrix very difficult.

Both TEM and SEM have been used extensively for size analysis of nanoparticles in nanotoxicity studies of zebrafish.

4.6. Atomic Force Microscopy

Atomic force microscopy (AFM) allows measurement of forces between a tip and a substrate (64). A cantilever (usually silicon nitride with a radius of curvature in the order of tens of nanometer) is approached to the sample using a step motor (or vice versa). When the distance is close enough for overlap of surface charges, the cantilever will be attracted or repelled by the substrate. The deflection is recorded by a photodetector through a laser beam focused on the cantilever tip. Forces between the cantilever and the substrate can be calculated from the deflection (x) and the spring constant of the cantilever (k) using Hooke's law ($F=-kx$).

Interparticle force measurements are possible by gluing a particle to the tip of the cantilever (65). AFM can give information on surface forces including interaction, adhesion, electrostatic, and van der Waals forces. The double layer forces can be used to estimate the surface potential of both surfaces in a symmetric system (same material for particles and substrate) or just one, provided that the surface potential of the other surface is known (66, 67). Both imaging and force measurement options of AFM instrumentation have strong potential in zebrafish toxicity studies. Force measurements between specific cells and particles can provide the means to understand the effects of factors such as size, shape, surface functionality, charge heterogeneity and physical roughness, on interactions between nanoparticles and specific cells. While TEM and SEM images have to be obtained in high vacuum and deionized water, AFM can provide high-resolution images under physiological conditions.

Puech et al. (68) measured adhesion forces between fibronectin—an abundant extracellular matrix component during gastrulation—adsorbed on glass and a single zebrafish mesendodermal progenitor cell, attached to the cantilever. Comparison with adhesion forces obtained for slb/wnt11 mutant embryos suggested that wnt11 was needed for the cell adhesion (Fig. 3).

Kreig et al. (69) used single-cell force spectroscopy to measure adhesion forces between isolated zebrafish germ-layer cells, one mounted on the cantilever and the other on a glass substrate. The authors concluded that mesoderm and endoderm progenitors were more cohesive than ectoderm cells and that the cohesion was mediated by E-Caherin. Forces between a glass bead mounted on

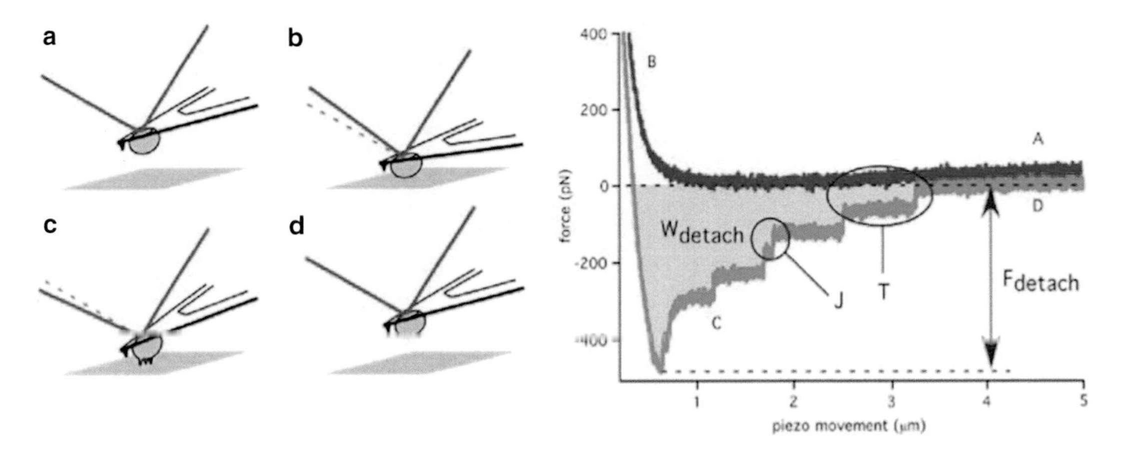

Fig. 3. Schematic diagram of an adhesion experiment and the corresponding typical force curves for approach (*red*) and retraction (*green*) from the surface. Figures (**a**)–(**d** show the approach (**a**), compliance (**b**) retraction, (**c**) and detachment (**d**) steps (from ref. 68) (Color figure online).

the cantilever and a progenitor cell mounted on the glass substrate were used to study progenitor cell-cortex tension.

So far the AFM force measurements are limited to the study of zebrafish gastrulating embryos. However, these pioneering studies show the potential of AFM force measurements in understanding the mechanisms involved in the nanoparticle uptake by various zebrafish cells.

4.7. Synchrotron Radiation-Based Analytical Techniques

Synchrotron facility radiation is emitted by electrons orbiting in a storage ring at velocities near the speed of light. When the electrons are accelerated, X-rays are emitted. Synchrotrons also provide ultra-bright ultraviolet light. Compared with conventional laboratory facilities, synchrotron-based X-ray sources have a wide energy range, high photon flux, and better beam coherence. Therefore, data acquisition and imaging times are much faster. The coherent beam allows for high contrast phase imaging with micrometer resolution (70). Synchrotron radiation-induced X-ray fluorescence (SRXRF) enables direct, multielemental analysis of solid samples with a detection limit of ng/g and submicron spatial resolution. Synchrotron radiation X-ray absorption spectroscopy can provide information on the local electronic and molecular structure around the atom of interest with sub-angstrom spatial resolution and detection limits in the order of parts per million (71). Other synchrotron radiation-based techniques include circular dichroism, X-ray diffraction, X-ray small angle scattering, and microcomputer tomography (72).

Neues et al. (73) used synchrotron-radiation microcomputer tomography (SRmCT), energy dispersive X-ray spectroscopy, along with optical and electron microscopy to study the dynamics of biomineralization in teeth and bones of zebrafish.

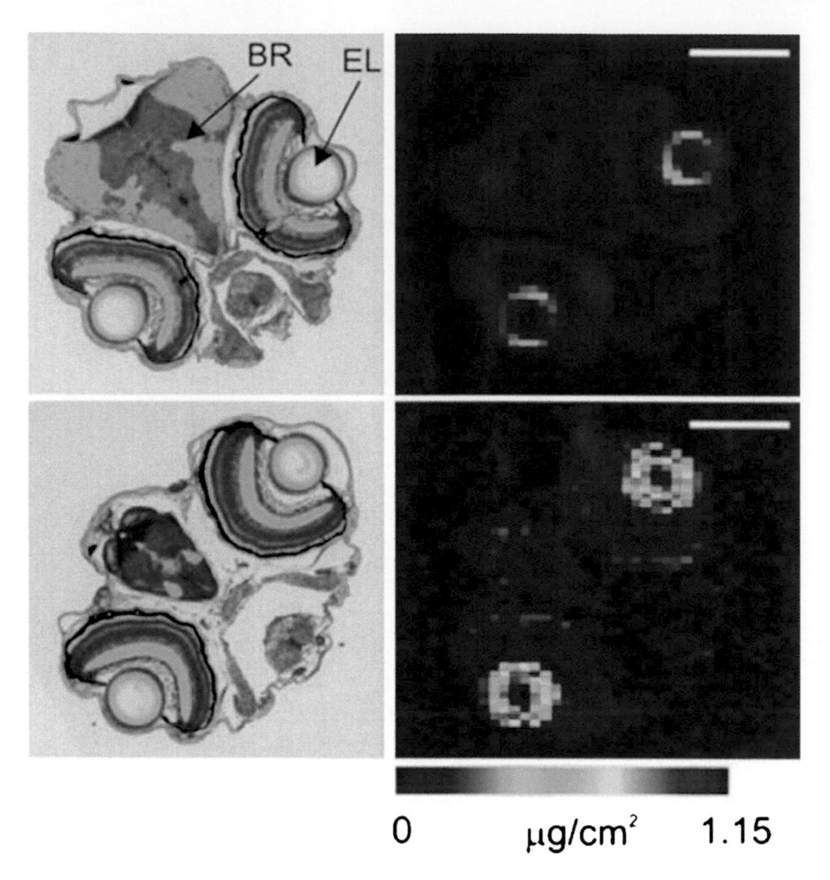

Fig. 4. Quantitative mercury distributions in zebrafish larvae after a 24-h exposure to 2 mM methylmercury l-cysteineate at 3.5 dpf (**a**), and followed by a 12-h recovery in fresh system water (**b**). Histological images (*left*) are compared with mercury distribution of the adjacent section (*right*). Scale bar 100 mm, *BR* brain, *EL* eye lens (from ref. 74). Copyright (2008) National Academy of Sciences, USA (Color figure online).

Korbas et al. (74) investigated the uptake and localization of organic mercury in zebrafish larvae using SRXRF imaging of live fish. The authors reported accumulation of methyl and ethyl mercury in the brain, gastrointestinal tract, and the highest accumulation rate was observed in the eye lens epithelium. Other areas such as the brain and the optic nerve were affected less. The mercury distribution was not homogenous and did not show any relation to the distribution of zinc or calcium. Similar patterns were observed in the experiments on sections obtained from zebrafish embryos (Fig. 4) (75). However, in the embryos the mercury was distributed homogenously throughout the lens, suggesting a discrepancy in the lens morphology at the two developmental stages. The authors reported redistribution of mercury from other organs to the lens in the post-exposure period and suggested that the organic mercury was carried into the lens epithelial cells by the large neutral amino acid carriers L or LAT.

5. Engineered Nanomaterials Under Nanotoxicology Investigation

5.1. Carbon Nanotubes

Carbon nanotubes (CNTs) are hollow hydrophobic graphite cylinders (76) capable of surface functionalization for varied applications. CNTs have a variety of uses ranging from biomedicine (targeting (76), antigen, gene and drug delivery (77), and tissue regeneration (78)) and construction materials with commercial and military applications to imbue enhanced mechanical, thermal, and electrical properties (79). It is estimated by 2011, CNTs will have a market value of approximately $1B with thousands of tons being produced annually (80). The nanotoxicity of CNTs is reported to be a function of several physical parameters including: length (81), aspect ratio (ratio of the longer dimension to the shorter), particle size (where CNTs have the potential to form large agglomerates in both fresh and saltwater at high pH (40, 82)), metal catalysts present due to synthesis (83, 84), and molecular weight (monodisperse, polydisperse, or aggregated (83, 85, 86)). Surface functionalization directly influences CNT nanotoxicity (87) with reports of bovine serum albumin (BSA) functionalization decreasing toxicity (88); moreover, PEGylation (conjugation of poly(ethylene glycol)) of CNTs has shown decreased cytotoxicity presumably by increasing the water-solubility of the CNTs (89). Oxidative treatments have been shown to modify CNT structure (85, 90), but there appears to be no change in particle structure with fluctuations in salinity or temperature (40, 91). Generally, the more soluble the particle, the smaller the toxic response induced (92) with a common nanotoxic result of deleterious generation of reactive oxygen species (ROS) (90). Carbon nanotubes also have differential toxicity based on whether they are single-walled (SWCNT) or multi-walled (MWCNT). MWCNTs have been shown to be toxic in a length-dependent manner in the mesothelial lining of mice chest cavities, while their SWCNT cousins exhibited no similar response (81). This difference in toxicity may be due to SWCNTs being easily phagocytosed and transported to local lymph nodes, eliminating their toxic threat, while MWCNTs accumulate and aggregate within larger tissue, disrupting its function (86). SWCNTs are more likely to form aggregates than MWCNTs due to smaller effective surface areas because of their relatively stronger van der Waals forces between their sidewalls (93). However, dispersed and water soluble CNTs may induce length-dependent toxicity since longer CNTs induce higher inflammatory responses believed from macrophages' inability to engulf CNTs properly (81). A summary of contemporary zebrafish nanotoxicology studies with CNTs can be found in (Table 3).

Evaluation of CNTs toxicity must be performed as they are known to act as bactericides, cause cell membrane damage (94),

Table 3
Carbon nanotube (CNT) zebrafish nanotoxicity studies and method(s) of physicochemical characterization

Nanoparticle	Size (avg) ($d \times l$)	Notes	Analysis method(s)	LC_{50}	Dosing Info.
SWCNT (54)	11 nm × 0.5–100 μm	Raw	EDS	n.d.	Waterborne 4 hpf
DWCNT (54)	n.d.			n.d.	
BSA-MWCNT (62)	19.9 nm × 0.8 μm	Both	S-TEM	n.d.	Micro-injection 72 hpf
PPEI-EI MWCNT (62)				n.d.	
Carboxyl acid-SWCNT (92)	4–5 nm × 500–1,500 nm	Doxorubicin added	TEM	n.d.	Micro-injection 48 hpf

SWCNT single-wall carbon nanotube, *DWCNT* double-wall carbon nanotube, *BSA* bovine serum albumin, *PPEI-EI* poly(propionylethylenimine-co-ethylenimine, *n.d.* not determined, *EDS* energy dispersive X-ray spectroscopy

mitochondrial DNA damage (95), granulomas (96), atherosclerotic lesions, and inhibit bacterial clearance from lung tissues (79, 83). Purportedly, SWCNTs have inherent strong antioxidant properties in zebrafish, and modification through functionalization with either different types of bonds or functional groups can alter the antioxidant ability (87). In 2007, Cheng et al. found that zebrafish embryos dosed with waterborne SWCNTs and double-walled carbon nanotubes (DWCNTs) exhibited no acute lethality or sublethality up to the end time point of96 hpf (hours post-fertilization), but hatching delay was observed (40). Both SWCNTs and DWCNTs formed aggregates in eggwater with sizes of 0.01–1.0 μm and were revealed on the surface of the chorion using SEM. Hatching delays of 52–72 hpf occurred more frequently with SWCNT exposed zebrafish (average length of 0.5–100 μm and diameter of 11 nm; concentrations > 120 mg/L) than DWCNT exposed fish. Hatching delays from 56 to 72 h were reported at concentrations >240 mg/L with assumed average lengths the same. Of note was the presence of adsorbed SWCNTs and DWCNTs on the chorion surface that appeared to be unable to cross the chorion barrier. Quantification of the chorion adsorbed ENMs was not reported. The basis for this hatching delay may be due to interference with nutrient and gas exchange of the chorion or residual Co/Ni catalysts used in the synthesis of the CNTs; 40 μg/L $NiSO_4$ and 3,840 μg/L $CoCl_2$ are the no-effect nominal concentrations, respectively; SWCNT at concentrations of 120 mg/L release approximately 3 mg/L and 9 mg/L, Ni^{2+} ions, respectively. DWCNTs are made with $Mg_{1-x}Co_xO$ catalysts and are known to encapsulate Co in its graphite shells (97). Ni is more specific and potent in inducing hatching delay (40) and potentially explains why DWCNTs have a less severe hatching delay. The data suggest that chemical synthesis waste products may be the primary mode of toxicity. A final possibility is CNTs synergistically interact with leftover metal catalysts to induce the hatching-inhibitive effect but this possibility was also not addressed (40).

Next, in 2009 Cheng et al. explored microinjection of CNTs into zebrafish; 2 ng of MWCNTs was microinjected into the zebrafish embryos at the one-cell stage. MWCNTs were acutely nontoxic, though they did observe an immune response and the treated fishes' offspring had significantly lower survival rates—suggesting an epigenetic nanotoxic response. Interestingly, BSA capping of the MWCNTs and labeling with the fluorescent dye fluorescein isothiocyanate (FITC) ameliorated toxicity but still produced an immune response as observed with whole mount in situ using the MMP-9 gene as a marker. Interestingly, FITC-BSA-MWCNTs entered blastoderm cells while BSA-FITC did not. MWCNTs preferentially accumulate in a swim bladder-like structure after going through the circulatory system. After 96 h post-injection, the majority of FITC-BSA-MWCNTs appeared to be gradually excreted out of the zebrafish larvae. There was no hatching delay. It is plausible that this result could be either the absence of aggregates on the chorion due to using microinjection exposures or attributed to different synthesis procedures since analysis with SEM (scanning electron microscopy), bright-field imagery, and TGA (thermal gravimetric analysis) showed no metal residues present (98). Chaudhuri et al. in 2010 microinjected 9.2 nL of SWCNTs 48 hpf and averaging 500–1,500 nm in length with diameters of 4–5 nm which did not induce lethality or gross morphological changes within the developing embryos though they did promote endothelial tubulogenesis. Endothelial tubulogenesis is a late step in angiogenesis, and it was correlated with an increase in $\alpha_v\beta_3$ integrin expression, which was observed by an immunofluorescence assay. There was weak cell internalization of the CNTs, even after PEGylation with OMe-PEG(2000)-NH_2 (76) which is known to increase blood circulation by decreasing opsonization and increasing solubility (99). Finally, representative of microinjected CNTs and associated nanotoxicity is the report by Asharani et al. in 2008. Here, Asharani et al. found that waterborne exposures of MWCNTs were acutely toxic to developing zebrafish embryos at concentrations greater than 60 μg/mL starting at 24 hpf until the endpoint 72 hpf. The mortality rate was $20 \pm 3\%$ at 70 μg/mL and increased to $60 \pm 2\%$ at 100 μg/mL; 100% mortality was observed at concentrations greater than 200 μg/mL. The study only characterized the waterborne MWCNTs as being 30–40 nm in diameter based on a characteristic absorption peak at 260 nm (100) from UV–vis spectroscopy and TEM microscopy images. Embyros microinjected with 5 ng/mL of solution had a mortality rate of 35%. Both exposure routes produced hatching delay, mucus production, shortened notochords, pericardial edema, and apoptosis. Bradycardia was witnessed only with waterborne exposures of the MWCNTs. MWCNTs appear to cause irritation and inflammation of the skin (101), similar to other particles (102), since at concentrations > 40 μg/mL there was a thick mucus secreted from the fish. DAPI analysis revealed kidney-shaped nuclei characteristic of macrophage. Mucus secretion is the first line of the body's defense against an irritant, functioning to trap and move the antigen from the body. Acridine orange (AO) assays used to identify apoptotic cells showed high levels of apoptosis and necrosis, with the highest response measured on the skin of treated embryos while controls did not exhibit the necessary orange fluorescence. Circulatory failures were observed at the specific site where the MWCNTs accumulated, suggesting that MWCNTs can aggregate in vivo and induce toxicity by blocking internal systems (101).

5.1.1. Extrapolations and Notes

Based on the aforementioned studies, size distribution does not appear to be a contributing factor to a zebrafish's nanotoxic response, but it may influence uptake (40, 76, 98). Other studies have found that larger CNTs (MWCNTs, lengths up to 56 μm and diameter of 165.02 ± 4.68 mm (81) and aggregates with mean values of 0.955 ± 0.064 μm (86)) more readily accumulate in body tissues and form larger aggregates or are readily excreted (81, 86). Along the same lines, the aspect ratio could be crucial in determining toxicity and would have important implications for biomedical uses. One study found in the CHO-k1 mammalian cytotoxicity assay that high-aspect ratio MWCNTs (length: ~10 μm diameter: 10–15 nm) had a lower 50% growth inhibition (GI_{50}) of 12.94 versus their low-aspect ratio counter parts (length: ~150 nm diameter: 10–15 nm), which had a GI_{50} of 60.2%. The same study concluded that both aspect ratios did not cause genotoxicity in bacterial reverse mutation tests, in vitro chromosome aberration tests, nor in vivo micronuclei tests (103). In normal human embryonic lung cells (WI-38), high-aspect ratio MWCNTs (length: 10,451 ± 8,422 nm diameter:10–15 nm) have statistically different decreases in cell viability than low-aspect ratio MWCNTs (length:192 ± 80 nm diameter:10–15 nm) at concentrations of 12.5, 25, 50, 100, and 200 μg/mL 24 h after dosing (104). Interestingly, in contrast to the previous two studies, a third study examining the effects of carbon-based nanomaterials on lung tumor cells in an MTT assay performed with three different human lung-tumor cell lines, H596, H446, and Calu-1 found that carbon black with an aspect ratio of one was more acutely toxic by inducing necrosis than MWCNTs with an aspect ratio of 30–40, though carbon nanofibers with an aspect ratio of 80–90 were more acutely toxic than MWCNTs but less than carbon black (105). Finally, a fourth study notes that in the human alveolar epithelial cell line

(A549) assay, dispersed SWCNTs did not induce significant decreases in cell viability, ROS generation, DNA fragmentation, nor cell count. However, aggregates induced granulomatous lung inflammation and fibrosis. Both the dispersed SWCNTs had aspect ratios relative to the aggregated SWCNTs suggesting aggregation rather than a high-aspect ratio as responsible for inducing toxicity (106). To complicate matters, other studies have found that there is toxicity with well-dispersed, high-aspect ratio MWCNTs aggregates of the same species decrease toxicity in mammalian cell lines (103). The discrepancies between these studies should be evaluated with further studies, along with testing in in vivo models in order to further elucidate the effect that aspect ratio has on CNT toxicity. Shape appears to affect uptake as CNTs are taken up less than their fullerene cousins (76), similar to rod-shaped gold nanoparticles, which are taken up less than spherical cousins (107). Synthesis remnants may be a critical factor inducing toxicity since one zebrafish study demonstrated that CNTs synthesized with Ni and Co catalysts instigate hatching delays (40) verifying in vitro studies with similar toxic responses due to catalysts (83). One such study found that CNTs, which were purified to remove synthesis catalysts, did not induce any toxic response (84). However, the in vitro studies testing the effects of different aspect ratios cited above went to great lengths to remove metal impurities from their CNTs through washing and then verifying through thermogravimetric analysis (TGA) found cell counts still decreased (103, 104). In vitro and in vivo studies have stated that nonfunctionalized CNTs tend to be more toxic (108), as their clearance rate is diminished, they more readily aggregate and become cytotoxic (90, 109), though this was not observed in the 2007 study. Interestingly, the only zebrafish study comparing MWCNTs and SWCNTs found that MWCNTs were not as toxic as SWCNTs (40) disagreeing with the studies done on mice (81, 86). However, another study found MWCNTs to be devastatingly toxic in zebrafish (101). Future studies are needed to more clearly understand the differential toxicity of SWCNTs and MWCNTs, if it exists, and to further elucidate the roles of size, synthesis procedure, aspect ratio, and surface functionalization.

5.2. Fullerenes, Fullerols (Fullerenols), and Metallofullerenes

5.2.1. Description

Carbon-based nanomaterials are the second most used nanomaterial in consumer products (110) and fullerenes, fullerols (or fullerenols), and metallofullerenes, with their characteristic 60-carbon cage structure, are the most prominent (111). Fullerenes, (C_{60}) also known as "bucky balls," are commonly utilized for drug delivery, groundwater redemption, food supplements, electronics, cosmetics, and fuel cells (79, 112). Studies have identified fullerenes' primary toxicity concern as ROS generation (113), correlating with the particles' surface modification (114) and synthesis scheme (115). Fullerene's unique 60-carbon structure adsorbs light, and in the presence of oxygen species transfers the energy to them and forms the highly reactive singlet oxygen state known as a reactive oxygen species (ROS) (116). C_{60} is insoluble in water (2×10^{-24} mol/L) (113) and naturally aggregates in river samples (11, 117, 118), but when modified to become soluble in aqueous solutions by adding surfactants or functional groups, it readily produces ROS (119) and becomes cytotoxic (113). Larger agglomeration sizes decrease uptake potential, and the rate at which C_{60} is able to mediate energy and electron transfer to form ROS (117–119). C_{60}'s surface is easy to modify, and attaching PEG, gamma-cyclodextrin, or other biocompatible molecules has been shown to decrease ROS generation and cytotoxicity (76, 114). Some studies suggest that soluble fullerols may be less toxic than its fullerene cousin (120), though it has been shown to generate dissolved inorganic carbon, i.e. CO_2, in solutions similar to rain and surface waters.(121) DF-1, another fullerene derivative, has been shown to be radioprotective in zebrafish due to its antioxidant properties (122, 123). Synthesis schemes should also be evaluated as toxic metals and nonmetals, such as barium and boron can leach from C_{60} (115). Interestingly, fullerenes have an affinity for accumulating in fatty tissues of fish, and other animals, even more so than the banned pesticide DDT (120). In vivo, it has been shown that C_{60}'s structure mimics clathrin, an endocytosis regulating structure, and thus can cross the external cellular membrane and preferentially localize within mitochondria (124), but can also cause lipid peroxidation through ROS generation (125). Fullerols were acutely nontoxic when microinjected (76) and found to internalize in lysosomal compartments (76). A summary of contemporary zebrafish nanotoxicology studies with fullerenes and their derivatives can be found in Table 4.

5.2.2. Zebrafish-Specific Studies

Zhu et al. in 2007 demonstrated that C_{60} induces toxicity in zebrafish embryos and becomes more lethal as exposure length increases at concentrations of 1.5 mg/L and greater. Water-soluble $C_{60}(OH)_{16-18}$, fullerols, did not exhibit toxic effects. The irregularly shaped C_{60} (avg size: 100 nm) were suspended using the known toxic chemicals tetrahydrofuran (THF) and benzene (126), and despite rigorous washing and purification, THF residue or degradation products accounted for 10% of the weight of the solution (43). THF controls did not induce toxicity, suggesting that the toxic response comes from C_{60} and its potential interactions. Generation of ROS is the foremost hypothesis on C_{60} toxicity and to test this hypothesis, C_{60} were co-exposed with 30 mg/L glutathione (GSH). GSH is known to mediate ROS in zebrafish since GSH's sulfide functional groups can capture unpaired electrons removing harmful free radicals (127). At first, co-exposed fish had less acute toxicity, but by the 96 hpf endpoint, acute toxicity was the same as fish exposed only to C_{60} suggesting that GSH can do

Table 4
Fullerenes, fullernols, and higher generation fullerenes zebrafish nanotoxicty studies and method(s) of physiochemical characterization

Nanoparticle	Size (avg) (nm)	Notes	Analysis (methods)	LC_{50}	Dosing Info.
C60 (43) $C_{60}(OH)_{18}$ (43)	100 nm n.d.	Irregular shape; THF; dechorionated	TEM	1.5 mg/L >50 mg/L	Waterborne 1.5 hpf
C60 (128) C70 (128) $C_{60}(OH)_{24}$ (128)	~175 nm (50 ppb) ~800 nm (200 ppb) ~500 nm (50 ppb) ~900 nm (200 ppb) ~100 nm (250 ppb) ~200 nm (1,000 ppb)	DMSO; dechorionated	PCS; TUNEL; Acridine Orange Staining	200 ppb 200 ppb 4,000 ppb	Waterborne 24 hpf Solution changed every 24 h
C60 (129) $C(OH)_{24}$ (129)	~175 nm (50 ppb) ~800 nm (200 ppb) ~100 nm (250 ppb) ~200 nm (1,000 ppb)	DMSO; Co-exposure: GSH, Vit-C, NAC, BSO, DEM, H_2O_2; dechorionated	PCS; Custom zebrafish gene arrays; Acridine orange;	DEM + BSO: 50 ppb	Waterborne 24 hpf; Solution changed every 24 h
C60 (39) C60-THF (39)	50–300 nm 50–300 nm	THF	Dark-field microscopy;gas chromatography–mass spectrometry	>0.025 mg/L ~0.000775 mg/L	Waterborne 75 hpf
$C60(OH)_{16-24}$ (76)	60–80 nm		TEM	n.d.	Microinjection 48 hpf
C60 (20) C70 (20) Higher Gen. (20)	n.d.	DMSO; dechorionated	LC/ESI-MS	130 μg/L	Waterborne 36 hpf

TEM transmission electron microscopy, *PCS* photon correlation spectroscopy, *DMSO* dimethyl sulfoxide, *THF* tetrahydrofuran, *TUNEL* terminal deoxynucleotidyl transferase dUTP nick end labeling, *LC/ESI-MS* liquid chromatography/electrospray ionization mass spectrometry, *GSH* l-glutathione, *NAC N*-acetylcy teine, *BSO* buthionine sulfoximide, *DEM* diethyl maleate, *n.d.* not determined little to reduce C_{60}-induced ROS toxicity, but that ROS generation is at least a partial factor for toxicity (43).

Usenko et al. in 2007 found that concentrations of C_{60} and C_{70} above 200 ppb resulted in 100% mortality while $C_{60}(OH)_{24}$ did not induce mortality until after concentrations exceeded 4,000 ppb. Sublethal effects observed from all species were pericardial edema, yolk sac edema, and pectoral fin malformations. All three species had increased aggregation as concentration increased. C_{60} had a size range of 100–500 nm at 50 ppb and 300–1,100 nm at 250 ppb. C_{70} had a size range of 100–700 nm at 50 ppb and 700–1,000 nm at 250 ppb. $C_{60}(OH)_{24}$ had a size range of up to 200 nm at 250 ppb and 100–500 nm at 1,000 ppb. DMSO was used to suspend C_{60}/C_{70} in media and interactions between the C_{60}/C_{70} and DMSO could induce toxicity, though the DMSO control by itself did not induce significant toxicity. AO found concentration-dependent cell death centralized in the head region, down the notochord to the site of caudal fin malformations. Interestingly, the apoptosis assay TUNEL had half as much fluorescence as AO for C_{60} dosed fish (128).

Usenko et al. in 2008 found DMSO-suspended C_{60} a powerful oxidant in the absence of functionalization. Four main points in the paper explain why C_{60} primary mode of action is through generation of ROS. The first is sublethal and cellular deaths were inhibited during dark exposures. The second is there is significant difference in sensitivity of embryonic zebrafish when C_{60} was co-administered with the known antioxidants GSH and ascorbic acid. Third, neither C_{60} nor $C_{60}(OH)_{24}$ had antioxidant properties when co-exposed with H_2O_2, a known oxidant, and thus had both increased lethal and sublethal effects. Fourth, a zebrafish microarray found that two genes directly related to glutathione activity GST-pi and GSH were up-regulated twofold over controls (129).

Henry et al. in 2007 found that THF-water from C_{60} washing, and C_{60} suspended with THF (size 50–300 nm) both had linear concentration (1–25%, v/v; 100 mL egg water) dependent lethal and sublethal effects (LC_{50} of 6.3% and 3.1%, respectively) but sonicated C_{60} eggwater had no significant lethal effects. THF-water and THF-C_{60} both had at least 124 genes that differed from controls with THF-C_{60} having 72% higher magnitude changes in gene expression over controls, including genes associated with GSH. The potential reason is that both THF-water and THF-C_{60} produce y-butyrolactone, a readily soluble colorless liquid, at approximately 10 and 5 mg/L, respectively, at their LC_{50} dosages. y-Butryolactone was found in higher concentrations in THF-C_{60} stocks than THF-water stocks (158 ppm vs. 103 ppm) suggesting that C_{60} acts as a catalyst to create more y-butyrolactone. y-Butyrolactone's LC_{50} value (.0047%; 47 mg/L) was far below the LC_{50} value (1.73%; 15.4 g/L) for THF with y-butyrolactone removed. (39) Y-butryolactone is readily converted into the neurotransmitter gamma amino butyric acid (GABA) (130) and GABA receptors may control ventilatory movements in early larval zebrafish development (131). The study's findings suggest that C_{60}'s reactive nature generates toxic organic chemicals that are the primary inducer of toxicity.

Isaacson et al. in 2007 found that DMSO solutions containing C_{60}, C_{70}, and higher order fullerenes up to C_{98} produced in dark laboratory conditions (to avoid ROS generation) had an LD_{50} of 130 μg/L. 90% of dosing materials were lost to adsorption onto the plate. Thus, doses given to fish were found to be significantly lower than calculated, with a 30–38% difference. After performing analysis with LC and ESI-MS, a more accurate LD_{50} of 0.079 μg/g of embryonic zebrafish was calculated. Fish uptake of the fullerene species increased as dosing concentration increased. Sublethal effects of delayed development, fin malformations, and pericardial edema were observed at concentrations lower or at 130 μg/L (20).

5.2.3. Extrapolations and Notes

Size or surface area is not a crucial factor for initiating fullerene's toxic response within zebrafish, as different sizes and higher generation fullerenes appeared to generate similar toxicity (128, 132). However, differing surface area or size may allow for greater generation of toxic chemicals (119). According to the studies it appears as if surface functionalization is the primary indicator of toxicity since functionalization can decrease the quantity of ROS produced, depending on what molecules are attached (43, 128). Three of the five zebrafish specific studies directly tested the effects of ROS (43, 128, 129) and found a clear link between ROS generation and zebrafish toxic response, although there is some disagreement on whether C_{60} itself generates ROS (128, 129) or the surfactants (39). Future studies should focus on surfactants and C_{60} synthesis schemes since current zebrafish-specific studies are lacking and in vivo mice studies have demonstrated that metals used in C_{60} may be the cause of the acute toxicity (133).

5.3. Quantum Dots

5.3.1. Description

Quantum dots (QDs) are commonly used in biomedical imaging, diagnostics (134), photovoltaics and electronics (135). They contain a semiconductor core (e.g. CdS) and are often encapsulated by a shell (e.g. ZnS) to enhance optical and electronic properties and their biointeraction (134, 136). Physicochemical considerations that influence toxicity are core and shell composition (137), metal leaching from QDs (138), protein interaction (139), their size (139, 140), and their net charge (139). Commonly produced QDs contain heavy metal chalcogenides and constitute a hazard to living organisms for both their toxic metal ion releases and their nanoscale properties that may make them more readily available (42). Similar to fullerenes, it has been found that QDs catalyze ROS that could lead to oxidative stress (139). In vitro studies using

Table 5
Quantum dot zebrafish nanotoxicty studies and method(s) of physiochemical characterization

Nanoparticles	Size (avg) (nm)	Notes	Analysis (methods)	LC_{50}	Dosing Info.
CdSe/ZnS–PLL (42)	9 nm	ζ 3.0	ICP-OES; UV–vis;	10×10^{11}	Waterborne
-350-OCH_3 (42)	7 nm	ζ 8.0	DLS; PALS;	n.d.	4–6 hpf
-5 k-COO^- (42)	14 nm	ζ 2.0		7×10^{12}	24 h renewal
-5 k-OCH_3 (42)	14 nm	ζ 5.0		10×10^{12}	
-5 k-NH_3 (42)	14 nm	ζ 2.0		15×10^{12}	
G3.5 (21)	n.d.			n.d.	Waterborne
G4.0 (21)	45 nm			0.4 μM	2 hpf
G3.5-RGD (21)	n.d.			n.d.	24 h renewal
G4.0-RGD (21)	n.d.			4.1 μM	

ICP-OES inductively coupled plasma-optical emission microscopy, *UV–vis* ultraviolet–visible spectroscopy, *DLS* dynamic light scattering, *PALS* phase-analysis light scattering, *PLL* poly-l-lysine, *RGD* arginine-glycine-aspartate, *n.d.* not determined

rat hepatocyte cell lines found differential toxicity based primarily on size (140), net charge, as well as ion release and surface modification (138, 140). Secondary coatings can protect against degradation and improve biocompatibility but the method of functionalization, e.g. electrostatics adsorption, multivalent chelation, and covalent bonding, etc., can change durability, stability, and bioreactivity (137). In vitro studies have suggested that size is the primary mechanism regardless of coating with QDs since smaller carboxyl-QDS were more toxic (141) potentially by being more bioavailable and having less powerful internal metallic bonds for QD stability. A study with rainbow trout found that QDs with different cores (CdS, CdTe or $CdSO_4$) reduce immune-competence but differential patterns of gene expression depending on composition (142). The QD has already been highly utilized within the zebrafish model to track development (143–146) and host interaction with viruses (147). Additionally, QDs have been shown to bond with viruses, and toxicological impact of this is still unknown (147). QDs have been shown to aggregate in vivo (145–147) and are toxic after the number of quantum dots exceeds 10^8 (144). A summary of contemporary zebrafish nanotoxicology studies with QDs can be found in (Table 5).

5.3.2. Zebrafish-Specific Study

King-Heiden et al. in 2009 demonstrated that different modified $CdSe_{core}/ZnS_{shell}$ QDs had differential toxicity. QDs functionalized with poly(l-lysine) (PLL), PEG_{5000}-NH_3, PEG_{5000}-COO^-, and PEG_{5000}-OCH_3 had 120 hpd (hours post dosing) LC_{50}'s of 7, 21, 28, and 42 μM, respectively, all considerably lower than $CdCl_2$ LC_{50} of 409 μM. Positively charged PLL-QDs and polyethylene glycol$_{350}$-OCH_3 aggregated quickly in suspension media (egg water) and were 42 and 85 times larger than their predicted diameter size of 9 and 7 nm, respectively. PEG_{350}-OCH_3 did not induce toxicity, most likely due to the lack of bioavailability. Negative zeta potential QDs did not aggregate, but were slightly larger than their predicted size. PLL-QD's higher mortality can largely be explained by PLL's nature as being lethally toxic itself, one thought being plasma membrane destabilization due to excess cationic charge. Both PLL-QDs and PLL exposures produced no significant sublethal effects. Other QD species induced the following sublethal effects: altered axial curvature, pericardial edema, ocular edema, submandibular edema, all similar to fish exposed to Cd ions, but also induced the following unique sublethal effects such as yolk sac edema, yolk sac, and tail malformations and opaque tissues. PEG_{5000}-OCH_3 had significantly less severe phenotypic malformations, since they lacked tail malformations and had less severe yolk malformations, necrosis, and opaque bodies. An additional difference between $CdCl_2$ and the QDs is the differential dose-dependent expression of the metallothionein (MT) genes. MT genes are induced by exposure to metals and influence the transportation and storage of metals within organisms. This appears to suggest that QDs are degraded in vivo, but surprisingly most of the species of QDs released less than 5% of the potential Cd, although (PLL) and PEG_{5000}-NH_3 released as much as 19% and 14% of their total Cd, respectively. Since QDs contain other heavy metals besides Cd, viz. Se and Zn—these metallic toxicities were also evaluated. Though Se was found to be toxic, the amount released in vivo or in solution would not be enough to induce its toxic effects. Zn appeared to be largely nontoxic (42).

5.3.3. Extrapolations and Notes

With only one study, it is difficult to understand all the different mechanisms that could contribute to QD toxicity, though negatively charged $CdSe_{core}/ZnS_{shell}$ QDs seem to be acutely toxic to zebrafish (42). Contrary to an in vitro study (138), there was no direct correlation between net surface charge and toxicity (42). Ion release could be another mechanism for inducing toxic responses, since other studies have found that QDs readily decompose in vivo and in the environment (137, 148), but almost all of the QDs remained relatively intact in the dosing medium and the zebrafish themselves (42). It should be clearly noted that the degree of degradation is surely a function of the surface coating and net charge (138). Surface coating appears to be the main mechanism of toxicity in zebrafish (42) and in vitro cell lines (138), although the differences between the surface coatings and their mode of inducing toxicity are largely unelucidated. Sadly, while the zebrafish study found Cd ions within the zebrafish embryos, it did not explore QD biodistribution, excretion, or bioavailability. Future studies should extrapolate on core composition and secondary coating since only $CdSe_{core}/ZnS_{shell}$ were used (42).

Table 6
Heavy metal ENMs zebrafish nanotoxicty studies and method(s) of physiochemical characterization

Nanoparticle	Size (avg)	Notes	Analysis (methods)	LC_{50}	Dosing Info.
Ni (44)	30 nm 60 nm 100 nm 540 nm	Quasi-spherical Dendritic Dechorionated	ICP-MS; FE-SEM; DLS	328 mg/L 361 mg/L 221 mg/L 115 mg/L	Waterborne 24 hpf
Ni (154) Co (154) Al (154)	6.1 nm 10.5 nm 41.7 nm	ζ 21.9 ζ 17.8 ζ 18.2	SEM; Zeta-Reader; DLS; ICP-MS	>10 mg/L >10 mg/L >10 mg/L	Waterborne <24 hpf for 48 h

ICP-MS inductively coupled plasma-mass spectroscopy, *FE-SEM* field emission-scanning electron microscopy, *DLS* dynamic light scattering, *SEM* scanning electron microscopy

5.4. Heavy Metals

5.4.1. Description

There are surprisingly few reports of toxicity of heavy metal nanoparticles and zebrafish. Currently, heavy metal nanoparticle composites, i.e. uranyl aluminate, have been manufactured (149, 150) and are being used in both synthesis and creation of other nanoparticles, such as quantum dots (141), in drug delivery (151), and groundwater remediation (152, 153). For example, Co has become a more widely used heavy metal, and though zebrafish studies have been conducted (154), researchers have not reported the effects of different shapes, surfactants, or magnetic properties, all of which can be easily controlled due to new advancements in material synthesis (155, 156). Problematically, studies have already found that nanoparticles such as uraninite (157) and other trace metals (158) are in the air and groundwater (159). Additionally, heavy metal ions may interact with silver nanoparticles (160) and potentially others (161) to create new conjugated species and/or change characteristics, i.e. increasing solubility (115, 161). Pertinent to zebrafish bioavailability is that heavy metals can leach from their nanoparticles, in vivo and in situ, and can switch from a solid state to a soluble form under normal pH conditions of typical surface waters and rainfall (115). CdTe nanoparticles have been shown to be toxic to mice and HepG2 cell lines (162). As mentioned previously, QDs regularly release their heavy metal cores and shells resulting in acute toxicity (42, 148). A summary of contemporary zebrafish nanotoxicology studies with heavy metal nanoparticles can be found in Table 6.

5.4.2. Zebrafish Specific Studies

Heavy Metal Nanoparticle Zebrafish Studies

Ispas et al. in 2009 compared sizes and shapes of different nickel nanoparticles (NiNPs) and their associated toxic effects. Zeta potential, ionic strength, and pH were all adjusted to be as close as possible in order to attribute toxic responses to a particular size or shape. Spherical NiNPs of different diameters (30, 60, and 100 nm) resulted in similar toxic responses in embryonic zebrafish, LD_{50} 328, 361, and 221 mg/L, respectively, and less toxic than soluble nickel's LD_{50} 221 mg/L. However, aggregated clusters of 60 nm NiNPs resulted in higher toxicity (LD_{50} 115 mg/L); 100 nm spherical NiNPs were less soluble (15.6 mg/L) than their 30 and 60 nm sphere and 60 nm dendrite (23 mg/L) counterparts. Interestingly, at pH > 8.5, 30 and 60 nm spherical nanoparticles switch from a positive to a negative surface charge possibly by formation of a nickel oxide outer layer. NiNP's sublethal effects were damage to skeletal fibers and the digestive tract dissimilar to soluble Ni's effects on the digestive tract (44).

Griffit et al. in 2008 found that the positively charged NiNPs were nontoxic to zebrafish embryos at concentrations up to 10 mg/L (6.1 nm; aggregated 446.1 nm). Positively charged, aggregated CoNPs are nontoxic to zebrafish at concentrations up to 10 mg/L (10.5 nm; aggregated 742 nm) (154).

Heavy Metal Ion Zebrafish Studies

Several heavy metal ion zebrafish-specific studies have been included in this review to demonstrate the potential toxic threat that their nanoparticle counterparts may have. Cd ions activate the stress gene heat shock protein 70 (HSP70) and are entirely lethal at 125 μM (163) arsenic (164) and arsenate (165) at concentrations of 2 ppb and 56 mg/L, respectively, have been shown to be toxic to zebrafish by decreasing the ability to remove viral loads and induced sensitivity to other toxicants. Hg induces hepatotoxicity and may cause metabolic disorders (166). ^{233}Uranium and depleted uranium in doses close to those found in the environment (20 μg/L) cause lethal and sublethal effects in zebrafish embryos and the isotopes have different bioavailabilities (167). Another uranium study found that different dosages of uranium caused different activation of genes, but interestingly lower doses of uranium caused more gene expression (54).

5.4.3. Extrapolations and Notes

The lack of studies concerning heavy metals is disconcerting since many heavy metal ions are toxic to zebrafish and are highly prevalent in the environment (54, 163–167). When additional studies are commissioned, care should be taken to insure ample characterization of physicochemical characteristics, since one study has shown that size and shape induces differential toxicity (44).

5.5. Gold

5.5.1. Description

Gold nanoparticles (AuNPs) are currently the most reported and well-characterized nanosystem in the literature (168). They have a wide range of uses for drug delivery vehicles (169, 170), gene therapy (171), biosensing (172), and cell imaging (173) due to their relative ease of characterization in in vivo systems. AuNPs also provide facile modification of their physicochemical characteristics such as size, shape, surface chemistry, monodispersity, and charge (107, 174). Gold's wide uses and ease of modification have led to a slew of studies that have identified size (175), shape (107), surfactants (176), and charge (177) as all influential to inducing toxicity. Though generally not cytotoxic due to their bio-inert properties (178, 179), AuNPs enter via the receptor-mediated endocytosis pathway, and there may be differing mechanisms for nonspecific adsorption of serum proteins that are size- and shape-dependent (107, 180). Once within cells or an organism, AuNPs have been shown to aggregate (179). Surface-modified AuNPs with a cationic charge have been shown to induce toxicity (177, 181) since the positive charge may disrupt the negative cell membrane by exploiting preexisting defects (177). Shape also induces differential effects, as worm-like AuNPs are more toxic, but taken up less than their spherical counterparts (107). Surfactant stabilization of worm-like AuNPs have utilized CTAB (cetyl trimethylammonium bromide), which is a well-known toxic agent indicative of the toxicity reported in human HeCat keratinocyte exposures (182, 183). Macromolecules can readily interact with AuNPs in the environment and change their solubility and aggregation state (184). Importantly, endotoxins have been shown to effectively bind to AuNPs during their synthesis and AuNPs augment the damage endotoxins cause (185), potentially by acting as a carrier in uptake pathways. High pH has been shown to aggregate AuNPs (186), potentially by weakening surface charge or allowing AuNPs to overcome their electrostatic repulsions. Within zebrafish, AuNPs are being used to create plasmonic bubbles, which have novel uses in noninvasive imaging (187, 188). A summary of contemporary zebrafish nanotoxicology studies with AuNPs can be found in (Table 7).

5.5.2. Zebrafish-Specific Studies

Bar-Ilan et al. in 2009 found that inert 3, 10, 50, and 100 nm non-aggregated spherical sized AuNPs induced less than 3% mortality and no sublethal effects at any concentration from 0.25 to 250 μM. AuNPs remained monodispersed for 24 h in eggwater, but aggregated after 5 days. To determine whether there was a difference in toxicity due to functionalization, triphenylphosphine monosulfonate (TPPMS) was functionalized onto gold nanoparticles, but did not differ in lethal or sublethal effects from their unfunctionalized counterparts. High concentrations of TPPMS (625 μg/mL) caused phenotypic malformations, although only 62.5 μg/mL was used to functionalize the AuNPs. Instrumental neutron activation analysis found that AuNPs did bioaccumulate in waterborne exposed zebrafish (47).

Browning et al. in 2009 found inert non-aggregated quasi-spherical AuNPs (average size: 11.6 nm) induced mortality, though not significantly different from water and supernatant controls; 2% of embryos treated with AuNPs had deformities at concentrations ranging from 0.025 to 1.2 nM, while no control fish had phenotypic changes. AuNPs were shown to diffuse with passive Brownian

Table 7
Precious metal ENMs zebrafish nanotoxicty studies and method(s) of physiochemical characterization

Nanoparticles	Size (avg)	Notes	Analysis (methods)	LC_{50}	Dosing Info.
Au (47) Ag (47)	3, 10, 50, 100 nm 3 nm 10 nm 50 nm 100 nm	Spherical	TEM; UV–vis; ICP-MS; INAA	n.d. 93.31 μM 125.66 μM 126.96 μM 137.26 μM	Waterborne 0–2 hpf 24 h renewal
Au(19)	11.6 nm	Spherical	DFOMS; DLA; UV–vis; HR-TEM	n.d.	Waterborne 0.75–2.25 hpf
MEE-Au (132) MES-Au (132) TMAT-Au (132)	0.8 nm 1.5 nm 0.8 nm 1.5 nm 0.8 nm 1.5 nm	Dechorionated		>250 ppm >250 ppm n.d. n.d. 1.8 ppm 30 ppm	Waterborne 8 hpf 24 h renewal
Ag (41)	5–20 nm	Spherical	TEM; Western Blot; TUNEL; RT-PCR; Protein assay; ICP-MS	250 mg/L	Waterborne
Ag-BSA (193) Ag-Potato Starch (193)	5–20 nm 5–20 nm	Roundish	TEM; UV–vis; Acridine Orange; DAPI staining	25–50 μg/L 25–50 μg/L	Waterborne
Ag (31)	11.6 nm	Spherical	SNOMS; DLS; HR-TEM	0.19 nM	Waterborne 0.75 hpf
Ag (154) Cu (154)	26.6 nm 26.7 nm	Spherical: ζ 27.0 ζ 0.69	SEM; Zeta-Reader; DLS; ICP-MS	7.07 mg/L 0.94 mg/L	Waterborne 0, 2, 24, and 48 h renewal

(continued)

Table 7 (continued)

Nanoparticles	Size (avg)	Notes	Analysis (methods)	LC_{50}	Dosing Info.
Ag (38) Cu (38)	26.6 nm 26.7 nm	Spherical: ζ 27.0 ζ 0.69	SEM; Zeta-Reader; CP-MS; Agilent 1 × 22 k zebrafish array	n.d. n.d.	Waterborne
Cu (37)	80 nm	Atypical Shapes	TEM; DLS; CPS	1.5 mg/L	Waterborne

TEM transmission electron microscopy, *UV–vis* ultraviolet–visible spectroscopy, *ICP-MS* inductively coupled plasma-mass spectroscopy, *INAA* instrumental neutron activation analysis, *DFOMS* dark-field optical microscopy, *DLS* dynamic light scattering, *HR-TEM* high resolution-transmission electron microscopy, *TUNEL* terminal deoxynucleotidyl transferase dUTP nick end labeling, *BSA* bovine serum albumin, *RT-PCR* real time-polymerase chain reaction, *SEM* scanning electron microscopy, *DAPI* 4′,6-diamidino-2-phenylindole, *SNOMS* near-field scanning optical microscopy, *SEM* scanning electron microscopy, *CPS* centrifugal particle sizing, *MEE* 2-(2-mercaptoethoxy)ethanol, *TMAT* *N,N,N*-trimethylammoniumethanethiol, *MES* 2-mercaptoethanesulfonate, *WB* Waterborne, ***n.d.*** not determined diffusion into the embryo via chorion pore canals, with higher concentrations being indicative of more AuNPs in the embryo (diffusion coefficient (*D*) range 2.8×10^{-11}–1.3×10^{-8} cm^2/s based on location in embryo). Larger AuNPs (avg size: 15 nm) had a lower $D = 1.3 \pm 1.0 \times 10^{-8}$ cm^2/s than smaller AuNPs (avg size: 9 nm, $D = 2.7 \pm 2.5 \times 10^{-8}$ cm^2/s). In murine tissue, smaller Au particles (10–50 nm) were more readily distributed than larger ones (189). The random stochastic walk of the AuNPs may lead to higher areas of AuNP accumulation in some embryos and induce sublethal effects within those areas explaining the seemingly random nature. Eggwater and the embryo interior exhibited the same LSPR (localized surface plasmon resonance) wavelength peaks (565 nm) suggesting little to no aggregation in the embryo. AuNPs stayed inside the embryo for the entirety of development (19).

Harper et al. in 2008 functionalized spherical AuNPs (sizes 0.8 and 1.5 nm) with: 2-(2-mercaptoethoxy)ethanol (MEE) to induce a neutral surface charge, *N*,*N*,*N*-trimethylammoniumethanethiol (TMAT) to induce a positive surface charge, and 2-mercaptoethanesulfonate (MES) to induce a negative surface charge. Zebrafish were then dosed from concentrations ranging from 16 ppb to 250 ppm. Positively charged AuNPs (TMAT-AuNPs) had a high mortality, negatively charged AuNPs (MES-AuNPs) had low mortality, and neutrally charged AuNPs (MEE-AuNPs) had no mortality; 1.5 nm sized AuNPs had higher toxicity in a concentration-dependent fashion, though the difference was less so than for differing surface charge (132).

5.5.3. Extrapolations and Notes

Neutrally charged AuNPs seem to cause no lethality in fish based on the culmination of the three aforementioned studies (19, 47, 132). Size distribution has conflicting evidence on whether or not it directly affects toxicity (47, 132), although the study stating that size caused differential toxicity used AuNPs differing in size by merely 7 Å (132, 175); 1.5 nm sized AuNPs were used in Harper's study, (132) while particles that are small were not used in either of the other two (19, 47). Moreover, 15 nm sized AuNPs induced no toxicity to cell lines (190) and 50 nm spheres were more readily taken up by endocytosis than smaller and larger sized ones (107). Future studies should focus on the synthesis scheme, acute attention to washing techniques, shape, and surfactants to evaluate their contributions to AuNP nanotoxicity.

5.6. Silver

5.6.1. Description

As of 2009, silver nanoparticles (AgNPs) were utilized twice as much over other nanomaterials (110). AgNPs have a wide variety of consumer and medical purposes such as antibacterials (191), burn treatments (silver sulfadiazine), sock components for odor elimination, detergents, soaps, water and air filters, washing machines, wet wipes, beddings, biomedical imaging, numerous biomedical uses, and industrial textiles (35). Alarmingly, an increasing number of studies are reporting extreme toxicity when compared to other nanomaterials (94), and the commonly utilized bactericidal properties could also be toxic to eukaryotic cells (35). Size appears to be the most important physicochemical aspect since smaller AgNPs can more readily release Ag^+ ions (192). Although the source of AgNP toxicity is still debated, it is generally believed that released Ag^+ ions cause DNA damage and instability, cell proliferation delay, and oxidative stress (193). AgNPs can remain dispersed in natural aqueous solutions due to a complex interaction between equilibrium with dissolved Ag^+ and high amounts of DOC (dissolved organic carbon) (11). It also appears as if high ionic strength, surface capping, and addition of environmentally relevant humic substances increased dispersity and stabilized them in solution making them potentially more bioavailable. Additionally, analysis of AgNPs in environmentally relevant conditions demonstrated a size discrepancy depending on the method of characterization used, i.e. DLS, TEM, etc. (194). A summary of contemporary zebrafish nanotoxicology studies with AgNPs can be found in (Table 7).

5.6.2. Zebrafish-Specific Studies

Bar-Ilan et al. in 2009 found waterborne exposed, nonaggregated spherical colloidal silver (cAg) particles capped with $NaBH_4$ and $Na_3C_6H_5O_7$ with sizes of 3, 10, 50, and 100 nm induced both sublethal and lethal effects. At 24 hpf and the highest concentration of 250 μM, smaller particles were more toxic than larger particles in a size linear fashion, though by 120 hpf all fish in the experiment were dead. The study showed cAg size-dependent toxicity is only for certain concentrations and time points (47). Size and surface area did not appear to be necessarily toxic indicators as suggested by other reports (36) since the same study compared cAg to cAu and found that cAu of approximately the same size and surface area was not toxic (47). At concentrations of 25 μM and lower, there was low incidence of lethal and sublethal effects. At the approximate LC_{50} (100 μM), numerous sublethal effects, such as opaqueness, nondepleted yolk, small head, jaw and snout malformations, stunted growth, circulatory malformations, tail malformations, body degradation, decaying tail issue, pericardial edema, and bent spine were observed to be induced by cAg at all sizes; 3 nm and 10 nm cAg differed from cAg sized 50 and 100 nm by inducing tail malformations, decaying tail tissue, pericardial edema, and bent spines. Instrumental neutron activation assays demonstrated that dosed fish readily took up silver, but could not differentiate between Ag^+ ions and AgNPs. Capping agents used to suspend cAg in solution induced no specific mortality or sublethal effects, suggest Ag^+ ions were the primary source of toxicity. $AgNO_3$ was used to test correlations between Ag^+ ions and cAg toxicity. $AgNO_3$ concentrations of 1 and 10 μM induced 100% toxicity in developing embryos, with one lone malformed survivor at 1 μM. In order to verify that toxic effects could be attributed to AgNPs instead of supernatants left from washing, fish were dosed with washings from each concentration. Only the washings from cAg10 induced similar sublethal effects as the corresponding concentration of AgNPs, while the rest of the washings ICP-MS found Ag^+ ions present in the first washing remnants with 3 nm cAg washings having significantly more Ag^+ ions (131 ppm vs. 17.2 ppm for the next highest, cAg10). This suggests that Ag^+ ions are more readily released from smaller cAg particles due to their higher surface-area-to-volume ratio (47). Additionally, since smaller sized AgNPs can more readily diffuse into zebrafish embryos (31), then perhaps more toxic Ag^+ ions are released within the embryo.

Griffit et al. in 2008 demonstrated that negatively charged aggregated AgNPs toxic response differed from dissolved Ag^+ ($AgNO_3$) in the gills of zebrafish. Interestingly, SEM showed more silver taken up by fish exposed to AgNPs than $AgNO_3$. AgNPs did not cause a morphological thickening of the gills as $AgNO_3$ did. Unfortunately, SEM cannot differentiate between the two species of silver, thus being unable to determine whether AgNPs released Ag^+ ions (38).

Choi et al. in 2009 found AgNP toxicity in zebrafish is not due to simply release of Ag^+ ions. Their findings with mono-dispersed spherical AgNPs sized 5–20 nm had an LC_{50} 250 mg/L. ICP-MS and UV–vis were used to verify that the deionized solutions contained no Ag^+ ions. TEM found that increasing concentrations accumulated more agglomerated AgNPs in liver tissue. A TUNEL assay found that AgNP induced apoptosis and was different from both eggwater and $AgNO_3$ dosed controls. Gene regulation analysis found MT2 mRNA, responsive to Ag^+ ions, was more up-regulated as dose increased suggesting that free Ag^+ ions were created in vivo. Additional gene regulation analysis of hepatic MDA (malondialdehyde) and GSH (glutathione) found that AgNPs cause oxidative damage in the liver, and western blot found that γ-H2AX (rabbit antibody), commonly found at the sites of double-strand breaks, was detected in liver cells exposed to AgNPs (41).

Lee et al. in 2007 found spherical nonaggregated AgNPs with an average size of 11.6 nm caused dose-dependent acute toxicity at concentrations above 0.19 nM. Supernatant fluids from washings did not induce mortality or morphological changes, suggesting that AgNPs are solely responsible for toxicity. AgNPs readily enter the chorion and embryo space through simple Brownian diffusion (D 3×10^{-9} cm^2/s) and through the chorion pore canals (0.5–0.7 μm in diameter). AgNPs spend a variable amount of time within the chorion pore canal ranging from 0.1 to 15 s suggesting adherence to the side of the canal. The majority of the AgNPs remained nonaggregated within the embryo space, as there was no shift in the eggwater LSPR (localized surface plasmon resonance) peak wavelength of 488 nm inside the embryo. AgNPs appeared to induce sublethal effects randomly at various concentrations, suggesting that the random diffusion of particles stochastically is the primary influence of toxicity (31). This would explain the aforementioned study that found differences in sublethal effects based on size (47), and may indicate that smaller sized particles have access and accumulate more readily in certain developing areas that larger ones do not. Different concentrations created different sublethal effects, i.e. from 0.04 to 0.71 nM finfold malformations were more common, and at 0.66–0.71 nM eye deformalities were observed. This seems to suggest that differences can be attributed to the concentration gradient of particles inside the embryo, and/or accumulated AgNPs can alter charge, diffusion or interaction of biomolecules, and/or the number of chorion canals blocked by aggregating nanoparticles. Interestingly, they found that AgNPs did not aggregate at standard eggwater with salt concentrations of 10 mM, but did begin to aggregate at 100 mM (31).

Asharani et al. in 2008 found organically capped AgNPs with either BSA or potato starch (used to suspend AgNPs in solution) did not induce linear mortality (LC_{50} 25–50 μg/mL). Increasing concentrations did, however, have an increase in hatching delay. They also found that later stage embryos were more resilient to AgNP treatment. TEM determined these AgNPs were quasi-spherical, 5–20 nm in size, were deposited on the inside of the nucleus of the cells, as well as on the skin and heart, and had aggregated within the body. UV–vis determined the particles did not aggregate in solution since it did not change from a peak wavelength of 424 nm after 5 days. Brown flakes were noticed on the inside of the chorion suggesting that the particles had diffused through it, and analysis with 4′,6-diamidino-2-phenylindole hydrate staining (DAPI) found AgNP-treated embryos differed from controls. AO staining found that both capped AgNPs induced apoptosis and necrosis. Surprisingly Ag^+ ions at the highest concentration of 20 nM did not have a similar mortality rate as AgNPs and did not induce hatching or sublethal effects (193).

Griffitt et al. in 2008 found negatively charged spherical AgNPs averaging in size of 26.6 nm, both aggregated in solution, and were acutely toxic to zebrafish adults and juveniles (48 h LC_{50} 7.07–7.20 mg/L). The LC_{50}'s differed from soluble silver ($AgNO_3$) with LC_{50} 0.022 and >10 mg/L, respectively (154).

5.6.3. Extrapolations and Notes

AgNPs are shown by every study aforementioned to be acutely toxic and to induce sublethal effects (31, 38, 41, 47, 154, 193). The prominent hypothesis for AgNP toxicity in zebrafish is the release of destructive Ag^+ ions in vivo, though six studies had different results in toxicity between AgNPs and $AgNO_3$ implicating that Ag^+ ions are not solely responsible for observed toxic responses (31, 38, 41, 47, 154, 193). Potentially, there could be a synergistic effect between the AgNPs and their released Ag^+ ions that culminates in the damage seen to embryos. AgNPs damage cell membranes, which lead to disruption in the ion efflux system, disabling the cell's ability to remove Ag^+ ions causing Ag^+'s well-documented DNA and ROS damage (195). Ag^+ ion dosed zebrafish have an LC_{50} 3 μM and at lower concentrations, hatching delays and similar black specs on the chorion are observed (196). Adding biocompatible surfactants, BSA, and potato starch had no effect on decreasing mortality (193). Size appears to be the primary physicochemical indicator of toxicity since smaller sized AgNP's toxicity is directly related to their ability to release more Ag^+ ions (195, 197, 198). A recent study found that AgNPs are nontoxic without Ag^+ ions being present in *Dapnia magna* (199). Future studies could focus on how AgNPs release Ag^+ ions in vivo for better understanding of silver's potent toxic abilities, as well as to evaluate other physicochemical characteristics' effect on toxicity.

5.7. Copper

5.7.1. Description

Copper nanoparticle (CuNP) toxicity is relatively unexplored, but is already widely used as a bactericide, air and liquid filtration, coatings on integrated circuits and batteries, and to increase thermal and electrical conductivity in coatings and sealants (200). CuNPs are highly reactive and produce large quantities of Cu ions in vivo in mice (201). CuNPs exhibit dose-dependent toxicity in human alveolar and macrophage cell lines (94) and dose- and size-dependent toxicity to rat neurons in vitro (202). Soluble copper is highly toxic to fish (203). Colloidal copper suspensions become increasingly toxic and aggregated with increasing amounts of dissolved organic carbon (DOC) (11). A summary of contemporary zebrafish nanotoxicology studies with CuNPs can be found in Table 7.

5.7.2. Zebrafish-Specific Studies

Again, Griffitt et al. in 2007 found negatively charged, jagged, unaggregated, and aggregated CuNPs caused toxicity in a dose-dependent linear fashion with 48 hpd (hours post dosing) LC_{50} 1.56 mg/L (size range: 18.8–450 nm; avg size of 50.2 nm). CuNPs were less toxic than soluble copper (dissolved $CuSO_4$, LC_{50} 0.25 mg/L). CuNPs release Cu ions in dosing media at all concentrations, with 0.09 mg/L released at the highest concentration representing 6% of the total CuNP weight. Based on the amount of Cu release only 16% of the mortality could be attributed to released Cu ions. Approximately 60% of the Cu from both CuNPs and sCu (soluble copper) was taken up by the fish, accumulating preferentially in the gills. Both species of copper caused damage to gill lamellae and proliferation of epithelial cells, though there was no significant evidence of injury to other major internal organs. Na^+/K^+-ATPase activity in the gills was decreased by 58% when exposed to the highest dose of CuNPs, which did not occur in fish exposed to sCu. In the gills, QPCR analysis found metal responsive genes, HIF-1, HSP70, and CTR, were significantly up-regulated when compared to sCu and control exposures. Up-regulation of those genes in the liver was found in both sCu and CuNP dosed fish and both differed significantly from the control (37).

Griffitt et al. in 2008 found a 48 hpd LC_{50} 1.56 mg/L in developing zebrafish embryos. The average size for the particles was 26.7 nm, but they still aggregated to sizes of 447.1 nm in media. Different organisms were dosed with CuNPs and the amount of material lost from the water column varied with species (154).

Griffitt et al. in 2009 found there was not a linear dose-dependent toxic response to CuNPs in adult zebrafish. CuNP dosed fish had significantly more thickening of the gill filament than sCu, potentially implicating a synergistic effect between CuNPs and their released Cu ions (204).

5.7.3. Extrapolations and Notes

CuNPs appear to selectively cause gill damage with little damage to other organs. Based off the zebrafish studies, size remains the only indicator of toxic response, as the smaller size in the second study (154) had a lower LC_{50} than the first study (37); nonetheless, this correlation cannot be assumed. Future studies should also explore the effects of modifying CuNP's physicochemical properties to determine the exact mechanism of toxicity and if there is a synergistic effect between CuNPs and their released Cu ions.

5.8. Platinum

To our knowledge, platinum nanoparticles, used in synthesis and biomedical products(205) have not been evaluated in the zebrafish model. Presumably, PtNPs represent a low threat since they have a small cytotoxic potential in human endothelial and lung cells and only induced a mild inflammatory response in rats (5).

5.9. Metal Oxides

5.9.1. Description

Metal oxides are a highly common ENM and manufacturing by-product (110) and their vast uses range from construction materials for flame proofing, paint, structural reinforcement, semiconductors, insulator, and antifogging (79), to cosmetics and sunscreen, to bactericides and biomedical uses such as fighting cancer or anthrax (15, 206). Their wide uses stem from the facile modification and control of physicochemical properties (207). MeOs (metal oxides) have been shown to have the following toxicological impacts, acute lethality, growth inhibition, apoptosis, DNA damage, lipid peroxidation, and inflammatory and immune responses (79). These modifiable characteristics may lead to differential toxicity based on charge (208), shape (209), size (210), or degree of dispersion (211). One study using bacteria found that semiconductor nano-oxides are more toxic than insulators in bulk (indicative of the quantum effect), and that different dopants can alter toxicity was well (208). Chemically, this may be due to a lower valence state being more capable of disrupting cell membranes (45). There is no correlation between surface area and size and MeO toxicity (94). Most MeOs regularly aggregate in media (15, 94);

however, several studies have found that several MeO aggregates are toxic (85, 94), and their difference in toxicity may be due to the degree that MeOs are monodisperse or polydisperse in media (82). Partially soluble particles, like ZnO are easier to disperse in media, but may be more toxic, by acting as a Trojan Horse to be unintentionally taken up by cells and disrupting cell function (212). Several factors can help disperse MeOs in media, such as BSA surfactants (88) or adsorbed dissolved organic carbon (213). Additionally, one study found that MeOs, even in aggregated form, are difficult to remove from ground water, as they regularly adhere to bacteria and sludge (214), therefore presenting a route for uptake by aquatic animals. A summary of contemporary zebrafish nanotoxicology studies with MeONPs can be found in Table 8.

5.9.2. ZnO

Description

Zinc oxide nanoparticles (ZnONPs) are common additives with a variety of applications and their ceramic nature allows them to function as a pigment and a semiconductor (15). ZnO was the most toxic of the metal oxide particles tested on bacteria (15, 45) facilitating its use as an antibacterial (215). ZnONPs have been speculated to create reactive oxygen species (15) and cause low amounts of DNA damage, DNA lesions, and nonviable cells in bacteria (211). Some studies suggested that increasing size might be the key toxic indicator after its semiconducting properties (208, 216). However, bioavailability may preclude ZnO from being dangerous to zebrafish since unmodified ZnO nanoparticles, though partially soluble in water, were not regularly taken up by *Oncorhynchus mykiss*, the rainbow trout, via gills, skin, or the digestive tract (36).

Zebrafish-Specific Studies

Zhu et al. in 2007 found aggregates of irregular shaped ZnONPs (size distribution of 50–360 nm) and mixed spherical and elliptical bulk ZnO (size distribution of 100–690 nm) were acutely toxic in a dose-dependent manner, but not significantly different from each other (LC_{50} at 96 h of 1.79 and 1.55 mg/L, respectively). Alarmingly, the manufacturer's published sizes for both ZnO species were significantly different from actual sizes as verified by DLS, TEM, and optical microscopy (43), and this was not the first incidence of this occurring (15). A zebrafish hatching delay was observed for both species of ZnO, though ZnONPs had tissue ulceration as the only sublethal effect. The similarity between the mortality rates and hatching delay suggests that Zn ions released may be the cause of toxicity as 10 mg/L ZnONPs released 2.67 ± 0.46 mg/L Zn ions. Zebrafish embryos dosed with 10 mg/L suspensions with ZnONPs removed using a 100 nm filter had a survival rate of 86.7%, which differed from the unfiltered ZnONP solution and the water control (43).

Zhu et al. in 2009 found that floccule-shaped ZnONP aggregates with a size distribution of 1,037–6,823 nm had an 84 h LC_{50}

Table 8
Metal oxide ENMs zebrafish nanotoxicty studies and method(s) of physiochemical characterization

Nanoparticle	Size (avg)	Notes	Analysis (methods)	LC_{50}	Dosing Info.
ZnO (43)	180 nm	Irregular shape		1.793 mg/L	
TiO_2 (43)	230 nm	Irregular shape	TEM; DLS	n.d.	Waterborne 1.5 hpf
Al_2O_3 (43)	930 nm	Potato-shape		n.d.	
ZnO (217)	2,196 nm	Floccules	SEM: DLS; GFAA	23.06 mg/L	Waterborne 1.5 hpf
CeO_2 (33)	14 nm (441 nm)[a] 20 nm (543 nm)[a] 28 nm (464 nm)[a]	Face-centered cubic crystal	SEM: NTA; Zetasizer 3000HSA;	>200 mg/L >200 mg/L >200 mg/L	Waterborne early stage
SiO_2 (14)	60 nm 200 nm	Spherical; labeled with: $[Ru(bpy)_3]Cl_2$	FCM; TEM; AFM; FTIR-ATR; FE-SEM;	n.d. n.d.	Waterborne 6 hpf
SiO_2 nanowire (235) Spheres (235)	55 nm × 2.1 μm 60 nm 200 nm	Wire functionalized with FITC Sphere functionalized with FITC, amine, and rhodamine	SEM; FCM; transgenic cell line for *shh*	110 pg/g embryo 20 ng/g embryo n.d.	Microinjection 0 hpf and 36 hpf
TiO_2 (223)	14.10 nm		HR-TEM; FTIR-S	30 ppm	Waterborne for 14 days

TEM transmission electron microscopy, *DLS* dynamic light scattering, *GFAA* graphite furnace atomic absorption, *SEM* scanning electron microscopy, *NTA* nanoparticle tracking analysis, *FCM* flow cytometry, *AFM* atomic force microscopy, *FTIR-ATR* Fourier transform infrared transmission spectroscopy-attenuated total reflectance, *FE-SEM* field emission-scanning electron microscopy, *shh* sonic hedgehodge, *HR-TEM* high resolution-transmission electron microscopy, *FTIR-S* fourier transform infrared transmission-spectroscopy $[Ru(bpy)_3]Cl_2$ tris(bipyridine)ruthenium(II) chloride, *FITC* fluorescein isothiocyanate, *n.d.* not determined

[a]Aggregate size in media

of 23.06 mg/L (217). Additionally, the hatching rate and pericardial edema both increased with increasing concentration. By 3 h 80% of the suspension had fallen to the bottom of the testing apparatus. These larger aggregates released only 1.01 mg of Zn ions. Flow cytometry evaluating a fluorogenic ROS indicator, H_2DCF-DA (2′,7′-dichlorodihydrofluorescein diacetate) found that there was ROS damage at treatment concentrations of 1–10 mg/L, increased with the dose, but was different from Zn ion's evaluated with the same test (217).

Extrapolations and Notes

Despite its potent acute toxicity, little has been done to elucidate the mechanism(s) for inducing mortality. Nano-ZnO (nZnO) aggregates could create ROS and/or compromise cellular antioxidant defense systems to induce toxicity. Size distribution may be the only plausible hypothesis, since smaller ZnONPs released more Zn ions (43, 217), but other mechanisms may be at work. Future studies could evaluate several variables of physicochemical characteristics to better understand ZnO's toxic capabilities.

5.9.3. TiO_2

Description

TiO_2 is currently used in an enormous number of consumer products and remains one of the fastest growing nanomaterials to date. Its uses range from construction materials, self-cleaning to antifogging properties (79), cosmetics and sunscreens as an opacifier and pigment (15, 211), and the biomedical field as a bactericide (79), or as an agent to fight cancer and anthrax (206). Compared to ZnO and other metal oxides, TiO_2 is the least toxic metal oxide but has shown potent cell toxicity in some cases (45). Its effects vary based on what cell line or tissue type to which it was exposed (85). It differs from its bulk counterpart by being internalized by mast cells and interfering with cell function (23), though different sizes do not appear to induce different toxic responses (218, 219). Purity (94) and shape (218) are potential indicators of toxic responses as stoichiometric TiO_2 is more toxic than non-stoichiometric TiO_2 (LC_{50} 432 μg/ml vs. 845.2 μg/L)(94). Pure anatase TiO_2 is much more toxic than its mixed, or rutile counterparts, perhaps by being more catalytically active (220, 221). Shape appears to influence toxicity as macrophage death occurred more frequently with dendritic TiO_2 particles than spherical or spindle morphologies (218, 222). In water suspensions, TiO_2 is more cytotoxic to bacteria species in the presence of light, suggesting ROS formation (15). One study found that $nTiO_2$ aggregates were taken up in the rainbow trout fish guts, but not in the skin, gill, liver, or blood (36).

Specific Zebrafish Studies

Zhu et al. in 2007 found amorphous blobs of $nTiO_2$ were acutely nontoxic to zebrafish (purity: 99.5% anatase) (16). Like the nZnO mentioned above, TiO_2 manufacturers' published size ranges of 100–500 nm were inaccurate with actual particle sizes <100 nm.

Two other studies using zebrafish, but different $nTiO_2$, found that although $nTiO_2$ readily aggregated (size range: 220.8–687.5 nm and approximately 10–900 nm, respectively) in eggwater media, produced no lethal or sublethal effects (purity: 20% rutile, 80% anatase; not reported for second) (38, 132). The former of the two studies found there was a clear effect on transcription patterns at 48 h, since genes controlling ribosomal structure and activity were down-regulated which could potentially be a source of long-term toxicity (38).

Palaniappan et al. in 2010 found sonicated $nTiO_2$ particles (average size of 14.10 ± 1.52 nm) had an LC_{50} 30 ppm which differed from bulk TiO_2's LC_{50} 100 ppm (223). The difference in toxicity between bulk TiO_2 and $nTiO_2$ may be due to size difference, since smaller NPs are likely to be more toxic due to their larger specific surface area since they would have greater bioavailability (15). Curve-fitting analysis on the amide I band in gill tissues found an increase in alpha-helical structure and a decrease in normal beta-sheet structure suggesting that in zebrafish (223), like rainbow trout (36), the gills regularly take up $nTiO_2$ that cause micro-phenotypic changes (223). The suggested mechanism for the changes in proteins is TiO_2 species regularly oxidize proteins resulting in added carbonyls (224) on select amino acids. Some of these carbonyls will reside adjacent to amines (225) explaining the rise in absorption peak in the FTIR (Fourier transform infrared) spectra for amide I (at 1,652 cm^{-1}: integrated area: 23.61 ± 0.98, 35.56 ± 0.90, and 31.61 ± 0.82 for control, $nTiO_2$ and bulk TiO_2, respectively) (223).

Extrapolations and Notes

Aggregated TiO_2 is nontoxic to fish, most likely because it is not taken up during waterborne exposures (16, 38, 132). Palaniappan et al. in 2010 verified the claim that 99.7% pure anatase TiO_2 is toxic to fish, but no studies have presented data on the toxic effects of nanoparticle shape in zebrafish. Further studies could attempt to verify claims by studies using bacteria that surfactants added to increase monodispersity cause greater toxicity (226).

5.9.4. SiO_2

Description

$nSiO_2$ is another common metal oxide with growing applications in both consumer products (110), and constructions materials (79), and biomedical applications such as drug delivery (170), diagnosis (227), targeting, and imaging (228, 229). Studies have indicated that surface area (230), size (231), and surface functionalization (7) may contribute to differential toxicity. $nSiO_2$ studies have demonstrated that $nSiO_2$'s unique surface chemistry and surface area can cause hemolysis, disturb exocytotic cell function, decrease cell viability, regularly aggregate within cells (7, 23), and damage the cell membrane (177). Water suspensions of bulk- and nano-TiO_2 have been shown to be toxic to bacteria (15). One study suggested that a decrease in surface area increased toxicity in mast cells (230) though others found no correlation between SA and toxicity in the A549 cell line (208). The difference between the two cell lines could be attributed to how much SiO_2 can influence the specific cell's metabolic activity, which has been shown to influence toxicity (232). Additionally, SiO_2 may be more readily taken up due to its unique hexagonal structure that allows it to act as "proton sponge" and be taken up by clathrin-coated vesicles due to favorable receptor-mediated interactions (209). However, other reports continue to suggest the primary indicator may be either size or surface functionalization (231).

Zebrafish-Specific Studies

Fent et al. in 2010 found spherical fluorescent core-shell silica nanoparticles of 60 nm and larger were not visibly taken up by the embryo and induced no toxic response. SEM showed monodisperse particles in dosing medium. Even though the $nSiO_2$ were similar sized to other toxic nanoparticles (14, 31, 42), and smaller than the 0.5 μm chorion pore canal (31), it seems as if the zebrafish chorion can selectively inhibit fluorescent silica nanoparticles. It is possible that the 3-aminopropyltriethoxysilane or $[Ru(bpy)_3]Cl_2$ dye attached to the $nSiO_2$ was interacting with the chorion (14). Other studies with hydras found specific regulation of particle uptake, and that positive fluorescent nanocrystals coated with the same surfactants were taken up by cell membranes in acidic pH (233). Dosing medium pH 7.4 for the zebrafish study and particles were found to only have a slight positive charge and may be under the necessary parameters to be taken up by the embryo (14).

Nelson et al. in 2010 found microinjecting raw or functionalized silica nanowires (surfactants: FITC, average size of 55 nm to 2.1 μm) had an LD_{50} 110 pg/g that differed from in vitro studies, which found an LD_{50} 7.2×10^7 pg/g (234). Surprisingly, they found LC_{50} was three orders of magnitude less than silica nanoparticles with a size of 50–200 nm, which did not differ from controls (235) (LD_{50} 2.0×10^4 pg/g). Surprisingly, all SiO_2 species were nontoxic during organogenesis (36 hpf), but nanowires were incredibly toxic during neuralation (10–14 hpf) and gastrulation (6 hpf) suggesting that they are selectively teratogenic. Deformities were specific to generation of body symmetry, including holoprosencephaly (incomplete separation of forebrain into hemispheres) or anophthalmia. Using the 2.2shh: GFP zebrafish transgenic line, researchers visualized the effects of $nSiO_2$ nanowires on the expression of *sonic hedgehog* (*ssh*) and found that there was diffuse expression of *ssh* in >60% of the embryos, suggesting that $nSiO_2$ nanowires caused a defect in *shh* expression or in the formation of the *shh*-expressing structures during gastrulation. *Shh* expression differed from both water and spherical $nSiO_2$ injected embryos, both of which displayed normal development (235).

Extrapolations and Notes

The bioavailability of SiO_2 needs to be further tested, since although they induce toxicity when microinjected (235), it is unclear whether they can pass through the chorion (14). Additionally, it appears that size, surface area, or surface functionalization is not a determining factor in causing a toxic response, and appears to be much more centered on shape and aspect ratios (235).

5.9.5. Al_2O_3

$nAl_2O_{3,}$ a commonly used additive in cosmetics and consumer products, and its bulk counterpart are acutely nontoxic to zebrafish, although as mentioned above, particle size was heavily aggregated (size range: 100–550 nm) and differed from the manufacturer's published size (20 nm) (16). In contrast, other studies have shown it to be cytotoxic, even in aggregated form, most likely through the generation of ROS (45, 85, 94).

5.9.6. CeO_2

$nCeO_2$ has been found to be acutely nontoxic to zebrafish embryos (33), although this may be due to adherence to the chorion since they aggregated to an average size of 425 nm from their much smaller sizes of 14, 20, and 29 nm. However, CeO_2 may be acutely nontoxic as it has been found to be neuroprotective in zebrafish(236) and nontoxic in vitro (208). Size appears to be an indicator of uptake, though not necessarily toxicity, since strangely, particles larger than 200 nm were more readily taken up in human lung fibroblasts than smaller CeO_2 particles (210), although the smaller particles regularly aggregated in media (10% fetal calf serum) into sizes greater than 200 nm. CeO_2 has been shown to be toxic to *E. coli* as its positive charge interacts with the negative cell membrane and destabilizes it (237). Another study found $nCeO_2$ did not accumulate in relatively large quantities in rainbow trout suggesting that $nCeO_2$ bioavailability to fish is fairly low (36). Surface functionalization and surfactants in CeO_2 formulations are valid areas for exploration, as CeO_2 has been shown to adsorb proteins which in some cases facilitates easier uptake into cells (238).

5.10. Dendrimers (PAMAM)

5.10.1. Description

Poly(amidoamine) (PAMAM) dendrimers are part of the next generation in biomedical materials because of their soluble ammonia core, and multivalent, multifunctional, and highly branched surfaces, which allow for selective manipulation to attach several targeting molecules, drugs, or stabilizing groups (239). These characteristics imbue PAMAM dendrimers with the ability to be used in drug delivery (240), gene delivery (241), and imaging and diagnostic (242). The hurdle for constructing these molecules for biological applications is that they are rapidly metabolized by peptidases and proteases, so modification of surface groups to prevent rapid systemic clearance is common, but this may consequentially also lead to increased toxicity (243). Due to their vast range of modifiable characteristics, they have been found to induce differing toxicity based on size (244), charge (245), and core structure (246). In *Daphnia magna, Vibrio fischeri, T. platyurus* and fish cell lines PLHC-1 (teleost liver cell line) and RTG-2 (fibroblastic-like cell line), PAMAM dendrimers induced toxicity correlated with increasing generation size and particle surface area, suggesting that more surface area leads to more area for biological interaction (177). Additionally, the change in the media's zeta potential was a statistically significant indicator of zebrafish toxicity (247). PAMAM-Gd (gadolinium) accumulates less readily in livers of mice than the less hydrophilic PPI diaminobutane suggesting the dendrimers core can influence bioaccumulation (244). PAMAM dendrimers have been shown to have membrane reducing potentials which could act as their primary source of toxicity by hindering the influx of Ca^{2+} ions (248) across the mitochondrial transmembrane, consequentially overloading the mitochondria with Ca^{2+} ions, causing ATP depletion and oxidative stress and leading to ischemia reperfusion and apoptosis (249).

5.10.2. Zebrafish-Specific Study

Heiden et al. in 2007 found zebrafish embryos treated with anionic G3.5 dendrimers at the highest tested concentration (200 μM) displayed no toxic response. Since bioaccumulation and bioavailability studies were not conducted, it is possible these dendrimers were not taken up by the zebrafish (42) especially since other negatively charged particles are not taken up by cell membranes (233). Cationic G4 dendrimers were toxic in a dose- and time-dependent manner (42), contradicting studies done on mice that low generation dendrimers were nontoxic (245). Cationic molecules, in general, can destabilize the negative cell membrane and result in cell lysis suggesting that charge is the primary indicator of toxicity (250). However, reducing damage to membranes by conjugating Arg-Gly-Asp (RGD), a recognition motif for integrin receptors located in the extracellular matrix on endothelial cells that assist in particle uptake (251) to G4 dendrimers reduced effects at low concentration, but were still acutely toxic at high concentrations (42). RGD-G3.5 dendrimers were nontoxic at all doses. Interestingly, embryos with chorions had a higher toxic response than those without, though the 120 hpf endpoint LC_{50} 0.7 μM was the same. Toxicity for G4 dendrimers differed based on what time point they were dosed, e.g. 24 hpf was significantly different than 6 hpf (LC_{50} at 120 hpf 0.7 μM vs. 0.4 μM, respectively). Sublethal effects induced by G4 dendrimers were reduced body growth, bent trunk, and smaller heads and eyes (42).

5.10.3. Extrapolations and Notes

Size and charge both appear to influence toxicity, with smaller G.35 dendrimers being less toxic in zebrafish (42) and other assays (239). The primary mode of toxicity would be cell membrane disruption culminating eventually in macro-morphological changes.

6. Conclusion

Our gathering of nanoparticle-specific studies on zebrafish implicate that the zebrafish model has potential to evolve into a whole-organism model of nanotoxicity. The intent of this review is to emphasize the ever-growing disparity in ENM designs and concomitant physicochemical characteristics. Ideally, nanotoxicology studies will one day parallel ENM design, synthesis, and implementation for predictive and/or correlative toxicity for rapid re-formulation and beneficial utility in the community.

References

1. Roco MC (2005) Environmentally responsible development of nanotechnology. Environ Sci Technol 39:106A–112A
2. Pitkethly MJ (2004) Nanomaterials-the driving force. Mater Today 7:20–29
3. Su DS, Chen XW (2007) Natural lavas as catalysts for efficient production of carbon nanotubes and nanofibers. Angew Chem Int Ed 46:1823–1824
4. Chaloupka K, Malam Y, Seifalian AM (2010) Nanosilver as a new generation of nanoproduct in biomedical applications. Trends Biotechnol 11:580–588
5. Elder A, Yang H, Gwiazda R, Teng X, Thurston S, He H, Oberdörster G (2007) Testing nanomaterials of unknown toxicity: an example based on platinum nanoparticles of different shapes. Adv Mater 19: 3124–3129
6. Oberdörster G, Oberdörster E, Oberdörster J (2005) Nanotoxicology: an emerging discipline evolving from studies of ultrafine particles. Environ Health Perspect 113:823–839
7. Brunner TJ, Wick P, Manser P, Spohn P, Grass RN, Limbach LK, Bruinink A, Stark WJ (2006) In vitro cytotoxicity of oxide nanoparticles: comparison to asbestos, silica, and the effect of particle solubility. Environ Sci Technol 40:4374–4381
8. Hower JC, Graham UM, Dozier A, Tseng MT, Khatri RA (2008) Association of the sites of heavy metals with nanoscale carbon in a Kentucky electrostatic precipitator fly ash. Environ Sci Technol 42:8471–8477
9. Gidney JT, Twigg MV, Kittelson DB (2010) Effect of organometallic fuel additives on nanoparticle emissions from a gasoline passenger car. Environ Sci Technol 44: 2562–2569
10. Wiesner MR, Lowry GV, Jones KL, Hochella MF Jr, Di Giulio RT, Casman E, Bernhardt ES (2009) Decreasing uncertainties in assessing environmental exposure, risk, and ecological implications of nanomaterials. Environ Sci Technol 43:6458–6462
11. Gao J, Youn S, Hovsepyan A, Llaneza VL, Wang Y, Bitton G, Bonzongo JCJ (2009) Dispersion and toxicity of selected manufactured nanomaterials in natural river water samples: effects of water chemical composition. Environ Sci Technol 43:3322–3328
12. Card JW, Magnuson BA (2010) A method to assess the quality of studies that examine the toxicity of engineered nanomaterials. Int J Toxicol 29:402–410
13. Choi JY, Ramachandran G, Kandlikar M (2009) The impact of toxicity testing costs on nanomaterial regulation. Environ Sci Technol 43:3030–3034
14. Fent K, Weisbrod CJ, Wirth-Heller A, Pieles U (2010) Assessment of uptake and toxicity of fluorescent silica nanoparticles in zebrafish (Danio rerio) early life stages. Aquat Toxicol 2:218–228
15. Adams LK, Lyon DY, Alvarez PJ (2006) Comparative eco-toxicity of nanoscale TiO2, SiO2, and ZnO water suspensions. Water Res 40:3527–3532
16. Zhu X, Zhu L, Duan Z, Qi R, Li Y, Lang Y (2008) Comparative toxicity of several metal oxide nanoparticle aqueous suspensions to Zebrafish (Danio rerio) early developmental stage. J Environ Sci Health Part A: Tox/ Hazard Subst Environ Eng 43:278–284
17. Donaldson K, Borm PJ, Oberdorster G, Pinkerton KE, Stone V, Tran CL (2008) Concordance between in vitro and in vivo dosimetry in the proinflammatory effects of

low-toxicity, low-solubility particles: the key role of the proximal alveolar region. Inhal Toxicol 20:53–62

18. Slikker WJ, Andersen ME, Bogdanffy MS, Bus JS, Cohen SD, Conolly RB, David RM, Doerrer NG, Dorman DC, Gaylor DW, Hattis D, Rogers JM, Setzer RW, Swenberg JA, Wallace K (2004) Dose-dependent transitions in mechanisms of toxicity: case studies. Toxicol Appl Pharmacol 201:226–294
19. Browning LM, Lee KJ, Huang T, Nallathamby PD, Lowman JE, Xu XHN (2009) Random walk of single gold nanoparticles in zebrafish embryos leading to stochastic toxic effects on embryonic developments. Nanoscale 1:138–152
20. Isaacson CW, Usenko CY, Tanguay RL, Field JA (2007) Quantification of fullerenes by LC/ESI-MS and its application to in vivo toxicity assays. Anal Chem 79:9091–9097
21. Heiden TC, Dengler E, Kao WJ, Heideman W, Peterson RE (2007) Developmental toxicity of low generation PAMAM dendrimers in zebrafish. Toxicol Appl Pharmacol 225: 70–79
22. Powers KW, Brown SC, Krishna VB, Wasdo SC, Moudgil BM, Roberts SM (2006) Research strategies for safety evaluation of nanomaterials. Part VI. Characterization of nanoscale particles for toxicological evaluation. Toxicol Sci 90:296–303
23. Jones CF, Grainger DW (2009) In vitro assessments of nanomaterial toxicity. Adv Drug Deliv Rev 61:438–456
24. Kroll A, Pillukat MH, Hahn D, Schnekenburger J (2009) Current in vitro methods in nanoparticle risk assessment: limitations and challenges. Eur J Pharm Biopharm 72:370–377
25. Fischer HC, Chan WCW (2007) Nanotoxicity: the growing need for in vivo study. Curr Opin Biotechnol 18:565–571
26. Langheinrich U (2003) Zebrafish: a new model on the pharmaceutical catwalk. Bioessays 25:904–912
27. Doerks T, Copley RR, Schultz J, Ponting CP, Bork P (2002) Systematic identification of novel protein domain families associated with nuclear functions. Genome Res 12:47–56
28. Pyati UJ, Look AT, Hammerschmidt M (2007) Zebrafish as a powerful vertebrate model system for in vivo studies of cell death. Semin Cancer Biol 17:154–165
29. Rubinstein AL (2003) Zebrafish: from disease modeling to drug discovery. Curr Opin Drug Discov Devel 6:218–223
30. Teraoka H, Dong W, Hiraga T (2003) Zebrafish as a novel experimental model for developmental toxicology. Congenit Anom (Kyoto) 43:123–132
31. Lee KJ, Nallathamby PD, Browning LM, Osgood CJ, Xu XH (2007) In vivo imaging of transport and biocompatibility of single silver nanoparticles in early development of zebrafish embryos. ACS Nano 1:133–143
32. Parng C (2005) In vivo zebrafish assays for toxicity testing. Curr Opin Drug Discov Devel 8:100–106
33. Van Hoecke K, Quik JTK, Mankiewicz-Boczek J, De Schamphelaere KAC, Elsaesser A, Van der Meeren P, Barnes C, McKerr G, Howard CV, Van De Meent D (2009) Fate and effects of CeO2 nanoparticles in aquatic ecotoxicity tests. Environ Sci Technol 43:4537–4546
34. Fujitani Y, Kobayashi T, Arashidani K, Kunugita N, Suemura K (2008) Measurement of the physical properties of aerosols in a fullerene factory for inhalation exposure assessment. J Occup Environ Hyg 5:380–389
35. Buzea C, Pacheco II, Robbie K (2007) Nanomaterials and nanoparticles: sources and toxicity. Biointerphases 2:17–71
36. Johnston BD, Scown TM, Moger J, Cumberland SA, Baalousha M, Linge K, van Aerle R, Jarvis K, Lead JR, Tyler CR (2010) Bioavailability of nanoscale metal oxides TiO(2), CeO(2), and ZnO to fish. Environ Sci Technol 44:1144–1151
37. Griffitt RJ, Weil R, Hyndman KA, Denslow ND, Powers K, Taylor D, Barber DS (2007) Exposure to copper nanoparticles causes gill injury and acute lethality in zebrafish (Danio rerio). Environ Sci Technol 41:8178–8186
38. Griffitt RJ, Hyndman K, Denslow ND, Barber DS (2009) Comparison of molecular and histological changes in zebrafish gills exposed to metallic nanoparticles. Toxicol Sci 107: 404–415
39. Henry TB, Menn FM, Fleming JT, Wilgus J, Compton RN, Sayler GS (2007) Attributing effects of aqueous C60 nano-aggregates to tetrahydrofuran decomposition products in larval zebrafish by assessment of gene expression. Environ Health Perspect 115: 1059–1065
40. Cheng J, Flahaut E, Cheng SH (2007) Effect of carbon nanotubes on developing zebrafish (Danio rerio) embryos. Environ Toxicol Chem 26:708–716
41. Choi JE, Kim S, Ahn JH, Youn P, Kang JS, Park K, Yi J, Ryu DY (2009) Induction of oxidative stress and apoptosis by silver nanoparticles in the liver of adult zebrafish. Aquat Toxicol 100:151–159

42. King-Heiden TC, Wiecinski PN, Mangham AN, Metz KM, Nesbit D, Pedersen JA, Hamers RJ, Heideman W, Peterson RE (2009) Quantum dot nanotoxicity assessment using the zebrafish embryo. Environ Sci Technol 43:1605–1611
43. Zhu X, Zhu L, Li Y, Duan Z, Chen W, Alvarez PJJ (2007) Developmental toxicity in zebrafish (Danio rerio) embryos after exposure to manufactured nanomaterials: Buckminsterfullerene aggregates (nC60) and fullerol. Environ Toxicol Chem 26:976–979
44. Ispas C, Andreescu D, Patel A, Goia DV, Andreescu S, Wallace KN (2009) Toxicity and developmental defects of different sizes and shape nickel nanoparticles in zebrafish. Environ Sci Technol 43:6349–6356
45. Hu X, Cook S, Wang P, Hwang HM (2009) In vitro evaluation of cytotoxicity of engineered metal oxide nanoparticles. Sci Total Environ 407:3070–3072
46. Darkow R, Groth T, Albrecht W, Litzow K, Paul D (1999) Functionalized nanoparticles for endotoxin binding in aqueous solutions. Biomaterials 20:1277–1283
47. Bar-Ilan O, Albrecht RM, Fako VE, Furgeson DY (2009) Toxicity assessments of multisized gold and silver nanoparticles in zebrafish embryos. Small 5:1897–1910
48. Oberdorster G, Maynard A, Donaldson K, Castranova V, Fitzpatrick J, Ausman K, Carter J, Karn B, Kreyling W, Lai D, Olin S, Monteiro-Riviere N, Warheit D, Yang H, ILSI Research Foundation/Risk Science Institute Nanomaterial Toxicity Screening Working (2005)) Principles for characterizing the potential human health effects from exposure to nanomaterials: elements of a screening strategy. Part Fibre Toxicol 2:8
49. Nölting B (2009) Methods in modern biophysics, 3rd edn. Springer, Berlin
50. Navin JK, Grass ME, Somorjai GA, Marsh AL (2009) Characterization of colloidal platinum nanoparticles by MALDI-TOF mass spectrometry. Anal Chem 81:6295–6299
51. Link V, Shevchenko A, Heisenberg CH (2006) Proteomics of early zebrafish embryos. BMC Dev Biol 6:1
52. Hanisch K, Kuster E, Altenburger R, Gundel U (2010) Proteomic signatures of the zebrafish (Danio rerio) embryo: sensitivity and specificity in toxicity assessment of chemicals. Int J Proteomics 10:1–13
53. Boyle D, Brix KV, Amlund H, Lundebye AK, Hogstrand C, Bury NR (2008) Natural arsenic contaminated diets perturb reproduction in fish. Environ Sci Technol 42:5354–5360
54. Lerebours A, Bourdineaud JP, van der Ven K, Vandenbrouck T, Gonzalez P, Camilleri V, Floriani M, Garnier-Laplace J, Adam-Guillermin C (2010) Sublethal effects of waterborne uranium exposures on the zebrafish brain: transcriptional responses and alterations of the olfactory bulb ultrastructure. Environ Sci Technol 44:1438–1443
55. Tadjiki S, Assemi S, Deering C, Veranth JM, Miller JD (2009) Detection, separation, and quantification of unlabeled silica nanoparticles in biological media using sedimentation field-flow fractionation. J Nano Res 11:981–988
56. Giddings JC, Yang FJ, Myers MN (1977) Flow field-flow fractionation: a new method for separating, purifying, and characterizing the diffusivity of viruses. J Virol 21:131–138
57. Giddings JC, Caldwell KD (1989) Field-flow fractionation. In: Rositer BW, Hamilton JF (eds) Physical methods of chemistry. Wiley, New York, pp 867–938
58. Assemi S, Tadjiki S, Donose BC, Nguyen AV, Miller JD (2010) Aggregation of fullerol C60(OH)24 nanoparticles as revealed using flow field-flow fractionation and atomic force microscopy. Langmuir 26:16063–16070
59. Giddings JC (1995) Measuring colloidal and macromolecular properties by FFF. Anal Chem 67:592A–598A
60. Ma PM, Buschmann MD, Winnik FM (2010) One-step analysis of DNA/chitosan complexes by field-flow fractionation reveals particle size and free chitosan content. Biomacromolecules 11:549–554
61. Kennedy AJ, Hull MS, Bedner AJ, Goss JD, Gunter JC, Bouldin JL, Vikesland PJ, Steevens JA (2010) Fractionating nanosilver: importance for determining toxicity to aquatic test organisms. Environ Sci Technol 44:9571–9577
62. Mühlfeld C, Rothen-Rutishauser B, Vanhecke D, Blank F, Gehr P, Ochs M (2007) Visualization and quantitative analysis of nanoparticles in the respiratory tract by transmission electron microscopy. Part Fibre Toxicol 4:11
63. Lin CL, Miller JD (1993) The development of a PC image-based on-line particle size analyzer. Miner Metallur Proc (SME) 29–35
64. Binnings G, Quate CF, Gerber C (1986) Atomic force microscope. Phys Rev Lett 56:930–933
65. Ducker WA, Senden TJ, Pashley RM (1991) Direct measurement of colloidal forces using an atomic force microscope. Nature 353:239–241
66. Veeramasuneni S, Yalamanchili MR, Miller JD (1996) Measurement of interaction forces

between silica and a-alumina by atomic force microscopy. J Colloid Interface Sci 184:594–600

67. Assemi S, Nalaskowski J, Miller JD, Johnson WP (2006) Isoelectric point of fluorite by direct force measurements using atomic force microscopy. Langmuir 22:1403–1405
68. Puech PH, Taubenberger A, Ulrich F, Krieg M, Muller DJ, Heisenberg CP (2005) Measuring cell adhesion forces of primary gastrulating cells from zebrafish using atomic force microscopy. J Cell Sci 118:4199–4206
69. Krieg M, Arboleda-Estudillo Y, Puech PH, Käfer J, Graner F, Müller DJ, Heisenberg CP (2008) Tensile forces govern germ-layer organization in zebrafish. Nat Cell Biol 10:429–436
70. Westneat MW, Socha JJ, Lee W-K (2008) Advances in biological structure, function, and physiology using synchrotron X-Ray Imaging. Ann Rev Physiol 70:119–142
71. Petibois C, Guidi MC (2008) Bioimaging of cells and tissues using accelerator -based sources. Anal Bioanal Chem 391:1599–1608
72. Wang B, Wang Z, Feng W, Wang M, Hu Z, Chai Z, Zhao Y (2010) New methods for nanotoxicology: synchrotron radiation-based techniques. Anal Bioanal Chem 398: 667–676
73. Neues F, Arnold WH, Fischer J, Beckmann F, Gaengler P, Epple M (2006) The skeleton and pharyngeal teeth of zebrafish (Danio rerio) as a model of biomineralization in vertebrates. Mat Sci Technol 37:426–431
74. Korbas M, Blechinger SR, Krone PH, Pickering IJ, George GN (2008) Localizing organomercury uptake and accumulation in zebrafish larvae at the tissue and cellular level. Proc Nat Acad Sci USA 105:12108–12112
75. Korbas M, Krone PH, Pickering IJ, George GN (2010) Dynamic accumulation and redistribution of methylmercury in the lens of developing zebrafish embryos and larvae. J Biol Inorg Chem 15:1137–1145
76. Chaudhuri P, Harfouche R, Soni S, Hentschel DM, Sengupta S (2010) Shape effect of carbon nanovectors on angiogenesis. ACS Nano 4:574–582
77. Liu J, Yang L, Hopfinger AJ (2009) Affinity of drugs and small biologically active molecules to carbon nanotubes: a pharmacodynamics and nanotoxicity factor? Mol Pharm 6:873–882
78. Bianco A (2004) Carbon nanotubes for the delivery of therapeutic molecules. Expert Opin Drug Deliv 1:57–65
79. Lee J, Mahendra S, Alvarez PJ (2010) Nanomaterials in the construction industry: a review of their applications and environmental health and safety considerations. ACS Nano 4:3580–3590
80. Thayer AM (2007) Carbon nanotubes by the metric ton. Chem Eng News 85:29–35
81. Poland CA, Duffin R, Kinloch I, Maynard A, Wallace WAH, Seaton A, Stone V, Brown S, MacNee W, Donaldson K (2008) Carbon nanotubes introduced into the abdominal cavity of mice show asbestos-like pathogenicity in a pilot study. Nat Nanotechnol 3:423–428
82. Velzeboer I, Hendriks AJ, Ragas AM, Van de Meent D (2008) Aquatic ecotoxicity tests of some nanomaterials. Environ Toxicol Chem 27:1942–1947
83. Panessa-Warren BJ, Maye MM, Warren JB, Crosson KM (2009) Single walled carbon nanotube reactivity and cytotoxicity following extended aqueous exposure. Environ Pollut 157:1140–1151
84. Pulskamp K, Diabate S, Krug HF (2007) Carbon nanotubes show no sign of acute toxicity but induce intracellular reactive oxygen species in dependence on contaminants. Toxicol Lett 168:58–74
85. Soto K, Garza KM, Murr LE (2007) Cytotoxic effects of aggregated nanomaterials. Acta Biomater 3:351–358
86. Fraczek A, Menaszek E, Paluszkiewicz C, Blazewicz M (2008) Comparative in vivo biocompatibility study of single-and multi-wall carbon nanotubes. Acta Biomater 4:1593–1602
87. Lucente-Schultz RM, Moore VC, Leonard AD, Price BK, Kosynkin DV, Lu M, Partha R, Conyers JL, Tour JM (2009) Antioxidant single-walled carbon nanotubes. J Am Chem Soc 131:3934–3941
88. Bihari P, Vippola M, Schultes S, Praetner M, Khandoga AG, Reichel CA, Coester C, Tuomi T, Rehberg M, Krombach F (2008) Optimized dispersion of nanoparticles for biological in vitro and in vivo studies. Part Fibre Toxicol 5:14
89. Heister E, Lamprecht C, Neves V, Tilmaciu C, Datas L, Flahaut E, Soula B, Hinterdorfer P, Coley HM, Silva SR, McFadden J (2010) Higher dispersion efficacy of functionalized carbon nanotubes in chemical and biological environments. ACS Nano 4:2615–2626
90. Liu Z, Davis C, Cai W, He L, Chen X, Dai H (2008) Circulation and long-term fate of functionalized, biocompatible single-walled carbon nanotubes in mice probed by Raman spectroscopy. Proc Nat Acad Sci USA 105:1410

91. Johnston DE, Islam MF, Yodh AG, Johnson AT (2005) Electronic devices based on purified carbon nanotubes grown by high-pressure decomposition of carbon monoxide. Nat Mater 4:589–592
92. Zhang LW, Zeng L, Barron AR, Monteiro-Riviere NA (2007) Biological interactions of functionalized single-wall carbon nanotubes in human epidermal keratinocytes. Int J Toxicol 26:103–113
93. Cui HF, Vashist SK, Al-Rubeaan K, Luong JH, Sheu FS (2010) Interfacing carbon nanotubes with living mammalian cells and cytotoxicity issues. Chem Res Toxicol 23:1131–1147
94. Lanone S, Rogerieux F, Geys J, Dupont A, Maillot-Marechal E, Boczkowski J, Lacroix G, Hoet P (2009) Comparative toxicity of 24 manufactured nanoparticles in human alveolar epithelial and macrophage cell lines. Part Fibre Toxicol 6:14
95. Karlsson HL, Cronholm P, Gustafsson J, Moller L (2008) Copper oxide nanoparticles are highly toxic: a comparison between metal oxide nanoparticles and carbon nanotubes. Chem Res Toxicol 21:1726–1732
96. Kim JE, Lim HT, Minai-Tehrani A, Kwon JT, Shin JY, Woo CG, Choi M, Baek J, Jeong DH, Ha YC, Chae CH, Song KS, Ahn KH, Lee JH, Sung HJ, Yu IJ, Beck GRJ, Cho MH (2010) Toxicity and clearance of intratracheally administered multiwalled carbon nanotubes from murine lung. J Toxicol Environ Health A 73:1530–1543
97. Flahaut E, Agnoli F, Sloan J, O'Connor C, Green MLH (2002) CCVD synthesis and characterization of cobalt-encapsulated nanoparticles. Chem Mater 14:2553–2558
98. Cheng J, Chan CM, Veca LM, Poon WL, Chan PK, Qu L, Sun YP, Cheng SH (2009) Acute and long-term effects after single loading of functionalized multi-walled carbon nanotubes into zebrafish (Danio rerio). Toxicol Appl Pharmacol 235:216–225
99. Immordino ML, Dosio F, Cattel L (2006) Stealth liposomes: review of the basic science, rationale, and clinical applications, existing and potential. Int J Nanomedicine 1:297
100. Elim HI, Ji W, Ma GH, Lim KY, Sow CH, Huan CHA (2004) Ultrafast absorptive and refractive nonlinearities in multiwalled carbon nanotube films. Appl Phys Lett 85:1799
101. Asharani PV, Serina NG, Nurmawati MH, Wu YL, Gong Z, Valiyaveettil S (2008) Impact of multi-walled carbon nanotubes on aquatic species. J Nanosci Nanotechnol 8:3603–3609
102. McLeish JA, Chico TJ, Taylor HB, Tucker C, Donaldson K, Brown SB (2010) Skin exposure to micro- and nano-particles can cause haemostasis in zebrafish larvae. Thromb Haemost 103:797–807
103. Kim JS, Lee K, Lee YH, Cho HS, Kim KH, Choi KH, Lee SH, Song KS, Kang CS, Yu IJ (2010) Aspect ratio has no effect on genotoxicity of multi-wall carbon nanotubes. Arch Toxicol 85(7):775–86
104. Kim JS, Song KS, Joo HJ, Lee JH, Yu IJ (2010) Determination of cytotoxicity attributed to multiwall carbon nanotubes (MWCNT) in normal human embryonic lung cell (WI-38) line. J Toxicol Environ Health A 73:1521–1529
105. Magrez A, Kasas S, Salicio V, Pasquier N, Seo JW, Celio M, Catsicas S, Schwaller B, Forro L (2006) Cellular toxicity of carbon-based nanomaterials. Nano Lett 6:1121–1125
106. Mutlu GM, Budinger GR, Green AA, Urich D, Soberanes S, Chiarella SE, Alheid GF, McCrimmon DR, Szleifer I, Hersam MC (2010) Biocompatible nanoscale dispersion of single-walled carbon nanotubes minimizes in vivo pulmonary toxicity. Nano Lett 10:1664–1670
107. Chithrani BD, Ghazani AA, Chan WC (2006) Determining the size and shape dependence of gold nanoparticle uptake into mammalian cells. Nano Lett 6:662–668
108. Sayes CM, Liang F, Hudson JL, Mendez J, Guo W, Beach JM, Moore VC, Doyle CD, West JL, Billups WE (2006) Functionalization density dependence of single-walled carbon nanotubes cytotoxicity in vitro. Toxicol Lett 161:135–142
109. Wang H, Wang J, Deng X, Sun H, Shi Z, Gu Z, Liu Y, Zhaoc Y (2004) Biodistribution of carbon single-wall carbon nanotubes in mice. J Nanosci Nanotechnol 4:1019–1024
110. Analysis Consumer Products Nanotechnology Project (2010) http://www.nanotechproject.org/inventories/consumer/analysis_draft/. Accessed 18 May 2011
111. Kroto H (2010) The 2009 Lindau Nobel Laureate Meeting: Sir Harold Kroto, Chemistry 1996. J Vis Exp
112. Loutfy RO, Lowe TP, Moravsky AP, Katagiri S (2002) Commercial Production of Fullerenes and Carbon Nanotubes. In: Osawa E (ed) Perspectives of fullerene nanotechnology. Kluwer Academic Publishers, Dordrecht, The Netherlands, pp 35–46
113. Irie K, Nakamura Y, Ohigashi H, Tokuyama H, YamagoS,NakamuraE(1996)Photocytotoxicity of water-soluble fullerene derivatives. Biosci Biotechnol Biochem 60:1359–1361
114. Zhao B, He YY, Chignell CF, Yin JJ, Andley U, Roberts JE (2009) Difference in phototoxicity of cyclodextrin complexed fullerene

[(gamma-CyD)2/C60] and its aggregated derivatives toward human lens epithelial cells. Chem Res Toxicol 22:660–667

115. Hull MS, Kennedy AJ, Steevens JA, Bednar AJ, Weiss CA Jr, Vikesland PJ (2009) Release of metal impurities from carbon nanomaterials influences aquatic toxicity. Environ Sci Technol 43:4169–4174
116. Arbogast JW, Darmanyan AP, Foote CS, Diederich FN, Whetten RL, Rubin Y, Alvarez MM, Anz SJ (1991) Photophysical properties of sixty atom carbon molecule (C60). J Phys Chem 95:11–12
117. Hou WC, Jafvert CT (2009) Photochemical transformation of aqueous C60 clusters in sunlight. Environ Sci Technol 43:362–367
118. Hou WC, Jafvert CT (2009) Photochemistry of aqueous C60 clusters: evidence of O2 formation and its role in mediating C60 phototransformation. Environ Sci Technol 43: 5257–5262
119. Lee J, Fortner JD, Hughes JB, Kim JH (2007) Photochemical production of reactive oxygen species by C60 in the aqueous phase during UV irradiation. Environ Sci Technol 41:2529–2535
120. Sayes CM, Fortner JD, Guo W, Lyon D, Boyd AM, Ausman KD, Tao YJ, Sitharaman B, Wilson LJ, Hughes JB (2004) The differential cytotoxicity of water-soluble fullerenes. Nano Lett 4:1881–1887
121. Kong L, Tedrow O, Chan YF, Zepp RG (2009) Light-initiated transformations of fullerenol in aqueous media. Environ Sci Technol 43:9155–9160
122. Daroczi B, Kari G, McAleer MF, Wolf JC, Rodeck U, Dicker AP (2006) In vivo radioprotection by the fullerene nanoparticle DF-1 as assessed in a zebrafish model. Clin Cancer Res 12:7086–7091
123. Brown AP, Chung EJ, Urick ME, Shield WP 3rd, Sowers AL, Thetford A, Shankavaram UT, Mitchell JB, Citrin DE (2010) Evaluation of the fullerene compound DF-1 as a radiation protector. Radiat Oncol 5:34
124. Foley S, Crowley C, Smaihi M, Bonfils C, Erlanger BF, Seta P, Larroque C (2002) Cellular localisation of a water-soluble fullerene derivative. Biochem Biophys Res Commun 294:116–119
125. Shinohara N, Matsumoto T, Gamo M, Miyauchi A, Endo S, Yonezawa Y, Nakanishi J (2009) Is lipid peroxidation induced by the aqueous suspension of Fullerene C60 nanoparticles in the brains of cyprinus carpio? Environ Sci Technol 43:948–953
126. Andrievsky G, Klochkov V, Derevyanchenko L (2005) Is the C 60 Fullerene Molecule Toxic?! Fuller Nanotub Car N 13:363–376
127. Lyon DY, Adams LK, Falkner JC, Alvarez PJJ (2006) Antibacterial activity of Fullerene water suspensions: effects of preparation method and particle size. Environ Sci Technol 40:4360–4366
128. Usenko CY, Harper SL, Tanguay RL (2007) In vivo evaluation of carbon fullerene toxicity using embryonic zebrafish. Carbon 45: 1891–1898
129. Usenko CY, Harper SL, Tanguay RL (2008) Fullerene C60 exposure elicits an oxidative stress response in embryonic zebrafish. Toxicol Appl Pharmacol 229:44–55
130. Bernasconi R, Mathivet P, Bischoff S, Marescaux C (1999) Gamma-hydroxybutyric acid: an endogenous neuromodulator with abuse potential? Trends Pharmacol Sci 20: 135–141
131. Turesson J, Schwerte T, Sundin L (2006) Late onset of NMDA receptor-mediated ventilatory control during early development in zebrafish (Danio rerio). Comp Biochem Physiol A Mol Integr Physiol 143:332–339
132. Harper S, Usenko C, Hutchison JE, Maddux BLS, Tanguay RL (2008) In vivo biodistribution and toxicity depends on nanomaterial composition, size, surface functionalisation and route of exposure. J Expl Nanosci 3:195–206
133. Cagle DW, Kennel SJ, Mirzadeh S, Alford JM, Wilson LJ (1999) In vivo studies of fullerene-based materials using endohedral metallofullerene radiotracers. Proc Natl Acad Sci USA 96:5182–5187
134. Michalet X, Pinaud FF, Bentolila LA, Tsay JM, Doose S, Li JJ, Sundaresan G, Wu AM, Gambhir SS, Weiss S (2005) Quantum dots for live cells, in vivo imaging, and diagnostics. Science (Washington DC, US) 307:538–544
135. Gratzel M (2007) Photovoltaic and photoelectrochemical conversion of solar energy. Philos Transact A Math Phys Eng Sci 365:993–1005
136. Nirmal M, Brus L (1999) Luminescence photophysics in semiconductor nanocrystals. Acc Chem Res 32:407–414
137. Hardman R (2006) A toxicologic review of quantum dots: toxicity depends on physicochemical and environmental factors. Environ Health Perspect 114:165
138. Derfus AM, Chan WCW, Bhatia SN (2004) Probing the cytotoxicity of semiconductor quantum dots. Nano Lett 4:11–18
139. Lovric J, Cho SJ, Winnik FM, Maysinger D (2005) Unmodified cadmium telluride quantum dots induce reactive oxygen species formation leading to multiple organelle damage and cell death. Chem Biol 12:1227–1234

140. Schipper ML, Iyer G, Koh AL, Cheng Z, Ebenstein Y, Aharoni A, Keren S, Bentolila LA, Li J, Rao J, Chen X, Banin U, Wu AM, Sinclair R, Weiss S, Gambhir SS (2009) Particle size, surface coating, and PEGylation influence the biodistribution of quantum dots in living mice. Small 5:126–134
141. Shiohara A, Hoshino A, Hanaki K, Suzuki K, Yamamoto K (2004) On the cyto-toxicity caused by quantum dots. Microbiol Immunol 48:669–675
142. Gagne F, Fortier M, Yu L, Osachoff HL, Skirrow RC, van Aggelen G, Gagnon C, Fournier M (2010) Immunocompetence and alterations in hepatic gene expression in rainbow trout exposed to CdS/CdTe quantum dots. J Environ Monit 12:1556–1565
143. Huang KS, Lin YC, Su KC, Chen HY (2007) An electroporation microchip system for the transfection of zebrafish embryos using quantum dots and GFP genes for evaluation, *Biomed.* Microdevices 9:761–768
144. Son SW, Kim JH, Kim SH, Kim H, Chung AY, Choo JB, Oh CH, Park HC (2009) Intravital imaging in zebrafish using quantum dots, *Skin Res.* Technol 15:157–160
145. Friedrich M, Nozadze R, Gan Q, Zelman-Femiak M, Ermolayev V, Wagner TU, Harms GS (2009) Detection of single quantum dots in model organisms with sheet illumination microscopy. Biochem Biophys Res Commun 390:722–727
146. Rieger S, Kulkarni RP, Darcy D, Fraser SE, Koster RW (2005) Quantum dots are powerful multipurpose vital labeling agents in zebrafish embryos. Dev Dyn 234:670–681
147. Wang CH, Wang WL, Wu JL, Peng CA (2010) Rapid antiviral assay using QD-tagged fish virus as imaging nanoprobe. J Virol Methods
148. Mahendra S, Zhu H, Colvin VL, Alvarez PJ (2008) Quantum dot weathering results in microbial toxicity. Environ Sci Technol 42:9424–9430
149. Chave T, Nikitenko SI, Scheinost AC, Berthon C, Arab-Chapelet B, Moisy P (2010) First synthesis of uranyl aluminate nanoparticles. Inorg Chem 49:6381–6383
150. Zhou Y, Itoh H, Uemura T, Naka K, Chujo Y (2002) Synthesis of novel stable nanometer-sized metal (M = Pd, Au, Pt) colloids protected by a $\alpha\beta$-conjugated polymer. Langmuir 18:277–283
151. Kester M, Heakal Y, Fox T, Sharma A, Robertson GP, Morgan TT, Altinoglu EI, Tabakovic A, Parette MR, Rouse S, Ruiz-Velasco V, Adair JH (2008) Calcium phosphate nanocomposite particles for in vitro imaging and encapsulated chemotherapeutic drug delivery to cancer cells. Nano Lett 8:4116–4121
152. Filip J, Zboril R, Schneeweiss O, Zeman J, Cernik M, Kvapil P, Otyepka M (2007) Environmental applications of chemically pure natural ferrihydrite. Environ Sci Technol 41:4367–4374
153. Elliott DW, Zhang WX (2001) Field assessment of nanoscale bimetallic particles for groundwater treatment. Environ Sci Technol 35:4922–4926
154. Griffitt RJ, Luo J, Gao J, Bonzongo JC, Barber DS (2008) Effects of particle composition and species on toxicity of metallic nanomaterials in aquatic organisms. Environ Toxicol Chem 27:1972–1978
155. Shukla N, Svedberg EB, Ell J, Roy AJ (2006) Surfactant effects on the shapes of cobalt nanoparticles. Materials Lett 60:1950–1955
156. Connolly J, Pierre TG, Rutnakornpituk M, Riffle JS (2004) Cobalt nanoparticles formed in polysiloxane copolymer micelles: effect of production methods on magnetic properties. J Phys D App Phys 37:2475
157. Utsunomiya S, Jensen KA, Keeler GJ, Ewing RC (2002) Uraninite and fullerene in atmospheric particulates. Environ Sci Technol 36:4943–4947
158. Utsunomiya S, Jensen KA, Keeler GJ, Ewing RC (2004) Direct identification of trace metals in fine and ultrafine particles in the Detroit urban atmosphere. Environ Sci Technol 38:2289–2297
159. Utsunomiya S, Kersting AB, Ewing RC (2009) Groundwater nanoparticles in the far-field at the Nevada Test Site: Mechanism for radionuclide transport. Environ Sci Technol 43:1293–1298
160. Bootharaju MS, Pradeep T (2010) Uptake of toxic metal ions from water by naked and monolayer protected silver nanoparticles: An X-ray Photoelectron Spectroscopic Investigation. J Phys Chem 114:8328–8336
161. O'Loughlin EJ, Kelly SD, Cook RE, Csencsits R, Kemner KM (2003) Reduction of uranium (VI) by mixed iron (II)/iron (III) hydroxide (green rust): formation of UO2 nanoparticles. Environ Sci Technol 37:721–727
162. Zhang Y, Chen W, Zhang J, Liu J, Chen G, Pope C (2007) In vitro and in vivo toxicity of CdTe nanoparticles. J Nanosci Nanotechnol 7:497–503
163. Blechinger SR, Warren JTJ, Kuwada JY, Krone PH (2002) Developmental toxicology of cadmium in living embryos of a stable

transgenic zebrafish line. Environ Health Perspect 110:1041–1046

164. Nayak AS, Lage CR, Kim CH (2007) Effects of low concentrations of arsenic on the innate immune system of the zebrafish (Danio rerio). Toxicol Sci 98:118–124
165. Liu F, Gentles A, Theodorakis CW (2008) Arsenate and perchlorate toxicity, growth effects, and thyroid histopathology in hypothyroid zebrafish Danio rerio. Chemosphere 71:1369–1376
166. Ung CY, Lam SH, Hlaing MM, Winata CL, Korzh S, Mathavan S, Gong Z (2010) Mercury-induced hepatotoxicity in zebrafish: in vivo mechanistic insights from transcriptome analysis, phenotype anchoring and targeted gene expression validation. BMC Genomics 11:212
167. Bourrachot S, Simon O, Gilbin R (2008) The effects of waterborne uranium on the hatching success, development, and survival of early life stages of zebrafish (Danio rerio). Aquat Toxicol 90:29–36
168. Grainger DW, Castner DG (2008) Nanobiomaterials and nanoanalysis: opportunities for improving the science to benefit biomedical technologies. Adv Mater 20:867–877
169. Powell AC, Paciotti GF, Libutti SK (2010) Colloidal gold: a novel nanoparticle for targeted cancer therapeutics, *Methods Mol.* Biol 624:375–384
170. Vivero-Escoto JL, Slowing II, Wu CW, Lin VS (2009) Photoinduced intracellular controlled release drug delivery in human cells by gold-capped mesoporous silica nanosphere. J Am Chem Soc 131:3462–3463
171. Thomas M, Klibanov AM (2003) Conjugation to gold nanoparticles enhances polyethylenimine's transfer of plasmid DNA into mammalian cells. Proc Natl Acad Sci USA 100:9138–9143
172. Pingarrón JM, Yáñez-Sedeño P, González-Cortés A (2008) Gold nanoparticle-based electrochemical biosensors. Electrochim Act 53:5848–5866
173. Copland JA, Eghtedari M, Popov VL, Kotov N, Mamedova N, Motamedi M, Oraevsky AA (2004) Bioconjugated gold nanoparticles as a molecular based contrast agent: implications for imaging of deep tumors using optoacoustic tomography. Mol Imag Biol 6:341–349
174. Jennings T, Strouse G (2007) Past, present, and future of gold nanoparticles. Adv Exp Med Biol 620:34–47
175. Tsoli M, Kuhn H, Brandau W, Esche H, Schmid G (2005) Cellular uptake and toxicity of Au55 clusters. Small 1:841–844
176. Simpson CA, Huffman BJ, Gerdon AE, Cliffel DE (2010) Unexpected toxicity of monolayer protected gold clusters eliminated by PEG-thiol place exchange reactions. Chem Res Toxicol 23:1608–1616
177. Leroueil PR, Berry SA, Duthie K, Han G, Rotello VM, McNerny DQ, Baker JRJ, Orr BG, Holl MM (2008) Wide varieties of cationic nanoparticles induce defects in supported lipid bilayers. Nano Lett 8:420–424
178. Khan JA, Pillai B, Das TK, Singh Y, Maiti S (2007) Molecular effects of uptake of gold nanoparticles in HeLa cells. Chembiochem 8:1237–1240
179. Wang Y, Seebald JL, Szeto DP, Irudayaraj J (2010) Biocompatibility and biodistribution of surface-enhanced raman scattering nanoprobes in zebrafish embryos: in vivo and multiplex imaging. ACS Nano 4:4039–4053
180. Hutter E, Boridy S, Labrecque S, Lalancette-Hebert M, Kriz J, Winnik FM, Maysinger D (2010) Microglial response to gold nanoparticles. ACS Nano 4:2595–2606
181. Goodman CM, McCusker CD, Yilmaz T, Rotello VM (2004) Toxicity of gold nanoparticles functionalized with cationic and anionic side chains. Bioconjug Chem 15:897–900
182. Wang S, Lu W, Tovmachenko O, Rai US, Yu H, Ray PC (2008) Challenge in understanding size and shape dependent toxicity of gold nanomaterials in human skin keratinocytes. Chem Phys Lett 463:145–149
183. Alkilany AM, Nagaria PK, Hexel CR, Shaw TJ, Murphy CJ, Wyatt MD (2009) Cellular uptake and cytotoxicity of gold nanorods: molecular origin of cytotoxicity and surface effects. Small 5:701–708
184. Diegoli S, Manciulea AL, Begum S, Jones IP, Lead JR, Preece JA (2008) Interaction between manufactured gold nanoparticles and naturally occurring organic macromolecules. Sci Total Environ 402:51–61
185. Vallhov H, Qin J, Johansson SM, Ahlborg N, Muhammed MA, Scheynius A, Gabrielsson S (2006) The importance of an endotoxin-free environment during the production of nanoparticles used in medical applications. Nano Lett 6:1682–1686
186. Nam J, Won N, Jin H, Chung H, Kim S (2009) pH-Induced aggregation of gold nanoparticles for photothermal cancer therapy. J Am Chem Soc 131:13639–13645
187. Wagner DS, Delk NA, Lukianova-Hleb EY, Hafner JH, Farach-Carson MC, Lapotko DO (2010) The in vivo performance of plasmonic nanobubbles as cell theranostic agents in zebrafish hosting prostate cancer xenografts. Biomaterials 31:7567–7574

188. Lukianova-Hleb EY, Santiago C, Wagner DS, Hafner JH, Lapotko DO (2010) Generation and detection of plasmonic nanobubbles in zebrafish. Nanotechnology 21:225102
189. Sonavane G, Tomoda K, Makino K (2008) Biodistribution of colloidal gold nanoparticles after intravenous administration: Effect of particle size, Coll. Surf. B 66:274–280
190. Pan Y, Neuss S, Leifert A, Fischler M, Wen F, Simon U, Schmid G, Brandau W, Jahnen-Dechent W (2007) Size-dependent cytotoxicity of gold nanoparticles. Small 3:1941–1949
191. Dai J, Bruening ML (2002) Catalytic nanoparticles formed by reduction of metal ions in multilayered polyelectrolyte films. Nano Lett 2:497–501
192. Santoro CM, Duchsherer NL, Grainger DW (2007) Antimicrobial efficacy and ocular cell toxicity from silver nanoparticles. Nanobiotechnology 3:55–65
193. Asharani PV, Wu YL, Gong Z, Valiyaveettil S (2008) Toxicity of silver nanoparticles in zebrafish models. Nanotechnology 19:255102
194. Cumberland SA, Lead JR (2009) Particle size distributions of silver nanoparticles at environmentally relevant conditions. J Chromatogr A 1216:9099–9105
195. Hwang ET, Lee JH, Chae YJ, Kim YS, Kim BC, Sang BI, Gu MB (2008) Analysis of the toxic mode of action of silver nanoparticles using stress-specific bioluminescent bacteria. Small 4:746–750
196. Powers CM, Yen J, Linney EA, Seidler FJ, Slotkin TA (2010) Silver exposure in developing zebrafish (Danio rerio): persistent effects on larval behavior and survival. Neurotoxicol Teratol 32:391–397
197. Carlson C, Hussain SM, Schrand AM, Braydich-Stolle LK, Hess KL, Jones RL, Schlager JJ (2008) Unique cellular interaction of silver nanoparticles: size-dependent generation of reactive oxygen species. J Phys Chem B 112:13608–13619
198. Liu W, Wu Y, Wang C, Li HC, Wang T, Liao CY, Cui L, Zhou QF, Yan B, Jiang GB (2010) Impact of silver nanoparticles on human cells: effect of particle size. Nanotoxicology 4:319–330
199. Kim J, Kim S, Lee S (2011) Differentiation of the toxicities of silver nanoparticles and silver ions to the Japanese medaka (Oryzias latipes) and the cladoceran Daphnia magna. Nanotoxicology 5(2):208–214
200. Cioffi N, Ditaranto N, Torsi L, Picca RA, De Giglio E, Sabbatini L, Novello L, Tantillo G, Bleve-Zacheo T, Zambonin PG (2005) Synthesis, analytical characterization and bioactivity of Ag and Cu nanoparticles embedded in poly-vinyl-methyl-ketone films. Anal Bioanal Chem 382:1912–1918
201. Meng H, Chen Z, Xing G, Yuan H, Chen C, Zhao F, Zhang C, Zhao Y (2007) Ultrahigh reactivity provokes nanotoxicity: explanation of oral toxicity of nano-copper particles. Toxicol Lett 175:102–110
202. Prabhu BM, Ali SF, Murdock RC, Hussain SM, Srivatsan M (2010) Copper nanoparticles exert size and concentration dependent toxicity on somatosensory neurons of rat. Nanotoxicology 4:150–160
203. Grosell M, Blanchard J, Brix KV, Gerdes R (2007) Physiology is pivotal for interactions between salinity and acute copper toxicity to fish and invertebrates. Aquat Toxicol 84:162–172
204. Gilbert B, Ono RK, Ching KA, Kim CS (2009) The effects of nanoparticle aggregation processes on aggregate structure and metal uptake. J Coll Interface Sci 339:285–295
205. Alexis F, Pridgen E, Molnar LK, Farokhzad OC (2008) Factors affecting the clearance and biodistribution of polymeric nanoparticles. Mol Pharmcol 5:505–515
206. Cai R, Kubota Y, Shuin T, Sakai H, Hashimoto K, Fujishima A (1992) Induction of cytotoxicity by photoexcited TiO2 particles. Cancer Res 52:2346–2348
207. Jolivet JP, Cassaignon S, Chanéac C, Chiche D, Durupthy O, Portehault D (2010) Design of metal oxide nanoparticles: control of size, shape, crystalline structure and functionalization by aqueous chemistry. Comp Rend Chim 13(1–2):40–51
208. Xu M, Fujita D, Kajiwara S, Minowa T, Li X, Takemura T, Iwai H, Hanagata N (2010) Contribution of physicochemical characteristics of nano-oxides to cytotoxicity. Biomaterials 31:8022–8031
209. Huang X, Teng X, Chen D, Tang F, He J (2010) The effect of the shape of mesoporous silica nanoparticles on cellular uptake and cell function. Biomaterials 31:438–448
210. Limbach LK, Li Y, Grass RN, Brunner TJ, Hintermann MA, Muller M, Gunther D, Stark WJ (2005) Oxide nanoparticle uptake in human lung fibroblasts: effects of particle size, agglomeration, and diffusion at low concentrations. Environ Sci Technol 39:9370–9376
211. Karlsson HL, Gustafsson J, Cronholm P, Moller L (2009) Size-dependent toxicity of metal oxide particles–a comparison between nano- and micrometer size. Toxicol Lett 188:112–118
212. Limbach LK, Wick P, Manser P, Grass RN, Bruinink A, Stark WJ (2007) Exposure of

engineered nanoparticles to human lung epithelial cells: influence of chemical composition and catalytic activity on oxidative stress. Environ Sci Technol 41:4158–4163

213. Keller AA, Wang H, Zhou D, Lenihan HS, Cherr G, Cardinale BJ, Miller R, Ji Z (2010) Stability and aggregation of metal oxide nanoparticles in natural aqueous matrices. Environ Sci Technol 44:1962–1967
214. Limbach LK, Bereiter R, Müller E, Krebs R, Gälli R, Stark WJ (2008) Removal of oxide nanoparticles in a model wastewater treatment plant: Influence of agglomeration and surfactants on clearing efficiency. Environ Sci Technol 42:5828–5833
215. Li LH, Deng JC, Deng HR, Liu ZL, Xin L (2010) Synthesis and characterization of chitosan/ZnO nanoparticle composite membranes. Carbohydr Res 345:994–998
216. Sawai J, Igarashi H, Hashimoto A, Kokugan T, Shimizu M (1996) Effect of particle size and heating temperature of ceramic powders on antibacterial activity of their slurries. J Chem Engin Jpn 29:251–256
217. Zhu X, Wang J, Zhang X, Chang Y, Chen Y (2009) The impact of ZnO nanoparticle aggregates on the embryonic development of zebrafish (Danio rerio). Nanotechnology 20:195103
218. Yamamoto A, Honma R, Sumita M, Hanawa T (2004) Cytotoxicity evaluation of ceramic particles of different sizes and shapes, *J. Biomed. Mater. Res.* Part A 68:244–256
219. Warheit DB, Webb TR, Colvin VL, Reed KL, Sayes CM (2007) Pulmonary bioassay studies with nanoscale and fine-quartz particles in rats: toxicity is not dependent upon particle size but on surface characteristics. Toxicol Sci 95:270–280
220. Sayes CM, Wahi R, Kurian PA, Liu Y, West JL, Ausman KD, Warheit DB, Colvin VL (2006) Correlating nanoscale titania structure with toxicity: a cytotoxicity and inflammatory response study with human dermal fibroblasts and human lung epithelial cells. Toxicol Sci 92:174–185
221. Braydich-Stolle LK, Schaeublin NM, Murdock RC, Jiang J, Biswas P, Schlager JJ, Hussain SM (2009) Crystal structure mediates mode of cell death in TiO 2 nanotoxicity. J Nanoparticle Res 11:1361–1374
222. Chen J, Zhou H, Santulli AC, Wong SS (2010) Evaluating cytotoxicity and cellular uptake from the presence of variously processed TiO2 nanostructured morphologies. Chem Res Toxicol 23:871–879
223. Palaniappan PL, Pramod KS (2010) FTIR study of the effect of nTiO2 on the biochemical constituents of gill tissues of Zebrafish (Danio rerio), *Food Chem.* Toxicol 48:2337–2343
224. Stadtman ER, Berlett BS (1991) Fenton chemistry. Amino acid oxidation. J Biol Chem 266:17201
225. Levine RL, Garland D, Oliver CN, Amici A, Climent I, Lenz AG, Ahn BW, Shaltiel S, Stadtman ER (1990) Determination of carbonyl content in oxidatively modified proteins. Methods Enzymol 186:464
226. Kwak SY, Kim SH, Kim SS (2001) Hybrid organic/inorganic reverse osmosis (RO) membrane for bactericidal anti-fouling: Preparation and characterization of TiO2 nanoparticle self-assembled aromatic polyamide thin-film-composite (TFC) membrane. Environ Sci Technol 35:2388–2394
227. Santra S, Wang K, Tapec R, Tan W (2001) Development of novel dye-doped silica nanoparticles for biomarker application. J Biomed Opt 6:160
228. Trewyn BG, Giri S, Slowing II, Lin VSY (2007) Mesoporous silica nanoparticle based controlled release, drug delivery, and biosensor systems. Chem Commun (Cambridge, UK) 2007:3236–3245
229. Slowing II, Trewyn BG, Giri S, Lin VSY (2007) Mesoporous silica nanoparticles for drug delivery and biosensing applications. Adv Funct Mater 17:1225–1236
230. Maurer-Jones MA, Lin YS, Haynes CL (2010) Functional assessment of metal oxide nanoparticle toxicity in immune cells. ACS Nano 4:3363–3373
231. Petushkov A, Intra J, Graham JB, Larsen SC, Salem AK (2009) Effect of crystal size and surface functionalization on the cytotoxicity of silicalite-1 nanoparticles. Chem Res Toxicol 22:1359–1368
232. Chang JS, Chang KLB, Hwang DF, Kong ZL (2007) In vitro cytotoxicitiy of silica nanoparticles at high concentrations strongly depends on the metabolic activity type of the cell line. Environ Sci Technol 41:2064–2068
233. Tortiglione C, Quarta A, Malvindi MA, Tino A, Pellegrino T (2009) Fluorescent nanocrystals reveal regulated portals of entry into and between the cells of Hydra. PLoS One 4(11):e7698
234. Adili A, Crowe S, Beaux MF, Cantrell T, Shapiro PJ, McIlroy DN, Gustin KE (2008) Differential cytotoxicity exhibited by silica nanowires and nanoparticles. Nanotoxicology 2:1–8
235. Nelson SM, Mahmoud T, Beaux Mn, Shapiro P, McIlroy DN, Stenkamp DL (2010) Toxic and teratogenic silica nanowires in developing vertebrate embryos. Nanomedicine 6:93–102

236. Schubert D, Dargusch R, Raitano J, Chan SW (2006) Cerium and yttrium oxide nanoparticles are neuroprotective. Biochem Biophys Res Commun 342:86–91
237. Thill A, Zeyons O, Spalla O, Chauvat F, Rose J, Auffan M, Flank AM (2006) Cytotoxicity of CeO2 nanoparticles for Escherichia coli. Physicochemical insight of the cytotoxicity mechanism. Environ Sci Technol 40:6151–6156
238. Patil S, Sandberg A, Heckert E, Self W, Seal S (2007) Protein adsorption and cellular uptake of cerium oxide nanoparticles as a function of zeta potential. Biomaterials 28:4600–4607
239. Duncan R, Izzo L (2005) Dendrimer biocompatibility and toxicity. Adv Drug Deliv Rev 57:2215–2237
240. Ihre HR, Padilla De Jesus OL, Szoka FCJ, Frechet JM (2002) Polyester dendritic systems for drug delivery applications: design, synthesis, and characterization. Bioconjugate Chem 13:443–452
241. Tang R, Palumbo RN, Nagarajan L, Krogstad E, Wang C (2010) Well-defined block copolymers for gene delivery to dendritic cells: probing the effect of polycation chain-length, *J.* Controlled Release 142:229–237
242. Margerum LD, Campion BK, Koo M, Shargill N, Lai JJ, Marumoto A, Christian Sontum P (1997) Gadolinium (III) DO3A macrocycles and polyethylene glycol coupled to dendrimers effect of molecular weight on physical and biological properties of macromolecular magnetic resonance imaging contrast agents. J Alloys Compd 249:185–190
243. Bracci L, Falciani C, Lelli B, Lozzi L, Runci Y, Pini A, De Montis MG, Tagliamonte A, Neri P (2003) Synthetic peptides in the form of dendrimers become resistant to protease activity. J Biol Chem 278:46590
244. Kobayashi H, Sato N, Hiraga A, Saga T, Nakamoto Y, Ueda H, Konishi J, Togashi K, Brechbiel MW (2001) 3D-micro-MR angiography of mice using macromolecular MR contrast agents with polyamidoamine dendrimer core with reference to their pharmacokinetic properties. Magn Res Chem 45:454–460
245. Roberts JC, Bhalgat MK, Zera RT (1996) Preliminary biological evaluation of polyamidoamine (PAMAM) Starburst™ dendrimers, *J. Biomed. Mater. Res.* Part A 30:53–65
246. Malik N, Wiwattanapatapee R, Klopsch R, Lorenz K, Frey H, Weener JW, Meijer EW, Paulus W, Duncan R (2000) Dendrimers: relationship between structure and biocompatibility in vitro, and preliminary studies on the biodistribution of 125I-labelled polyamidoamine dendrimers in vivo. J Control Release 65:133–148
247. Naha PC, Davoren M, Casey A, Byrne HJ (2009) An ecotoxicological study of poly-(amidoamine)-dendrimers-toward quantitative structure activity relationships. Environ Sci Technol 43:6864–6869
248. Labieniec M, Gabryelak T (2008) Preliminary biological evaluation of poli(amidoamine) (PAMAM) dendrimer G3.5 on selected parameters of rat liver mitochondria. Mitochondrion 8:305–312
249. Vergun O, Reynolds IJ (2005) Distinct characteristics of Ca(2+)-induced depolarization of isolated brain and liver mitochondria. Biochim Biophys Acta 1709:127–137
250. Rittner K, Benavente A, Bompard-Sorlet A, Heitz F, Divita G, Brasseur R, Jacobs E (2002) New basic membrane-destabilizing peptides for plasmid-based gene delivery in vitro and in vivo. Mol Ther 5:104–114
251. Pasqualini R, Koivunen E, Ruoslahti E (1997) αv-integrins as receptors for tumor targeting by circulating ligands. Nature Biotechnol 15:542–546
252. Zhu X, Wang J, Zhang X, Chang Y, Chen Y (2010) Trophic transfer of TiO(2) nanoparticles from daphnia to zebrafish in a simplified freshwater food chain. Chemosphere 79: 928–933

Chapter 20

Application of Embryonic and Adult Zebrafish for Nanotoxicity Assessment

Jiangxin Wang, Xiaoshan Zhu, Yongsheng Chen, and Yung Chang

Abstract

As an emerging model for toxicological studies, zebrafish has been explored for nanotoxicity assessment. In addition to endpoint examination of embryo/fish mortality and/or developmental disorders, molecular analyses of differential gene expression have also been employed to evaluate toxic effects associated with the exposure to nanomaterials. Here, we describe zebrafish-based assays, including both embryo and adult, for evaluation of nanotoxicity caused by metal oxide nanoparticles (NPs), in particular, zinc oxide (ZnO) and titanium oxide (TiO_2) nanoparticles.

Key words: Zebrafish embryos, Nanoparticles, Nanotoxicology, Microarray, Acute, Chronic toxicity

1. Introduction

Zebrafish is an attractive model to study environmental toxicity, attributed to their inherent advantages: small size and relatively rapid life cycle for easy breeding and housing in large numbers, as well as their feasibility for high-throughput but cost-effective screening of numerous environmental toxic pollutants. In addition, optically transparent embryos allow direct observation of phenotypic changes in live organisms (1). Furthermore, the expression microarrays across the whole genome (gene chips) are commercially available and are being broadly applied in ecotoxicity study (2). Thus, zebrafish has been employed for toxicity assessments of a wide array of nanomaterials, including metal NPs, carbon-based nanomaterials, and polymers (3–8). While embryos are frequently used for assessing acute and developmental toxicity of NPs, adult zebrafish has been exploited for the study of NP fate, transport and bioaccumulation as well as chronic toxicity (9).

Joshua Reineke (ed.), *Nanotoxicity: Methods and Protocols*, Methods in Molecular Biology, vol. 926,
DOI 10.1007/978-1-62703-002-1_20,

In this chapter, we described experimental methods utilized to assess acute and chronic nanotoxicity in zebrafish embryos and adult fish, after their being exposed to ZnO and TiO_2 nanoparticles, respectively.

2. Materials

2.1. Chemicals

1. Uncoated nanoscale ZnO (nZnO, 99.7 + % purity) and anatase TiO_2 ($nTiO_2$, 99 + % purity) were obtained from Sigma-Aldrich (Sigma-Aldrich, Inc., St. Louis, MO, USA). The particle sizes of nZnO and $nTiO_2$ from the vendor were 20 and 60 nm, respectively.
2. All the stock solutions of NPs were prepared with DI water (i.e. deionized water with purity up to 18 MΩ cm at 25°C). They were stored at room temperature in the dark (unless indicated otherwise) (see Note 1).
3. Nitric acid (*Trace*ELECT®, for trace analysis, ≥69.0%, Sigma-Aldrich, Inc., St. Louis, MO, USA).
4. Sulfuric acid–ammonium sulfate solution: 400 g ammonium sulfate in 700 mL concentrated sulfuric acid, boiled.
5. 20 nm filters (Whatman International Ltd, Maidstone, UK).
6. Fish culture medium: 64.75 mg/L $NaHCO_3$, 5.75 mg/L KCl, 123.25 mg/L $MgSO_4 \cdot 7H_2O$, and 294 mg/L $CaCl_2 \cdot 2H_2O$ (Sigma-Aldrich, Inc., St. Louis, MO, USA) prepared according to the ISO (International Organization for Standardization, Geneva, Switzerland) standard 7346-3:1996. The medium is stable at room temperature. The fish culture medium specified here is used in the step 3 of Subheading 3.1.
7. Sediment stock: standard formulated sediment containing 19% sand and 2% organic carbon was prepared according to US EPA guidelines (EPA 600/R-99/064). Specifically, 219 g sand (White Quartz #1 dry), 1242 g silt–clay mixture (ASP 400), 77.3 g alpha cellulose, 0.15 g humic acid, and 7.5 g dolomite.
8. 2,7-Dichlorodihydrofluorescein diacetate (H_2DCF-DA) fluorescent probe (Sigma-Aldrich, Inc., St Louis, MO, USA).
9. Porcine trypsin, ethylenediaminetetraacetic acid, (EDTA) and benzocaine (ethyl-*p*-amino benzoate) were purchased from Sigma-Aldrich (Sigma-Aldrich, Inc., St Louis, MO, USA).
10. Collagenase H was purchased from Roche Diagnostics, Mannheim, Germany.
11. TRIzol® reagent was obtained from Invitrogen, Carlsbad, CA.

12. M-MuLV reverse transcriptase and oligo (dT) were obtained from New England Biolabs, Inc., Boston, MA.
13. 1× M-MuLV reverse transcriptase reaction buffer: 50 mM Tris–HCl, 75 mM KCl, 3 mM $MgCl_2$, 10 mM dithiothreitol. pH 8.3 at 25°C.
14. Agilent 4×44k zebrafish microarray (Santa Clara, CA) with two-color Amplification Kit and Gene Expression Hybridization Kit (Agilent), Feature Extraction 9.5 (Agilent).

2.2. Equipment

1. Micro-flow imaging (MFI) (DFA 4100, Brightwell Technologies Inc., Ottawa, ON, Canada) (see Note 2).
2. Dynamic light scattering device (DLS) (i.e., BI9000AT, Brookhaven Instrument Corporation, Holtsville, NY).
3. Transmission electron microscope (TEM, Tecnai G2T20ST, Philips, Holland) with energy dispersive X-ray (EDX).
4. Atomic force microscope (AFM, a molecular imaging 5500 AFM, Agilent) with silicon probes coated with alumina reflex coating (Tap300Al, Budgetsensors, Redding, CA).
5. Inductively coupled plasma mass spectrometry (ICP-MS; ELEMENT2, Thermo Fisher, Dubuque, IA).
6. Graphite furnace atomic absorption spectroscopy (GFAAS) (Varian SpectrAA 400, Agilent).
7. Inverse microscope (Olympus, Japan) equipped with a digital camera.
8. 2-mL glass/glass tissue grinder (Potter-Elvehjem-type; Braun Biotech, Sartorius, Goettingen, Germany) with a defined pestle/wall distance of 50–70 μm.
9. Flow cytometry (FACScalibur; BD Bioscience, San Jose, CA).
10. NanoDrop sensor (Coleman Technologies Inc., Chadds Ford, PA).
11. Agilent's bioAnalyzer 2100 (Agilent).
12. G2505 B microarray scanner (Agilent).

2.3. Fish and Embryos

1. Adult zebrafish (*Danio rerio*) of an AB strain were raised and maintained in a closed flow-through culture system (Aquatic Eco-Systems, Inc., FL), housed in a fish room of The Biodesign Institute at Arizona State University. The room is set up at 28±0.5°C with a photoperiod of 14 h light and 10 h dark. All experiments were carried out in compliance with the guidelines stipulated by the Arizona State University Institutional Animal Care and Use Committee (IACUC). The zebrafish were fed twice a day with TetraMin dry flakes (Tetra, Melle, Germany). For breeding, one female and one male were set up in a 2 L breeding tank in the afternoon. The spawning was triggered

next morning once the light was turned on and was complete within 30 min. At 4–5 h post-fertilization (hpf), embryos were collected and rinsed several times with fish culture medium to remove residues on the egg surface. Healthy embryos at blastula stage were then selected for subsequent experiments.

3. Methods

3.1. Preparation of NPs Aqueous Suspensions

1. Add 10 mg nZnO or $nTiO_2$ to 100 mL of nanopure water, respectively, to prepare the NPs stock solutions.
2. Disperse the above nanoparticle solutions by stirring solutions vigorously in fish culture medium using a magnetic agitator at room temperature for 2 h (nZnO), or by ultrasonication (50 W/L at 40 kHz) at room temperature for 0.5 h ($nTiO_2$).
3. Test solutions were prepared immediately prior to use by diluting the above stocks of NPs with fish culture medium (see item 6 in Subheading 2.1 for composition).
4. During the preparation of diluted solution, continuously stir the stock solutions with a magnetic stirrer to maintain the suspension at as stable of a concentration as possible.

3.2. Size Distribution, Characterization, and Actual Concentration of NPs in Medium Solution (See Note 3)

1. Analyze the sizes of nZnO or $nTiO_2$ in the prepared suspensions.
   (a) Measure larger particles by MFI.
   (b) Determine the number of pixels that compose the image of a particle.
   (c) Count the number of particles within different size ranges, size distribution, shape, and particle number/mL.
   (d) Analyze size distribution of smaller particles using a dynamic light scattering device after separation of the larger particles using sedimentation (see Note 4).
   (e) Calculate the hydrodynamic diameters of particles from the diffusion coefficients in terms of the Stokes–Einstein equation. The mean particle size measured by DLS is the so-called "*z*-average" diameter, which is weighted by the intensity of light scattered by the particles.
   (f) Obtain the particle size distribution by using the non-negatively constrained least square (NCLS) approach to solve the light scattering intensity autocorrelation function of the particles; this method is built into the DLS measurement software.

2. Examine the structures of $nZnO/nTiO_2$ or their aggregates using TEM and AFM.
   (a) TEM samples were prepared by drying several drops of NPs suspension onto 300-mesh copper grids.
   (b) AFM samples were prepared by dropping 2.5 μL of NP suspension on the clean silicon wafer and air dried for 5 min on a hotplate at a temperature of 35 °C. The images are acquired in acoustic alternating current (AAC) mode with a scanning speed of 2,000–5,000 nm/s, a drive frequency of 8 kHz, and a drive amplitude of 0.1 V (10).
3. Determine the concentration of nZnO or $nTiO_2$ in water samples.
   (a) Transfer 10 mL nZnO test solution to triangular flasks, dry, and then digest nZnO by nitric acid at room temperature.
   (b) Use sulfuric acid–ammonium sulfate solution to decompose the dried $nTiO_2$ into Ti^{4+} ion, where $nTiO_2$ in sulfuric acid–ammonium sulfate solution were heated until $nTiO_2$ were dissolved and the suspensions became clear.
   (c) Analyze the mass concentrations of Zn^{2+} and Ti^{4+} concentrations in the digested samples using ICP-MS.
   (d) Calculate the concentrations of these NPs (nZnO and $nTiO_2$) in the suspension based on the Zn^{2+} and Ti^{4+} concentrations measured from ICP-MS, respectively.

3.3. Determination of Dissolved Metal Ions Concentration Released from NPs and Metal Ions-Mediated Toxicity (See Note 5)

1. Filter 5 mL water samples taken from the test solutions immediately after a defined exposure time through a 20 nm filter to separate the dissolved Zn ions from the nZnO aggregates that range from a few hundred nanometers to several microns in diameter.
2. Determine the concentration of released Zn ions in the filtrates by GFAAS after filtration.
3. Examine the dissolved Zn ion solutions (collected in step 2) for their toxicity to zebrafish embryos (see Subheading 3.5).

3.4. Formulated Sediments (See Note 6)

1. Divide the whole dry sediment equally and transfer into six-well plates at 2 g/well.
2. Mix the sediment with 10 mL fish culture medium (as specified in item 6, Subheading 2.1, and used as a control) in a well, or with 10 mL fish culture medium along with various doses of nanoparticle (0.5–10 mg/L per well).
3. Stir the formulated sediments and fish culture medium gently with a small glass stick, and allow the mixture to settle for 1 h before the addition of 10 zebrafish embryos per well.
4. Test ten replicates (wells) per nanoparticle concentration and control ($n = 100$).

3.5. Embryo Toxicity Test

For the assessment of embryonic toxicity, we adopted the procedure developed by Schulte and Nagel (11) with some modifications:

1. Transfer selected embryos (blastula stage) into 24-well plates at 1 embryo/well, in which 20 wells contain 2 mL nZnO test solution (as treatment) and four wells contain 2 mL of fish culture medium (as control) per well.
2. Select a wide range of nZnO concentrations for the assessment, i.e., 500, 100, 50, 10, 5, 1, 0.5, 0.1 mg/L. Place one embryo per well in 2-mL nZnO suspension at a defined concentration, and 60 embryos in total three 24-well plates are treated per each treatment, and 12 embryos are included as the control.
3. Monitor the treated embryos and larvae daily for their viability, hatching and developmental abnormality using an inverse microscope and document photographically at specified time points (t=6, 12, 24, 36, 48, 60, 72, 84, and 96 h post fertilization (hpf)). The endpoints for documenting developmental toxicity include embryo/larvae survival and embryo hatching rate.
4. Record malformations (see Note 7) (e.g., pericardial edema, see Note 8) observed in nZnO treated embryos, which have been reported in the zebrafish embryos exposed to other toxins (12, 13).
5. Compare the incidence of malformation among various treatment groups.
6. Collect water samples for water quality assessment and for the filtrate test at the end of the experiment.
7. To determine whether light influences the toxic potential of some light sensitive NPs, perform a similar experiment as described above in a dark room, where the NP stock solution is prepared and kept in the dark and the 24-well plates containing embryos and NPs are protected from the light during the entire exposure period (96 h).

3.6. Cellular Reactive Oxygen Species (ROS) (See Note 9)

1. Set up 45–50 embryos per group for various treatments, i.e., 1.0 and 10 mg/L nZnO, 1 mg/L Zn^{2+} control (this concentration was selected based on the measurement of dissolved Zn^{2+} in the test solutions) and fish culture medium control.
2. Collect embryos 4 days after nZnO treatment, and rinse quickly with fish culture medium and phosphate-buffered saline (PBS).
3. Euthanize zebrafish embryos in 250 mg/L benzocaine in PBS buffer.
4. Conduct enzymatic digestion to prepare a single cell suspension, following the methods of Kosmehl et al. (14).
   (a) After rinsing in water, transfer 45–50 embryos to 1.5-mL Eppendorf tubes containing 1 mL 0.02–2% (w/v) collagenase in PBS followed by 0.05% (w/v) trypsin/EDTA solution (0.5 g/L porcine trypsin, 0.2 g/L EDTA and 4 Na).

(b) Incubate for 30 min for the digestion by collagenase and 5 min for the digestion by trypsin.

(c) Stop enzymatic digestion by addition of 1 mL 10% (v/v) fetal calf serum in PBS.

(d) Alternatively, prepare single cell suspension by mechanical agitation to disintegrate the tissues using 2-mL glass/glass tissue grinder.

5. Pass the cell suspension through 70 μm gauze into 1.5-mL Eppendorf tubes to separate undigested tissues from isolated cells.
6. Spin-down the cells at 200 × *g* and 4°C for 10 min and re-suspend the cell pellet with 1 mL PBS, centrifuge the cell suspension at 180 × *g* and 4°C for 7 min.
7. Incubate the obtained cell suspension with 1.25 mg/L H2DCF-DA at 28.5°C for 20 min in the dark.
8. Wash and re-suspend the cells in PBS.
9. Analyze the cells by the flow cytometry.
10. Collect and analyze a total of 10,000 cells/sample using Cell Quest 3.2 software. The fluorescence intensity of the cells is displayed in histogram.

3.7. Exposure of Adult Zebrafish to NPs and Reproduction Assessment

Adult zebrafish (more than 3 months old; 0.33 ± 0.09 g) are used for this study.

1. Prior to $nTiO_2$ treatment, acclimate fish to experimental conditions, including daily manipulation (fish handling and water changes) and weekly breeding for at least 4 weeks to ensure their reproductivity.

   (a) Feed all fish daily with TetraMin dry flakes (Tetra, Melle, Germany) containing no $nTiO_2$.

   (b) Keep fish in fish tanks or breeding tanks containing 1 L standard fish culture medium throughout the experiment, including the breeding period.

   (c) Observe and record the health status of the fish daily.

2. Expose reproductively active fish (12 females in two tanks and 12 males in two tanks) to one of the following treatments daily for 13 weeks using a semi-static exposure: 0.1 mg/L, 1.0 mg/L $nTiO_2$ or fish water only, in which the exposure medium was changed every 24 h with a new aliquot of appropriate concentrations of NPs or water for the control group.
3. Set up one male and one female each week from the same group in a breeding tank in the afternoon (one pair of fish in one breeding tank; 12 pairs for one treatment group). Trigger spawning next morning when the light is turned on and completed within 30 min.

4. Record numbers of spawning fish and eggs from individual breeding tanks.
5. Transfer eggs from each breeding pair into petri dish with 4 mL fish medium, and keep them in the fish room at 28 ± 0.5°C with an automatic switch between 14 h light and 10 h dark. At 48 hpf, record live embryos and calculate the embryo survival rate (live over the total eggs per spawning fish).
6. Monitor the water pH, temperature, hardness and dissolved oxygen concentration throughout the experiments.

3.8. Accumulation of NPs in Fish Tissues (See Note 10)

1. Section the tissues from control and $nTiO_2$-treated fish, in particular, gills and ovaries in our study.
2. Digest the dissected tissues in 5 mL of sulfuric acid–ammonium sulfate solution to decompose TiO_2 into titanium ion (Ti^{4+}) using a microwave digestion system (i.e., Mars 5, CEM, USA) with a four-stage digestion protocol (5 min at 150°C, 5 min at 190°C, 10 min at 230°C, and 10 min at 240°C).
3. Evaporate samples to dryness at 120°C for several hours after the digestion.
4. Convert $nTiO_2$ to Ti^{4+} using sulfuric acid–ammonium sulfate as aforementioned.
5. Determine the Ti^{4+} concentration by ICP-MS. In our study, $nTiO_2$ recovery in biological samples ranged from 80% to 90%.
6. Calculate the concentrations of $nTiO_2$ in the tissue based on the Ti^{4+} concentrations measured from ICP-MS in step 5.

3.9. Tissue Histology Analysis to Assess Oocytes Maturation

Perform the histology examination following the protocol described previously (15).

1. Fix freshly isolated ovary tissues in freshly prepared 4% paraformaldehyde in PBS at 4°C overnight.
2. Embed the tissue in paraffin after a stepwise dehydration in ethanol and xylene, and then prepare 5 μm transverse sections, deparaffinize in xylene, and stain with hematoxylin and eosin.
3. Collect the sections for further analyses from the tissue 1 mm below the ovary surface.
4. Examine eight to ten sections of each ovary, and count and assign more than 100 follicles from 1 to 3 sections of each ovary (total 5 ovaries for each group) to their appropriate stages based on the size and morphology as defined by Selman et al. (16).
5. Calculate the distribution of stages I–IV in each ovary as the percentage of total follicles enumerated.
6. Present the averages of different stages of follicles among five individual ovaries in each experimental group.

3.10. Gene Expression Analysis for Toxicity Assessment

3.10.1. Isolation of RNA by TRIZOL Following the Manufacture's Procedure with Slight Modification

1. Euthanize embryos with 0.25 mg/mL tricaine methanesulfonate and rinse thoroughly. Remove excess water from the samples, and transfer embryos to 1.5 mL Eppendorf tube. For each group, use total 15–30 embryos.
2. Add 1 mL TRIzol® reagent to the Eppendorf tube and homogenize embryos using a pestle on ice. If not used immediately, store the samples at −80°C until processing.
3. Thaw homogenates samples on ice once all samples were collected.
4. Vortex each sample at the highest speed for 30 s and incubate the homogenized samples at room temperature for 5 min.
5. Add 300 μL chloroform into each samples and vortex vigorously for 15 s.
6. Incubate the mixtures at room temperature for 3 min.
7. Centrifuge the homogenates at 12,000 × *g* at 4°C for 10 min.
8. Transfer the supernatants to a clean RNAse free Eppendorf tube.
9. Add 500 μL isopropanol per sample and incubate at room temperature for 10 min.
10. Centrifuge the samples at 12,000 × *g* at 4°C for 12 min.
11. Remove all liquid in the sample tubes and wash the pellets twice with 70% ethanol. Air-dry the pellets for about 5 min and resuspend pellets in 30 μL diethylpyrocarbonate (DEPC) treated RNase free water.
12. Quantify RNA using NanoDrop sensor and determine the RNA quality using Agilent's bioanalyzer.
13. Perform reverse transcription immediately. Or, flash freeze RNA samples using liquid nitrogen and stored at −80°C.

3.10.2. qRT-PCR

Reverse transcription is performed immediately following the protocol reported previously (8).

1. Reversely transcribe 2 μg of total RNA by M-MuLV reverse transcriptase and oligo-dT (New England Biolabs, Inc., USA) in 20 μL 1× M-MuLV reverse transcriptase reaction buffer (New England Biolabs, Inc.).
2. Amplify cDNA by real-time PCR with gene specific primers using the SYBR Green procedure on an ABI-7000 system.
3. Determine the relative abundance of β-actin and use it as an internal standard.
4. Perform all calculations and statistical analyses as described in the ABI 7000 sequence detection system User Bulletin 2.

3.10.3. Fish Microarray

1. For microarray analysis, construct the control RNA by pooling equal amounts of RNA from four individual control ovary samples. For treated groups, use both pooled (equal amounts

of RNA, 250 ng each, from four individual treated ovary samples) and two individual ovary RNA samples in the microarray analysis.

2. Conduct cRNA synthesis and transcription using a two-color Amplification Kit and Gene Expression Hybridization Kit, according to the manufacturer's protocols.
3. Hybridize the labeled samples to the arrays for 17 h at 65°C.
4. Wash, and immediately scan the arrays using a G2505 B microarray scanner.
5. Extract individual spots using Feature Extraction 9.5. For identification of genes that are differentially expressed between exposed and control zebrafish, calculate *p*-values of Student's 1-sample *t*-test and mean fold changes between the exposed and the control samples. Use the threshold based on both fold change and *p*-value (fold change > 2 and *p*-value <0.05) to select probes that respond to the exposure.
6. Select genes with consistent expression patterns under different $nTiO_2$ treatment conditions to control the false positive rate.
7. Use hierarchical clustering analysis to search for common gene expression patterns. Apply our recently developed integrative framework (17) to address the multiple testing situations that arise in the context with the simultaneous observation of more than 40,000 probes in the microarray, fusing different types of knowledge and data sources to comprehensively evaluate the biological relevance of the genes.
8. Obtain human homologues of zebrafish genes from the ZFIN consortium (http://zfin.org) and use pathway analysis with the IPA (Ingenuity Pathways Analysis) system to identify the significantly affected biological processes and the involved genes.
9. Assign zebrafish genes to GO categories according to the annotations from the ZFIN consortium. Use GO enrichment analysis to identify significantly affected GO terms (from biological process, molecular function, and cellular component catalogs) and their related genes.
10. Use web search with GenBank (http://www.ncbi.nlm.nih.gov/Genbank/) and references from PubMed (http://www.ncbi.nlm.nih.gov/pubmed/) to understand the functions of the genes.
11. Download gene expression data from other studies that address disrupted ovary development, such as TCDD-treated zebrafish, from Gene Expression Omnibus (GEO, http://www.ncbi.nlm.nih.gov/geo/), and use cross-array analysis to analyze the expression patterns of the selected genes in other studies to help us understand the mode of action of $nTiO_2$-mediated perturbation of zebrafish reproduction.

12. Integrate different evaluation criteria to generate a final list of genes that were differentially expressed.
13. Verify the gene expression of some important and interesting genes on the list by qRT-PCR.

3.11. Statistical Analysis

All exposure experiments are repeated three times independently, and data should be presented as the mean ± standard error of the mean (SEM). For the embryo/larvae bioassays, a one-way analysis of variance (ANOVA) with Tukey's multiple comparisons should be used to identify significant differences between the treatments and controls. The lowest observed effect observed effect concentrations (LOECs) can be determined if the treatments that were statistically different from controls ($p < 0.05$). LOECs were calculated for sublethal concentrations with >50% survival.

4. Notes

1. Stock solutions are usually stable for at least 1 month.
2. A digital camera is used to capture images of particles as they flow through the sensing zone of the micro-flow imager.
3. There is a growing consensus that properly characterized NPs are essential to the assessment of the potential toxicity of NPs in biological systems (18, 19), which is attributed to the following considerations: (1) consistent physicochemical properties of NPs are critical to the identification of their contribution to the biological effects; (2) commercial NPs in solution have a tendency to agglomerate or form multiple single particle clusters during storage, which often behave differently from primary nanoparticles; (3) to minimize variations, NPs with defined characteristics are usually synthesized or obtained from the same source until standard reference NP become available. Consequently, we routinely perform detailed characterization of NPs prepared in our fish culture medium.
4. Sedimentation is conducted at room temperature for half hour.
5. It has been recognized that a fraction of nZnO could be dissolved as metal ions (20). Thus it is important to distinguish the toxicity inflicted by nZnO from the one by dissolved Zn^{2+}.
6. NPs were frequently observed to form much larger aggregates and settle out of the water column in aquatic environments (8), implicating a scenario of their accumulation in and interactions with sediments. Thus, the influence of sediments on the bioavailability and/or toxicity of NPs was evaluated here as well.

7. Malformation: including deformation of head and tail, reduced mouth and jaw formation, lordosis and edema, etc.
8. Pericardia edema: is swelling caused by fluid retention—excess fluid is trapped in the pericardial chamber of embryos, which is often indicative of a problem in the heart.
9. Oxidative stress, represented by reactive oxygen species (ROS), has been implicated in NP-induced toxicity both in vitro and in vivo (8, 21). To assess the ROS level generated in the zebrafish embryos upon being exposed to NP, we employ a fluorescence-based assay, using a cell-permanent ROS indicator, H2DCF-DA fluorescent probe (8). The acetates of colorless H2DCF-DA are cleaved by intracellular esterases to be converted to fluorescent (dichloro) fluorescein (DCF) in the presence of ROS, which can be detected by a flow cytometer or fluorescence microscopy.
10. In addition to the knowledge on the concentrations and size distribution of NPs in the environment and their exerted toxicity to biota, reliable ecological risk assessments of NPs also require information on the fate and transport of NPs, i.e., their bioconcentration, bioaccumulation, and biomagnifications in the aquatic environment. Thus, the examination of NP accumulation in fish tissues is critical to the fate assessment of NPs in biota.

Acknowledgments

This study was supported by the U.S. Environmental Protection Agency Science to Achieve Results Program (Grant # RD831713) (to Y.S. Chen and Y. Chang). Article contents are sole responsibility of the authors and do not represent official views of the sponsors.

References

1. Segner H (2009) Zebrafish (*Danio rerio*) as a model organism for investigating endocrine disruption. Comp Biochem Physiol C Toxicol Pharmacol 149:187–195
2. Schirmer K, Fischer BB, Madureira D, Pillai S (2010) Transcriptomics in ecotoxicology. Anal Bioanal Chem 397:917–923
3. Fako VE, Furgeson DY (2009) Zebrafish as a correlative and predictive model for assessing biomaterial nanotoxicity. Adv Drug Deliv Rev 61:478–486
4. Velzeboer I, Hendriks AJ, Ragas AM, van de Meent D (2008) Aquatic ecotoxicity tests of some nanomaterials. Environ Toxicol Chem 27:1942–1947
5. Griffitt RJ, Luo J, Gao J, Bonzongo JC, Barber DS (2008) Effects of particle composition and species on toxicity of metallic nanomaterials in aquatic organisms. Environ Toxicol Chem 27:1972–1978
6. Zhu XS, Zhu L, Duan Z, Qi R, Li Y, Lang YJ (2008) Comparative toxicity of several metal oxide nano-particle aqueous suspensions to zebrafish (*Danio rerio*) early developmental stage. J Environ Sci Health A 43: 278–284
7. Griffitt RJ, Hyndman K, Denslow ND, Barber DS (2009) Comparison of molecular and histological changes in zebrafish exposed to metallic nanoparticles. Toxicol Sci 107:404–415

8. Zhu XS, Wang JX, Zhang XZ, Chang Y, Chen YS (2009) The impact of ZnO nanoparticle aggregates on the embryonic development of zebrafish (Danio rerio). Nanotechnology 20: 195103
9. Zhu XS, Wang JX, Zhang XZ, Chang Y, Chen YS (2010) Trophic transfer of TiO2 nanoparticles from daphnia to zebrafish in a simplified freshwater food chain. Chemosphere 79: 928–933
10. Balnois E, Wilkinson KJ, Lead JR, Buffle J (1999) Atomic force microscopy of humic substances: effects of pH and ionic strength. Environ Sci Technol 33:3911–3917
11. Schulte C, Nagel R (1994) Testing acute toxicity in the embryo of zebrafish, *Brachydanio rerio*, as an alternative to the acute fish test-preliminary results. Atla-Alternatives Lab Anim 22:12–19
12. Hill AJ, Bello SM, Prasch AL, Peterson RE, Heideman W (2004) Water permeability and TCDD-induced edema in zebrafish early-life stages. Toxicol Sci 78:78–87
13. Osman AGM, Wuertz S, Mekkawy IA, Exner H-J, Kirschbaum F (2007) Lead induced malformations in embryos of the African catfish *Clarias gariepinus* (Burchell, 1822). Environ Toxicol 22:375–89
14. Kosmehl T, Hallare AV, Reifferscheid G, Manz W, Braunbeck T, Hollert H (2006) A novel contact assay for testing genotoxicity of chemicals and whole sediments in zebrafish embryos. Environ Toxicol Chem 25:2097–2106
15. Wilson-Rawls J, Hurt CR, Parsons SM, Rawls A (1999) Differential regulation of epaxial and hypaxial muscle development by paraxis. Development 126:5217
16. Selman K, Wallace RA, Sarka A, Qi XP (1993) Stages of oocyte development in the zebrafish, *Brachydanio rerio*. J Morphol 218:203–224
17. Zhao Z, Wang JX, Sharma S, Agarwal N, Liu H, Chang Y (2010) An integrative approach to identifying biologically relevant genes. In: Proceedings of SIAM international conference on data mining (SDM), Columbus, April 29–May 1, 2010
18. Hood E (2004) Nanotechnology: looking as we leap. Environ Health Perspect 112:A740–A749
19. The Royal Society (2004) Nanoscience and nanotechnologies: opportunities and uncertainties. RS Policy document 19/04
20. Zhang Y, Chen YS, Westerhoff P, Hristovski K, Crittenden JC (2008) Stability of commercial metal oxide nanoparticles in water. Water Res 42:2204–2212
21. Zhu X, Zhu L, Lang Y, Chen Y (2008) Oxidative stress and growth inhibition in the freshwater fish Carassius auratus induced by chronic exposure to sublethal fullerene aggregates. Environ Toxicol Chem 27(9):1979–85

Chapter 21

Applications of Subsurface Microscopy

Laurene Tetard, Ali Passian, Rubye H. Farahi, Brynn H. Voy, and Thomas Thundat

Abstract

Exploring the interior of a cell is of tremendous importance in order to assess the effects of nanomaterials on biological systems. Outside of a controlled laboratory environment, nanomaterials will most likely not be conveniently labeled or tagged so that their translocation within a biological system cannot be easily identified and quantified. Ideally, the characterization of nanomaterials within a cell requires a nondestructive, label-free, and subsurface approach. Subsurface nanoscale imaging represents a real challenge for instrumentation. Indeed the tools available for high resolution characterization, including optical, electron or scanning probe microscopies, mainly provide topography images or require taggants that fluoresce. Although the intercellular environment holds a great deal of information, subsurface visualization remains a poorly explored area. Recently, it was discovered that by mechanically perturbing a sample, it was possible to observe its response in time with nanoscale resolution by probing the surface with a micro-resonator such as a microcantilever probe. Microcantilevers are used as the force-sensing probes in atomic force microscopy (AFM), where the nanometer-scale probe tip on the microcantilever interacts with the sample in a highly controlled manner to produce high-resolution raster-scanned information of the sample surface. Taking advantage of the existing capabilities of AFM, we present a novel technique, mode synthesizing atomic force microscopy (MSAFM), which has the ability to probe subsurface structures such as non-labeled nanoparticles embedded in a cell. In MSAFM mechanical actuators (PZTs) excite the probe and the sample at different frequencies as depicted in the first figure of this chapter. The nonlinear nature of the tip–sample interaction, at the point of contact of the probe and the surface of the sample, in the contact mode AFM configuration permits the mixing of the elastic waves. The new dynamic system comprises new synthesized imaging modes, resulting from sum- and difference-frequency generation of the driving frequencies. The specific electronics of MSAFM allows the selection of individual modes and the monitoring of their amplitude and phase. From these quantities of various synthesized modes a series of images can be acquired. The new images contain subsurface information, thus revealing the presence of nanoparticles inside the cells.

Key words: Nanotoxicology, Subsurface microscopy, Ultrasonic microscopy, Atomic force microscopy, Mode synthesizing atomic force microscopy, Nanoparticles, Nanomaterials, Intracellular

Joshua Reineke (ed.), *Nanotoxicity: Methods and Protocols*, Methods in Molecular Biology, vol. 926,
DOI 10.1007/978-1-62703-002-1_21, © Springer Science+Business Media, LLC 2012

1. Introduction

An increasing array of engineered nanomaterials is under development for diverse industrial, consumer, and medical applications (1–4). In parallel, safety and health issues related to nanoparticles exposure continue to be debated (5–10). The new area of research to address potential safety concerns consists of toxicity studies, development of novel methods, screenings and characterization of nanomaterials and their effect on biological systems is now referred to as nanotoxicology (10–13). Assessing potential toxicity is complicated by a number of factors (14). One of the most significant challenges is the ability to track the fate of nanomaterials as they enter the body and potentially penetrate biological tissue, which requires the ability to noninvasively monitor the response of the system with high spatial resolution. Furthermore, a number of studies (6, 9, 14, 15) suggest that particle-specific toxicity studies are necessary, with a careful screening for key characteristics such as the size (12), the shape, the crystallinity, the chemical composition, the surface properties (area, chemistry or charge) (16) to reach a full understanding of the mechanisms of the nanoparticle–biomaterial interaction. Imaging techniques capable of localizing and visualizing nanoparticles with nanometer resolution while keeping the cellular structure intact are therefore of tremendous importance (2). The development of analytical techniques to assess biodistribution, translocation, and cellular uptake, both in vivo and in vitro, is a key component for the advancement of the field. The ability to track nanoparticles uptake at the level of a single cell would allow mechanistic studies of particle translocation and distribution within a cell, which are central to dissecting potential biological effects. Further, with respect to screening for exposures, such a technique would create a tool to examine a surrogate biological sample such as peripheral blood that could be collected with minimal invasion. We focus here on the visualization of nanoparticles inside animal cells using a novel technique called mode synthesizing atomic force microscopy (MSAFM) (17–20). We also review the various technologies, which have been used to determine the presence of nanoparticles in cells.

1.1. Optical Imaging

Confocal microscopy is a powerful optical imaging technique commonly used in biology. Confocal imaging has the ability to noninvasively investigate an organism and reconstruct a 3D image of the system, providing insights into the structure and behavior of the system, an important attribute for nanotoxicology studies. Many studies have reported the subcellular localization of nanoparticles in cells using confocal fluorescence microscopy (15, 21, 22). Fluorescence measurements are carried out by labeling the nanoparticles before permitting them to interact with the cell cultures. Although this process allows recognition and tracking of the nanoparticles in the different organelles of the cell, the dyes used as labels make it difficult to separate the biological properties of the nanoparticles from the action of the dyes. With techniques such as confocal microscopy, bright field or dark field microscopy, it is only possible to observe phenomena (delivery, cellular uptake, etc.) of nanomaterials larger than the diffraction limit of light. It is then necessary to turn to other complementary techniques for advanced studies involving lower concentration of nanoparticles and their localized interaction with biological tissues (21).

1.2. Electron Microscopy

Electron microscopy is an imaging technique commonly used in nanotoxicology studies, owing to its high spatial resolution and depth of field (23, 24) capability. Scanning electron microscopy (SEM) and transmission electron microscopy (TEM) provide important insight into nanomaterial characterization. SEM is usually used for morphology inspection of the cells before and after exposure to the nanoparticles. Attempts to use TEM to locate nanotubes within alveolar macrophages positively identify only aggregates (24, 25). Porter et al. reported the observation of carbon nanotubes in lysosomes (25). However, the composition of cells is rich in carbon. One of the difficulties encountered when using electron microscopy to trace nanoparticles inside a cell is the low contrast between the cell structures and the nanomaterials, in particular for unlabelled carbon-based particles such as carbon nanotubes (25). Staining the sample sections prior to TEM imaging can be performed in order to increase the contrast between nanomaterials and cell structures. However, it is time consuming, labor intensive, and requires the use of heavy metal (osmium tetroxide, uranyl acetate, lead citrate) that may disrupt the biological sample. Other electron microscope-based approaches, such as electron energy loss (EEL) or energy-dispersive X-ray spectroscopies have also been used to trace nanoparticle uptake in cells (25). The information extracted from the bulk plasmon peak in the EEL (26) data revealed the presence of carbon nanotubes in the lysosomes of the cell after 2 days.

Confocal fluorescence and electron microscopy are best used together to get a good representation of material uptake and localization. However, in many cases, for instance when unlabeled carbon nanoparticles are involved, it remains difficult to locate and identify single nanoparticles within a cell. It is even more complex and challenging to operate on living cells, especially in the high-vacuum conditions of electron microscopy.

1.3. Scanning Probe Microscopy

A range of techniques in scanning probe microscopy (SPM) are available to noninvasively characterize materials and access physical, structural, and chemical properties with high spatial resolution (27). Atomic force microscopy (AFM) is now widely used in life sciences (28–30). It is advantageous because of simpler sample preparation and the ability for ambient imaging. However, conventional AFM typically only probes the surface of a specimen, making it difficult to analyze structures within a cell (31). AFM imaging is generally performed using contact or tapping mode. In contact mode, the tip of the cantilever probe is in direct contact with the sample during the image scan, exerting a force of 50–100 pN on the sample at all times. In tapping mode, the cantilever is mechanically excited close to its resonance frequency. The tip and the sample contact only at the lowest position of the oscillation, significantly decreasing the forces applied on the sample. Tapping mode is very popular in biological applications (28, 29), especially for imaging of soft samples or samples that are not well attached to the substrate, which is generally the case in liquid environments. The measurements performed in contact or tapping mode provide some insight on the surface and interfacial properties on the system of interest. However, subsurface imaging with nanoscale resolution remains an unexplored area (32–35). One may envision this limitation as being due to a lack of mechanical excitation of the sample resulting in the inability of the probe to gather the subsurface dynamic attributes (36–39). MSAFM (19, 20) offers a potential circumvention that is capable of subsurface imaging of soft samples and thus is a potential method for probing cellular uptake of nanoparticles. The phase information in the coupled modes may be detected at various frequencies corresponding to a synthesized mode (20), which can be displayed as a function of the spatial location of the scanning cantilever tip (Fig. 1). The resultant amplitude or phase map may then reveal any contrasts due to acoustic impedance variation that results from nanoscale heterogeneity in a volume of the cell directly underneath the AFM tip and amounts to a mapping of the cell's elastic response. We explored this viability by using MSAFM to determine the cellular fate of single-wall carbon nanohorns (SWCNH) and silica nanospheres using a mouse model to detect and visualize particles within lavage macrophages and erythrocytes (blood cells) (18, 19) as described in Fig. 2.

2. Materials

2.1. Sample Preparation

1. Isoflurane anesthesia to sacrifice the mice.
2. Bell jar or other air-tight container for anesthesia.
3. Blunt 22-gauge needle for bronchoalveolar lavage (BAL) fluid collection.
4. Cold sterile phosphate buffered saline (PBS).
5. Sterile tubes (1.5 ml).
6. Cytospin.
7. Hematology stain (e.g., Hema3, Fisher Scientific).
8. Cleaved mica (see Note 1).

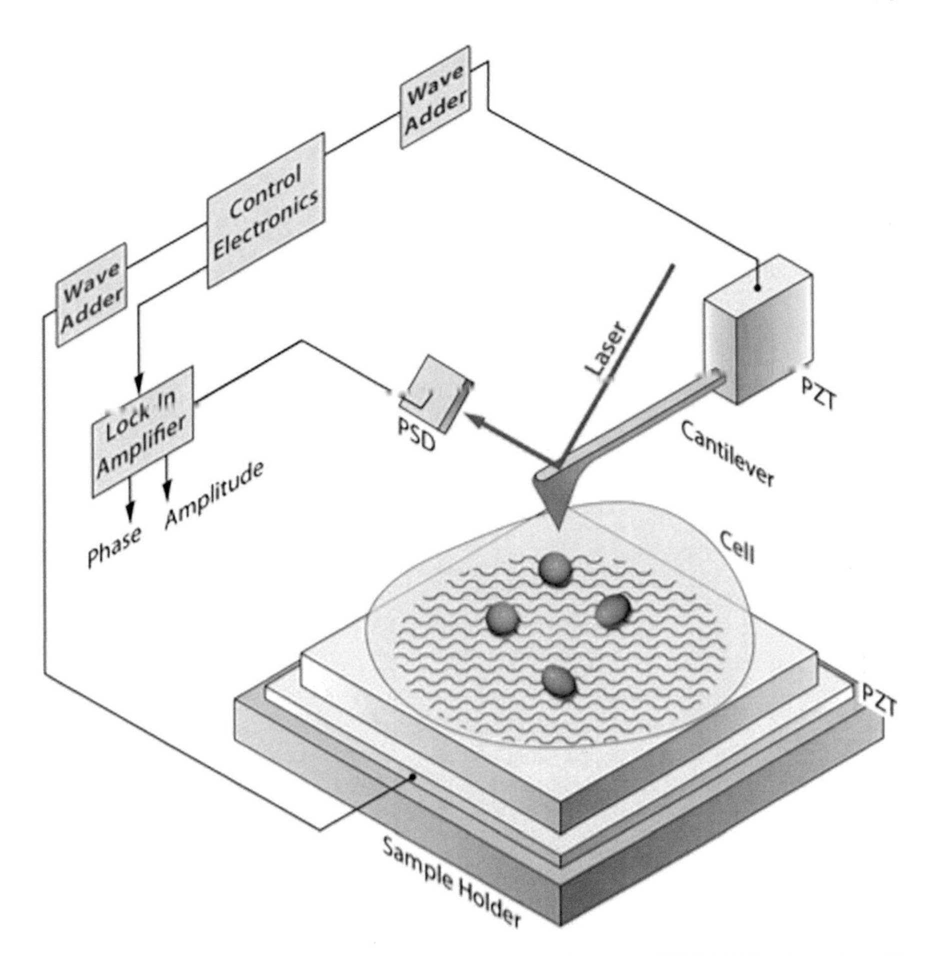

Fig. 1. Mechanism of operation of the mode synthesizing atomic force microscope (MSAFM). The dynamics of the cantilever is monitored using an optical readout system (laser). The signal is detected by the lock-in amplifier using the difference frequency as reference. The images emerging from MSAFM reveal the presence of nanoparticles below the cell surface.

9. Methanol.
10. Heparinized capillary tubes for blood collection.
11. Mica substrates for the samples.

3. Methods

3.1. Sample

In order to study the fate of the nanoparticles in biological system, it is important to choose a route of exposure corresponding to a real life situation. For instance, pharyngeal aspiration has been reported as an efficient method for exposure and distribution of nanoparticles throughout the alveolar regions of the lungs. In the case of carbon nanoparticles, the OSHA-permissible limit of exposure across twenty 8-h days or twenty days of 8-h days is equivalent to exposing a mouse to 30 μg of carbon nanohorns SWCNH.

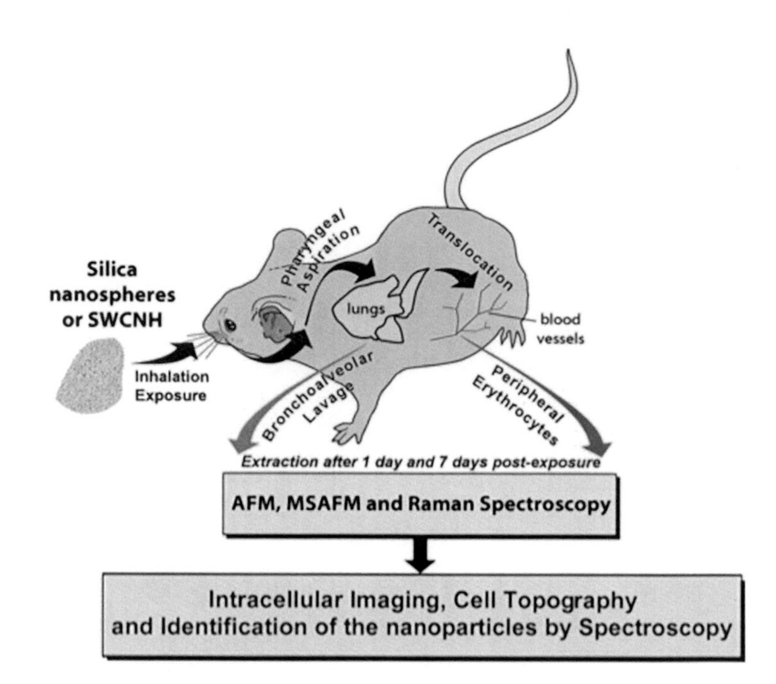

Fig. 2. Protocol for the mice exposure to SWCNH or silica nanospheres. Each mouse was exposed to 30 μg of SWCNH. Control and SWCNH-exposed mice were sacrificed 24 h and 7 days post-exposure. Mouse pharyngeal aspiration was used for carbon nanohorn administration. Mice were sacrificed by isoflurane overdose in a bell jar, and BAL was performed according to standard protocols. For light microscopy analysis, cytospin slides were stained with a Hema3 kit (Fisher Scientific). For AFM analysis, cells were centrifuged onto freshly cleaved mica using a cytospin and fixed with methanol. In addition, peripheral blood was collected using heparinized capillaries from the abdominal aorta. Blood samples were diluted in PBS, centrifuged onto freshly cleaved mica using a cytospin, and fixed with methanol. The samples were characterized using atomic force microscopy (AFM), mode synthesizing force microscopy (MSAFM) and micro-Raman spectroscopy.

The nanoparticles were administered by pharyngeal aspiration after anesthetizing the mice with isoflurane. The solution of nanoparticles or the control solution was applied in a bolus on the back of the tongue, which is extended using forceps. The tongue was released after the fluid was fully aspired into the lungs. The mouse was allowed to recover in its home cage after exposure and was anesthetized at time points of choice for collection of BAL fluid and/or blood, as follows:

1. Mice are sacrificed by isoflurane overdose in a bell jar.
2. Bronchoalveolar lavage (BAL) fluid is collected according to standard protocols (40).
3. The trachea is exposed and a blunt 22-gauge needle inserted into the trachea. After securing the needle with sutures, lavage is performed five times with cold sterile PBS.
4. Fluid is gently aspirated while massaging the chest. The first lavage is performed with 0.6 ml of PBS and was kept separate for analysis for another study.

5. The second and third lavages are performed with 1.0 ml of PBS and are pooled in sterile tubes, centrifuged, and resuspended in PBS.
6. For light microscopy analysis, cytospin slides are stained with a Hema3 kit.
7. For AFM analysis, cells are centrifuged onto freshly cleaved mica (see Note 1) using a cytospin and fixed with methanol (see Notes 2 and 3).
8. Peripheral blood is collected from the abdominal aorta into heparinized capillary tubes.
9. Blood samples are diluted ~10-fold in PBS, centrifuged onto freshly cleaved mica using a cytospin, and fixed with methanol (see Notes 2 and 3).
10. The mica substrate is affixed onto the sample actuator with glue.

3.2. Mode Synthesizing Atomic Force Microscopy

The MSAFM was developed on a modified multimode AFM (Nanoscope III, Digital Instrument). MSAFM operates on the basis of expanding the dynamic attributes of the probe and the sample of an AFM to boost the frequency content of the system. Actuators, located at the base of the cantilever and the sample (Fig. 1), exerted mechanical forces on the probe and the substrate, respectively (see Notes 4 and 5), of the order of few kilohertz to several tens of megahertz. The acoustic waves, created at the bottom of the sample, transmitted through the sample and modified the position of the tip of the cantilever that is in contact with the sample (see Note 6). At the tip–sample interface, the nanomechanical interaction resulted in a nonlinear force that leads to difference and sum frequency generation yielding a series of synthesized oscillations via higher order coupling in the system. The evolution of the probe's state in time is monitored using the position sensitive detector of the AFM. The resulting signal $S(t)$ is interpreted as a contrast map, representative of the forces at each point of the scanned area. In Fourier space, $S(t)$ takes the form of a rich spectrum, in which the synthesized modes created in the system are engaged for amplitude or phase imaging. An MSAFM session can be established from the following instructions:

1. The laser of the optical read-out system is aligned with the cantilever.
2. The sample is mounted on the AFM sample holder.
3. Using the MSAFM module, the actuators are connected to waveform generators.
4. The approach is executed by engaging the cantilever for contact imaging (see Note 2).
5. The excitation frequencies are tuned to generate a set of frequencies interesting to probe the sample (see Note 7).

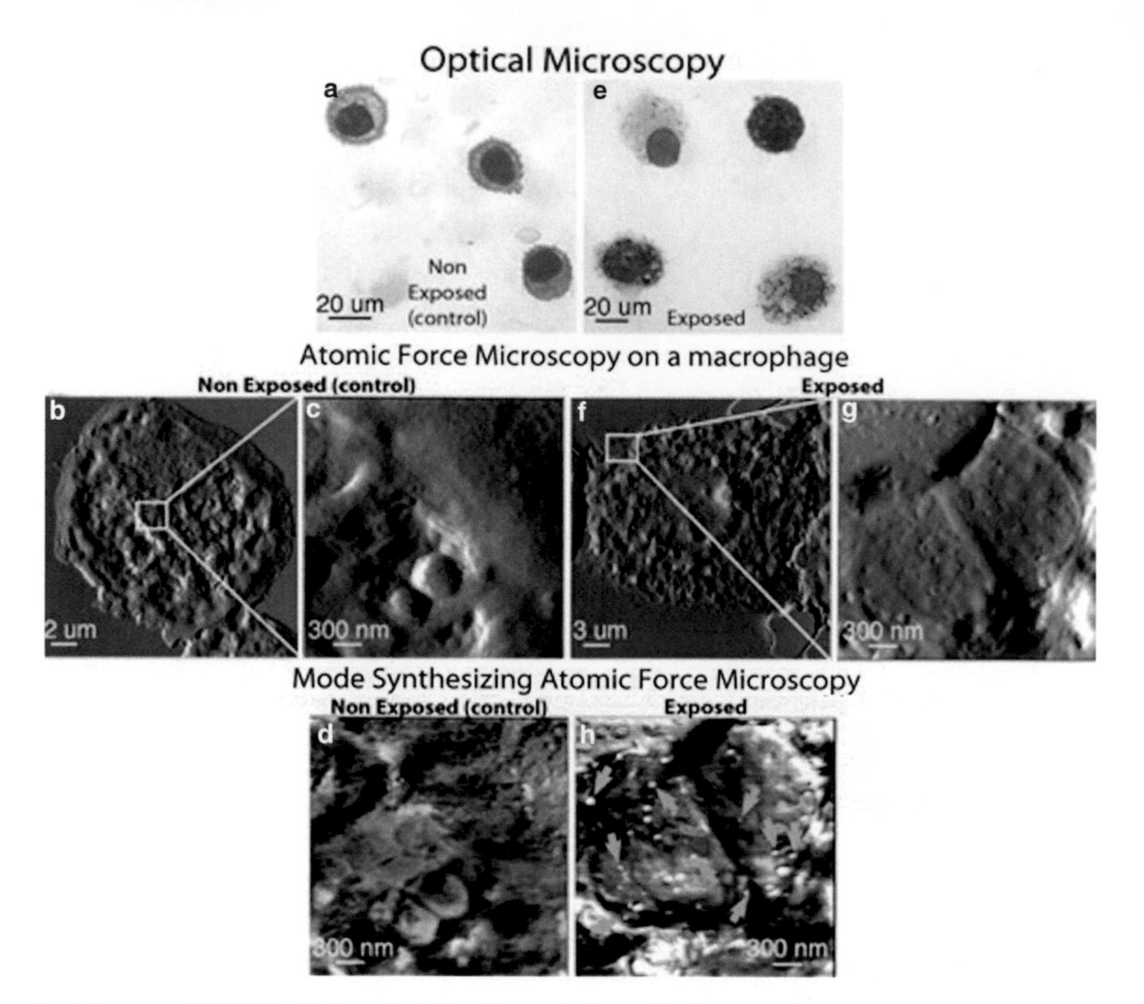

Fig. 3. Presence of SWCNH inside the cells obtained from mice lungs. Alveolar macrophages from mice exposed to vehicle control (**a–d**) and SWCNH after 7 days (**e–h**). Representative optical (**a**, **e**), AFM topography (**b**, **c**, **f**, **g**), and MSAFM phase (**d**, **h**) images from control and treated mice. *White dots* in (**h**) (some indicated by the *green arrows*) correspond to SWCNH (Color figure online).

6. The frequencies of the modes of interest are selected. These frequencies will be used as reference in the lock-in measurement stage.
7. The signal input $S(t)$ is sent as input for lock-in measurements. The resulting signals (amplitudes or phases) are recorded.
8. The mode of images to be displayed (mode, amplitude, or phase) is selected. A standard AFM topography image is obtained along with amplitude and phase images extracted from the MSAFM module.

The resulting images, presented in Fig. 3, display the surface and subsurface information of the sample, including embedded particles (see Note 8). The morphology of the nanoparticles alone was

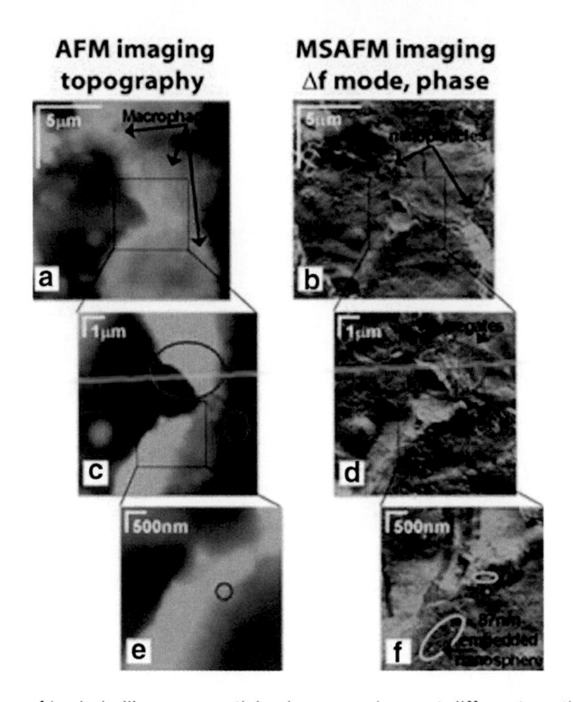

Fig. 4. Images of buried silica nanoparticles in macrophage at different spatial resolution (15, 8, and 4 μm scans). (*Left*) AFM images. (*Right*) MSAFM phase images obtained by locking on the difference mode Δf. $f_c = 4.248$ MHz, $a_c = 5.4$Vpp, $f_s = 3.950$ MHz, $a_s = 2.2$Vpp (Color figure online).

examined using standard AFM imaging. Similar measurements, presented in Fig. 4, were carried out for silica nanospheres (17, 18). Silica and carbon have very different elastic properties (stiffness and density).

3.3. Spectroscopy Measurements

Although the MSAFM images reveal the presence of nanoscale features inside the cell, their chemical composition remains unknown. The portfolio of tools capable of imaging cell structures for chemical recognition with high resolution is very limited. For example, spectroscopic measurements are diffraction limited.

We carried out Raman spectroscopy measurements on the samples in order to corroborate the presence/absence of carbon nanohorns inside the cell (19) (see Note 9). The results are presented in Fig. 5. The Raman spectra were obtained from a Renishaw micro-Raman spectrometer, equipped with a 785 nm Near Infrared laser and a Leica Raman Imaging microscope. In micro-Raman spectroscopy, the Near IR laser is used in conjunction with an optical microscope to excite the sample, and the inelastic scattering resulting from the light impinging the sample is monitored using a monochromator and photon sensitive detector. Raman effect results from the relaxation of the excited molecules accompanied

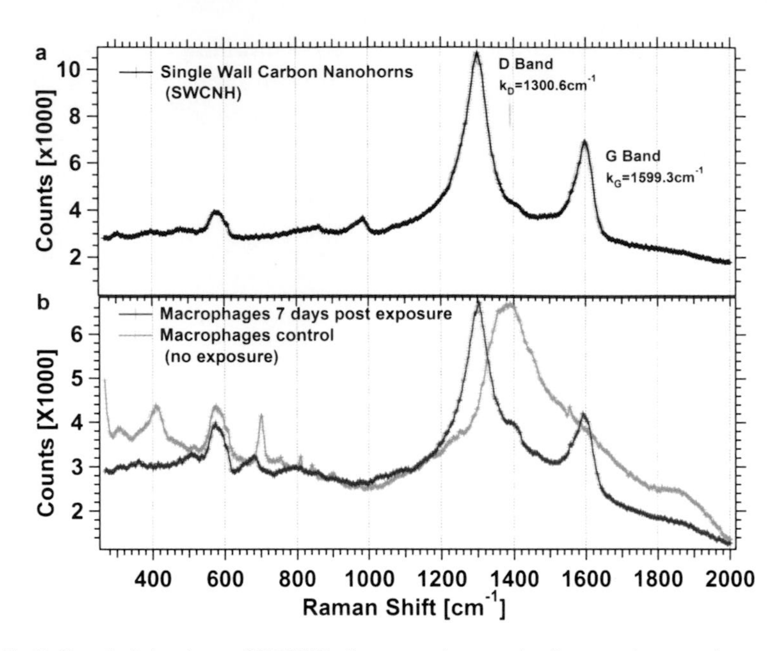

Fig. 5. Chemical signatures of SWCNH in the macrophages using Raman micro-spectroscopy. Raman spectrum of the solution of SWCNH used for the aspiration (**a**) and of the exposed and non-exposed macrophages (**b**). The measurements presented were acquired on the macrophages.

by the photon emission while the molecules return to a lower energy state. The following steps can be followed to carry out the Raman measurements:

1. Prepare the sample of a substrate that is transparent to the excitation light source; here we used glass.
2. Perform a background scan to ensure that the focus of the microscope is optimized.
3. Insert the sample in the beam path in the sample compartment.
4. Perform control measurements on the as-produced nanomaterial.
5. Perform the measurements on the cell samples. Depending on the model of the microspectrometer, it is possible to obtain spectra at various locations of the sample by manual translation or taking advantage of an automated X–Y stage.

This procedure is important in order to identify the regions where nanoparticles are present. One of the major drawbacks is the lack of spatial resolution and sensitivity for lower concentration of nanomaterials. However, the observations made with MSAFM were strengthened by the Raman spectroscopy measurements, confirming the presence of carbon nanohorns in the macrophages (17, 19).

4. Notes

1. The surfaces usually used for AFM sample preparation include silicon (wafer), mica, glass (slide), or gold surfaces. Special precautions should be taken to ensure the surface properties of the substrate (hydrophobic/hydrophilic, electrostatic, etc.) and those of the sample are compatible. Mica substrates should be used because the glass slides readily respond to ultrasonic waves.
2. The measurements were carried out on dried samples in ambient conditions. At the scale of the tip, there exists a thin film of water at the surface of the sample. Because of the capillary forces created by the meniscus between the tip and the sample, the AFM resolution is not optimum when operated in contact mode. Imaging in liquid, although minimizing capillary and hydrophobic/hydrophilic forces, would also present new challenges due to the new liquid environment (28, 29).
3. Imaging in liquid should be considered for better preservation of the cell structure and minimization of capillary and hydrophobic/hydrophilic forces.
4. In the case of our study the actuators that created the mechanical stimulus to the probe and sample were piezoelectric crystals.
5. Improvements can be made in the contact between sample and sample actuator.
6. The microcantilever properties, such as spring constant, length and probe tip shape, determine the quality of the images. Softer cantilevers (i.e. less stiff) should be used to preserve the cells. One should have a precise idea of the conditions of operation for a given measurements, such as the forcing amplitude and frequencies, in order to assess the best suited microcantilever probe.
7. The new (synthesized) modes result from the difference and sum frequency generation of the external forces applied on the system (20). The amplitude of the new modes generally reaches a maximum when matching one of the resonance frequencies of the tip–sample system (20). However, the optimization of the settings for imaging does not necessarily correspond to the maximum amplitude of the mode and remains a complex process (17).
8. Salt create crystals on the sample, making it difficult to locate engineered embedded nanoparticles. Contamination is a serious problem in the preparation of samples for AFM studies, since AFM resolves nanometer scale features.
9. There is a need for chemical characterization to ensure that the features revealed by the images are the nanoparticles. One of the major drawbacks remains the lack of sensitive and specific high spatial resolution chemical identification techniques.

Acknowledgments

The study was supported by the Oak Ridge National Laboratory (ORNL) Technology Maturation Fund. ORNL, Oak Ridge, Tennessee, 37831-6123, is managed by UT, Battelle, LLC for the Department of Energy under Contract No. DE-AC05-0096OR22725.

References

1. Fan X, Tan J, Zhang G, Zhang F (2007) Isolation of carbon nanohorn assemblies and their potential for intracellular delivery. Nanotechnology 18:195103–195108
2. Stone V, Donaldson K (2006) Nanotoxicology—signs of stress. Nat Nanotechnol 1:23–24
3. Sadik OA, Zhou AL, Kikandi S, Du N, Wang Q, Varner K (2009) Sensors as tools for quantitation, nanotoxicity and nanomonitoring assessment of engineered nanomaterials. J Environ Monit 11:1782–1800
4. Strassert CA, Otter M, Albuquerque RQ, Hone A, Vida Y, Maier B, De Cola L (2009) Photoactive hybrid nanomaterial for targeting, labeling, and killing antibiotic-resistant bacteria. Angew Chem Int Ed Engl 48: 7928–7931
5. Colvin V (2003) The potential environmental impact of engineered nanomaterials. Nat Biotechnol 21:1166–1170
6. Holsapple M, Farland W, Landry T, Monteiro-Riviere N, Carter J, Walker N, Thomas K (2005) Research strategies for safety evaluation of nanomaterials. Part II. Toxicological and safety evaluation of nanomaterials, current challenges and data needs. Toxicol Sci 88: 12–17
7. Kreyling WG, Semmler-Behnke M, Moeller W (2006) Health implications of nanoparticles. J Nanopart Res 8:543–562
8. Medina C, Santos-Martinez MJ, Radomski A, Corrigan OI, Radomski MW (2007) Nanoparticles: pharmacological and toxicological significance. Br J Pharmacol 150: 552–558
9. Nel A, Xia T, Madler L, Li N (2006) Toxic potential of materials at the nanolevel. Science 311:622–627
10. Oberdorster G, Oberdorster E, Oberdorster J (2005) Nanotoxicology: an emerging discipline evolving from studies of ultrafine particles. Environ Health Perspect 113:823–839
11. Myllynen P (2009) Nanotoxicology: damaging DNA from a distance. Nat Nanotechnol 4: 795–796
12. Dawson KA, Salvati A, Lynch I (2009) Nanotoxicology: nanoparticles reconstruct lipids. Nat Nanotechnol 4:84–85
13. Donaldson K, Poland CA (2009) Nanotoxicology: new insights into nanotubes. Nat Nanotechnol 4:708–710
14. Donaldson K, Aitken R, Tran L, Stone V, Duffin R, Forrest G, Alexander A (2006) Carbon nanotubes: a review of their properties in relation to pulmonary toxicology and workplace safety. Toxicol Sci 92:5–22
15. Verma A, Uzun O, Hu Y, Hu Y, Han H-S, Watson N, Chen S, Irvine DJ, Stellacci F (2008) Surface-structure-regulated cell-membrane penetration by monolayer-protected nanoparticles. Nat Mater 7:588–595
16. Auffan M, Rose J, Bottero J-Y, Lowry GV, Jolivet J-P, Wiesner MR (2009) Towards a definition of inorganic nanoparticles from an environmental, health and safety perspective. Nat Nanotechnol 4:634–641
17. Tetard L, Passian A, Farahi RH, Thundat T (2010) Atomic force microscopy of silica nanoparticles and carbon nanohorns in macrophages and red blood cells. Ultramicroscopy 6: 586–591
18. Tetard L, Passian A, Lynch RM, Voy BH, Shekhawat G, Dravid VP, Thundat T (2008) Elastic phase response of silica nanoparticles buried in soft matter. Appl Phys Lett 93:133113
19. Tetard L, Passian A, Venmar KT, Lynch RM, Voy BH, Shekhawat G, Dravid VP, Thundat T (2008) Imaging nanoparticles in cells by nanomechanical holography. Nat Nanotechnol 3:501–505
20. Tetard L, Passian A, Thundat T (2010) New modes for subsurface atomic force microscopy through nanomechanical coupling. Nat Nanotechnol 5:105–109
21. Panyam J, Sahoo SK, Prabha S, Bargar T, Labhasetwar V (2003) Fluorescence and electron microscopy probes for cellular and tissue uptake of poly(,-lactide-co-glycolide) nanoparticles. Int J Pharm 262:1–11

22. Zhang Y, Kohler N, Zhang M (2002) Surface modification of superparamagnetic magnetite nanoparticles and their intracellular uptake. Biomaterials 23:1553–1561
23. Geiser M, Rothen-Rutishauser B, Kapp N, Schurch S, Kreyling W, Schulz H, Semmler M, Hof V, Heyder J, Gehr P (2005) Ultrafine particles cross cellular membranes by nonphagocytic mechanisms in lungs and in cultured cells. Environ Health Perspect 113:1555–1560
24. Worle-Knirsch J, Pulskamp K, Krug H (2006) Oops they did it again! Carbon nanotubes hoax scientists in viability assays. Nano Lett 6:1261–1268
25. Porter AE, Gass M, Muller K, Skepper JN, Midgley PA, Welland M (2007) Direct imaging of single-walled carbon nanotubes in cells. Nat Nanotechnol 2:713–717
26. Daniels HR, Brydson R, Brown A, Rand B (2003) Quantitative valence plasmon mapping in the TEM: viewing physical properties at the nanoscale. Ultramicroscopy 96:547–558
27. Lang HP, Gerber C (2009) Up close & personal with atoms & molecules. Mater Today 12:18–25
28. Allison DP, Mortensen NP, Sullivan CJ, Doktycz MJ (2010) Atomic force microscopy of biological samples. Wiley Interdiscip Rev Nanomed Nanobiotechnol 2:618–634
29. Kim S, Kihm KD, Thundat T (2010) Fluidic applications for atomic force microscopy (AFM) with microcantilever sensors. Exp Fluids 48:721–736
30. Mao YD, Sun QM, Wang XF, Ouyang Q, Han L, Jiang L, Han D (2009) In vivo nanomechanical imaging of blood-vessel tissues directly in living mammals using atomic force microscopy. Appl Phys Lett 95(1):13704
31. Garcia R, Margerle R, Perez R (2007) Nanoscale compositional mapping with gentle forces. Nat Mater 6:405–411
32. Chung J, Kim K, Hwang G, Kwon O, Lee JS, Park SH, Choi YK (2010) Nanoscale range finding of subsurface structures by measuring the absolute phase lag of thermal wave. Rev Sci Instrum 81(5):053701
33. Shekhawat G, Srivastava A, Avasthy S, Dravid V (2009) Ultrasound holography for noninvasive imaging of buried defects and interfaces for advanced interconnect architectures. Appl Phys Lett 95(26):263101
34. Shekhawat GS, Avasthy S, Srivastava AK, Tark SH, Dravid VP (2010) Probing buried defects in extreme ultraviolet multilayer blanks using ultrasound holography. IEEE Trans Nanotechnol 9:671–674
35. Valdes O, Cuberes MT (2009) Characterization of a new scaffold formed of polyelectrolyte complexes using atomic force and ultrasonic force microscopy. J Biomed Nanotechnol 5: 716–721
36. Rabe U, Arnold W (1994) Acoustic microscopy by atomic force microscopy. Appl Phys Lett 64:1493–1495
37. Kolosov OV, Martin RC, Marsh CD, Briggs GA, Kamins TI, Williams RS (1998) Imaging the elastic nanostructures of ge islands by ultrasonic force microscopy. Phys Rev Lett 81:1046
38. Martinez NF, Patil S, Lozano JR, Garcia R (2006) Enhanced compositional sensitivity in atomic force microscopy by the excitation of the first two flexural modes. Appl Phys Lett 89:153115
39. Sahin O (2008) Scanning below the cell surface. Nat Nanotechnol 3:461–462
40. Lynch RM, Naswa S, Rogers GL, Kania SA, Das S, Chesler EJ, Saxton AM, Langston MA, Voy BH (2010) Identifying genetic loci and spleen gene coexpression networks underlying immunophenotypes in BXD recombinant inbred mice. Physiol Genomics 41:244–253

Chapter 22

Application of ICP-MS for the Study of Disposition and Toxicity of Metal-Based Nanomaterials

Mo-Hsiung Yang, Chia-Hua Lin, Louis W. Chang, and Pinpin Lin

Abstract

Many nanomaterials, such as quantum dots, nano-gold, nano-silver, nano-ZnO, etc., consist of metal components. When these metal-based nanomaterials are used for biological applications, their biological safety must be evaluated. The biological disposition (ADME: absorption, deposition, metabolism, and elimination) of these nanomaterials need to be evaluated. Such evaluation can be made via tracking of the metallic constituents of the nanoparticles in various tissues and organs after exposure. Although atomic absorption (AA) spectrometry is traditionally used for metal analyses, inductively couple plasma mass spectrometry (ICP-MS) is a more modern and preferred technique for metal analyses. ICP-MS has distinct advantages over the traditional AA technique by being much more sensitive, efficient, and effective. Because the metallic contents in nanomaterials are usually of very minute amounts, the use of ICP-MS for their tracking is recommended. Specifics of applications and detailed technical protocols for ICP-MS analyses are provided. Some study results on quantum dots (QDs) and nano-gold (AuNP) with ICP-MS are also illustrated.

Key words: ICP-MS, Nanomaterials, Disposition, Toxicity, Metals

1. Introduction

Nanotechnology development is probably one of the most revolutionary industrial advancements in the twenty-first century. The potential uses and applications of engineered nanomaterials (NMs) are broad, ranging from simple cosmetics, sunscreens to electronics, sporting goods, and biomedical applications (diagnoses, drug delivery, imaging, etc.). Whilst the many potential uses and applications of NMs are promising and exciting (1), concerns on various undesirable health effects of some of the NMs have been raised (2–8). The bio-reactivity of the NMs is closely related to and may vary with the chemico-physical contents, properties, and characteristics of the NMs (2). In this chapter, we will only

Joshua Reineke (ed.), *Nanotoxicity: Methods and Protocols*, Methods in Molecular Biology, vol. 926,
DOI 10.1007/978-1-62703-002-1_22, © Springer Science+Business Media, LLC 2012

focus on the assessments and analyses of those NMs containing metals, such as cadmium, lead, aluminum, etc. Many of these metals are known to be highly toxic to humans. The most fundamental approach to assess an NM after entering into an intact biological system is to understand the deposition, including the absorption, distribution, metabolism, and excretion (ADME) of the NM in vivo. Such information will be critical for one to predict, to assess, and to correlate the toxic effects of the NM in question.

For metal-containing NMs, the tracking (qualitative analyses) and measurement (quantitative analyses) of the metal(s) from the NMs will be important. In the past, AA spectrophotometry was used to identify and to quantify metals in the environment and in tissue specimens. While the AA method is acceptable, it is slow (one metal at a time), relatively insensitive (needs larger specimen and higher metal concentrations), time consuming, and more laborious. We would like to introduce here a more contemporary approach in metal analysis via ICP-MS. ICP-MS has unique characteristics to provide very sensitive, accurate, and precise determinations of multiple metallic elements at the same time (9). The advantages of ICP-MS over AA analyses are many: (1) high sensitivity: much smaller sample size is required and it can measure very low levels (for most elements, normally in the sub-ppb range) of metals in the specimen; (2) excellent efficiency: it can analyze up to 70% of the elements of the periodic table and can determine up to 17 elements simultaneously from one sample if needed (10); (3) desirable effectiveness: it can provide speciation analyses of the metals if desired; (4) better time and labor efficiency.

Many metals are used in the fabrication of engineered NMs (Table 1), including silver, gold, titanium oxide, copper, zinc and cerium oxides as well as cadmium containing quantum dots. Potential toxicities of these metals are of concern to many investigators (11–15). ICP-MS can be used to quantify elements in these metal-containing NMs before they were administrated into the biological systems. It can also locate, quantify, and assess the distribution, deposition, relocation (redistribution), bio-accumulation, disintegration, elimination, and excretion of the NMs in vivo via tracking of the key metal(s) in the NMs. Toxic metals frequently induce disruption of homeostasis of essential trace elements and metals, such as copper, iron, zinc, calcium, magnesium, selenium, manganese, etc., in the biological system. Alterations in these essential elements and metals as a result of exposures to NMs can also be assayed via ICP-MS analyses. Knowing the disposition (ADME) of the toxic metals in tissues and organs as a result of NM exposure, one can then evaluate the safety or risks of NMs in question. We have used ICP-MS very successfully in many of our nanotoxicity and safety studies in the past years and we would like to share our experience here.

Table 1
The isobaric elemental interferences and isobaric polyatomic ion interferences in ICP-MS

Material	Element	Stable Isotopes	Abundance	Isobaric elemental interferences	Isobaric polyatomic ion interferences
QD705	Cd	111	12.8		$^{95}Mo^{16}O^+$, $K_2{}^{16}O_2{}^1H^+$
		112	24.1	^{112}Sn	$^{40}Ca_2{}^{16}O_2$, $^{40}Ar_2{}^{16}O_2$, $^{96}Ru^{16}O^+$, $^{95}Mo^{16}O^1H^+$,$^{96}Mo^{16}O^+$
		114	28.7	^{114}Sn	$^{97}Mo^{16}O^1H^+$,$^{98}Mo^{16}O$, $^{98}Ru^{16}O^+$
	Te	126	18.95	^{126}Xe	$^{110}Pd^{16}O^+$
		128	31.69	^{128}Xe	$^{96}Ru^{16}O_2{}^+$
		130	33.80	^{130}Xe	$^{98}Ru^{16}O_2{}^+$
AuNP	Au	197	100		$^{181}Ta^{16}O^+$
Trace elements in biological sample	Mn	55	100		$^{40}Ar^{14}N^1H^+$, $^{39}K^{16}O^+$, $^{37}Cl^{18}O^+$, $^{40}Ar^{15}N^+$, $^{38}Ar^{17}O^+$
	Cu	63	69.1		$^{40}Ar^{23}Na^+$,$^{31}P^{16}O_2{}^+$,$^{47}Ti^{16}O^+$, $^{23}Na^{40}Ca^+$, $^{46}Ca^{16}O^1H^+$
		65	30.9		$^{32}S^{16}O_2{}^1H^+$, $^{40}Ar^{25}Mg^+$, $^{40}Ca^{16}O^1H^+$, $^{36}Ar^{14}N_2{}^1H^+$
	Zn	64	48.89	^{64}Ni	$^{40}Ar^{24}Mg$,$^{32}S^{16}O_2{}^+$,$^{32}S_2{}^+$,$^{31}P^{16}O_2{}^1H^+$, $^{48}Ca^{16}O^+$, $^{32}S_2{}^+$
		66	27.81		$^{34}S^{16}O_2{}^+$, $^{33}S^{16}O2^1H^+$, $^{32}S^{16}P^{18}O^+$, $^{32}S^{17}O_2{}^+$, $^{33}S^{16}O^17O^+$
		68	18.57		$^{36}S^{16}O_2{}^+$,$^{34}S^{16}O^{18}O^+$, $^{40}Ar^{14}N_2{}^+$, $^{35}Cl^{16}O^{17}O^+$,$^{34}S^{2+}$
	Se	77	7.58		$^{40}Ar^{37}Cl^+$, $^{36}Ar^{40}Ar^1H^+$, $^{38}Ar_2{}^1H^+$,$^{12}C^{19}F^{14}N^{16}O_2{}^+$
		78	23.52		$^{40}Ar^{38}Ar^+$, $^{38}Ar^{40}Ca^+$
		82	9.19		$^{12}C^{35}Cl_2{}^+$,$^{34}S^{16}O_3{}^+$,$^{40}Ar_2{}^1H_2{}^+$, $^{81}Br^1H^+$

Engineered nanomaterials, such as QDs and AuNPs, have great potential in a variety of biomedical applications, including medical imaging, diagnosis, and targeted delivery of therapeutic drugs. QDs are inorganic nanocrystals with a unique autofluorescent property and great potential for applications in medical diagnosis and drug delivery (16–18). Because some QDs contain toxic metals, such as cadmium, their safety is of great concern. AuNPs are being developed for drug delivery, cancer therapy, and treatment of rheumatoid arthritis (19–22). Because Au is not present in normal biological specimens, the biodistribution of AuNPs can be tracked easily and has been investigated by several laboratories (11, 23, 24). In this chapter, we use cadmium-based QD and AuNP as examples for the applications of ICP-MS analyses and in our study protocols.

2. Materials

2.1. Nanomaterials Studied

1. QD705 (Invitrogen, Inc., Hayward, CA, USA): QD705 contained the Cd/Se/Te core covered with a ZnS shell and modified with methoxy-PEG-5000 coating. Each tube of this product contained 200 μL of 2 μM solution in 50 mM borate buffer, pH 8.3 (see Note 1).
2. AuNPs were synthesized by reducing hydrogen tetrachloroaurate with sodium citrate (see Note 2).

2.2. Reagents and Biological Sample Components

1. Isoflurane.
2. Borate buffer: boric acid and sodium hydroxide (NaOH) in 100 mL double distilled H_2O, pH 8.3. The pH of borate buffer was adjusted by adding HCl.
3. Tissue samples (ICR mice): plasma, red blood cells (RBCs), liver, lungs, kidneys, spleen, muscle, thymus, fat, brain, skin, and bones.
4. Cultivated mammalian cells: mouse leukemic monocyte macrophage RAW 264.7 cell line (American Type Culture Collection, Manassas, VA, USA).
5. Phosphate buffer: NaH_2PO_4 and Na_2HPO_4 in double distilled H_2O, pH 7.4.
6. Culture medium for RAW264.7 cells: Cells were maintained in DMEM supplemented with 10% FBS, 2 mMl-glutamine, 4.5 g/L glucose, 10 mM HEPES, 1.0 mM sodium pyruvate, and antibiotics including 100 U/mL penicillin and 100 μg/mL streptomycin.
7. Methanol.

2.3. Microwave Digestion Components

1. Nitric acid (HNO_3): Ultrex Ultra II Pure HNO_3 (J.T. Baker, Phillipsburg, NJ, USA).
2. Hydrochloric acid (HCl): analytical grade HCl (37%).
3. Aqua regia: freshly mixing concentrated HNO_3 and concentrated HCl (volumetric ratio of 1:3).

2.4. Metal Analysis Components

1. Standard metals (Cd, Fe, Cu, Mn, Se, and Zn) solutions (multi-element solution 2, Claritas PPT, Spex, USA).
2. Standard metals (Au and Te) solutions (multi-element solution 3, Claritas PPT, Spex, USA).
3. Certified reference material (CRM) for Cd: Trace Element in lyophilized Bovine Liver CRM 185R, European commission (Promochem, Molsheim, France) (see Note 3).
4. Certified reference material for Au: Seronorm™ Trace Elements Serum, Level 2 (Seronorm, Cheshire, UK) (see Note 4).

5. Mass spectrometer tuning solution: PerkinElmer Pure Elan 6100 Setup/Stab/Masscal solution (PerkinElmer, CA, USA) (see Note 5).
6. Washing solution (10% HNO_3): 100 mL HNO_3 was added to 1 L ddH_2O.

3. Methods

3.1. Preparation of Nanomaterials

3.1.1. Determine the Hydrodynamic Diameters and Zeta Potentials of Nanomaterials by the Zetasizer Nano System (Zetasizer Nano ZS, Malvern Instruments, Worcestershire, UK)

1. Suspend nanomaterials in working solution (see Note 6) and sonicate up to 30 min (see Note 7).
2. Put the suspension into a cuvette at 25°C to enable particle size and zeta potential analysis.

3.1.2. Determination and Confirmation of the Shape and Size of Nanomaterials by Transmission Electron Microscopy (TEM) (H-7650, Hitachi, Japan)

1. Suspend nanomaterials in methanol.
2. Drip nanomaterial solutions on copper grids for TEM.
3. Preserve all copper grids in a dry cabinet.
4. Characterize the particle shape and size of nanomaterials by TEM.

3.2. Administration of Nanomaterials in Biological Systems

3.2.1. In Vivo

1. Suspend nanomaterials in working solution and sonicate up to 30 min.
2. Administer to mice via different routes (see Note 8) with nanomaterial suspensions of different volumes per mouse (see Note 9).

3.2.2. In Vitro

1. Suspend nanomaterials in cell culture medium (see Note 10) and sonicate up to 30 min.
2. Incubate cells (from human or rodents) with nanomaterial suspensions (see Note 11).

3.3. Sample Preparation

Blood and tissue samples are collected at various time intervals according to specific study design.

3.3.1. Blood and Tissue Preparation

1. In vivo, sacrifice animals (under isoflurane anesthesia) after nanomaterial dosing.
2. Carefully remove, collect, and weigh blood and tissues.
3. Dry tissues with a lyophilizer (Dura-Dry corrosion resistant freeze-dryer, FTS system, USA) at −50°C for 24–48 h.

4. Ash carcasses with furnace (Thermolyne Type 48000 Muffel Furnace, Taylor Scientific, St. Louis, MO, USA) at 450°C for 6 h.
5. In vitro, harvest cells after nanomaterials dosing. RAW 264.7 cells were washed with PBS and subsequently harvested with cell scrapers.

3.3.2. Microwave Digestion of the Sample

Prior to measurement with ICP-MS, except for blood and urine, the collected biological samples should be further treated with acid digestion to destroy or remove the sample matrix. Blood and urine samples may be subjected directly to ICP-MS analysis after a simple dilution process without acid digestion (25). This procedure will help to reduce or eliminate undesirable interferences. It will also ensure all NMs are converted into ionic species for analysis. The use of a microwave digestion system with closed vessels has the advantage of faster reaction rates because of the high temperature and pressure attained inside the sealed containers. The percentage distribution of NMs in different organs may vary greatly and some elements to be measured may lie in trace and ultratrace levels. Thus, strict control to avoid contamination in the entire analytical procedure should be made. The use of highly pure reagents, acid clean vessels, and dust-free laboratory are prerequisite for accurate measurements.

1. Transfer 1 mL of blood, 30–70 mg/dry weight of tissue and carcass to cleaned and dried digestion tubes (see Note 12).
2. In vitro, transfer cells to digestion tubes.
3. Liquefy the QD705- and AuNP-treated biological samples with 0.5–2.5 mL of HNO_3 (see Note 13) and aqua regia (see Note 14), respectively.
4. Prepare the appropriate sample blanks and CRM with addition of HNO_3 or aqua regia.
5. Subject the samples to microwave digestion systems for digestion (Fig. 1) (Multiwave 3000, Anton Paar GmbH, Graz, Austria). Program the microwave digestion temperature to increase from 30°C to 75°C at a rate of 7.5°C/min with a 1-min hold at 75°C, and up to 130°C at a rate of 11°C/min with a hold of 30 min at 130°C.
6. After cooling the digestion tubes, transfer the solutions to polystyrene tubes.
7. Diluted the digested solution to 15–50 mL with deionized water. The final acid solution of the major components of NMs are, respectively, Cd: 3.5% HNO_3; Te: 3.5% HNO_3/3.7% HCl; Au: 3.5% HNO_3/10.5% HCl, and the essential elements such as Cu, Mn, Se, and Zn are in 3.5% HNO_3 solution.
8. Carry out all measurements of samples with the ICP-MS (PE Elan 6100, CA, USA).

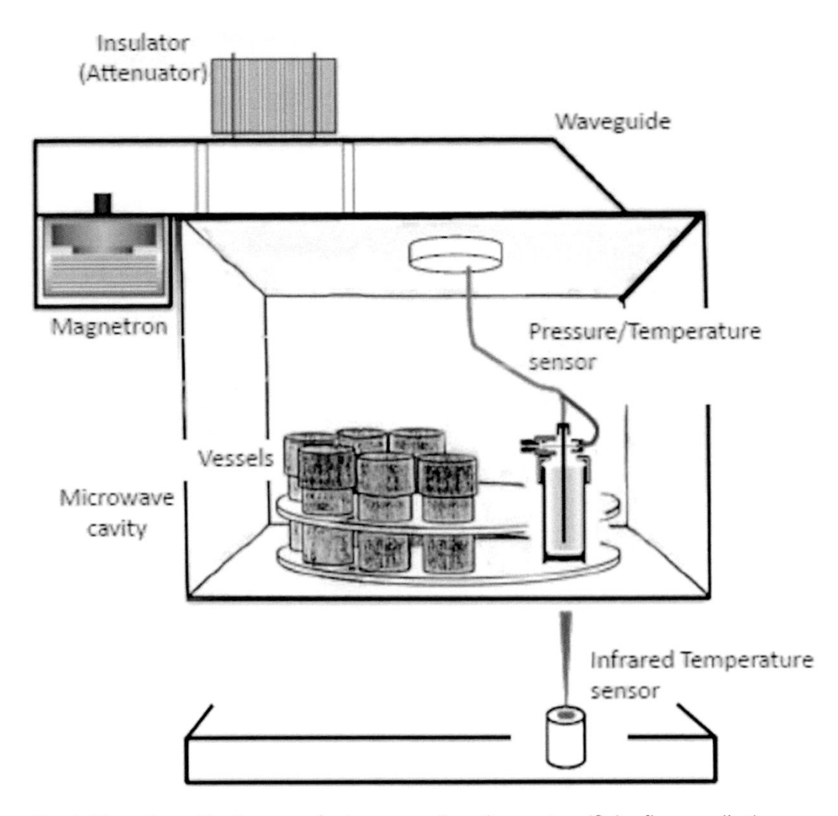

Fig. 1. The schematic diagram of microwave digestion system (Color figure online).

3.4. ICP-MS Analysis

ICP-MS is currently the most frequently applied inorganic spectrometry technique for fast multi-element determinations in the trace and ultratrace concentration levels for both life science and environmental research. Since the plasma is operated at temperatures of 6,000–8,000 K, each injected sample is completely dissociated, atomized, and ionized independently of the structure of the original substance. The ions generated in the plasma are then introduced into the mass analyzer, where the ions are separated in accordance with their mass-to-charge ratio (m/z). As to the mass analyzers, though double-focusing sector field (SF) and time-of-flight (TOF) analyzers can be operated at higher resolution, thus overcoming typical isobaric interferences of atomic ions of analytes and polyatomic ions. Quadrupole mass analyzers are now still most commonly used in the ICP-MS system because of its low cost and simple operation.

As shown in Fig. 2, a quadrupole-based ICP-MS instrument consists of four main parts: the sample introduction system; the ion source for the evaporation, atomization, and ionization of the sample; the quadrupole mass analyzer for the mass separation of ion beams that are extracted from the ion source; and the ion detection system. Although quadrupole ICP-MS is considered to be the ideal technique for elemental analysis, it is not exempt from limitation of

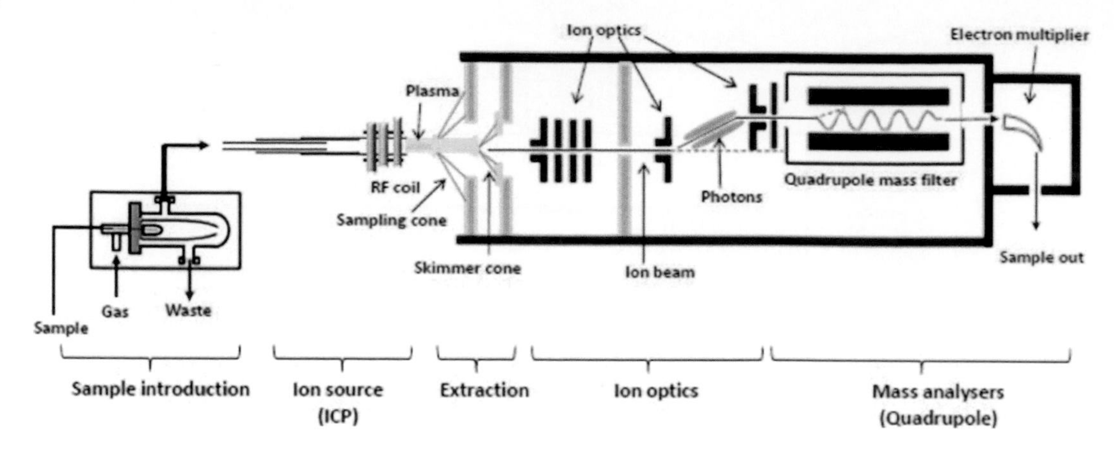

Fig. 2. The schematic diagrams of Quadrupole ICP-MS system (Color figure online).

polyatomic interferences from argon, atmospheric gases and the biological matrices and isobaric elemental interferences. Table 1 shows typical isobaric interferences that may be encountered in the determination of major components of QD705 (Cd and Te) and AuNP (Au), and trace elements in biological samples in this work. Most of those interferences can now be successfully eliminated using collision and reaction cell ICP-MS instruments which are now commercially available (26, 27).

3.4.1. Measurement with ICP-MS

1. The quadrupole ICP-MS system used is a PerkinElmer ELAN 6100 Model (Massachusetts, USA) equipped with a standard sample-introduction device (Cross flow nebulizer; scott double pass spray chamber; platinum interface cones).
2. Set the main instrumental settings as follows: RF power 1.3 kW; plasma gas flow: 15 L/min; auxiliary gas flow: 1.01 L/min; nebulizer gas flow: 1.03 L/min; lens voltage, 11.95 V; dwell time/amu: 100 ms; repetition: four times.
3. Aspirate the washing solution through the ICP-MS system for 30 min before tuning of the instrument.
4. Daily performance is made using the mass spectrometer tuning solution (see Note 5) to calibrate the ICP-MS over the entire mass spectrum.
5. Before analysis of the samples, construct the calibration curves of the respective elements of major components of NMs (Cd, Te, and Au) and trace constituents in tissue samples (Mn, Cu, Se and Zn) (see Note 15). Isotopes used for measurement are ^{111}Cd, ^{128}Te, ^{197}Au, ^{55}Mn, ^{65}Cu, ^{66}Zn, and ^{82}Se, respectively.
6. All measurements of standard and samples are carried out with the ICP-MS system.
7. After sample analysis, aspirate the system with the washing solution for 3 min to remove the memory effect from previous samples (see Note 16).

8. Regularly check the performance of the analytical system with a quality control protocol including linearity of calibration curve, blank control, duplicate analysis, spike recovery and quality control sample analysis after completing a cycle of every ten sample analyses.

3.4.2. Accuracy and Quality Control of ICP-MS

1. For evaluation of data reliability, two certified reference materials (Trace Elements in lyophilized Bovine Liver CRM 185R and Seronorm™ Trace Elements Serum) are used. The results found in this work are in good agreement with the certified values. The recoveries for Cd and Au are 97 and 95% ($n=3$) respectively. Other essential elements are also found in good agreement with the certified values (see Note 17).
2. The method detection limits based on the analyte concentration that gives a signal which is three times the standard deviation of the procedure blank ($n=7$) are estimated to be 0.018, 0.017, and 0.045 μg/L for Cd, Te, and Au respectively. Other trace elements are also conducted in the same way, and the results are shown (see Note 18).
3. To ensure routine analytical performance, quality control programs are regularly conducted which include linearity of calibration curve, blank control test, and analysis of duplicate, spike, and quality control sample on the basis of every ten sample as a batch. The results of nonspectroscopic interferences caused by matrix effects are recorded (see Note 19).

3.5. Data

3.5.1. Tissue Distribution

1. Tissue distribution of QD705 in ICR mice.

   Our PBPK model simulations, in comparison with experimental data on tissue concentrations of QD705, for the entire 6-month experimental period are presented in Fig. 3. On the basis of the results of our overall pharmacokinetic studies and PBPK modeling, the following conclusions may be drawn: (1) QD705 accumulates and persists in the spleen, liver, and kidneys through time-dependent redistribution from other tissues of the body; (2) principal mechanisms of tissue distribution of QD705 may involve its deposition, the binding of protein or other receptors, endo-, trans-, and phagocytosis, and/or transport through cellular gaps and openings; (3) possible metabolism and/or excretion of QD705 occurs slowly and gradually about 1 month after dosing; and (4) the whole body half-life for QD705 in the mice is very long, on the order of several months.
2. Tissue distribution of gold in mice (23).

   Comparative study on average colloidal gold distribution of various sizes after 24 h of gold NP administration in vivo was reported by Sonavane et al. (23) (Table 2). In this study, gold NP with 15, 50, and 100 nm sizes, the highest concentration was found in the liver followed by lungs, kidneys, brain, and

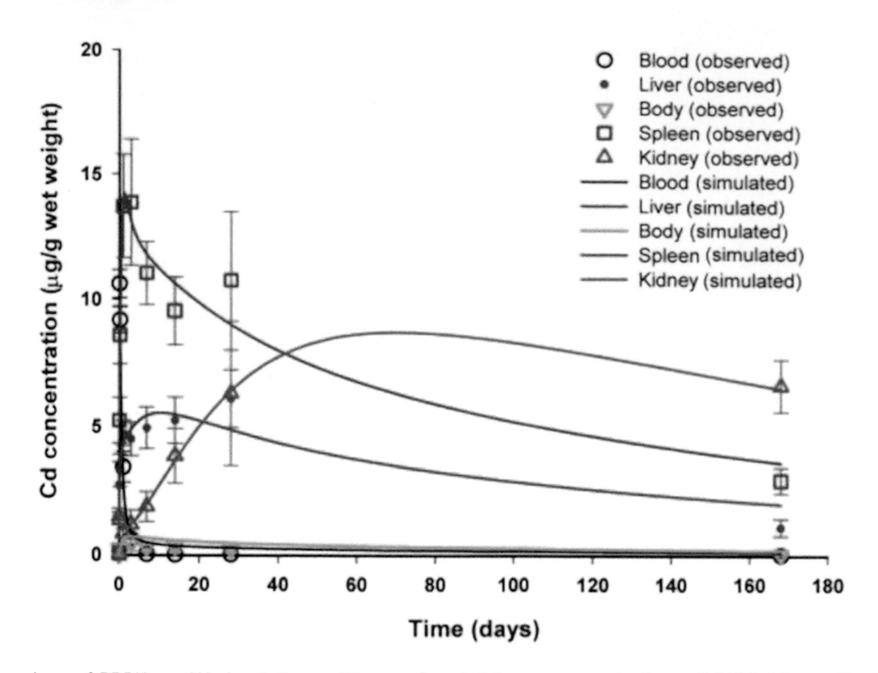

Fig. 3. Comparison of PBPK model simulations with experimental tissue concentrations of QD705 6 months following an intravenous dosing (30) (Color figure online).

spleen (Table 2). Pancreas also showed trace amount of gold (Table 2). In case of 200 nm size gold NP, highest concentration of gold was observed in liver followed by spleen, lung, and kidney. Only small amounts were observed in the brain, pancreas, stomach, and blood. It was apparent that particles of larger size may not cross the blood–brain barrier to enter the brain.

3.5.2. Degradation of Nanoparticles In Vivo

The degradation of QD705 can be estimated via tracking on the changes in the ratios of the two metals, Cd and Te, associated with QD705. The molar ratio of Cd/Te in QD705 is approximately 69:1. Both Cd and Te are non-biologically essential metals. In rodents the biological half-life of Cd is much longer (10–15 times) than that of Te (28). If QD705 is indeed disintegrated in vivo, the released Te will be excreted from the body at a much more rapid rate than the released Cd. Therefore, an increased Cd/Te ratio (>69) in tissues over time would reflect time-corresponded increase in disintegration of the Cd/Se/Te complex. When the Cd/Te ratios in the spleen, liver, and kidneys were calculated, no significant change in Cd/Te ratios were observed in the spleen and liver (Fig. 4). However, sharp and time-corresponded increase in Cd/Te ratio was found in kidneys tissues (Fig. 4). These findings signified that QD705 disintegration of QD705 particle or Cd/Se/Te complex may be organ or tissue specific. While there was no significant QD705 disintegration (elevation in Cd/Te ratio) in spleen and liver, kidneys appeared to have the capability for such breakdown.

Table 2
Comparative average colloidal gold distribution at the organ/tissue level of albino mice after 24 h of dose administration (23)

Tissues	15 nm gold NP			50 nm gold NP			100 nm gold NP			200 nm gold NP			Control[a]
	µg Au/g of tissue	% of given dose	Number of gold NP distribution	µg Au/g of tissue	% of given dose	Number of gold NP distribution	µg Au/g of tissue	% of given dose	Number of gold NP distribution	µg Au/g of tissue	% of given dose	Number of gold NP distribution	
Lung	32.27 ± 1.94	0.244	9.03×10^{18}	18.65 ± 3.94	0.073	1.34×10^{17}	15.23 ± 1.5	0.059	1.36×10^{16}	19.40 ± 4.50	0.047	1.98×10^{15}	ND[a]
Liver	52.26 ± 3.47	0.395	1.46×10^{19}	21.25 ± 5.47	0.084	1.52×10^{17}	27.11 ± 4.2	0.106	2.43×10^{16}	58.78 ± 10.4	0.143	6.00×10^{15}	ND
Kidney	25.48 ± 2.14	0.193	7.13×10^{18}	3.75 ± 1.14	0.014	2.69×10^{16}	1.29 ± 0.4	0.005	1.15×10^{15}	9.35 ± 3.65	0.022	9.55×10^{14}	ND
Spleen	5.46 ± 1.26	0.041	1.52×10^{18}	11.53 ± 2.26	0.045	8.39×10^{16}	12.92 ± 3.4	0.050	1.17×10^{15}	28.89 ± 8.54	0.070	2.95×10^{15}	ND
Heart	1.05 ± 0.26	0.007	2.93×10^{17}	0.97 ± 0.28	0.003	6.98×10^{15}	3.24 ± 0.8	0.012	2.91×10^{15}	2.86 ± 0.44	0.006	2.92×10^{14}	ND
Brain	9.95 ± 1.25	0.075	2.78×10^{18}	9.12 ± 3.25	0.036	6.56×10^{16}	6.01 ± 1.4	0.023	5.40×10^{15}	0.15 ± 0.03	0.0003	1.53×10^{13}	ND
Blood	0.56 ± 0.26	0.004	1.56×10^{17}	0.59 ± 0.23	0.002	4.24×10^{15}	ND[a]	ND[a]	ND[a]	0.11 ± 0.02	0.0002	1.12×10^{13}	ND
Stomach	3.57 ± 0.86	0.027	9.99×10^{17}	0.25 ± 0.06	0.0009	1.79×10^{15}	0.80 ± 0.2	0.003	7.19×10^{14}	0.15 ± 0.03	0.0003	1.53×10^{13}	ND
Pancreas	NC[b]	NC[b]	NC[b]	1.73 ± 0.45	0.006	1.24×10^{16}	7.52 ± 1.5	0.029	6.76×10^{16}	0.11 ± 0.02	0.016	6.91×10^{14}	ND

[a]Not detectable
[b]Not calculated

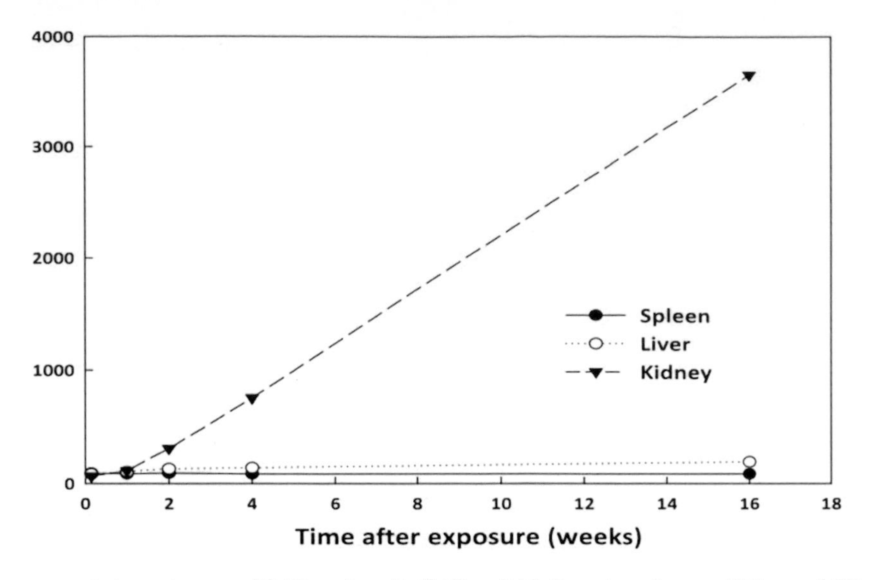

Fig. 4. A time course study on changes of Cd/Te molar ratio (Cd/Te ratio) in the spleen, liver, and kidneys of ICR mice treated with 40 pmol of QD705 (31).

4. Notes

1. QD705 analyses in our laboratory revealed that in our QD705 samples, there were 56.16 ± 8.54% Cd, 5.80 ± 3.60% Se, and 0.92 ± 0.20% Te. QD705 were spherical-shaped nanoparticles with sizes of about 12.3 ± 5.2 nm. The averaged hydrodynamic diameter of QD705 in borate buffer was 143.3 nm and the size distribution was between 10 nm and 250 nm. The zeta potential of QD705 in borate buffer was −24.8 ± 5.1 mV. Aggregates of QD705 were insoluble and remained dispersed in borate buffer for at least 1 h. The maximal emission of QD705 is about 705 nm (λ_{ex} = 350 nm).
2. The method for AuNP generation was modified from Turkevich et al. (29), and the detailed protocol was described in Sonavane et al. (23).
3. The certified values for Cd, Cu, Mn, Se, and Zn were 0.544 ± 0.017, 277 ± 5, 11.07 ± 0.29, 1.68 ± 0.14, and 138.6 ± 2.1 mg/kg, respectively.
4. The certified value for Au is 1,956 μg/L.
5. The "Mass spectrometer tuning solution" (PerkinElmer Pure Elan 6100 Setup/Stab/Masscal solution) contains 10 μg/L of Mg, Cu, Rh, Cd, Ln, Ba, Ce, Pb, and U in 1% HNO_3 solution.
6. In vivo: QD705 was suspended in borate buffer and AuNPs were suspended in H_2O or phosphate buffer; In vitro: both QD705 and AuNPs were suspended in RAW 264.7 cell culture medium.

7. The nanoparticle dispersion was sonicated with a Bransonic 8,510 40 kHz sonicator (BRANSON, Danbury, CT, USA). The size distribution and degree of agglomeration depended not only on the employed methodology but also on the physicochemical property of nanomaterials.
8. Mice were given nanomaterials through different routes including oral, intratracheal, dermal, and intravenous delivery. Exposure route of nanomaterial is determined by the primary route of human exposure. For example, the IV route for QD705 injection was chosen to mimic potential human medical imaging applications and provide cleaner pharmacokinetic profiles without the complication of the absorption phase.
9. The volume of nanomaterial suspension used for administration to rodents was in accordance with the concentration of nanomaterials expected for a given application or exposure.
10. Different culture media were selected based on the type of cells.
11. RWA 264.7 cells were seeded for 24 h, changed to serum-free media for additional 24 h incubation, and then incubated with vehicle and nanomaterial suspensions at indicated doses (0.1 nm to 20 μM) for 6–48 h.
12. Digestion tubes are washed before use by adding 10 mL HNO_3 and performing digest cycle in the microwave oven.
13. The volume of digestion acid depends on the metal species and the weight of biological samples.
14. The digestion procedure varies slightly with the metal species under consideration.
15. The calibration curves for Cd, Cu, Mn, Se, and Zn are prepared in the concentration range of 0.1, 0.5, 1, 5, 10, 50, 100, and 500 μg/L in 3.5% HNO_3 solution; where Te and Au were, respectively, in 3.5% HNO_3/3.7% HCl and 3.5% HNO_3/10.5% HCl solution.
16. The duration time of this washing step varies on the species and the concentration of metals in the samples. For example, the memory effect of Au is more pronounced, requiring longer duration for washing the ICP-MS system.
17. The agreement of some essential elements was found satisfactory with that of CRM, the recoveries for Cu, Mn, Se, and Zn were 89, 93, 92, and 93% ($n = 3$).
18. The method detection limits for Cu, Mn, Se, and Zn were 0.066, 0.090, 0.24, and 7.2 μg/L, respectively.
19. The nonspectroscopic interferences caused by matrix effect were investigated. The results expressed as spike recovery were 102, 102, 104, 105, 102, 104, 106% for Cd, Te, Au, Cu, Mn, Se, and Zn ($n = 6$), respectively.

Acknowledgments

We would like to thank Dr. Yu-Teh Chung for his assistance in gathering information in ICP-MS and microwave digestion. This work was supported by grant NM-097-PP08 and NM-098-PP08 from the Center for Nanomedicine Research, National Health Research Institutes, Taiwan.

References

1. Michalet X, Pinaud FF, Bentolila LA, Tsay JM, Doose S, Li JJ, Sundaresan G, Wu AM, Gambhir SS, Weiss S (2005) Quantum dots for live cells, in vivo imaging, and diagnostics. Science 307:538–544
2. Hardman R (2006) A toxicologic review of quantum dots: toxicity depends on physicochemical and environmental factors. Environ Health Perspect 114:165–172
3. Derfus AM, Chan WCW, Bhatia SN (2004) Probing the cytotoxicity of semiconductor quantum dots. Nano Lett 4:11–18
4. Nel A, Xia T, Madler L, Li N (2006) Toxic potential of materials at the nanolevel. Science 311:622–627
5. Male KB, Lachance B, Hrapovic S, Sunahara G, Luong JH (2008) Assessment of cytotoxicity of quantum dots and gold nanoparticles using cell-based impedance spectroscopy. Anal Chem 80:5487–5493
6. Chang E, Thekkek N, Yu WW, Colvin VL, Drezek R (2006) Evaluation of quantum dot cytotoxicity based on intracellular uptake. Small 2:1412–1417
7. Zhang LW, Yu WW, Colvin VL, Monteiro-Riviere NA (2008) Biological interactions of quantum dot nanoparticles in skin and in human epidermal keratinocytes. Toxicol Appl Pharmacol 228:200–211
8. Mortensen LJ, Oberdorster G, Pentland AP, Delouise LA (2008) In vivo skin penetration of quantum dot nanoparticles in the murine model: the effect of UVR. Nano Lett 8: 2779–2787
9. Becker JS, Jakubowski N (2009) The synergy of elemental and biomolecular mass spectrometry: new analytical strategies in life sciences. Chem Soc Rev 38:1969–1983
10. Bettmer J, Montes Bayon M, Encinar JR, Fernandez Sanchez ML, de la Campa F, Mdel R, Sanz Medel A (2009) The emerging role of ICP-MS in proteomic analysis. J Proteomics 72:989–1005
11. De Jong WH, Hagens WI, Krystek P, Burger MC, Sips AJ, Geertsma RE (2008) Particle size-dependent organ distribution of gold nanoparticles after intravenous administration. Biomaterials 29:1912–1919
12. Liu W, Wu Y, Wang C, Li HC, Wang T, Liao CY, Cui L, Zhou QF, Yan B, Jiang GB (2010) Impact of silver nanoparticles on human cells: effect of particle size. Nanotoxicology 4:319–330
13. Liu Y, Gao Y, Zhang L, Wang T, Wang J, Jiao F, Li W, Li Y, Li B, Chai Z, Wu G, Chen C (2009) Potential health impact on mice after nasal instillation of nano-sized copper particles and their translocation in mice. J Nanosci Nanotechnol 9:6335–6343
14. Wang J, Liu Y, Jiao F, Lao F, Li W, Gu Y, Li Y, Ge C, Zhou G, Li B, Zhao Y, Chai Z, Chen C (2008) Time-dependent translocation and potential impairment on central nervous system by intranasally instilled TiO(2) nanoparticles. Toxicology 254:82–90
15. Xia T, Kovochich M, Liong M, Madler L, Gilbert B, Shi H, Yeh JI, Zink JI, Nel AE (2008) Comparison of the mechanism of toxicity of zinc oxide and cerium oxide nanoparticles based on dissolution and oxidative stress properties. ACS Nano 2:2121–2134
16. Gao X, Cui Y, Levenson RM, Chung LW, Nie S (2004) In vivo cancer targeting and imaging with semiconductor quantum dots. Nat Biotechnol 22:969–976
17. Obonyo O, Fisher E, Edwards M, Douroumis D (2010) Quantum dots synthesis and biological applications as imaging and drug delivery systems. Crit Rev Biotechnol 30(4):283–301
18. Mulder WJ, Strijkers GJ, Nicolay K, Griffioen AW (2010) Quantum dots for multimodal molecular imaging of angiogenesis. Angiogenesis 13:131–134
19. Maysinger D (2007) Nanoparticles and cells: good companions and doomed partnerships. Org Biomol Chem 5:2335–2342
20. Huang X, Jain PK, El-Sayed IH, El-Sayed MA (2007) Gold nanoparticles: interesting optical properties and recent applications in cancer diagnostics and therapy. Nanomedicine (Lond) 2:681–693

21. Cho K, Wang X, Nie S, Chen ZG, Shin DM (2008) Therapeutic nanoparticles for drug delivery in cancer. Clin Cancer Res 14: 1310–1316
22. Tsai CY, Shiau AL, Chen SY, Chen YH, Cheng PC, Chang MY, Chen DH, Chou CH, Wang CR, Wu CL (2007) Amelioration of collagen-induced arthritis in rats by nanogold. Arthritis Rheum 56:544–554
23. Sonavane G, Tomoda K, Makino K (2008) Biodistribution of colloidal gold nanoparticles after intravenous administration: effect of particle size. Colloids Surf B Biointerfaces 66:274–280
24. Fent GM, Casteel SW, Kim DY, Kannan R, Katti K, Chanda N (2009) Biodistribution of maltose and gum arabic hybrid gold nanoparticles after intravenous injection in juvenile swine. Nanomedicine 5:128–135
25. Fong BM, Siu TS, Lee JS, Tam S (2007) Determination of mercury in whole blood and urine by inductively coupled plasma mass spectrometry. J Anal Toxicol 31:281–287
26. Becker JS, Dietze HJ (2000) Inorganic mass spectrometric methods for trace, ultratrace, isotope, and surface analysis. Int J Mass 197:1–35
27. Wang M, Feng WY, Zhao YL, Chai ZF (2010) ICP-MS-based strategies for protein quantification. Mass Spectrom Rev 29:326–348
28. Nordberg G (2007) Handbook on the toxicology of metals, 3rd edn. Academic, Amsterdam
29. Turkevich J, Stevenson PC, Hillier J (1951) A study of the nucleation and growth processes in the synthesis of colloidal gold. Discuss Faraday Soc 11:55–75
30. Lin P, Chen JW, Chang LW, Wu JP, Redding L, Chang H, Yeh TK, Yang CS, Tsai MH, Wang HJ, Kuo YC, Yang RS (2008) Computational and ultrastructural toxicology of a nanoparticle, Quantum Dot 705, in mice. Environ Sci Technol 42:6264–6270
31. Lin CH, Chang LW, Chang H, Yang MH, Yang CS, Lai WH, Chang WH, Lin P (2009) The chemical fate of the Cd/Se/Te-based quantum dot 705 in the biological system: toxicity implications. Nanotechnology 20:215101

Chapter 23

Quantitative Nanoparticle Organ Disposition by Gel Permeation Chromatography

Abdul Khader Mohammad and Joshua Reineke

Abstract

Gel permeation chromatography (GPC) also known as size exclusion chromatography (SEC) is a highly valuable tool for the purification, separation, and characterization of synthetic and natural polymers. In this technique, the analyte (usually a polymer) is dissolved in a suitable solvent and is injected into columns made of porous inert material. The columns separate the analyte based on its hydrodynamic size rather than the molecular weight. GPC systems typically have an RI detector, UV detector, or light scattering unit attached to the columns. With advanced detection systems coupled to the GPC, we can obtain important information about polymers including their molecular weight distribution, average molecular mass, degree of branching in the polymers, etc. In addition to the separation of polymers, GPC allows for the separation of enzymes, nucleic acids, polysaccharides, and hormones. With regards to nanotoxicity, GPC can be used for the quantitative determination of tissue deposition of polymer nanoparticles after in vivo exposure. Understanding the organ specific exposure to a nanomaterial is helpful in choosing appropriate toxicity assays, interpreting data from other toxicity assessments, and in determining potential risk.

Key words: Gel permeation chromatography, Refractive index, Polymer, Nanoparticle, Nanotoxicity, Biodistribution

1. Introduction

Gel permeation chromatography (GPC) is a technique that separates molecules based on their hydrodynamic volume or size (1). This technique is extensively used to analyze synthetic and natural polymers (2). It is the most widely accepted technique to determine the molecular weight distribution and molecular weight average of polymers (1, 3). The main operational unit in a GPC is a set of columns made of porous gels or rigid inorganic particles through which the molecules travel and undergo separation based on their pore permeation (4). The polymer is dissolved in a solvent and the injection needle injects the polymer solution into the columns.

Joshua Reineke (ed.), *Nanotoxicity: Methods and Protocols*, Methods in Molecular Biology, vol. 926,
DOI 10.1007/978-1-62703-002-1_23, © Springer Science+Business Media, LLC 2012

The mobile phase, usually the same solvent used to dissolve the polymer, passes through the columns at a fixed flow rate. As the mobile phase moves forward in the column, the molecules that are too large to penetrate the pores of the packing elute in the interstitial or void volume, i.e., they pass right through the column. However, molecules with size less than the pore size of the packing material penetrate into the pores thereby traveling a much longer path and as a result elute at a later time; therefore resulting in separation based on their size. Since GPC is a relative and not absolute molecular weight technique, the columns must be calibrated with polymer standards of known molecular weight (1) for molecular weight determination of unknown samples. This technique allows us to separate analytes over a broad molecular mass ranging from a simple monomer unit to several millions of grams per mole (4) dependent on the column bank used.

Typically, GPC is used to obtain information about the polymer characteristics like molecular weight distribution, average molecular weight, branching, etc. However, it has been shown that GPC in conjunction with a refractive index (RI) detector can also be used in biodistribution studies of polymer nanoparticles to determine the amount of the polymer and therefore nanoparticles present in the tissues (5–8). We frequently utilize a GPC for quantitative biodistribution studies of polymer nanoparticles apart from typical usage including the determination of molecular weight.

Given the tremendous importance of polymers in drug delivery, the information obtainable from GPC with regards to polymer concentration and polymer molecular weight makes it a very powerful analytical tool in the quantitative study of the biodistribution of nondegradable polymer nanoparticles including polystyrene. Though operating a GPC involves tedious maintenance procedures and close monitoring of the instrument, it is a very powerful analytical tool in the field of drug delivery and nanotoxicity applications for the clinically relevant information it generates. The materials and methods sections of this chapter mainly deal with the use of GPC for determining the amount of polymer nanoparticles present in tissues following in vivo administration. The use of the term GPC Max refers to the Viscotek GPC system VE 2001.

2. Materials

2.1. Sample Preparation

1. Tetrahydrofuran (THF) stabilized with Butylated hydroxytoluene (BHT) (or) unstabilized THF (see Notes 1 and 2).
2. Nylon syringe filters, 0.2 μm pore size.
3. Conical vials, 1.5 ml volume (see Note 3).

2.2. GPC Components

1. UV protected solvent tank.
2. Solvent filter (Malvern Instruments, USA) (see Note 4).
3. Frit for fill port (Malvern Instruments, USA) (see Note 5).
4. Light scattering filter unit (Malvern Instruments, USA) (see Note 6).
5. Nylon filter for light scattering filter unit (Malvern Instruments, USA) (see Note 7).
6. Guard column (Malvern Instruments, USA).
7. Analytical columns (Malvern Instruments, USA) (see Notes 8 and 9).

3. Methods

The methods section has been written for the determination of polymer (using GPC) in animal tissues after administration. In a typical nanoparticle biodistribution study, the animals are exposed to the nanoparticles of interest followed by which the tissues are harvested from the animal at predetermined time points post nanoparticle administration. The harvested tissues are then homogenized, frozen, and lyophilized to obtain the dry homogenized tissue. Following this, the polymer in the tissues is extracted by addition of a suitable solvent for the polymer and subsequent extraction for 4 days. At the end of the extraction period, the tissue sample is filtered using a 0.2 μm filter. The filtrate contains polymer and/or tissue extracts dissolved in the solvent used for extraction. This filtrate is frozen and lyophilized; following which it is ready for analysis by GPC.
It is recommended that the whole process of sample preparation for GPC analysis be carried out in a clean biosafety hood in order to eliminate any chances of contamination to the sample. This indirectly helps prolong the life of the columns as a result of elimination of particulate matter and other contaminants into the column.

3.1. Preparation of Sample

1. Add 0.5 ml of THF to the lyophilized tissue extract sample and shake gently in order to dissolve the vial contents (see Note 10).
2. Place this vial on a rotating mixer for about an hour to ensure complete dissolution of the contents of the vial (see Note 11).
3. Using a syringe, draw out all the solution and attach a 0.2 μm nylon syringe filter to the end of the syringe while holding the syringe in upright position.
4. Filter the contents of the syringe into a conical GPC vial carefully and cap the vial (see Note 12).
5. After all the samples are filtered in the above manner, place all the conical vials in order in the sample rack of the GPC Max and make a note of the order.

3.2. Preparing the GPC Max Before Running the Samples

1. Before beginning to run the samples, set the flow rate to 1 ml/min (see Note 13).
2. In the sample sequence, add a couple samples with just the blank THF before the actual experimental samples (see Note 14).
3. Make sure the column bank is set at a constant temperature in the range of 25–35 °C (see Note 15).
4. Enlist the sequence in the software based on the number of samples you are analyzing (see Note 16).
5. Make sure that the sequence number coincides with the appropriate vial. For example, sequence #1 belongs to vial 1, sequence #2 belongs to vial 2, and so on.
6. Check to make sure that the number of sequences set in the software is equal to the number of samples present in the GPC Max (see Note 17).
7. Wash the injection needle trice by using the wash command in the OmniSEC software.
8. Purge the RI detector for about 5 min.
9. Start running the samples by clicking the data acquisition button on the OmniSEC software.

3.3. Data Analysis

The resulting chromatograms for each sample should be analyzed individually and manually (see Note 18) for the total peak area under the peak of interest (see Note 19). The peak area can be linearly related to peak areas of a standard curve of known concentrations. This will yield a concentration for each sample that, once corrected for the known total sample volume, can be used to calculate the total amount of polymer in your sample. We have mostly performed this analysis on whole tissue homogenates allowing the direct, quantitative determination of the amount of polymer (amount of nanoparticles) in a specific tissue (see Note 20). This data can be reported as a percentage of the total exposed dose or as a per gram of tissue if tissue mass is recorded.

4. Notes

1. It is advised to use BHT stabilized THF which helps prevent the formation of peroxides in the solvent. However, if the detection method is based on UV-absorption, then unstabilized THF should be used as it is transparent for UV-absorption. Additionally, it is better to use fresh solvent instead of one that has been in storage for long periods of time (several months to years). If unstabilized THF is being used for UV-based analysis, purge the solvent bottle with an inert gas (such as helium, argon, or nitrogen) each time the bottle is opened in order to reduce

the chances of peroxide formation. Peroxides not only make the THF explosive, but also have strong UV absorbance thereby leading to interference in the UV-based analysis methods.

2. Solvent containers should be tightly capped and the cap covered with aluminum foil after use. To reduce the risk of cross contamination, solvent containers that are to be used on the GPC should be designated as such and not used for any other experiments. It is advised to use a new pipette each time in order to draw out the solvent from the bottle for sample preparation.
3. Using vials with a conical shape at the bottom helps improve the sensitivity of the assay as the sample can be prepared in as small a volume of solvent as feasible. The conical shape raises the level of the sample in the vial as opposed to that in a vial with a flat base. This raised sample level allows the sample to be drawn by the injection needle even with a lesser volume present.
4. Changing the filter in the solvent tank every 2–3 months ensures that the solvent going into the columns from the tank is free from particulate impurities thereby extending the life of the columns.
5. Change the frit in the fill port every 2 months or more frequently depending on the usage.
6. We found that installing a filter unit before the guard column helps tremendously in maintaining the pressure in the whole column bank unit apart from improving the life span of the guard column. It also keeps the viscous tissue extracts from building up in the main column bank.
7. If tissue samples are being analyzed, changing the nylon filter in the light scattering filter unit twice every month (or more frequently depending on usage) helps prevent the tissue extracts from building up in the main column.
8. Using a single column often does not give sufficient separation as all the molecules larger than the pore size of the column elute at the same time. Therefore, using columns of different size exclusion limits gives much better separation as compared to that of a single column. Often column banks of three or more columns are used.
9. If there is a need to remove one or more of the columns, always keep the columns closed on both ends by using screw caps. When removing a column from the column bank, first fill up the inlet end of the column with the solvent and then cap the inlet. Following this, close the outlet of the column with a screw cap and store it properly.
10. Prior to dissolution in THF, the samples should be completely dry without any trace of water. Any presence of water is detrimental to the columns that use an organic solvent as the mobile phase.

To increase the sensitivity of the assay, the concentration of sample can be increased by preparing the sample in a volume of less than 500 μl of THF keeping in mind the solubility of the sample.

11. We have found that viscous tissue extracts of tissues including small intestine, liver, and kidneys especially do not dissolve readily on the addition of solvent. However, after being placed on the mixer for an hour, a clear solution forms. This step also helps in ensuring the complete dissolution of polymer present in the extracts.
12. Do not apply excessive pressure on the syringe while filtering. This will cause the sample to spill out from the edges of the syringe filter. Specifically, for viscous tissue samples like liver, the application of high pressure may rupture the syringe filter resulting in loss of the sample and possible spillage onto the person performing the procedure. Once all the contents are filtered, close the vial carefully to make sure it forms a tight closure. If the cap has a septum, make sure it is not open. Open cap septa may allow access to external impurities into the vial. Furthermore, if the sample prepared is among the last in a sequence of multiple samples (for example in a sequence of 50 samples), most of the sample may have evaporated by the time it is reached in the sequence. This could result in the incorporation of air into the columns resulting in irreversible damage.
13. High flow rates (greater than 1 ml/min) may damage the columns. Do not increase the flow rate suddenly from the standby flow rate (usually as low as 0.2 ml/min). For instance, if the standby flow rate is 0.2 ml/min, increase it to 0.5 ml/min first and let it run for a minute and then increase it to 0.7 ml/min, and again after a minute raise it to 1 ml/min. This will ensure that the columns do not receive large pulses of pressure.
14. We found that running blank THF samples in the sequence preceding the actual experimental samples helps in stabilizing the baseline of the experimental samples.
15. Storing the columns in a chamber with temperature control and keeping the columns at a constant temperature helps impart stability and reproducibility in the chromatograms.
16. Running the samples in the expected increasing order of polymer concentration (if known) will greatly minimize any contamination from the high concentration samples into those with low polymer concentration.
17. In cases where the number of samples in the sequence (in software) is more than the actual number of samples present in the GPC Max, and if these samples are running overnight, then the flow rate will continue to be the same until corrected manually (usually at the end of the sequence, the flow rate automatically goes back to standby flow rate). In cases where the

sample sequence is programmed to end, this small error will cause all of the solvent to go to waste overnight and could be deleterious to the columns if all of solvent is emptied thereby resulting in air entrapment into the columns.

18. In our experience, relying on peak detection software omits lower values in the data that otherwise can very readily be visualized in the chromatogram. Setting limits manually can achieve better sensitivity. However, any manual analysis must be applied across the entire data set to maintain objectiveness (i.e., set peak analysis for a retention time range on all chromatograms).

19. Depending on the GPC software analysis program you are running it may be easier to do peak analysis by exporting data into a graphical analysis software program.

20. Theoretically, one could sample a known weight percentage of a tissue and extrapolate the results to a whole tissue concentration in order to free tissue for other analytical methods. However, this assumes equal distribution within a tissue, which may not be the case. Additionally, by taking a portion of a tissue it may reduce the quantity below the detectable limit yielding false results. Therefore, we feel that the data has the most utility if it is from a whole tissue.

References

1. Mori S, Barth HG (1999) Size exclusion chromatography. Springer, New York
2. Gaborieau M, Nicolas J, Save M et al (2008) Separation of complex branched polymers by size-exclusion chromatography probed with multiple detection. J Chromatogr A 1190:215–223
3. Kostanski LK, Keller DM, Hamielec AE (2004) Size-exclusion chromatography – a review of calibration methodologies. J Biochem Biophys Methods 58:159–186
4. Striegel AM (2008) Size-exclusion chromatography: smaller, faster, multi-detection, and multidimensions. Anal Bioanal Chem 390:303–305
5. Jani P, Halbert GW, Langridge J et al (1990) Nanoparticle uptake by the rat gastrointestinal mucosa: quantitation and particle size dependency. J Pharm Pharmacol 42:821–826
6. Florence AT, Hillery AM, Hussain N et al (1995) Nanoparticles as carriers for oral peptide absorption: studies on particle uptake and fate. J Control Release 36:39–46
7. Jani P, Halbert GW, Langridge J et al (1989) The uptake and translocation of latex nanospheres and microspheres after oral administration to rats. J Pharm Pharmacol 41: 809–812
8. Hillery AM, Jani PU, Florence AT (1994) Comparative, quantitative study of lymphoid and non-lymphoid uptake of 60 nm polystyrene particles. J Drug Target 2:151–156

Chapter 24

Physiologically Based Pharmacokinetic Modeling for Nanoparticle Toxicity Study

Mingguang Li and Joshua Reineke

Abstract

This chapter introduces the principles and development procedures for physiologically based pharmacokinetic (PBPK) models, and their application for nanoparticle toxicity studies. PBPK models describe the concentration–time or mass–time profiles of chemicals or nanoparticles in individual tissues and organs within the body. They have been used mostly for toxicology and pharmacology studies of small molecules, and their application for nanoparticles are in the early stages. Due to the biodistribution differences between nanoparticles and small molecules, modification may be necessary to build PBPK models for nanoparticles. PBPK models for nanoparticles may be applied to biodistribution predictions, data extrapolation, and property–biodistribution relationships, and, thus, can be a powerful tool in toxicity evaluation.

Key words: Blood flow limited, Membrane limited, Biodistribution, ADME, Toxicology, Pharmacokinetics, PBPK

1. Introduction

Nanoparticles have shown great potential in medical and pharmaceutical applications. However, before any nanoparticles can be safely applied to clinical applications, the toxicology of nanoparticles needs to be carefully evaluated. There are various methods for nanoparticle toxicity evaluation, including in vitro cell viability and mechanistic tests, tissue histology and hematology, and biomarker measurements (1). However, the reliability and predictive ability of these methods are limited by the fact that they are not closely related to in vivo nanoparticle concentrations in individual tissues. Toxicity of nanoparticles is often determined by their properties (size, surface charge, chemical composition, etc.) and tissue exposure (concentration and duration); therefore, a thorough

Joshua Reineke (ed.), *Nanotoxicity: Methods and Protocols*, Methods in Molecular Biology, vol. 926,
DOI 10.1007/978-1-62703-002-1_24, © Springer Science+Business Media, LLC 2012

understanding of in vivo nanoparticle distribution is necessary for toxicity evaluation.

Much effort has been made, especially over the past 10 years, to gain understanding of nanoparticle biodistribution, including nanoparticle–cell interaction, mechanisms of nanoparticles crossing biobarriers, and tissue distribution after various exposure routes. These works significantly advanced the knowledge of in vivo nanoparticle behavior. However, there is a lack of methodologies that can comprehensively interpret experimental data and compare results from studies of various designs including nanoparticle composition, animal models, exposure routes, and doses.

Physiologically based pharmacokinetic (PBPK) models utilize physiological information to simulate and predict the concentration–time profiles of chemicals and recently nanoparticles. These models list tissues and organs as compartments and mass transportation among them is described by mathematical equations. Compared to traditional pharmacokinetic models, PBPK have two major advantages. First, they can describe the nanoparticle concentration–time profiles in individual tissues and organs, which is necessary for evaluation of the toxicity and potential applications. Second, PBPK models enable data extrapolation enabling prediction of experimental results based on known data, including inter-species, inter-routes, inter-formulations, and inter-doses extrapolation (2).

PBPK models have been utilized mostly for small molecules, and occasionally for macromolecules such as antibodies. They have only recently been applied to nanoparticle applications in the past several years. Increasing efforts are being made to utilize PBPK models for the advancement of nanoparticle research, and their advantages have been recognized. PBPK models have been listed as one of the current quantitative support tools for investigation of nanoparticle hazard assessment as specified in the Organization for Economic Cooperation and Development (OECD) guideline and the new European Union regulatory framework REACH (Registration, Evaluation and Authorization of Chemicals) (3).

ADME behaviors of small molecules are likely much different from that of nanoparticles. The ADME of small molecules includes processes of diffusion, active transport, enzymatic metabolism, and excretion. While for nanoparticles, physiological processes could be much more complex, including some that are not usually relevant to small chemical molecules (4), such as opsonization in the blood, cellular recognition and internalization, enzymatic degradation, physical property changes, etc. Due to these differences, the principles of building PBPK models need to be modified based on thorough understanding of the physiological processes of nanoparticles.

2. Principles of PBPK Model Development

Although the concept of PBPK modeling has been well established and accepted, there are no formal definitions to distinguish it from traditional pharmacokinetic models. Traditional pharmacokinetic models simply use one, two, or three compartmental structures to estimate the one that best fits experimental data, according to some statistical criterion. The final structures and parameters are determined (and dependent) solely on, and do not extend beyond, experimental data. However, a PBPK model is built based on realistic physiological elements including tissues, organs, and body fluids (such as blood and lymph), which are independent to the chemicals or nanoparticles under study. The characteristics of chemicals or nanoparticles under study will determine the values of the coefficients in the mathematical equations describing the model.

Nearly all of the PBPK models developed to date are used to describe the kinetics within the whole body of humans or animals. These models are called "whole body" physiologically based pharmacokinetic (WBPBPK) models (5, 6). These models are generally built centering around the blood circulation, which connects other organs or tissues into a system. A typical PBPK model structure is shown in Fig. 1 (7). Whole body PBPK models are generally divided into two groups according to the transportation mechanisms between blood and tissues: blood flow (or perfusion) limited and membrane (or permeability) limited (8), as shown in Fig. 2.

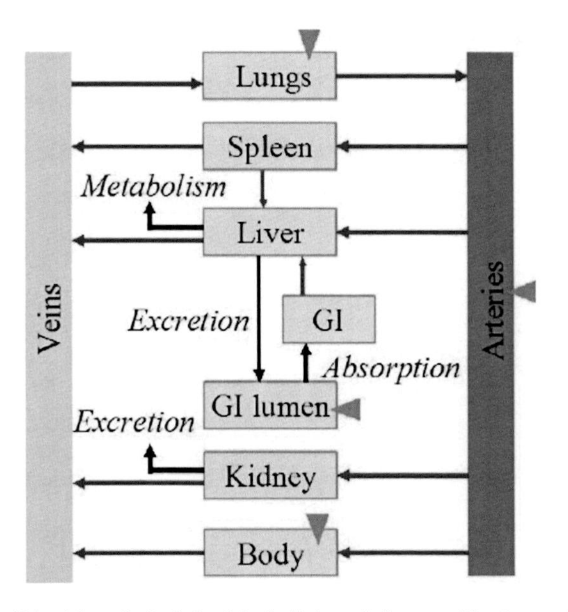

Fig. 1. A typical blood flow-limited physiologically based pharmacokinetics model structure. *Arrows* indicate the transportation of drugs or nanoparticles. *Triangles* show administration routes. Reproduced from (7) with permission from (ACS publications).

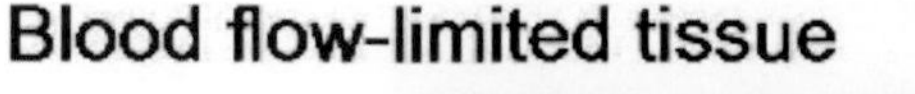

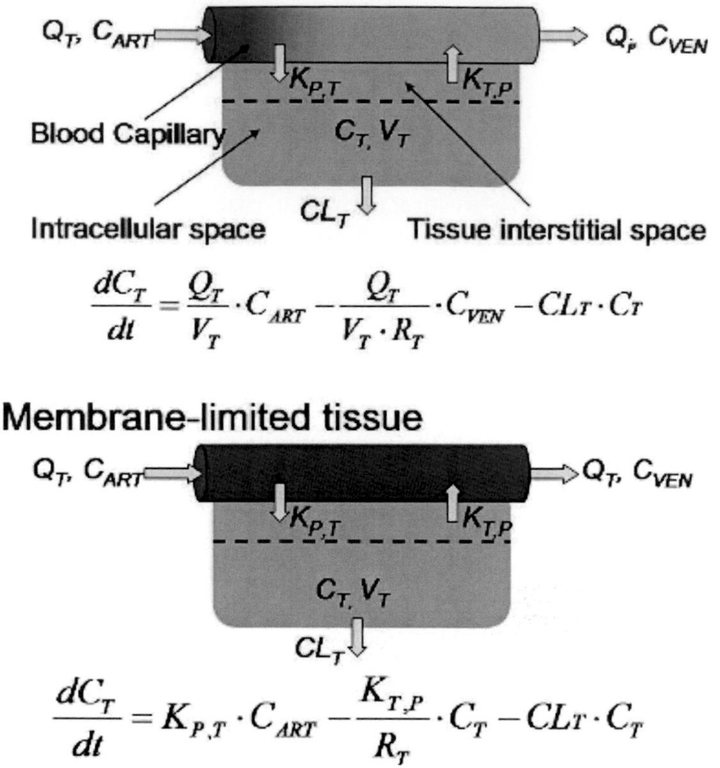

Fig. 2. Diagrams and equations for a blood flow-limited tissue (*upper panel*) and a membrane-limited tissue with the vascular membrane as the limiting membrane (*lower panel*). **C** concentration, **CL** clearance, **R** tissue-**to-**plasma partition coefficient, **Q** blood flow, **V** volume, **K** transportation coefficient, subscripts **ART**, **VEN, P**, **T**, and **V** indicate arterial, venous, plasma, and tissue, respectively. For tissues without elimination, the CL will be zero. Reproduced from (7) with permission from (ACS publications).

In the blood flow-limited model, it is assumed that the drug transportation into tissues is very fast and equilibrium between blood and tissue is reached instantly. The limitation of transportation from blood to tissues is the blood flow rates, which are easily obtained. For the membrane-limited model, the transportation from blood into tissues across the tissues specific biological barriers is the rate-limiting step of distribution.

PBPK models can also be used to describe pharmacokinetics within part of the whole body or a single organ (9). The organ can be divided into a number of compartments and the mass transfer among them described by mathematical equations, similar to whole body models. These models are often termed "partial" PBPK models in comparison with whole body models, and a blood compartment is not necessarily included. For "partial" PBPK models, all the transportation rate constants should be determined individually. "Partial" PBPK models can also be treated as part of a whole body model to describe the local mass transfer kinetics (6).

Similar to traditional pharmacokinetics, PBPK models are also utilized to describe available information of kinetic processes in organisms. Based on available information, including in the area of general physiology and in the area of chemical or nanoparticle specifics, the development of a PBPK model can be divided into the following steps.

2.1. Specification of the Model Structure

The structure of the model is generally determined by the available experimental data, the study design, and the distribution properties of the substances (drugs or nanoparticles) tested (10). No specific structures are required for PBPK models, such as the number of compartments, and which tissues or organs should be listed. However, some general rules may need to be followed. A relatively large number of compartments are generally needed to produce precise simulation. For whole body PBPK models, no major tissues or organs should be missing from the model structure to ensure the mass conservation of the model. A generally adapted method is to include a single compartment, such as "body" or "carcass," that consists of all the tissues and organs not individually included as compartments. Additionally, multiple tissues and organs can be combined into a single compartment based on similar properties such as fast perfusing or slow perfusing. Although it is ideal to choose a particular structure based on structure sensitivity tests, this may be too costly in terms of time and resources. More work is needed to determine optimal PBPK model structures, as this will have far-reaching consequences for model analysis and experimental design.

Excretion and metabolism should be included if they have significant influence on the disposition of nanoparticles. The most common excretion routes include the urine and bile. The kinetics of these excretion routes can be estimated through including experimental data obtained in urine and GI lumen (except when exposure in the GI tract is tested). Metabolism has not yet been included in any PBPK models of nanoparticles. This is likely because nanoparticle metabolism/degradation will result in changed nanoparticle properties, which in turn will cause the drifting of kinetic model parameters. The models currently developed cannot address this issue.

After the whole structure is determined, the next step is to determine the nanoparticle transportation mechanism at the tissue level. In most current reports, a blood flow-limited mechanism was used and assumed that each tissue or organ was one well-stirred compartment. This model assumes that nanoparticles from the incoming blood flow distribute into the tissue very fast and equilibrium is reached instantly. In this scenario, the rate-limiting step of nanoparticle distribution into a tissue is solely dependent on the blood supply. Such a model has been applied to most small molecules and simulated experimental data quite well. However, its application to nanoparticles has resulted in a mixture of good and poor simulation. In fewer reports, a permeability-limited model was adopted and assumed that there are membranes limiting the transportation of nanoparticles from the blood into the tissue. The limiting membrane could be at the capillary wall or at the cell membrane, or both. Even more complicated mechanisms could be suggested given sufficient information available regarding the transportation of nanoparticles. The above-mentioned mechanisms can also be combined into one model to represent different situations in various tissues.

2.2. Equation Writing and Programming into Software

Equation writing should be according to the mass transfer routes and transportation kinetics. In PBPK models, linear ordinary differential equations are often used under the assumption that linear processes are involved. When nonlinearity is assumed, such as the saturation of nanoparticle uptake by cells (11), nonlinear differential equations may apply. In fewer cases, algebraic equations are used for static processes and partial differential equations for dispersion models (12). A combination of these equations may be used within the same model. Their selection is based on the transportation mechanisms.

Equation writing should follow the rule of mass conversation; meaning the total mass of nanoparticles (including excretion and metabolism if applicable) should remain constant and equal the dose of administration throughout the entire study time course. Any missing portion will result in biased simulation. This requirement also stipulates that the experiments should analyze nanoparticles within the whole body and any possible elimination such as urine and feces. In reality, tissues of less importance can be combined as one tissue for analysis. Tissues with very large masses such as muscles may be analyzed in a portion and then extrapolated to the whole tissue based on mass (although the related assumptions may introduce inaccuracies).

After equations are written and validated, they need to be programmed into software for parameter estimation and data simulation. There are many software products used in the literature, including Advanced Continuous Simulation Language (ACSL), MATLAB, and STELLA, to name a few. There are a number of software products developed for general pharmacokinetic modeling that can also be used for PBPK model development, such as ACSL Tox, SAAM II, and NONMEM.

2.3. Estimation of Model Parameters

Once the structure and the mechanisms are determined, the parameters of the equations need to be estimated. There are two groups of parameters for PBPK models. The first group is parameters related to animal anatomical structures and physiological processes. These include: body/tissue/fluid weights and volumes, blood/lymph/bile/urine flow rates, and other necessary information. These parameters are generally assumed to be independent to the nanoparticles studied, and, thus, can be borrowed from studies not related to

Table 1
Flow of blood, through the major organs, and other fluids in the mouse, rat, rabbit, monkey, dog, and human. Reproduced from (13) with permission from (Springer)

	Mouse (0.02 kg)	Rat (0.25 kg)	Rabbit (2.5 kg)	Monkey (5 kg)	Dog (10 kg)	Human (70 kg)
Blood flows (ml/min)						
Brain	–	1.3	–	72	45	700
Liver	1.8	13.8	177	218	309	1,450
Kidneys	1.3	9.2	80	138	216	1,240
Heart	0.28	3.9	16	60	54	240
Spleen	0.09	0.63	9	21	25	77
Gut	1.5	7.5	111	125	216	1,100
Muscle	0.91	7.5	155	90	250	750
Adipose	–	0.4	32	20	35	260
Skin	0.41	5.8	–	54	100	300
Hepatic artery	0.35	2	37	51	79	300
Portal vein	1.45	9.8	140	167	230	1,150
Cardiac output	8	74	530	1,086	1,200	5,600
Urine flow(ml/day)	1	50	150	375	300	1,400
Bile flow (ml/day)	2	22.5	300	125	120	350
GFR (ml/min)	0.28	1.31	7.8	10.4	61.3	125

PBPK modeling. The general physiological parameters of mice and human needed for PBPK modeling are listed in Table 1 (13).

The second group is parameters that relate to the nanoparticles under study. These parameters include: permeability through membranes, interaction with blood and tissue cells, traffic within tissues and cells, and others dependent on the nanoparticles. Some of these parameters may be obtained from separate studies, but in most cases, they need to be estimated through the model under development.

There are two methods in the literature for parameter estimation. The first one is a step-wise expansion from a simple model structure to a more complex structure (14). The estimation procedure starts with the simplest, a two-compartment model for example, with blood and all other remaining tissues as the physiological compartments. Parameters involved in this simple model are then estimated based on experimental data. Then, one tissue is separated out from the remaining tissues compartment, forming a three-compartment model. The parameters obtained from the two-compartment model are used for the estimation of the parameters of the third compartment. Such a process can be repeated until a sufficient number of compartments are reached. Note that because in each state the remaining tissue compartment differs from its predecessor, its parameter values have to be fitted anew together with those for the newly introduced compartment.

In another method (15) the concentration–time profile in blood circulation is fitted with an empirical pharmacokinetic model, for example, a bi-exponential function as shown below:

$$C(t) = A_1 e^{-\lambda_1 t} + A_2 e^{-\lambda_2 t}$$

This model is used as an input "forcing" function to estimate the parameters of each parallel peripheral tissue separately from others. The parameters obtained are then used as starting values for final estimation, normally termed a "close loop" method, which simultaneously fit all the equations of the model to the experimental data.

2.4. Model Implementation, Validation, and Data Extrapolation

With the equations written and parameters estimated, the model can be implemented using software as mentioned above. It needs to be pointed out that the model's ability to simulate the experimental data largely depends on the expertise of the researcher and the quality of the experimental data, instead of on the software selection. Development of a PBPK model requires physiological knowledge of the animal model, the characteristic of nanoparticles and their in vivo behaviors, the mathematical description of dynamic processes, and computational programming of the software. It takes much effort to optimize the model structure, the equations, the parameters, and the organization of experimental data to obtain the best simulation. There is also a possibility that the model cannot be implemented or parameters have significant uncertainty, if the model structures or the experiments are not designed well.

When implementing the model, experimental results may be described in a dynamic manner, providing more information beyond the original data. One of the most important questions concerning in vivo nanoparticle studies is the transportation kinetics within the body; that is, to what extent and rate nanoparticles distribute among various tissues. Simulation of the experimental data can fill gaps in experimental data, including limitations in measurement, limitations in time and tissues that are not feasibly detected (16). The model can be extended to a time scale that allows prediction of how long it takes to eliminate all nanoparticles from the body, for example.

Another important application of the model is data extrapolation among species, tissues, exposure routes, nanoparticles, and doses (17, 18). Experimental data from one study may then be used to predict the results under different experimental conditions. The inter-species extrapolation may be of the highest significance. The final purpose of most nanoparticle studies are their application and toxicity to humans, but in vivo studies are generally carried out solely on animal models. To interpolate data from one animal model to another, the model parameters need to be adjusted, the physiological parameters in most cases, according to the differences between them. However, the model structures normally are required to remain constant.

It is ideal to validate the model to assure that the model is properly built and sufficient for use of data prediction and extrapolation. Validation of the model is normally performed by taking experimental data, external from the data used for model development, and establishing model fit. For example, to validate the extrapolation from mice to rats, the concentration–time profiles in rat tissues should be compared with simulated results by the model based on experimental results from mice. If a good agreement is achieved, the model can be considered successful for such an interpolation. However, in some cases, particularly when extrapolating to humans, the validation is difficult due to the lack of human tissue study availability.

2.5. Evaluation of the Model and Sensitivity Analysis

After parameter optimization, the mass–time profiles of nanoparticles are simulated by the software, and then used to evaluate how well the model can describe the experimental data. One method is by using of the index R^2 (19) which is calculated by:

$$R^2 = \frac{\left[\sum_i (x_i y_i) - \left(\sum_i x_i \sum_i y_i\right) / n\right]^2}{\left[\left(\left(\sum_i x^2 - \left(\sum_i x\right)^2 / n\right)\left(\sum_i y^2 - \left(\sum_i y\right)^2 / n\right)\right)^{1/2}\right]^2}$$

where x is the observed value of nanoparticle concentration and y is the predicted value of the same tissues. An R^2 value close to unit indicates the best prediction of the experimental data by the model.

As one of the several steps in model building, determination of parameters is the most influential on model results. Sensitivity analysis of transportation coefficients is not only critical to model validation, but also serves to guide future research efforts. Parameters that have the greatest impact on the concentration–time profiles need to be stringently evaluated through experimental work.

Parameter sensitivity can be denoted by the sensitivity coefficient, the ratio of the change in output to the change in input while all other transportation coefficients remain constant. There are many methods to determine the sensitivity coefficient. It is reported that the sensitivity coefficient values obtained from different analysis methods are just slightly different from each other, thus selection of analysis methods would not change the sensitivity evaluation significantly.

The most straightforward and commonly used sensitivity analysis is a local sensitivity analysis method, also referred to as "one-at-a-time" sensitivity measure. This is performed by increasing each parameter by a given percentage while leaving all others constant, and quantifying the change in model output. For example, the value of each parameter was increased by 1 %, the model simulations were repeated, and the blood AUC recalculated. The relative sensitivity coefficients for significant parameters were calculated using the following equation (15):

$$\text{Sentitivity} = \frac{\mathrm{d}C \,/\, \mathrm{d}P}{C \,/\, P}$$

i.e., the percentage change in the AUC divided by the percentage change in the parameter value (P).

3. Special Consideration for Nanoparticle PBPK Models

Due to the difference between the ADME behaviors of nanoparticles from that of small molecules, there are some additional factors that need to be considered when building PBPK models for nanoparticles. PBPK models have only been applied to nanoparticles over the past few years.

Appropriate descriptions of transportation mechanisms are the foundation for prediction of nanoparticle biodistribution. A thorough understanding of nanoparticle transportation within the living systems is needed for selection and modification of transportation mechanisms. Various mechanisms have been proposed for PBPK models including blood flow-limited and permeability-limited mechanisms. Modification of these basic mechanisms may be needed to better suit specific situations.

Although a blood flow-limited model works well for small molecules in most cases, its preliminary application to nanoparticles was a mix of good and poor prediction of experimental data (20, 21). For example, Lee et al. (21) tried to validate whether a blood flow-limited, whole body PBPK model could be applied across various types of quantum dots. Classical tissue-to-blood partition coefficients were used in this work. The results indicated that the model used was not sufficient to produce satisfactory simulation of quantum dot biodistribution, especially at the early time points. Unsatisfactory simulation may result from the utilization of an incorrect transportation mechanism. Additionally, as hypothesized by the authors, the lymphatic system, which is not included in the model, could have a significant influence on the simulation accuracy.

Membrane-limited or more complicated models have yet to be applied to nanoparticles, possibly because of poor understanding of nanoparticle transportation mechanisms. Limited knowledge available regarding nanoparticles extravasation (22), lymphatic washout from tissue interstitial space and returning to blood circulation (23), binding to cell surfaces (especially for active targeting of nanoparticles) (24), and internalization into cells. It is very challenging to mathematically describe the processes of nanoparticle trafficking from blood circulation into tissues, within tissue interstitial space, and within cells. Similar difficulties exist for macromolecules and relative successful PBPK models were developed employing a two-pore model mechanism of transportation (15). Nanoparticle PBPK modeling may benefit from these works.

There is one very important parameter in PBPK models: the tissue/blood partition coefficient, which is defined as the ratio of drug or nanoparticle concentration in the tissue to that in the emergent venous blood of the tissue. Since partition coefficients depend on the properties of the drugs or nanoparticles under study, they need to be estimated individually. Both in vitro and in vivo methods were proposed for partition coefficients estimation for small molecules (25) and were adopted in some PBPK modeling works of nanoparticles (21). By using such a parameter, it is assumed that there are the same transfer kinetics of nanoparticles from blood into tissue as from tissue back into blood circulation, and equilibrium of concentrations between blood and tissue exists. These assumptions may not be appropriate for nanoparticles. Some researchers have already questioned the suitability of partition coefficients for nanoparticles PBPK modeling (26, 27).

Most of the PBPK modeling works do not included the lymphatic system, except a few developed for macromolecules (antibodies, etc.) (15, 18). However, significant amounts of nanoparticles could traverse into the lymphatic system, especially when given through administration routes other than IV (such as pulmonary (28), oral (29), intradermal injections (30), and others). If the lymphatic system were excluded from the model, in these scenarios, significant error in simulation of the experimental data would likely occur and the parameters obtained would be biased.

Modeling of the metabolism of nanoparticles could also be different from that of small molecules. Metabolism of small molecules consists of a series of chemical reactions. Each of the metabolites has a distinct chemistry and ADME profiles from that of the parent molecule. PBPK models built for each of the metabolites, given sufficient information, can be developed, and then be connected to that of the parent molecule through the major metabolizing compartment (i.e., liver) (31). However, metabolism of nanoparticles, in most cases, is a gradual process (32, 33). This could be explained by the fact that nanoparticles are relatively large clusters of molecules or atoms. Any change of an individual molecule only changes the nanoparticle by a small fraction, and it may often require many such changes before properties and ADME profiles are altered. Metabolism may result in drafting of the transportation parameters for nanoparticles. The changes depend upon mechanisms of degradation involved for particular particles. Exploring the mechanism of nanoparticle degradation might assist in solving this issue.

4. Application of PBPK Models for Nanoparticle Toxicity Studies

Toxicity of chemicals and nanoparticles is directly related to their local concentration and time of exposure. Understanding the distribution kinetics will greatly help in toxicity management and reduction. Although in vitro cellular models have been widely used to predict in vivo toxicity, it has often been difficult to relate the in vitro data to the in vivo event. One reason for this is that the concentration of exposure of drug or nanoparticles in a cell-based assay can be difficult to relate to the in vivo exposure. Histology/histopathology still is the most reliable way for toxicity evaluation, but it is difficult to be predictive without being correlated with nanoparticle concentration–time profiles. Toxicity biomarkers also need to be connected to nanoparticle distribution to be used as quantitative criterion for toxicity.

The most common application of PBPK models is for toxicity evaluation. By PBPK modeling, the ADME of nanoparticles could be systematically described and understood, providing more insights into their in vivo behavior, and then toxicology. PBPK models could aid in identifying factors that influence the nanoparticle distribution and further their toxicity management. The concentration–time profiles of individual tissues and organs can be generated by the model. Based on these results, the toxicity studies can be more rationally designed and the results better interpreted. The powerful extrapolation ability of PBPK models make it possible to support the toxicity evaluation under specific conditions by utilizing data from wider sources such as various animal models, nanoparticle formulations, administration routes, and doses.

PBPK models have been used to improve toxicity studies of small molecules in many ways. Predictive toxicology tools have been proposed through the integration of physiologically based pharmacokinetic/pharmacodynamic (PBPK/PD) and quantitative structure–activity relationship (QSAR) modeling with focused mechanistically based experimental toxicology (34). Another example is to derive the toxicity reference values such as doses for chemical molecules through data extrapolation, making it possible to calculate and predict internal dose metrics with reasonable scientific certainty (35). Lastly, PBPK models have even been coupled with toxicity models to quantitatively predict the toxic responses after chemical exposure, by integrating the target organ concentrations and the associated area-under-curve (AUC)-based toxicological dynamics (36). Although the applications to nanoparticles are still very limited, the same methods and theories can be readily applied.

References

1. Marquis BJ, Love SA, Braun KL et al (2009) Analytical methods to assess nanoparticle toxicity. Analyst 134:425–439
2. Ings RM (1990) Interspecies scaling and comparisons in drug development and toxicokinetics. Xenobiotica 20:1201–1231
3. Seaton A, Tran L, Aitken R, Donaldson K (2010) Nanoparticles, human health hazard and regulation. J R Soc Interface 7(Suppl 1): S119–129
4. Riviere JE (2009) Pharmacokinetics of nanomaterials: an overview of carbon nanotubes, fullerenes and quantum dots. Wiley Interdiscip Rev Nanomed Nanobiotechnol 1:26–34
5. Nestorov I (2007) Whole-body physiologically based pharmacokinetic models. Expert Opin Drug Metab Toxicol 3:235–249
6. Willmann S, Hohn K, Edginton A et al (2007) Development of a physiology-based whole-body population model for assessing the influence of individual variability on the pharmacokinetics of drugs. J Pharmacokinet Pharmacodyn 34:401–431
7. Li M, Al-Jamal KT, Kostarelos K et al (2010) Physiologically-based pharmacokinetic modeling of nanoparticles. ACS Nano 4(11):6303–17
8. Kwon KI (1987) Development of physicological pharmacokinetic model. Arch Pharm Res 10:250–257
9. Sturm R (2007) A computer model for the clearance of insoluble particles from the tracheobronchial tree of the human lung. Comput Biol Med 37:680–690
10. Nestorov IA, Aarons LJ, Arundel PA et al (1998) Lumping of whole-body physiologically based pharmacokinetic models. J Pharmacokinet Biopharm 26:21–46
11. Wilhelm C, Gazeau F, Roger J et al (2002) Interaction of anionic superparamagnetic nanoparticles with cells: kinetic analyses of membrane adsorption and subsequent internalization. Langmuir 18:8148–8155
12. Nestorov I (2003) Whole body pharmacokinetic models. Clin Pharmacokinet 42:883–908
13. Davies B, Morris T (1993) Physiological-parameters in laboratory-animals and humans. Pharm Res 10:1093–1095
14. Lankveld DP, Oomen AG, Krystek P et al (2010) The kinetics of the tissue distribution of silver nanoparticles of different sizes. Biomaterials 31:8350–8361
15. Davda JP, Jain M, Batra SK et al (2008) A physiologically based pharmacokinetic (PBPK) model to characterize and predict the disposition of monoclonal antibody CC49 and its single chain Fv constructs. Int Immunopharmacol 8:401–413
16. Hagens WI, Oomen AG, de Jong WH et al (2007) What do we (need to) know about the kinetic properties of nanoparticles in the body? Regul Toxicol Pharmacol 49:217–29
17. Evans MV, Dowd SM, Kenyon EM et al (2008) A physiologically based pharmacokinetic model for intravenous and ingested dimethylarsinic acid in mice. Toxicol Sci 104:250–260
18. Baxter LT, Zhu H, Mackensen DG et al (1995) Biodistribution of monoclonal antibodies: scale-up from mouse to human using a physiologically based pharmacokinetic model. Cancer Res 55:4611–4622
19. MacCalman L, Tran CL, Kuempel E (2009) Development of a bio-mathematical model in rats to describe clearance, retention and translocation of inhaled nanoparticles throughout the body. J Phys: Conf Ser 151:012028
20. Gerlowski LE, Jain RK (1983) Physiologically based pharmacokinetic modeling: principles and applications. J Pharm Sci 72:1103–1127
21. Lee HA, Leavens TL, Mason SE et al (2009) Comparison of quantum dot biodistribution with a blood-flow-limited physiologically based pharmacokinetic model. Nano Lett 9:794–749
22. Pegaz B, Debefve E, Ballini JP et al (2006) Effect of nanoparticle size on the extravasation and the photothrombic activity of meso(p-tetracarboxyphenyl)porphyrin. J Photochem Photobiol B 85:216–222
23. Nishioka Y, Yoshino H (2001) Lymphatic targeting with nanoparticulate system. Adv Drug Deliv Rev 47:55–64
24. Tassa C, Duffner JL, Lewis TA et al (2010) Binding affinity and kinetic analysis of targeted small molecule-modified nanoparticles. Bioconjug Chem 21:14–19
25. Lin JH, Sugiyama Y, Awazu S et al (1982) In vitro and in vivo evaluation of the tissue-to-blood partition coefficient for physiological pharmacokinetic models. J Pharmacokinet Biopharm 10:637–647
26. Lin P, Chen JW, Chang LW et al (2008) Computational and ultrastructural toxicology of a nanoparticle, Quantum Dot 705, in mice. Environ Sci Technol 42:6264–670
27. Stern ST, Hall JB, Yu LL et al (2010) Translational considerations for cancer nanomedicine. J Control Release 146:164–174
28. Videira MA, Botelho MF, Santos AC et al (2002) Lymphatic uptake of pulmonary delivered radiolabelled solid lipid nanoparticles. J Drug Target 10:607–613

29. Hussain N, Jaitley V, Florence AT (2001) Recent advances in the understanding of uptake of microparticulates across the gastrointestinal lymphatics. Adv Drug Deliv Rev 50:107–142
30. Reddy ST, van der Vlies AJ, Simeoni E et al (2007) Exploiting lymphatic transport and complement activation in nanoparticle vaccines. Nat Biotechnol 25:1159–1164
31. Hofmann AF, Molino G, Milanese M et al (1983) Description and simulation of a physiological pharmacokinetic model for the metabolism and enterohepatic circulation of bile-acids in man - cholic-acid in healthy man. J Clin Invest 71:1003–1022
32. Okon E, Pouliquen D, Okon P et al (1994) Biodegradation of magnetite dextran nanoparticles in the rat—a histologic and biophysical study. Lab Invest 71:895–903
33. Anderson JM, Shive MS (1997) Biodegradation and biocompatibility of PLA and PLGA microspheres. Adv Drug Deliv Rev 28:5–24
34. Yang RS, Thomas RS, Gustafson DL et al (1998) Approaches to developing alternative and predictive toxicology based on PBPK/PD and QSAR modeling. Environ Health Perspect 106(Suppl 6):1385–1393
35. Lu Y, Rieth S, Lohitnavy M et al (2008) Application of PBPK modeling in support of the derivation of toxicity reference values for 1,1,1-trichloroethane. Regul Toxicol Pharmacol 50:49–60
36. Liao CM, Liang HM, Chen BC et al (2005) Dynamical coupling of PBPK/PD and AUC-based toxicity models for arsenic in tilapia Oreochromis mossambicus from blackfoot disease area in Taiwan. Environ Pollut 135:221–233

Chapter 25

Biophysical Methods for Assessing Plant Responses to Nanoparticle Exposure

Tatsiana A. Ratnikova, Ran Chen, Priyanka Bhattacharya, and Pu Chun Ke

Abstract

As nanotechnology rapidly emerges into a new industry—driven by its enormous potential to revolutionize electronics, materials, and medicine—exposure of living species to discharged nanoparticles has become inevitable. Despite the increased effort on elucidating the environmental impact of nanotechnology, literature on higher plants exposure to nanoparticles remains scarce and often contradictory. Here we present our biophysical methodologies for the study of carbon nanoparticle uptake by *Allium cepa* cells and rice plants. We address the three essential aspects for such studies: identification of carbon nanoparticles in the plant species, quantification of nanotransport and aggregation in the plant compartments, and evaluation of plant responses to nanoparticle exposure on the cellular and organism level. Considering the close connection between plant and mammalian species in ecological systems especially in the food chain, we draw a direct comparison on the uptake of carbon nanoparticles in plant and mammalian cells. In addition to the above studies, we present methods for assessing the effects of quantum dot adsorption on algal photosynthesis.

Key words: Uptake, Adsorption, Fullerene, Quantum dots, Hydrophobicity, Cell membrane, Plant cells, Mammalian cells

1. Introduction

Nanotoxicity studies over the past two decades have been focused predominately on animal subjects. Few studies so far have been addressing the impact of nanoparticles on the ecosystems of plants, bacteria, fish, and wildlife. Among the studies available for aquatic species, it has been reported that uncoated fullerenes exerted oxidative stress and caused severe lipid peroxidation in fish brain (1). *Daphnia magna*, or water flea, ingested lipid-coated carbon nanotubes through normal feeding and utilized the lipid coating as a food source. Consequently, nanotubes were no longer water soluble

Joshua Reineke (ed.), *Nanotoxicity: Methods and Protocols*, Methods in Molecular Biology, vol. 926,
DOI 10.1007/978-1-62703-002-1_25, © Springer Science+Business Media, LLC 2012

post ingestion and readily accumulated on the external surfaces of the organisms to compromise their mobility (2).

In comparison with the aquatic studies, research eliciting the behaviors of nanoparticles in higher plants has been less than consistent. Lu et al. (3) reported that a mixture of nano-SiO_2 and nano-TiO_2 at low concentrations increased nitrate reductase activity, enhanced the water and nutrient uptake, stimulated antioxidant system, and hastened germination and growth of soybean (*Glycine max*). Zheng et al. (4) found that nano-TiO_2 (at 0.25 %) increased seed germination, plant dry weight, chlorophyll production, and the RuBP activity and rate of photosynthesis of spinach (*Spinacia oleracea*), while nano-TiO_2 of concentrations greater than 0.4 % were detrimental to plant growth. Govorov and Carmeli (5) found that chlorophyll-bound gold and silver nanoparticles enhanced production of excited electrons in the photosynthetic complex. Khodakovskaya et al. (6) showed that multiwalled carbon nanotubes (MWNTs) of 10–40 μg/mL penetrated tomato seeds and enhanced their germination and growth rates. In contrary, Yang and Watts (7) reported that uncoated alumina nanoparticles (at 2,000 mg/L) reduced the root growth of corn, carrot, cucumber, soybean, and cabbage seedlings while alumina nanoparticles coated with phenanthrene had no effect on root growth. They attributed the protective effect of coated alumina nanoparticles to their ability to scavenge free radicals and prevent oxidative damage. Lin and Xing (8) showed that toxicity varied among different types of nanoparticles and plant species. At 2,000 mg/L, aluminum, alumina, and MWNT suspensions did not affect seed germination, but Zn and ZnO suspensions inhibited germination of rye and corn seeds. Contrary to the report by Khodakovskaya et al. (6), this study found no effect of MWNT suspension on root growth. Lin and Xing (8) further showed that alumina suspensions reduced corn root growth, but had no effect on other crops, and aluminum suspensions had no effect on cucumber roots, but promoted root growth of radish and rape seedlings and retarded root growth of rye and lettuce seedlings. The toxicities induced by Zn and ZnO nanoparticles were attributed to the direct adsorption of nanoparticles on root surface; not due to ion dissolution. In a follow-up study, Lin and Xing (9) reported that ZnO nanoparticles damaged root tip, entered root cells and inhibited seedling growth, thus reducing biomass. However, translocation of ZnO from plant root to shoots was found minimal in their study.

We have studied the uptake of carbon nanoparticles, namely, fullerene C_{70} suspended in natural organic matter (NOM) and fullerene derivative $C_{60}(OH)_{20}$ in rice plants (10) and *Allium cepa* cells (11). We have characterized the biodistribution of the C_{70}-NOM in the roots, stems, leaves, and seeds of the rice plants and discovered transfer of the nanoparticles through the progeny of the plants (10). We have also demonstrated the differential uptake of C_{70}-NOM and $C_{60}(OH)_{20}$ by plant cells resulting from the contrasting hydrophobicity of these nanoparticles (11). These fundamental studies—though still preliminary—cast a light on the implications of nanomaterials in plant systems. In the following we introduce the protocols and methodologies used for these biophysical studies. These methods are effective in acquiring high-resolution information unavailable from the routine biochemical and toxicological methods in the literature.

2. Materials

1. Fullerene C_{70} (SES Research, purity: 99 %), purchased from SES Research.
2. Fullerene Derivative $C_{60}(OH)_{20}$, purchased from Bucky USA.
3. Nordic NOM, purchased from IHSS, MN.
4. C_{70}-NOM: form an NOM solution of 100 mg/L in Milli-Q water. Prepare a C_{70}-NOM stock suspension of 1,000 mg/L by suspending C_{70} in the NOM. Dissolve $C_{60}(OH)_{20}$ directly into Milli-Q to obtain a stock suspension of 1,000 mg/L. Dilute the nanoparticle suspensions in Milli-Q to concentrations of 10–110 mg/L.
5. Rice seeds and MS germination buffer: soak newly harvested rice seeds (*Oryza sativa* L. ssp. *japonica*, cv Taipei 309) in 70 % ethanol for 30 s, surface sterilized twice in 10 % (v/v) Clorox® bleach plus two drops of Tween-20™ (Polysorbate 20), and stir for 30 min. Mix rice germination buffer (12) (half-strength MS basal salts and vitamins and 7.5 g/L sucrose, pH = 5.7) with C_{70}-NOM and autoclaved at 120 °C for 20 min.
6. *Allium cepa* cells: obtain *Allium cepa* (onion) cells directly from produce quality onion bulbs. Remove storage leaves of area 1 cm^2 and collect laminar cells from the inner layers of the plant tissue. Immerse samples in $C_{60}(OH)_{20}$ and C_{70}-NOM suspensions to obtain final concentrations of 10–110 mg/L in MS buffer.
7. HT-29 human colonic adenocarcinoma cell lines (ATCC)—mammalian cell line.
8. The Plant Cell Viability Assay Kit (PA0100-KT): designed for the differential viability staining of plant cells. The kit employs a dual color fluorescent staining system (propidium iodide or PI, and fluorescein diacetate or FD) to highlight viable and nonviable cells. This procedure has been used to stain intact plant tissue, callus tissue, cell suspension culture, and protoplasts. Specifically, the PI (Ex/Em: 535/617 nm) dye is membrane impermeable and is therefore generally excluded by

viable cells. The FD dye is optimal for staining viable plant cells. It does not photobleach as quickly as calcein AM and produces much less background fluorescence than carboxyfluorescein diacetate in plant cells.

9. Quantum dots (for Algal Adsorption and Photosynthesis): Yellow fluorescent CdSe/ZnS core/shell quantum dots (QDs) (Ex: <550 nm; Em: 570–585 nm) from NN-Laboratories, LLC. These QDs are rendered water soluble by coating mercaptoundecanoic acid (MUA) ligands on their surfaces. The average hydrodynamic diameter of the QDs can be determined by dynamic light scattering (Zetasizer S90) as ~10 nm. The dimensions of dried QDs can also be determined by transmission electron microscopy (TEM) as 5–9 nm.
10. Fresh *Chlamydomonas* sp., the spherically shaped algal cells can be harvested from greenhouse or other natural resources, or purchased from the Carolina Biological Supply Company.
11. A bicarbonate indicator solution (0.2 g of thymol blue, 0.1 g of cresol red, in 0.01 M $NaHCO_3$) is used to monitor the depletion of CO_2 by the algae.
12. To measure oxygen production of algae in the presence of QDs, an Oxyg32 system (Hansatech Instruments) is used for a fixed amount of algal cells treated with various dosages of the QDs.

3. Methods

3.1. Characterization of Nanoparticle Suspensions

3.1.1. C_{70}-NOM Stability by UV–Vis Spectrophotometry

1. Measure the absorbance of C_{70}-NOM at 400 nm immediately after probe sonication, using a spectrophotometer (Biomate 3).
2. Incubate samples at room temperature for 9 h, as used in all plant and mammalian cell experiments.
3. Plot the absorbance readings vs. nominal C_{70} concentration. The decrease in absorbance indicates C_{70} precipitation, especially at high concentrations (e.g. 110 mg/L) (see Note 1).

3.1.2. $C_{60}(OH)_{20}$ Stability by Ultracentrifugation

1. Measure the absorbance of freshly prepared $C_{60}(OH)_{20}$ and $C_{60}(OH)_{20}$ stored for 9 h at room temperature using a spectrophotometer (252 nm, Biomate 3).
2. Plot absorbance values vs. nominal $C_{60}(OH)_{20}$ concentration before and after ultracentrifugation (10,000 × *g* RCF, for 5 min, Eppendorf, Centrifuge 5810 R).

3.1.3. C_{70}-NOM Size Distribution by Dynamic Light Scattering

1. Prepare C_{70}-NOM samples of low and high nominal concentrations (10 and 110 mg/L) and measure their size distributions using a dynamic light scattering device (Malvern Nanosizer S90).

2. For C_{70}-NOM of 10 mg/L, filter the suspension through a 20 nm pore size Anotop 10 (Whatman) filter to suppress scattering from large particles which masks the presence of the small nanoparticles (see Note 2). For C_{70}-NOM of 110 mg/L, filter the suspension through a 0.45 mm pore size Nalgene filter to remove dust particles from the ambient air. The size distribution of C_{70}-NOM of 10 mg/L ranges between 15 and 45 nm, while the size of C_{70}-NOM of 110 mg/L ranges between 25 and 100 nm (see Notes 3 and 4).

3.1.4. $C_{60}(OH)_{20}$ Size Distribution by Dynamic Light Scattering

1. Prepare $C_{60}(OH)_{20}$ samples of low and high concentrations (10 and 110 mg/L) and measure their size distributions using a dynamic light scattering device (Nanosizer S90). Apply no filtration to the suspensions since these nanoparticles are highly hydrophilic. The size distribution of $C_{60}(OH)_{20}$ at 10 mg/L ranges between 1 and 2 nm, while the size distribution of $C_{60}(OH)_{20}$ at 110 mg/L ranges between 15 and 25 nm.

3.1.5. Mammalian Cell Culture

1. Culture HT-29 human colonic adenocarcinoma cell lines in DMEM with 1 % penicillin streptomycin, 1 % sodium pyruvate, and 10 % fetal bovine serum.
2. Seed approximately 5,000 HT-29 cells in each well (200 mL) of an 8-chamber glass plate and allow the cells to attach overnight at 37 °C with 5 % CO_2.
3. Add C_{70}-NOM and $C_{60}(OH)_{20}$ to each chamber glass well after the cells reach a 60 % confluence to obtain nominal nanoparticle concentrations of 10, 30, 50, 70, 90, and 110 mg/L.
4. After 9 h incubation, rinse the cells thoroughly three times using PBS buffer to remove dead cells and un-bound nanoparticles.

3.2. Nanoparticle Uptake by Cells

In this section we present methods for a parallel study of carbon nanoparticle uptake by plant and mammalian cells (11). Specifically, *Allium cepa* and HT-29 human colonic adenocarcinoma cell lines are used as model plant and mammalian systems and are exposed to C_{70}-NOM and $C_{60}(OH)_{20}$. NOM is a collective term for the heterogeneous organic substances derived from decomposed living species. The use of NOM is justified because of its abundance in the natural water sources and soil and its likelihood to interact with discharged nanoparticles (10, 13). This study demonstrates that variations in nanoparticle size and hydrophobicity as well as structural differences between plant and mammalian cells underlie nanoparticle–cell interaction and cell damage.

3.2.1. Nanoparticle Uptake by Plant Cells

1. Prepare laminar *Allium cepa* cells (see Note 5).
2. Incubate the cells separately with C_{70}-NOM and $C_{60}(OH)_{20}$ of 10–110 mg/L for 9 h.

3. Acquire bright field and fluorescence images of the plant cells using a fluorescence microscope (Imager A1, Zeiss). The bright field images illustrate *Allium cepa* cell morphology, while the orange PI fluorescence indicates loss of membrane integrity in the presence of $C_{60}(OH)_{20}$ and C_{70}-NOM (see Note 6). Specifically, a number of orange fluorescent spots should be observed for plant cells exposed to $C_{60}(OH)_{20}$ at 30–70 mg/L, resulting from considerable cell damage. The bright green fluorescence regions signify hydrolysis of FD by intracellular esterases, which are indicative of viable cells.
4. To confirm location of cell damage, add mannitol (0.8 M) to the cells and incubate for 15 min. Note these cells have been pre-incubated with C_{70}-NOM of all concentrations for 9 h. The mannitol gradient across the cell surfaces induces an osmotic pressure, which in turn splits plant cell walls from their underlining plasma membranes. C_{70} aggregates should be revealed by the osmosis assay as mostly adsorbed on or trapped within the hydrophobic cellulose matrices of the plant cell walls.
5. Cell damage is calculated by counting percent of nonviable cells in the PI channel, while the FD channel can be used as a reference due to its susceptibility to crosstalk from the PI channel and cell autofluorescence (see Notes 7 and 8).

3.2.2. Nanoparticle Uptake by Mammalian Cells

1. Incubate HT-29 cells with C_{70}-NOM and $C_{60}(OH)_{20}$ of all concentrations for 9 h in 8-chamber wells.
2. Thoroughly wash the samples to remove dead cells and unbound nanoparticles.
3. Examine the cells using a confocal microscope (LSM510, Zeiss). Due to their structural differences from plant cells, especially with the absence of the cell wall, mammalian cells show distinctly different responses to nanoparticles exposure.
4. Excite sample cells with an Argon laser of 488 nm. Acquire 10 images (900 mm × 900 mm) for each sample condition using a 10× objective.
5. Analyze the images and count the cells of each sample using LSM Image Browser (free to download from the Zeiss website) (see Notes 9 and 10).

3.2.3. Transmission Electron Microscopy Imaging of Allium cepa Uptake of Nanoparticles

1. Fix thin layers of Allium cepa cells in 3.5 % glutaraldehyde overnight and dehydrate the samples in a graded series of ethanol.
2. Embed the dehydrated samples in LR white resin overnight at 40 °C and section the samples into thin films approximately 200 nm thick using an Ultracut E Microtome. No osmium tetroxide should be added in order to avoid artifacts.
3. Acquire TEM images using a Hitachi H7600 microscope operated at 80 and 100 kV. Capture the lattice structures of

C_{70}-NOM and $C_{60}(OH)_{20}$ using a high-resolution Hitachi H9500 microscope operated at 150 kV.

4. Analyze the lattice spacings of the nanoparticles in *Allium cepa* by performing Fast Fourier Transform (FFT) of the TEM images, using the "Diffractogram" software.

3.3. Uptake of Nanoparticles by Rice Plants

In this section we introduce experimental strategies on the uptake, accumulation, and generational transmission of NOM-suspended carbon nanoparticles in rice plants, the staple food crops of over half the world's population (10). This study facilitates our assessment of the potential impact of nanoparticle exposure on plant development and the food chain.

3.3.1. Rice Plant Germination, Regeneration, and Exposure to Nanoparticles

1. Incubate newly harvested rice seeds in Petri dishes that contain 15 mL of different concentrations of C_{70}-NOM in rice germination buffer.
2. After germination at 25 ± 1°C for 2 weeks transplant the seedlings to soil in big pots and grow in a green house to maturity without addition of nanoparticles. Maintain five pots of plants per sample condition for analysis. Refer to these plants as the first generation. Refer to the plants grown in the germination buffer as the control. Use identical amounts of NOM for C_{70}-NOM of all concentrations.
3. Harvest mature seeds from the control plants and C_{70}-treated plants 6 months after germination. Choose 60 seeds of similar size for each plant and sterilize the seeds.
4. Plant 10 seeds in each Petri dish filled with rice germination buffer and keep at 25 ± 1 °C for 2 weeks. Refer to the germinated plants without addition of nanomaterials as the second generation.

3.3.2. Bright Field Imaging of Nanoparticle Uptake by Plant Compartments

1. Take tissues of rice plant roots, stems, and leaves at various developmental stages.
2. Thoroughly wash the samples using distilled water.
3. Cut and section the samples to make thin layers.
4. Image the samples on glass slides using a bright field microscope (Imager A1, Zeiss) (see Note 11).

3.3.3. Fourier Transform Infrared Spectroscopy for Nanoparticle Biodistribution

1. Collect Fourier transform infrared (FTIR) spectra at room temperature for both the first- and second-generation rice plant tissue samples. Compare the spectra with the IR-spectral finger prints available in the literature (14) to confirm the uptake and transmission of C_{70} in the plants.
2. Use a Bruker Fourier transform infrared spectrometer (model IFS 66v/s) equipped with a deuterated triglycine sulfate detector to collect the infrared absorption spectra of the selected

samples in the range of 400–4000 cm^{-1}. Use ~3 mg of the root, leaf, or stem mixed with ~50 mg of KBr powder and pressed into a ~5 mm diameter. To eliminate interfering IR absorption by water vapor and CO_2 present in the ambient atmosphere evacuate the sample chamber down to 0.002 mbar.

3. Fit each of the C_{70} peaks to a Lorentzian line shape and calculate the area under the peak (integrated intensity) via equation: $I = A\pi / \Gamma$, where A is the amplitude, and Γ is the FWHM. Convert this area into a percent uptake of C_{70} by dividing it by the total area of all the combined samples for the roots, stems, leaves, and seeds. This plot represents the biodistribution of C_{70} in the whole rice plant (10) (see Notes 12 and 13).

3.3.4. Electron Microscopy Imaging of Rice Plant Uptake of Nanoparticles

1. For scanning electron microscopy (SEM) imaging, coat ten samples of the plant roots and leaves evenly with a thin film of platinum (~5 nm) using a Hummer® 6.2 sputtering system.
2. Acquire SEM images using an FESEM, Hitachi 4800, microscope operating at 5 kV. This procedure offers clues on the adsorption of nanoparticles on plant tissue (especially roots).
3. For TEM imaging, place ten samples of the roots and leaves of rice plants in 3.5 % glutaraldehyde.
4. Fix the samples in osmium tetroxide.
5. Dehydrate the samples in a graded series of ethanol.
6. Embed the samples in LR White embedding media.
7. Polymerize the sample overnight.
8. Section the samples using an Ultracut E microtome. Cut the tissues of plant roots and leaves at 60–90 nm. Acquire TEM images using a Hitachi H7600 microscope operated at 80 and 100 kV.
9. Analyze the lattice spacing of C_{70} particles by performing FFT of the TEM images, using the software "Diffractogram." This procedure provides information on the rice tissue and cell distributions of the nanoparticles (10).

3.4. Plant Gene Amplification in the Presence of Nanoparticles

In this section we present experimental and molecular dynamics (MD) simulation methods on the amplification of a plant heat shock transcription factor (HSTF) gene by polymerase chain reaction (PCR) in the presence of $C_{60}(OH)_{20}$ (15). The experimental study suggests that the inhibition of DNA amplification is mainly due to the binding of $C_{60}(OH)_{20}$ with Taq DNA polymerase; the binding of $C_{60}(OH)_{20}$ with free dNTPs, primers, and DNA products also occur but do not impact DNA amplification for conventional PCR stoichiometry. The atomistic MD simulations show a clear tendency for hydrogen-bond-mediated binding between $C_{60}(OH)_{20}$ and the dNTP and ssDNA components of the PCR reaction.

3.4.1. Polymerase Chain Reaction

1. The PCR primers used in this study are designed from genomic DNA sequence of soybean HSTF gene and synthesized by IDT (Coralville). The gene is localized within the region assigned to the Linkage Group A on the soybean genetic molecular map.
2. Conduct physical detection of this gene-rich region using bacterial artificial chromosome (BAC) library constructed from the ancestral germplasm PI 437654. Identify BAC clone containing HSTF gene and digest it with restriction enzyme *Spe*I.
3. Isolate a 7 kb DNA fragment corresponding to HSTF and subclone it into the pBlueScriptII plasmid. The pBlueScriptII-HSTF is maintained in *Escherichia coli* DH5alpha and used as a template for DNA amplification in PCR in the presence of $C_{60}(OH)_{20}$.
4. For PCR, the primer sequences are HSTF1F 5′-TATTCTTTG TGGGCGTTTAT-3′ and HSTF1R 5′-TTTAACTGTTCTCC AAGACA-3′. For real-time PCR, the primer sequences are HS TF2F 5′-TCCGCCAGCTCAATACCTA.
5. C-3′ and HSTF2R 5′-CAGCTCAGTGCCAATATCCA-3′.
6. Suspend $C_{60}(OH)_{20}$ in Milli-Q, filter and sterilize the nanoparticle suspension using Anotop 10 filters (0.2 mm, Whatman).
7. Prepare PCR (25 μL), each containing 1 ng pBlueScriptII-HSTF DNA, 10 pmol primer, 5 nmol dNTPs, 1 unit (1 U) AmpliTaq Gold DNA polymerase in 1× PCR buffer II (Applied Biosystems), 2 mM $MgCl_2$, and 10 μL $C_{60}(OH)_{20}$ of a final concentration of 0.2×10^{-4}, 0.6×10^{-4}, 1.0×10^{-4}, 1.6×10^{-4}, and 4.0×10^{-4} mM, respectively.
8. Perform PCR using the following protocol: denaturation at 94 °C for 5 min, 30 cycles at 94 °C for 30 s, 58 °C for 30 s and 72 °C for 1 min, and extension at 72 °C for 10 min.
9. Obtain amplified products through size-fractionation by gel electrophoresis (1 % agarose) and visualize the products by staining with ethidium bromide.

3.4.2. Nanosizer Measurement of $C_{60}(OH)_{20}$ and Taq DNA Polymerase Binding

1. Probe the binding between $C_{60}(OH)_{20}$ and Taq DNA polymerase at room temperature using a nanosizer (Malvern, S90).
2. Prepare the concentrations of $C_{60}(OH)_{20}$ and Taq DNA polymerase at 4.0×10^{-3} mM and 10 U, and repeat the measurement three times. An upper shift in the peak size indicates the hydrogen binding between the $C_{60}(OH)_{20}$ and the Taq DNA polymerase.

3.4.3. Real-Time PCR in the Presence of $C_{60}(OH)_{20}$

1. To monitor individual HSTF gene amplification cycles, carry out this experiment using an iCycler iQ™ Real-Time PCR Detection System with iQ SYBR Green Supermix (BioRad).

2. Prepare the reaction as the following: a 25 μL volume reaction containing 1 ng pBlueScriptII-HSTF DNA, 10 pmol each primer, and 10 μL $C_{60}(OH)_{20}$ of 0.2×10^{-4} to 4.0×10^{-4} mM.
3. Run real-time PCR to amplify HSTF gene, following this protocol: denaturation at 95 °C for 2 min, 40 cycles at 95 °C for 30 s and 58 °C for 30 s, and extension at 72 °C for 30 s.

3.5. Adsorption of Quantum Dots by Algae

In this section we use *Chlamydomonas* sp., the single-celled green algae, as a model system for examining the interaction of QDs with plant species. Like most high plants and bacteria, algae possess a cell wall outside their cell membrane. However, unlike mammalian cells, algae do not show robust endocytosis when exposed to foreign materials. Previous spectroscopic and electron microscopic results suggest that small QDs of less than 5 nm in diameter, when aided by light, can enter bacteria possibly by means of oxidative damage to the cell wall and the cell membrane (16). Since functionalized QDs are typically larger than 5 nm, we focus on addressing the effects of QDs adsorption on algae photosynthesis.

3.5.1. Incubation of Algae with QDs

1. Concentrate algal cells by low-speed centrifugation (11,700 RCF/12 960 rpm) for 3 min.
2. Incubate the concentrated algae with QDs of 0.05–5 ppm at room temperature for 2 h.
3. Prepare four samples of each concentration to ensure experimental repeatability and establish error bars.

3.5.2. Bright Field Imaging

1. Image algae incubated with QDs under the bright field mode of a fluorescence microscope (Imager A1, Zeiss).
2. Flow 10 μL of algae/QDs solution into a sample channel sandwiched between a glass substrate and a cover glass prior to imaging.

3.5.3. Confocal Fluorescence Imaging

1. Incubate algal cells with QDs in an eight-well chamber glass overnight prior to confocal fluorescence imaging (LSM 510, Zeiss).
2. Excite the samples with an argon ion laser at 488 nm and capture the fluorescence images with a BP 570–590 filter set and a 40× oil immersion objective.

3.5.4. Transmission Electron Microscopy

1. Air-dry a small volume of QDs suspension (10 μL, 0.1 mg/mL) directly onto a TEM grid prior to imaging.
2. Perform imaging with a Hitachi H-9500 high-resolution transmission electron microscope, under a 100 kV accelerating voltage.

3.5.5. Quantification of QDs Adsorption

1. Use a UV–vis spectrophotometer (Biomate 3) to quantify the amount of QDs adsorbed on the algae. Record absorbance at

545 nm, where QDs show strong absorption, before and after adding QDs of various concentrations into the algal growth medium. Denote the differences as total concentrations of the QDs.

2. After 2 h of incubation, add 10 μL of NaOH in the algae/QDs solution before filtering through membranes with a pore size of 0.45 μm (Nalgene). The introduction of NaOH is to prevent aggregation of negatively charged QDs in the weakly acidic environment of the algal growth medium (pH 6.45).
3. After filtration, block all algal cells by the membranes because of their large sizes, while the absorbance of the filtrate indicates the amount of free (or unadsorbed) QDs. The amount of adsorbed QDs can be calculated by Eq. 1:

$$\mathrm{Abs}_{\mathrm{adsorbed}} = (\mathrm{Abs}_{\mathrm{QDs}} + \mathrm{algae} - \mathrm{Abs}_{\mathrm{algae}}) - \mathrm{Abs}_{\mathrm{filtered}} \quad (1)$$

where $\mathrm{Abs}_{\mathrm{filtered}}$ and $\mathrm{Abs}_{\mathrm{QDs+algae}}$ denote the absorbance of the algae/QDs solution before and after filtration, and $\mathrm{Abs}_{\mathrm{algae}}$ is the absorbance of the algae alone. Establish an adsorption curve by varying the QDs concentration from 0.1 to 5 ppm.

4. To better understand the physical adsorption of QDs to algae, use the Freundlich model to fit the adsorption isotherms. The Freundlich model (17) is a modification of the Langmuir adsorption scheme and is appropriate for describing rough inhomogeneous adsorbent (i.e., algae) surfaces with multiple adsorption sites. Considering the adsorbate–adsorbate interactions for QDs–QDs, we express the empirical Freundlich equation in Eq. 2:

$$q_{\mathrm{eq}} = kC_{\mathrm{eq}}^{n} \quad (2)$$

where k is a coefficient indicating the affinity of QDs for algae, and n is a constant characteristic of the adsorption system and is related to the binding efficiency. An n value of less than 1 indicates a favorable adsorption, while an n value higher than 1 reflects a weak adsorption (18). The parameters C_{eq} and q_{eq} represent the concentrations of nonadsorbed QDs and the QDs adsorbed on the algae at equilibrium, respectively.

3.5.6. Analysis of Algal Photosynthesis

1. The standard photosynthetic reaction is described by Eq. 3:

$$6CO_2 + 12H_2O + \text{photons} = C_6H_{12}O_6 + 6O_2 + 6H_2O \quad (3)$$

To evaluate the effects of QDs absorption on algal bioactivities, we measure both oxygen production and carbon dioxide depletion of the algae incubated with the QDs.

2. Use a bicarbonate indicator solution to monitor the depletion of CO_2, the activities of which are depicted by Eq. 4:

$$HCO_3^- + H^+ = H_2O + CO_2 \quad (4)$$

3. Mix the algae/QDs solution with the indicator solution and tightly seal the samples to prevent gas exchange. Algae consume CO_2 over time during photosynthesis, causing the pH value of the indicator solution to increase accordingly. A color change from yellow to purple indicates a transition from acidic to basic condition, which is accompanied by an increase of absorbance at 574 nm.
4. Calculate the depletion rates of CO_2 based on the increase of absorbance values for different sample concentrations.
5. To measure oxygen production of the algae in the presence of the QDs, use an Oxyg32 system (Hansatech Instruments) for a fixed amount of algal cells treated with various dosages of QDs. All measurements are conducted at room temperature under identical lighting conditions.

4. Notes

1. For $C_{60}(OH)_{20}$ stability calibration, a straight line should be obtained for absorbance vs. nominal concentration prior to ultracentrifugation, while a bending curve should be observed for the sample after ultracentrifugation. The differences between the two curves indicate gradual aggregation of $C_{60}(OH)_{20}$ with increased concentration.
2. For nanoparticle size distribution measurement using dynamic light scattering note here "size" refers to the hydrodynamic diameter of a Rayleigh particle. According to Rayleigh's approximation, the intensity of scattering is proportional to the sixth power of the diameter of a particle. In other words the presence of small quantity of nanoparticle aggregates could easily block the readout for the presence of large number of small nanoparticles. The proper use of Anotop filters (e.g., 100, 200, or 500 nm pore size) holds the key to obtaining good experimental data.
3. For dynamic light scattering the instrument (e.g. Malvern Nanosizer S90) resolution typically ranges between 1 nm up to a few micrometers. In principle the decimal readings less than 1 nm can be ignored.
4. Some high-end dynamic light scattering devices have the capability of measuring zeta potential of colloidal and nanoparticle suspensions. Results from such zeta potential measurements are important for interpreting the stability/aggregation of nanoparticle suspensions and for describing the interaction between nanoparticles and (charged) biological systems (e.g. cell membranes, organelles, and biomacromolecules).

5. The following steps are proceeded in order minimize unintended damage to the *Allium cepa* cells. First, the outermost storage leaves are removed to expose inner leaves for sample collecting. The inner leaves are then detached from the bulb, placed on a clean glass substrate with the inner surface facing up, and cut into cubes with a top surface area of 1 cm^2. Laminar cells from the inner surfaces are then peeled off from the cubes using tweezers. It is not recommended to peel off the whole laminar cell layer from the inner surface of the leaves before cutting it into smaller pieces. The laminar cells layer is usually peeled from a corner of the cube using sharp-tipped metal tweezers; once the layer is detached at the corner, it can be completely peeled off using a pair of flat-tipped plastic tweezers. The laminar cell layer usually wraps around itself once it is peeled off the storage leaf, so the layer has to be unwrapped using the sharp-tipped tweezers on a glass slide. A piece of cover slide is placed on top of the laminar cell layer to keep it from wrapping. Practice is needed before one can perform this procedure. Damaged cells near the edges of the layer should be neglected when the sample is observed under a fluorescence microscope; those damages are most likely introduced by human errors during the peeling and cutting processes.
6. For the *Allium cepa* assay, a comparison between the bright field and fluorescence images should yield a good correlation between damaged membranes (osmosis assay with mannitol), cells of impaired viability (PI and FD fluorescence channels), and cells of altered morphology (bright field).
7. For the *Allium cepa* assay, the appearances of contagious non-viable cells suggests that upon $C_{60}(OH)_{20}$ uptake cells undergo necrosis, which is typically invoked by abnormal environmental conditions and viruses.
8. For the *Allium cepa* assay, C_{70}-NOM should cause minimal or no plant cell damage at all concentrations. This phenomenon is attributed to the large size and hydrophobicity of the C_{70}-NOM, which tends to block the porous plant cell wall and form clusters therein through hydrophobic interactions. $C_{60}(OH)_{20}$, in comparison, should trigger a steady rise in cell damage, causing a maximum ~5 % more damage than the control at 70 mg/L. Due to their small size and good solubility, $C_{60}(OH)_{20}$ particles readily permeate through the plant cell wall driven by a concentration gradient and are mostly excluded by the plasma membrane due to their hydrophilicity, mutual electrostatic repulsion, and hydrogen-bonding with water. Under capillary and *van der Waals* forces, these nanoparticles are confined between the cell wall and the plasma membrane and accumulate under the concentration gradient to protrude the plasma membrane. Since $C_{60}(OH)_{20}$ have been shown as

relatively inert in creating reactive oxygen species (ROS) (19), the loss of membrane integrity is therefore inferred as a result of mechanical damage exerted by nanoparticle aggregation. Such damage may impinge on membrane fluidity and the transport of nutrients and ions between the plant cell and its extracellular space, further stressing the physiological state of the cell and its neighboring cells. $C_{60}(OH)_{20}$ clusters are expected to occasionally appear near the plasma membrane within the cytoplasm, due to membrane damage and a low-level steady state endocytosis (20, 21). The ease of cell damage at high concentrations (e.g., 90 and 110 mg/L) can be attributed to the aggregation of $C_{60}(OH)_{20}$ at these concentrations, which hinders nanoparticle uptake.

9. For the HT-29 assay, the number/density of viable HT-29 cells is expected to decrease continuously with increased C_{70}-NOM concentration up to 70 mg/L, and then level off at higher concentrations due to nanoparticle aggregation. The cell morphology should also change from the healthy elongated form to the less viable more spherical shapes at higher C_{70}-NOM concentrations, showing abundant nanoparticle aggregates bound to/imbedded in the cell membranes. Cell lysis may be visible at times, due to extensive endocytosis and occurrence of necrosis in the damaged cells.
10. For the HT-29 assay, no cell damage is expected for $C_{60}(OH)_{20}$ of all concentrations because of the low affinity of $C_{60}(OH)_{20}$ for mammalian cell membranes. This assay suggests that C_{70}-NOM impacts on mammalian cells similar to C_{60}, possibly due to the hydrophobicity and dissociation of C_{70}-NOM to facilitate C_{70} interacting with the fatty acyl chains in the lipid bilayer. Such hydrophobic interaction, when coupled with the ROS production by C_{70}, could result in cytotoxicity and cell lysis, especially at high nanoparticle concentrations (11). Unlike C_{70}-NOM, $C_{60}(OH)_{20}$ is more hydrophilic and, therefore, is largely excluded by mammalian cells due to the same reasons discussed for plant cell membranes.
11. For the rice plant assay, the experimenter is expected to spot black aggregates appearing frequently in the seeds and roots, and less frequently in stems and leaves. This specific biodistribution is due to the sequence of nanoparticle uptake, which occurs first in the plant seeds and roots and then in the plant stems and leaves. When imaging for rice stems, the experimenter is expected to spot black aggregates mostly in and near the stem's vascular system, since the transport of C_{70} occurs simultaneously with the uptake of water and nutrients in the plant xylem. Black aggregates may also be spotted in the leaf tissues of the second-generation plants, though much less frequently (10).

12. For the FTIR assay, expect a prevalent C_{70} particle distribution in the roots, stems and leaves of the 2-week-old plants, and little or no concentration dependence.
13. For the FTIR assay, for mature (6-month-old) plants, expect to find C_{70} predominantly in or near the stems' vascular systems, less in the leaves, and even less in the seeds due to the multiplied uptake rates. Furthermore, there should be little or no C_{70} left in the roots of the mature plants due to the robust transport of the nanoparticles in conjunction with the transpiration of water from the roots to the leaves (10).

Acknowledgments

Ke acknowledges the support of NSF CAREER grant #CBET-0744040 and US EPA grant #R834092. Bhattacharya acknowledges a COMSET graduate fellowship. The authors thank Sijie Lin, Matthew Stone, JoAn Hudson, Junjun Shang, and Halina Knap for their valuable contributions cited in this presentation.

References

1. Oberdorster E (2004) Manufactured nanomaterials (fullerenes, C60) induce oxidative stress in the brain of juvenile large-mouth bass. Environ Health Perspect 112:1058–1062
2. Roberts AP, Mount AS, Seda B, Qiao R, Lin S, Ke PC, Rao AM, Klaine SJ (2007) In vivo biomodification of lipid-coated carbon nanotubes by *Daphnia magna*. Environ Sci Technol 41:3025–3029
3. Lu CM, Zhang CY, Wen JQ, Wu GR, Tao MX (2002) Research of the effect of nanometer materials on germination and growth enhancement of Glycine max and its mechanism. Soybean Sci 21:168–172
4. Zheng L, Hong F, Lu S, Liu C (2005) Effect of nano-TiO_2 on strength of naturally aged seeds and growth of spinach. Biol Trace Elem Res 104:83–91
5. Govorov AO, Carmeli I (2007) Hybrid structures composed of photosynthetic system and metal nanoparticles: plasmon enhancement effect. Nano Lett 7:620–625
6. Khodakovskaya M, Dervishi E, Mahmood M, Xu Y, Li Z, Watanabe F, Biris AS (2009) Carbon nanotubes are able to penetrate plant seed coat and dramatically affect seed germination and plant growth. ACS Nano 3:3221–3227
7. Yang L, Watts D (2005) Particle surface characteristics may play an important role in phytotoxicity of alumina nanoparticles. Toxicol Lett 158: 122–132
8. Lin D, Xing B (2007) Phytotoxicity of nanoparticles: inhibition of seed germination and root growth. Environ Pollut 150:243–250
9. Lin D, Xing B (2008) Root uptake and phytotoxicity of ZnO nanoparticles. Environ Sci Technol 42:5580–5585
10. Lin S, Reppert J, Hu Q, Hudson JS, Reid ML, Ratnikova T, Rao AM, Luo H, Ke PC (2009) Uptake, translocation and transmission of carbon nanomaterials in rice plants. Small 5: 1128–1132
11. Chen R, Ratnikova TA, Stone MB, Lin S, Lard M, Huang G, Hudson JS, Ke PC (2010) Differential uptake of carbon nanoparticles by plant and mammalian cells. Small 6:612–617
12. Murashige T, Skoog F (1962) A revised medium for rapid growth and bioassays with tobacco tissue cultures. Physiol Plant 15:473–497
13. Hyung H, Fortner JD, Hughes JB, Kim JH (2007) Natural organic matter stabilizes carbon nanotubes in the aqueous phase. Environ Sci Technol 41:179–184
14. Eklund PC, Rao AM, Zhou P, Wang Y, Holden JM (1995) Photochemical transformation of C_{60} and C_{70} films. Thin Film Solids 257:185–203
15. Shang J, Ratnikova TA, Anttalainen S, Salonen E, Ke PC, Knap HT (2009) Experimental and

simulation studies of real-time polymerase chain reaction in the presence of a fullerene derivative. Nanotechnology 20:415101
16. Kloepfer JA, Mielke RE, Nadeau JL (2005) Uptake of CdSe/ZnS quantum dots into bacteria *via* purine-dependent mechanisms. Appl Environ Microbiol 71:2548–2557
17. Weber WJ Jr (1985) Adsorption technology: a step-by-step approach to process evaluation and application. Dekker, New York
18. Ribeiro MHL, Lourenco PAS, Monteiro JP, Ferreira-Dias S (2001) Kinetics of selective adsorption of impurities from a crude vegetable oil in hexane to activated earths and carbons. Eur Food Res Technol 213:132–138
19. Sayes CM, Fortner JD, Guo W, Lyon D, Boyd AM, Ausman KD, Tao YJ, Sitharaman B, Wilson LJ, Hughes JB, West JL, Colvin V (2004) The differential cytotoxicity of water-soluble fullerenes. Nano Lett 4:1881–1887
20. Šamaj J, Baluska F, Voigt B, Schlicht M, Volkmann D, Menzel D (2004) Endocytosis, actin cytoskeleton, and signaling. Plant Physiol 135:1150–1161
21. Etxeberria E, Gonzalez P, Baroja-Fernandez E, Romeo JP (2006) Fluid phase endocytic uptake of artificial nano-spheres and fluorescent quantum dots by sycamore cultured cells: evidence for the distribution of solutes to different intracellular compartments. Plant Signal Behav 1:196–200

Chapter 26

In Vivo Nanotoxicity Assays in Plant Models

Mamta Kumari, Vinita Ernest, Amitava Mukherjee, and Natarajan Chandrasekaran

Abstract

Increasing application of silver nanoparticles (SNPs) and zinc oxide nanoparticles (nZnO) in consumer products like textiles, cosmetics, washing machines and other household products increases their chance to reach the environment. Intensive research is required to assess the nanoparticles' toxicity to the environmental system. The toxicological effect of nanoparticles has been studied at the miniscule scale and requires intensive research to be conducted to assess its unknown effects. Plants are the primary target species which need to be included to develop a comprehensive toxicity profile for nanoparticles. So far, the mechanisms of toxicity of nanoparticles to the plant system remains largely unknown and little information on the potential uptake of nanoparticles by plants and their subsequent fate within the food chain is available. The phytoxicological behaviour of silver and zinc oxide nanoparticles on *Allium cepa* and seeds of *Zea mays* (maize), *Cucumis sativus* (cucumber) and *Lycopersicum esculentum* (tomato) was done. The in vitro studies on *A. cepa* have been done to check the cytotoxicological effects including mitotic index, chromosomal aberrations, vagrant chromosomes, sticky chromosomes, disturbed metaphase, breaks and formation of micronucleus. In vitro and in vivo studies on seed systems exposed to different concentration of nanoparticles dispersion to check phytotoxicity end point as root length, germination effect, adsorption and accumulation of nanoparticles (uptake studies) into the plant systems. In vivo studies in a seed system was done using phytagel medium. Biochemical studies were done to check effect on protein, DNA and thiobarbituric acid reactive species concentration. FT-IR studies were done to analyze the functional and conformational changes in the treated and untreated samples. The toxicological effects of nanoparticles had to be studied at the miniscule scale to address existing environment problems or prevent future problems. The findings suggest that the engineered nanoparticles, though having significant advantages in research and medical applications, requires a great deal of toxicity database to ascertain the biosafety and risk of using engineered nanoparticles in consumer products.

Key words: Silver nanoparticles, Nano zinc oxide particles, Cytotoxicity, Genotoxicity, *Allium cepa* root cells, Toxicity to seeds, Biochemical, FTIR, Adsorption, Uptake studies

Joshua Reineke (ed.), *Nanotoxicity: Methods and Protocols*, Methods in Molecular Biology, vol. 926,
DOI 10.1007/978-1-62703-002-1_26, © Springer Science+Business Media, LLC 2012

1. Introduction

Due to the rapid development in the field of nanotechnology, it has resulted in a vast array of nanoparticles with varying size, shape, surface charge chemistry; coating and solubility behaviour. The term "nanotechnology" refers to technology of the very small, with dimensions in the range of nanometers. Nanoparticles are defined as particles less than 100 nm in one dimension at atomic, molecular and macromolecular scales (1, 2). The nanoparticle differs from its own bulk form in its physical properties (3, 4) and more toxic than its bulk form (5, 6). Nanotechnology has wide applications in various industries, thereby enhancing the economy of a country. On the other hand, it also creates negative impacts to human, non-human biota and to the ecosystem defined (7).

There are nearly 800 consumer products where nanoparticles are being used (8). The antimicrobial properties of silver nanoparticles are being increasingly exploited in consumer products like deodorants, clothing materials, bandages, and also in cleaning solutions and sprays (9, 10). The USEPA will now move to regulate products that contain nano silver and that claims to act as an "anti-bacterial" as pesticides, including Samsung's "Nano Silver" range of appliances (washing machine, refrigerator, vacuum cleaner and air conditioner). Samsung's "Nano Silver" washing machine releases nano silver directly into waste water systems. Samsung's own advertising claims that its nano silver products will "sterilize over 650 types of bacteria" (11). There is a real risk that effluent containing nano silver would kill beneficial bacteria and disrupt ecosystem functioning. Therefore stringent regulatory measures are needed for nanoparticles usage before entering the market. In the near future there may be a risk and enhanced bioavailability of the nanoparticles in the environmental components.

2. Materials

1. Silver nanoparticles (<100 nm), surface area (5.0 m^2/g).
2. Zinc oxide nanoparticles (<100 nm) and surface area (15–25 m^2/g) (Nanoparticles were procured from Sigma Aldrich, USA).
3. Silver nitrate (<10 micron), surface area (10 m^2/g).
4. Zinc oxide (ZnO) (<5 μm) and surface area (1.2 m^2/g).
5. *Allium cepa* (onion).
6. *L. esculentum* (tomato).
7. *C. sativus* (cucumber).
8. *Z. mays* (maize) (see Note 1).

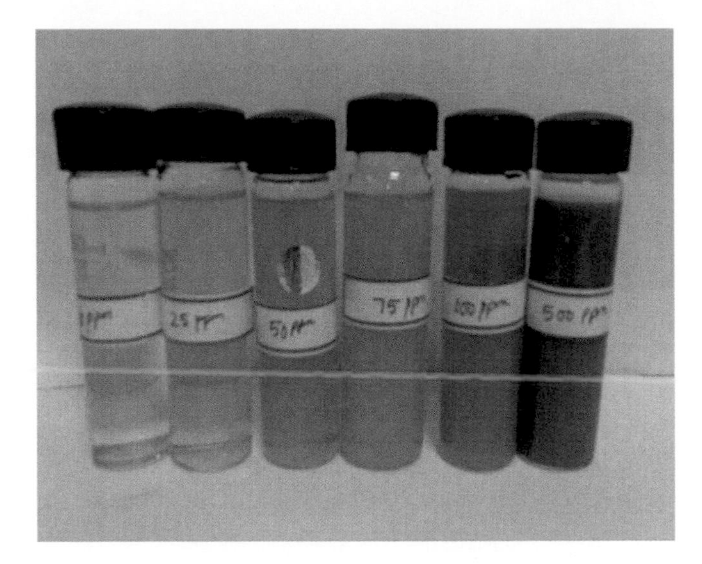

Fig. 1. Dispersion of silver nanoparticles (Color figure online).

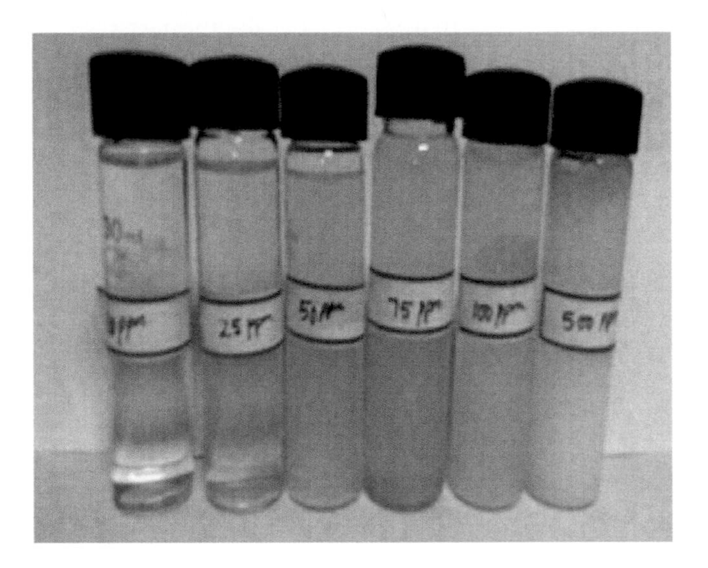

Fig. 2. Dispersion of zinc oxide nanoparticles (Color figure online).

3. Methods

3.1. Nanoparticle Dispersion

1. The engineered silver nanoparticles (SNPs) and zinc oxide nanoparticles (nZnO particles) are dispersed in deionized (MilliQ) water as shown in Figs. 1 and 2.
2. It was then sonicated using ultrasonic vibrations (Sonics Vibracell ultrasonicator, 130 W, 20 kHz) for 30 min to produce five different concentrations such as 10, 25, 50, 100 and 500 μg/ml for both nanoparticles (see Note 2).

Table 1
Bioavailable concentration of nanoparticles in dispersion

	Available concentration (mean ± SE)	
Added conc.(μg/ml)	SNPs particles	nZnO particles
10	4.04 ± 0.95	4.04 ± 0.95
25	9.5 ± 1.15	9.5 ± 1.15
50	22.2 ± 1.6	20.7 ± 1.6
75	27.9 ± 0.91	31.9 ± 0.91
100	41 ± 1.03	43.7 ± 1.03
500	168.7 ± 0.98	159.7 ± 0.98

3.1.1. Availability of Nanoparticles in Dispersion

When nanoparticles are dispersed in aqueous media, it is important to check its availability in the dispersion. i.e., the real concentration that is available for the test system. For example *A. cepa* root tip test. After dispersion, there is a possibility of agglomeration and settling down of particles. Therefore, the real concentration that was taken during the experiment will not be reflected. Thus, the bioavailable concentration is taken into account. Another reason is that when nanoparticles (silver and zinc oxide) are dispersed in water, the chances of ions getting released from the nanoparticles into the medium is most likely. Some examples of bioavailable concentrations for nanoparticles are shown in Table 1.

1. Centrifuge the nanoparticle dispersion at 14,000×*g* for 10 min.
2. Filter through 0.22 μm Anapore membrane disc and carefully collect the clear filtered supernatant in a boiling tube.
3. Add 2 ml of 1% nitric acid and analyze using atomic absorption spectrophotometer (12, 13) (see Note 3).

3.2. Characterization of Nanoparticles (Analytical Studies of Nanoparticles)

The characteristic peaks of the silver and zinc oxide nanoparticles were identified using UV–vis double beam spectrophotometer (Systronics 2201). The peaks were at 424 and 374 nm for silver and zinc oxide nanoparticles, respectively. The morphological features of the nanoparticles were characterized using Transmission Electron Microscope (Technai10, Philips). Silver nanoparticles showed spherical to oval shape while zinc oxide nanoparticles showed spherical to hexagonal shaped particles. Particle size distribution, effective diameter and polydispersity were assessed by 90Plus Particle Size Analyzer (Brookhaven Instruments Corporation). Fourier Transform-Infrared spectroscopy (FT-IR) (Thermonicolar, Avatar-330, USA) was done for surface characterization and to show presence of functional groups like carboxylic

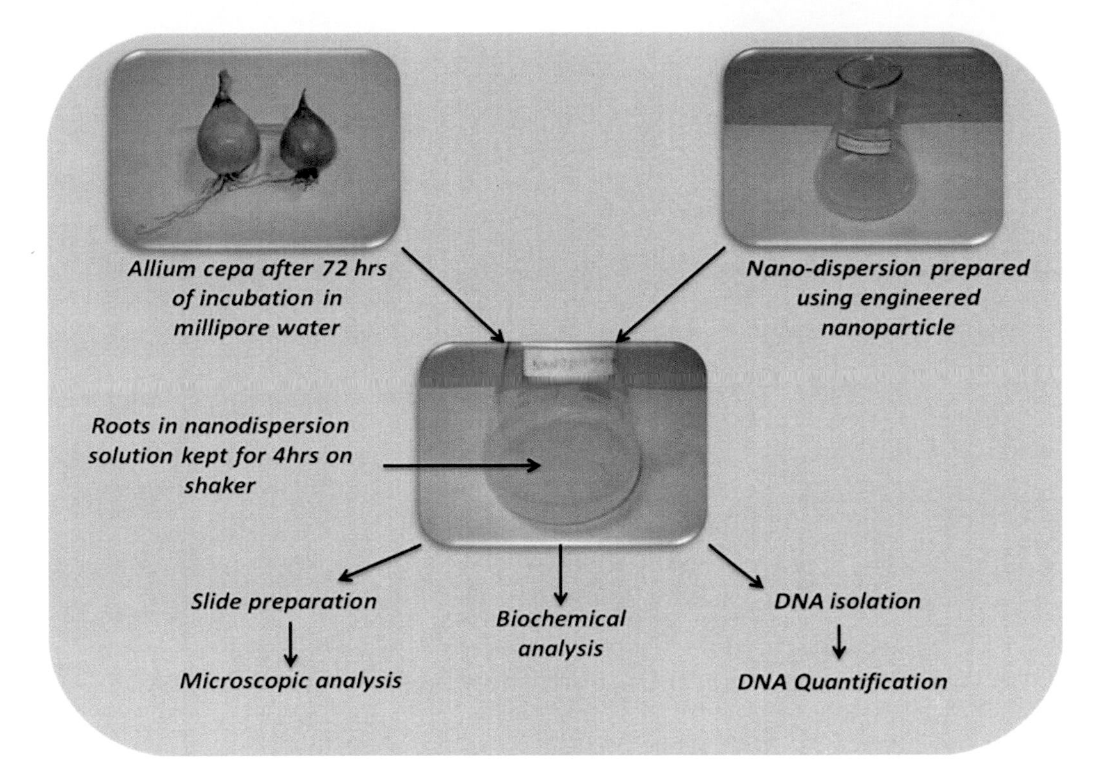

Fig. 3. Root cells of *Allium cepa* treated with nanoparticle dispersion (Color figure online).

ester group, amide stretching bands, and strong fingerprint region for nanoparticles in control and test. Detailed methods of these analytical techniques are described in the previous chapter. Atomic Force Microscopic analyses were done to find the surface morphology and shape of the nanoparticles as described below.

3.2.1. Atomic Force Microscopy

1. 10 μg/ml of nanoparticles were dispersed in deionized water by sonication for 15 min using, a 750 W (20 kHz) ultrasonic processor, (Sonics Corp., USA).
2. A drop of nanoparticle dispersion was placed onto the cover slip and spread evenly in order to get a thin film.
3. It was dried in hot air oven at 60°C for 30 min.
4. The slides were subjected to microscopy analysis (see Note 4).

3.3. Phytotoxicity Assessment of Nanoparticles

3.3.1. In Vitro Studies on A. cepa (Root Test Assay)

A. cepa was used for this study, as it is the suitable bio-indicator for testing toxicity of materials. It is also easy to analyze the cellular and chromosomal deformation caused due to novel materials because of its low chromosomal number ($2n = 16$). An overview of the methodology is shown in Fig. 3.

1. Grow four healthy onion bulbs (20–25 g) in dark in a cylindrical glass beaker at room temperature (28 ± 0.5°C) and renew water supply every 24 h.

2. When the roots reached 2–3 cm in length, treat them with different concentrations of nanoparticles suspension for 4 h (14, 15).
3. Five replicates were used for each concentration (see Note 5).

3.3.2. Microscopic Examination

1. Use five bulbs and eight new root tips for each concentration of SNP and nano zinc oxide particles in dispersion.
2. Prepare the slides for each concentration and control following Saffranin squash technique.
3. Keep the root tips in 1 M HCl for 6 min followed by staining with 40–45% saffranin.
4. Continue staining for 5–6 min.
5. Analyze the slides at 1,000× magnification for cytological changes. The mitotic index is calculated as the number of dividing cells per number of 1,000 observed cells (14).
6. Note the number of aberrant cells per total cells scored at each concentration (16).
7. Calculate the MN index as mentioned below (17) (see Note 6).

3.3.3. Effect of $AgNO_3$, $ZnCl_2$, $Zn(NO_3)_2$, $ZnSO_4$ on Mitotic Index and Micronucleus Index

Ionic effect of $AgNO_3$, $ZnCl_2$, $Zn\ (NO_3)_2$, $ZnSO_4$ on mitotic index is done using the below-mentioned formulae.

$$\text{Mitotic index (MI)} = \text{TDC} / \text{TC} \times 100 \quad (1)$$

$$\text{Phase index (PI)} = \text{TC} / \text{TDC} \times 100 \quad (2)$$

$$\text{Total percentage of abnormal cells} = T_{\text{abn}} / \text{TDC} \times 100 \quad (3)$$

$$\text{MN index (\%)} = T_{\text{MN}} / T_{\text{BN}} \times 100 \quad (4)$$

where *TDC*—total no. of dividing cells; *TC*—total no. cells observed; T_{abn}—total no. of abnormal cells; T_{MN}—total no. of micronucleus observed; T_{BN}—total no. of binucleated cells observed.

3.3.4. In Vitro Studies: Seed Experiment

Exposure of Silver nanoparticles and nano zinc oxide particles to the seeds of *C. sativus*, *Z. mays* and *L. esculentum*

Seed Germination Test

This test is conducted following the standard method (18). The test was performed on three seeds (*C. sativus, L. esculentum* and *Z. mays*). The Relative Seed Germination rate (RSG) and Relative Root Growth (RRG) were calculated using the Eqs. 1 and 2. Germination Index (GI) was also determined using Eq. 3. In addition, 50% effective concentration and its 95% confidence level of nanoparticles (IC_{50}) was determined by using probit software computer program (US-EPA, 1994).

$$\text{Relative Seed Germination rate} = (\text{Ss} / \text{Sc}) \times 100 \quad (1)$$

$$\text{Relative Root Growth} = (\text{Rs} / \text{Rc}) \times 100 \quad (2)$$

$$\text{Germination Index} = (\text{RSG} \times \text{RRG}) / 100 \quad (3)$$

where Ss is the no. of seed germinated in sample; Sc is the no. of seed germinated in control; Rs is the average root length in sample; Rc is the average root length in control.

Seedling Exposure

1. Check the seeds for viability by suspending them in deionized water.
2. Select the seeds which settle to the bottom for further study.
3. Soak the seeds for 10 min in 10% sodium hypochlorite solution, which acts as a surface sterilizing agent (19).
4. After surface sterilization, rinse the seeds in deionized water thrice and then stir for 2 h in nanoparticles dispersion (10, 100, and 500 μg/ml) using a magnetic stirrer.
5. Place Whatman No.1 filter paper into each Petri dish (100 mm × 15 mm) and add 5 ml of the respective particle suspensions using a Pasteur pipette.
6. Transfer the seeds to the Petri dish, with ten seeds per dish placed equidistant from one another.
7. Cover and seal the dishes with sealing tape and place in dark conditions.
8. The end points of the experiment are when at least 80% of the control seeds have germinated; 80–85 h for *L. esculentum*, 52 h for *C. sativus* and 48–50 h for *Z. mays*, carry out the experiment in triplicates (18).

3.4. In Vivo Studies: Seed Experiment

3.4.1. Seedling Exposure Using Phytagel

In the present study plants are used to access the soil ecotoxicity and bioavailability of silver nanoparticles. *C. sativus* and *Z. mays* are selected as the test species because of their importance of valuable plant species, both the plant species are selected based on germination time. The test species are selected because of less germination time, sensitive to toxicity, routine use in phytotoxicity tests and their importance as in food plants. *L. esculentum* was not selected as test species for in vivo studies because of very small size and high germination time. Most nanoparticles are hardly water soluble so a plant agar test has been employed to avoid precipitation of nanoparticles and to distribute nanoparticles evenly in test species.

Preparation of Culture Dispersion

1. Culture dispersion of nanoparticles is achieved by adding phytagel powder (Sigma-Aldrich, USA, melting point – 90°C) and the needed amount of nanoparticles to deionized water.
2. Shake the dispersions sufficiently after sonication to break up agglomerates.
3. Prepare each nanoparticle treatment concentration separately without dilution to avoid agglomeration in the dispersion and sonicate for 30 min.
4. Melt agar separately in deionized water.

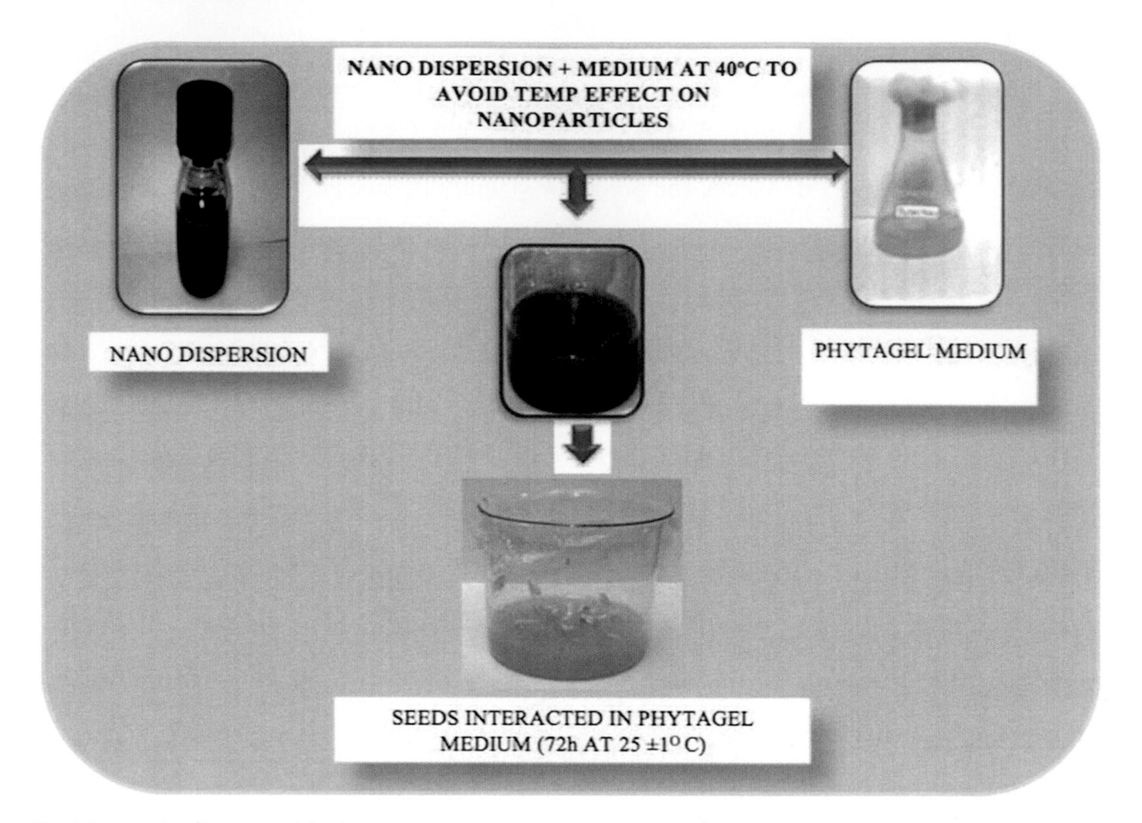

Fig. 4. Interaction of nanoparticles in phytagel with seed system (Color figure online).

5. Add nano dispersion to the melted agar at an optimum temperature of 40–45°C to avoid temperature effect on nanoparticles.
6. Range the concentration of nanoparticles from 0 to 500 μg/ml. The agar culture media has the advantage of easy dispersion of nanoparticles without precipitation (see Notes 7 and 8).

Acute Toxicity Test Using Plant Agar Method

7. Sterilize seeds in a 10% sodium hypochlorite solution for 10 min.
8. Rinse thoroughly with deionized water and subsequently place in wet cotton at a controlled temperature of 25 ± 1°C in the dark.
9. After 24 h check seeds for the germination. Use seeds that have sprouted in the test.
10. Conduct the toxicity tests in beakers. Each test unit contains 30 ml of 1.5% of agar media with a specific concentration of nanoparticles and immediately hardened in a freezer to avoid the possible precipitation of nanoparticles.
11. Place test plant seedlings just above the surface of the agar in the test units (Fig. 4).

12. Keep the test units for incubation at a controlled temperature 25 ± 1 °C in dark. The exposure period is for 72 h. After an incubation period of 3 days, separate the plants from the agar media, and measure seedling growth.
13. Calculate IC_{50} using probit analysis (20).

3.4.2. Biochemical Analysis-Determination of Protein and DNA Content

The biochemical assay involves determination of the total protein content of the samples using the Lowry's method (21) and DNA content of the sample by standard protocol of DNA isolation and quantification (22, 23). Lipid peroxidation assay is done following a standard protocol.

Lipid Peroxidation

Lipid peroxidation is determined by measuring the amount of thiobarbituric acid reactive species (TBARS). Each experiment is run with three replications.

1. Cut a total of 0.2 g of root tissues from control and treated plants into small pieces.
2. Homogenize by the addition of 1 ml of 5% trichloroacetic acid (TCA) solution.
3. Transfer the homogenates into fresh tubes and centrifuged at 12,000 rpm for 15 min at room temperature.
4. Add equal volumes of supernatant and 0.5% thiobarbituric acid (TBA) in 20% TCA solution (freshly prepared) into a new tube and incubate at 96°C for 25 min.
5. Transfer the tubes into an ice bath and then centrifuge at $10{,}000 \times g$ for 5 min.
6. Record the absorbance of the supernatant at 532 nm and correct for non-specific turbidity by subtracting the absorbance at 600 nm (see Note 9).

3.4.3. Adsorption Studies

Adsorption studies are done by washing the exposed seeds thrice in distilled water followed by quantification of nanoparticles in the supernatant using atomic absorption spectroscopy (Varian-AA 240). This study shows the metal adsorbed onto the surface of the seeds. This could be due to the greatest surface site densities (positively or negatively charged sites) and cation exchange capacities (negatively charged sites only) on to the surface of nanoparticles and seeds and also because of the extent of metal adsorption depends on the total metal concentration and the pH.

3.4.4. Accumulation of Nanoparticles in Root

To determine silver and zinc oxide nanoparticles accumulation in plant root tissue after 72–82 h:

1. Wash all plants thoroughly with distilled water to remove the test medium.

2. Dissolve 0.1 g of sample in 2 ml of conc. HNO_3 and 2 ml of deionized water at 90°C for 2 h.
3. Filter the solution through a glass frit.
4. Make it up to a specified volume in a volumetric flask.
5. The concentration is measured using an atomic absorption spectrophotometer (see Note 10).

4. Notes

1. Selection of Test System has been done based on US Environment Protection Act, 2007.

   A. cepa

   - Best bio-indicator to check environmental pollution and toxicant
   - Easily available
   - Low chromosome number
   - Sensitivity to toxicity

   Seeds (*L. esculentum*, *C. sativus* and *Z. mays*)

   - Easily obtainable
   - Long shelf-life
   - Sensitivity to toxicity

2. The stability of nanoparticles in aqueous dispersion is an important factor in studying its effects to the test system. When nanoparticle concentration is high in the aqueous dispersion, the possibility of it getting agglomerates is high, causing least effect to the test system.
3. Total available concentration was determined since nanoparticles agglomerate very fast and settled down in the dispersion. When nanoparticles concentration is high in the aqueous dispersion, the possibility of it getting agglomerated is high, causing least effect to the test system. For chemical analysis of Zn^{2+} and Ag^{2+} in the dispersion, the suspension was filtered through 0.2 μm filter to get rid of aggregated particles over size 200 nm (if any). The filtrate was acid digested and analyzed using AAS. The ion concentration reported is essentially the total Zn^{2+} and Ag^{2+} dissolved from all the ZnO and Ag nanoparticles (below 200 nm size) present in the filtrate. The result gives an idea about maximum Zn^{2+} and Ag^{2+} ions which would be dissolved from the particles in suspension.

4. Sample preparation for Atomic Force Microscopy (AFM) analysis should be done carefully by spreading a drop of sample on cover slip, oven dry should not be done above 60 °C since shape distortion can occur at high temperature.
5. *A. cepa* is used for genotoxicity assay since it is one of the frequently used bioindicators. *A. cepa* has been regarded as a favorable indicator to assess the chromosome damages and disturbances in the mitotic cycle due to the presence of good chromosome condition such as large chromosomes and reduced in number ($2n = 16$).
6. For a confirmatory test of micronucleus, treat root cells of *A. cepa* with 0.6 µM of micronucleus inducing chemical Mitomycin C ($C_{18}H_{15}N_4O_5$). Use it as a positive control for the observed micronucleus. Colchicine (0.05%) treatment can be done on root cells of the *A. cepa*; and used as positive control for the confirmatory test of the mutagenic effect on the root cells of the *A. cepa*.
7. Most nanoparticles are hardly water soluble, so phytagel a plant agar test system has been used to prepare culture medium to avoid the precipitation of nanoparticles and to distribute nanoparticles evenly in culture medium. Phytagel is water soluble hence it could not bind nanoparticles.
8. No solvent should be used to improve solubility of nanoparticles, since solvent enhanced solubilization of nanoparticles, but is not an environmentally favorable method because of toxicity of the solvent that remains.
9. 0.5% TBA in 20% TCA solution was used as the blank. TBARS content was determined using the extinction coefficient of 155 mM^{-1} cm^{-1}.
10. Total silver concentrations are reported here as a weight percentage on a dry plant tissue basis.

References

1. Osterberg R, Persson D, Bjursell G (1984) The condensation of DNA by chromium (III) ions. J Biomol Struct Dyn 2:285–290
2. The Royal Society and Royal Academy of Engineering, UK (2004) Nanoscience and nanotechnology, opportunities and uncertainties. Available at http://www.nanotech.org.uk/finalReport.htm
3. Munzuroglu O, Geckil H (2002) Effects of metals on seed germination, root elongation, and coleoptile and hypocotyl growth in *Triticum aestivum* and *Cucumis sativus*. Arch Environ Contam Toxicol 43:203–213
4. Oberdorster G, Oberdorster E, Oberdorster J (2005) Nanotoxicology: an emerging discipline evolving from studies of ultrafine particles. Environ Health Perspect 113:823–839
5. Anastasio C, Martin ST (2001) Atmospheric nanoparticles. Rev Miner Geochem 44: 293–349
6. Nel A, Xia T, Madler L, Li N (2006) Toxic potential of materials at the nanolevel. Science 311:622–627
7. Agency for toxic substances and Disease Registry (1990) Toxicological profile for silver prepared by clement international corporation under contract 205-88-0608, U.S. Public Health Service. ATSDR/TP-90-24
8. Maynard AD, Aitken RJ, Butz T, Colvin V, Donaldson K, Oberdörster G, Philbert MA,

Ryan J, Seaton A, Stone V, Tinkle SS, Tran L, Walker NJ, Warheit DB (2006) Safe handling of nanotechnology. Nature 444:267–269

9. Chen X, Schluesener HJ (2008) Nanosilver: a nanoproduct in medical application. Toxicol Lett 176(Pt 1):1–12
10. Tripathy A, Chandrasekran N, Raichur AM, Mukherjee A (2008) Antibacterial applications of silver nanoparticles synthesized by aqueous extract of Azadirachta indica (Neem) leaves. J Biomed Nanotechnol 4:1–6
11. USEPA (2009) European Agency for Safety and Health Report
12. Wiesner MR, Lowry GV, Alvarez P, Dionisiou D, Biswas P (2006) Assessing the risks of manufactured nanomaterials. Environ Sci Technol 15:4336–4345
13. Xia T, Kovochich M, Brant J, Hotze M, Sempf J, Oberley T, Sioutas C, Yeh JI, Wiesner MR, Nel AE (2006) Comparison of the abilities of ambient and manufactured nanoparticles to induce cellular toxicity according to an oxidative stress paradigm. Nano Lett 6:1794–1807
14. Hsin Y, Chen C, Huang S, Shih T, Lai P, Chueh PJ (2008) The apoptotic effect of nanosilver is mediated by a ROS- and JNK-dependent mechanism involving the mitochondrial pathway in NIH3T3 cells. Toxicol Lett 179:130–139
15. Franklin NM, Rogers NJ, Apte SC, Batley E, Gadd E, Casey PS (2007) Comparative toxicity of nanoparticulate ZnO, bulk ZnO, and ZnCl to a freshwater microalga (*Pseudokirchneriella subcapitata*): the importance of particle solubility. Environ Sci Technol 41:8484–8490
16. USEPA (2007) Nanotechnology White Paper. Science Policy Council, Washington
17. Bakare AA, Mosuro AA, Osibanjo O (2000) Effect of simulated leachate on chromosomes and mitosis in roots of *Allium cepa* (L). J Environ Biol 21:263
18. U.S. Environmental Protection Agency Ecological Effects Test Guidelines (OPPTS 850.4200), Seed Germination/Root Elongation Toxicity Test (1996). http://www.epa.gov/opptsfrs/publications/OPPTS_Harmonized/850_Ecological_Effects_Test_Guidelines/Drafts/850-4200.pdf
19. Leme DM, Marin-Morales MA (2008) Chromosome aberration and micronucleus frequencies in *Allium cepa* cells exposed to petroleum polluted water—a case study. Mutat Res 650:80–86
20. USEPA Data Quality Objectives Decision Error Feasibility Trials *(DQO/DEFT)*, Version 4.0, EPA QA/G-4D (1994) Washington
21. Lowry OH, Rosbrough NJ, Farr AL, Randall RJ (1951) Protein measurement with the folin phenol reagent. J Biol Chem 193:267–275
22. Hoisington D, Khairallah M, Gonzales de leon D (1994) Laboratory protocols. CIMMYT Applied Biotechnology Center, Mexico
23. Zhu H, Han J, Xiao JQ, Jin Y (2008) Uptake, translocation and accumulation of manufactured iron oxide nanoparticles by pumpkin plants. J Environ Monit 10:713–717

Index

Joshua Reineke (ed.), *Nanotoxicity: Methods and Protocols*, Methods in Molecular Biology, vol. 926, DOI 10.1007/978-1-62703-002-1, © Springer Science+Business Media, LLC 2012

T

U

V

W

X

Z

KT-452-749

THE RAINS CAME

BY THE SAME AUTHOR

A MODERN HERO

TWENTY-FOUR HOURS

LILLI BARR

THE STRANGE CASE OF MISS ANNIE SPRAGG

EARLY AUTUMN

POSSESSION

THE GREEN BAY TREE

THE FARM

HERE TODAY AND GONE TOMORROW

THE MAN WHO HAD EVERYTHING

IT HAD TO HAPPEN